Edited by
Yoshiki Chujo

Conjugated Polymer Synthesis

Related Titles

Elias, H.-G.

Macromolecules

4 volumes

2009

ISBN: 978-3-527-31171-2

Dubois, P., Coulembier, O., Raquez, J.-M. (eds.)

Handbook of Ring-Opening Polymerization

2009

ISBN: 978-3-527-31953-4

Matyjaszewski, K., Advincula, R. C., Saldivar-Guerra, E., Luna-Barcenas, G., Gonzalez-Nunez, R. (eds.)

New Trends in Polymer Sciences

2009

ISBN-13: 978-3-527-32735-5

Barner-Kowollik, C. (ed.)

Handbook of RAFT Polymerization

2008

ISBN: 978-3-527-31924-4

Schnabel, W.

Polymers and Light

Fundamentals and Technical Applications

2007

ISBN: 978-3-527-31866-7

Freund, M. S., Deore, B. A.

Self-Doped Conducting Polymers

2007

ISBN: 978-0-470-02969-5

Hadziioannou, G., Malliaras, G. G. (eds.)

Semiconducting Polymers

Chemistry, Physics and Engineering

2007

ISBN: 978-3-527-31271-9

Edited by
Yoshiki Chujo

Conjugated Polymer Synthesis

Methods and Reactions

WILEY-VCH Verlag GmbH & Co. KGaA

The Editor

Prof. Yoshiki Chujo
Kyoto University
Graduate School of Engineering
Department of Polymer Chemistry
Katsura Nishikyo-ku
Kyoto 615-8510
Japan

Library of Congress Card No.: applied for

British Library Cataloguing-in-Publication Data
A catalogue record for this book is available from the British Library.

Bibliographic information published by the Deutsche Nationalbibliothek
The Deutsche Nationalbibliothek lists this publication in the Deutsche Nationalbibliografie; detailed bibliographic data are available on the Internet at http://dnb.d-nb.de.

Typesetting Thomson Digital, Noida, India
Printing and Binding Fabulous Printers Pte Ltd, Singapore
Cover Design Adam-Design, Weinheim

Printed in Singapore
Printed on acid-free paper

ISBN: 978-3-527-32267-1

Contents

Conjugated Polymer Synthesis. Edited by Yoshiki Chujo

ISBN: 978-3-527-32267-1

Preface

This book "Conjugated Polymer Synthesis – Methods and Reactions" aims to summarize the major developments in the topics of synthesis of new conjugated polymers, novel methodologies for the preparation of conjugated polymers, and inorganic-elements containing mainchain-type conjugated polymers. These new compounds and materials are all set to be very important in the fields of electrical, optical, and magnetic applications.

Much effort has been devoted to the design and synthesis of a wide variety of conjugated polymers such as polyacetylene, polythiophene, polypyrrole, poly(*p*-phenylene), poly(*p*-phenylenevinylene), poly(*p*-phenylene-ethynylene), polyfluorene and their derivatives. This is not only due to an academic interest in their particular characteristic properties, but also owing to various industrial utilizations. Potential applications here include organic light-emitting diodes, flat panel displays, sensory materials, semiconductors, field-effect transistors, photovoltaic cells, and so on. To achieve high performance in these applications, it is now necessary to explore new, more conjugated polymers. These new conjugated systems might be expected to show high emission efficiency, fine-tunability of their band-gaps, processability, dramatic improvements in durability, thermal-, air- and photo-stabilities, and other important characteristic properties.

In each of the chapters, all of them written by internationally acclaimed experts, the book covers the whole spectrum of the synthesis of new conjugated polymers from fundamentals to material science applications. I hope that the readers will enjoy this new chemistry and methodology developed throughout the book.

August 2010 *Yoshiki Chujo*

Conjugated Polymer Synthesis. Edited by Yoshiki Chujo

ISBN: 978-3-527-32267-1

List of Contributors

Kazuo Akagi
Kyoto University
Department of Polymer Chemistry
Katsura
Kyoto 615-8510
Japan

Yoshiki Chujo
Kyoto University
Graduate School of Engineering
Department of Polymer Chemistry
Katsura, Nishikyo-ku
Kyoto 615-8510
Japan

Derek P. Gates
University of British Columbia
Department of Chemistry
2036 Main Mall
Vancouver, British Columbia
Canada V6T 1Z1

Martin D. Hager
Friedrich-Schiller-University Jena
Laboratory of Organic and
Macromolecular Chemistry
Humboldtstr. 10
07743 Jena
Germany

Masahiko Iyoda
Department of Chemistry
Graduate School of Science
Tokyo Metropolitan University
Hachioji
Tokyo 192-0397
Japan

Jacky W.Y. Lam
The Hong Kong University of
Science & Technology
Department of Chemistry
Clear Water Bay
Kowloon, Hong Kong
China

Jianzhao Liu
The Hong Kong University of
Science & Technology
Department of Chemistry
Clear Water Bay
Kowloon, Hong Kong
China

Richard D. McCullough
Carnegie Mellon University
Department of Chemistry
4400 Fifth Ave.
Pittsburgh, PA 15213
USA

Conjugated Polymer Synthesis. Edited by Yoshiki Chujo

ISBN: 978-3-527-32267-1

Yasuhiro Morisaki
Kyoto University
Graduate School of Engineering
Department of Polymer Chemistry
Katsura, Nishikyo-ku
Kyoto 615-8510
Japan

Atsushi Nagai
Kyoto University
Graduate School of Engineering
Department of Polymer Chemistry
Katsura, Nishikyo-ku
Kyoto 615-8510
Japan

Kensuke Naka
Kyoto Institute of Technology
Graduate School of Science and Technology
Department of Chemistry and Materials Technology
Goshokaido-cho, Matsugasaki, Sakyo-ku
Kyoto 606-8585
Japan

Itaru Osaka
Graduate School of Engineering
Department of Applied Chemistry
1-4-1 Kagamiyama, Higashi-hiroshima
Hiroshima 739-8527
Japan

Ulrich S. Schubert
Friedrich-Schiller-University Jena
Laboratory of Organic and Macromolecular Chemistry
Humboldtstr. 10
07743 Jena
Germany

Paul W. Siu
University of British Columbia
Department of Chemistry
2036 Main Mall
Vancouver, British Columbia
Canada V6T 1Z1

Masayoshi Takase
Tokyo Metropolitan University
Graduate School of Science
Department of Chemistry
Hachioji
Tokyo 192-0397
Japan

Ben Zhong Tang
The Hong Kong University of Science & Technology
Department of Chemistry
Clear Water Bay
Kowloon, Hong Kong
China

and

Zhejiang University
Department of Polymer Science & Engineering
Hangzhou 310027
China

Andreas Wild
Friedrich-Schiller-University Jena
Laboratory of Organic and Macromolecular Chemistry
Humboldtstr. 10
07743 Jena
Germany

Andreas Winter
Friedrich-Schiller-University Jena
Laboratory of Organic and
Macromolecular Chemistry
Humboldtstr. 10
07743 Jena
Germany

Takakazu Yamamoto
Tokyo Institute of Technology
Chemical Resources Laboratory
4259 Nagatsuta, Midori-ku
Yokohama 226-8503
Japan

Tsutomu Yokozawa
Kanagawa University
Department of Material and Life
Chemistry
Rokkakubashi, Kanagawa-ku
Yokohama 221-8686
Japan

1
Organometallic Polycondensation for Conjugated Polymers

Takakazu Yamamoto

1.1
Basic Organometallic C–C Coupling

Diorganonickel(II) complexes NiR_2L_m undergo reductive coupling (or reductive elimination) reactions to give R–R (Eq. (1.1)) [1]. The controlling factors of this coupling reaction have long been studied by our research group and by others (L = neutral ligand such as 2,2′-bipyridyl (bpy) and tertiary phosphine):

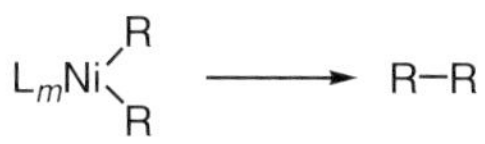

$$L_mNi\begin{matrix}R\\R\end{matrix} \longrightarrow R{-}R \quad (1.1)$$

This basic C–C coupling on Ni introduced the concept of "reductive elimination" to the field of organometallic chemistry [1a–1p] and the first experimental support [1a] for the concept of "back-donation to an olefin [1q,1r]" was given during the study. The coordination of molecules leading to back-donation (e.g., electron-accepting olefin [1a] or aromatic compound [1k,1s]) to the central metal facilitates the reductive elimination of R–R, and the concepts of reductive elimination and back-donation are now widely accepted in chemistry.

The Ni–R bond in NiR_2L_m is considered to be polarized as $Ni^{\delta+}$–$R^{\delta-}$, whereas the reductive elimination produces an electrically neutral R–R molecule. Consequently the reductive elimination is assumed to involve electron migration from the R group to Ni, and this electron migration is considered to be enhanced by coordination of an electron-withdrawing olefin (e.g., $CH_2{=}CHCN$ and $CH_2{=}CHBr$) and an aromatic compound (e.g., C_6H_5CN, C_6H_5Br, and C_6F_6) [1a,1k,1s,1t] (Figure 1.1). For $NiEt_2(bpy)$ (bpy = 2,2′-bipyridyl), the enhancement effect is as large as 10^{10}–10^{13} [1t].

This enhancement effect is similar to that of an electron-withdrawing group on the acid dissociation of substituted benzoic acid (Hammett's effect), however, the enhancement effect on the reductive elimination is much larger than the Hammett's

Conjugated Polymer Synthesis. Edited by Yoshiki Chujo

ISBN: 978-3-527-32267-1

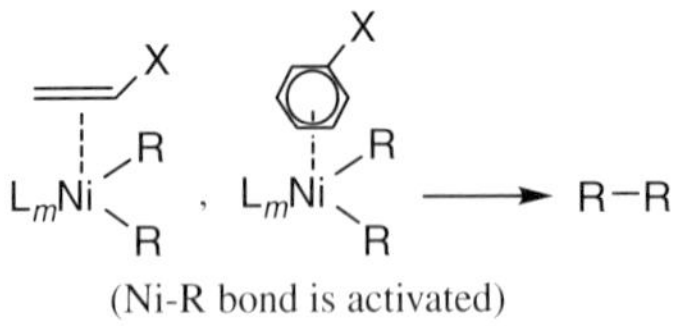

X X R R L_mNi , L_mNi ⟶ R–R R R

(Ni-R bond is activated)

Figure 1.1 Activation of Ni–R bond by coordination of an electron-accepting olefin and aromatic compounds.

effect on the acid dissociation. Because the electron withdrawing ability of the R group increases in the order:

$$\text{Pr(propyl)} < \text{Et} < \text{Me} < \text{Ph}$$

The stability of the Ni–R bond is considered to increase in this order. Actually thermal stability of NiR_2(bpy) increases in the order:

$$NiPr_2(bpy) < NiEt_2(bpy) < NiMe_2(bpy),$$

and insertion of CO into an Ni–Et bond is usually easier than into an Ni–Me bond [1u,1v]. However, the Ni–Ph bond in $NiPh_2L_m$ seems to be less stable and undergoes reductive elimination to give Ph–Ph. Attempts to isolate $NiPh_2L_m$ have not been successful, and they usually give the reductive elimination product Ph–Ph. When the Ph group has a strongly electron-withdrawing substituent(s) as in C_6F_5, the $NiPh_2L_m$ type complex (e.g., $Ni(C_6F_5)_2$(bpy)) can be isolated, and its molecular structure suggests the presence of electronic interaction between the two aromatic ligands through π-electrons in the two aromatic units [1s] (Figure 1.2):

The presence of such an electronic interaction between the two aromatic groups accounts for the ease of reductive elimination of Ph–Ph from $NiPh_2L_m$. Because of (i) the enhancement effect of electron-accepting aromatic compounds on the reductive elimination and (ii) the ease of the reductive elimination from the $NiPh_2L_m$ type complex, organometallic dehalogenative polycondensation is considered to be especially suited to polymerization of dihaloaromatic monomers, X–Ar–X, which are considered to behave as typical electron-accepting ligands to Ni.

The basic coupling reaction (reductive elimination) is a key step in Ni-promoted organic syntheses (e.g., RMgX + R′X → R–R′; 2RX + Zn → R–R; 2RX + Ni(0)

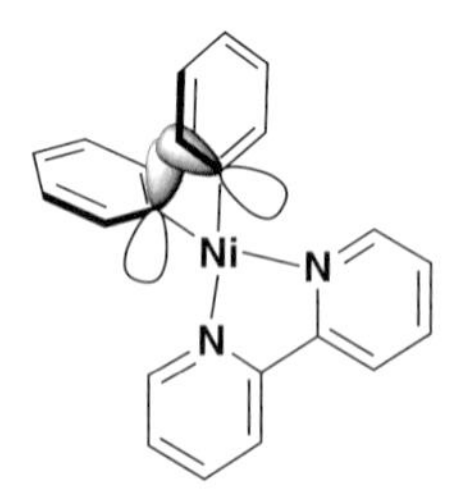

Figure 1.2 Electronic interaction between two aromatic groups through π-electrons in the aromatic ligands. For aromatic unit = C_6F_5, $Ni(C_6F_5)_2$(bpy) can be isolated [1s].

complex → R–R; X = halogen) [2]. We have developed further the utilization of this coupling reaction for the polycondensation of dihaloaromatic compounds:

$$nX{-}Ar{-}X + nMg \rightarrow n[X{-}Ar{-}MgX] \xrightarrow{\text{Ni-complex}} {-}(Ar)_n{-} \quad [3] \tag{1.2}$$

$$nX{-}Ar{-}X + nNi(0)L_m \rightarrow {-}(Ar)_n{-} + nNiX_2L_m \quad [4] \tag{1.3}$$

In some cases, $Ni(0)L_m$ formed *in situ* by chemical (e.g., by Zn) or electrochemical reduction of Ni(II)-compounds are also usable in this polycondensation, thus providing the following catalytic reactions (Eq. (1.4)) [5,6a–e]. It was reported that NaH and hydrazine hydrate could also be used as the reducing agents [6f,6g]

$$nX{-}Ar{-}X + nZn\,(\text{or } 2e^-) \xrightarrow{\text{Ni-complex}} {-}(Ar)_n{-} + nZnX_2(\text{or } 2X^-) \quad [5,6] \tag{1.4}$$

Polyarylenes can be prepared by the organometallic polycondensation as well as by chemical and electrochemical oxidation of aromatic compounds, and books and reviews have been published concerning the preparation and properties of polyarylenes [1t,7].

Organopalladium(II) complexes also undergo C–C coupling on Pd [8]. We applied C–C coupling to the following polycondensation [9a–e] which is based on Pd-promoted synthetic reactions of arylacetylenes [8b,8c,10a,b]. Acetylenic ligands of Cu complexes can migrate to Pd [8b,8c], and this migration reaction seems to occur in the C–C coupling reaction and the polycondensation to give PAE (poly(aryleneethynylene)) type polymers.

$$nX{-}Ar{-}X + nHC{\equiv}C{-}Ar'{-}C{\equiv}CH \xrightarrow{\text{Pd-Cu}} \underset{\text{PAE-type polymer}}{{-}(Ar{-}C{\equiv}C{-}Ar'{-}C{\equiv}C)_n{-}} \quad [9] \tag{1.5}$$

Successful polycondensation usually requires highly effective basic coupling reactions. However, the polycondensations expressed by Eqs. (1.2)–(1.5) give polymers with high molecular weights even when the basic C–C coupling reaction is not so effective. One of the reasons for the successful polycondensation seems to be an energetic advantage of the polycondensation leading to poly(arylene)s which seem to be stabilized by forming the extended π-conjugation system along the polymer chain. In relation to this, it was reported that polymerization of propylene to give crystalline stereoregular poly(propylene) proceeded at a much faster velocity than that giving amorphous stereo-irregular poly(propylene), presumably due to the stabilization energy attained by forming the crystal in the stereoregular polymerization [10c].

As described above, the basic C–C coupling reaction in the polycondensation is considered to proceed well, especially when the dihalo compounds, such as X–Ar–X, come (or coordinate) to the propagating species ($(\text{polymer})_a{-}NiL_m{-}(\text{polymer})_b$) to produce $(\text{polymer})_a{-}(\text{polymer})_b$ in the polycondensation. A concept that the propagation reaction proceeds selectively when the monomer comes to the propagating species also explains the smooth polycondensation and the high molecular weight polymer obtained by the polycondensation.

Because organometallic polycondensations can give the π-conjugated aromatic polymers effectively, various analogous polycondensations have been developed. For example, organostannanes and organoborons undergo similar Pd-catalyzed C–C coupling reactions [1t,7,11a–d]:

$$n\text{X}-\text{Ar}-\text{X}+n\,\underset{\text{m:SnR}_3\text{ or B(OR)}_2}{\text{m}-\text{Ar}'-\text{m}} \rightarrow -(\text{Ar}-\text{Ar}')_n- \quad [12] \tag{1.6}$$

Pd-promoted coupling reactions between ArX and olefin are also known [11e,11f]. They have also been applied to the polymerization [12]. The polymerization expressed by Eq. (1.2) is applicable to dihaloalkanes (e.g., $X-(CH_2)_n-X$) by using Cu catalyst [3g]. Use of C–OY (Y = tosyl, etc.) compounds, instead of C–X compounds (OY = leaving group or pseudo-halogen), is also possible for the polycondensation [3g,5e], which is considered to proceed through oxidative addition of C–OY to a transition metal as studied previously [13a–c].

$$\text{BrMg}-(\text{CH}_2)_m-\text{MgBr}+\text{TsO}-(\text{CH}_2)_n-\text{OTs} \xrightarrow{\text{Cu}} -(\text{CH}_2)_{m+n}- \tag{1.7}$$

$$n\text{TsO}-\text{Ar}-\text{OTs} \xrightarrow{\text{Ni(0)L}_m} -(\text{Ar})_n- \quad [5e] \qquad \text{Ts} = \text{Tosyl} \tag{1.8}$$

When the polymerization is carried out using $Ni(0)L_m$(Eq. (1.3)), the polymerization is considered to proceed through the following fundamental reactions [2f,4c];

$$\text{Ni(0)L}_m + \text{X}(\text{Ar})_i\text{X} \xrightarrow{\text{oxidative addition}} \underset{\text{Complex I}}{\text{L}_m\text{Ni}(\text{X})(\text{Ar})_i\text{X}} \tag{1.9}$$

$$\underset{\text{Complex I}}{\text{L}_m\text{Ni}(\text{X})(\text{Ar})_i\text{X}} + \underset{\text{Complex I'}}{\text{L}_m\text{Ni}(\text{X})(\text{Ar})_j\text{X}} \xrightarrow{\text{disproportionation}} \text{L}_m\text{NiX}_2 + \underset{\text{Complex II}}{\text{L}_m\text{Ni}[(\text{Ar})_i\text{X}][(\text{Ar})_j\text{X}]} \tag{1.10}$$

$$\text{Complex II} \xrightarrow{\text{reductive elimination}} \text{L}_m\text{Ni} + \underset{\text{P}_{i+j}}{\text{X}-(\text{Ar})_{i+j}-\text{X}} \tag{1.11}$$

The oxidative addition of C–X [13d–f] and C–OY [13a–c] to $Ni(0)L_m$ (Eq. (1.9)) is well known, and the disproportionation reaction [1s,13g] is also known. When the Ni–C bond has high stability, the Complexes I [2e] and II [1k,l], as well as a complex of type $L_m(X)Ni–Ar–Ni(X)L_m$ [1g] can be isolated. Thus, the basic concepts (reductive elimination, back-donation, oxidative addition) and the basic reactions in organometallic chemistry studied by us support the organometallic polycondensation.

1.2 Syntheses of π-Conjugated Polymers

By using the organometallic polycondensations expressed by Eqs. (1.2)–(1.6), various π-conjugated poly(arylene)s have been prepared. Figure 1.3 shows examples of the

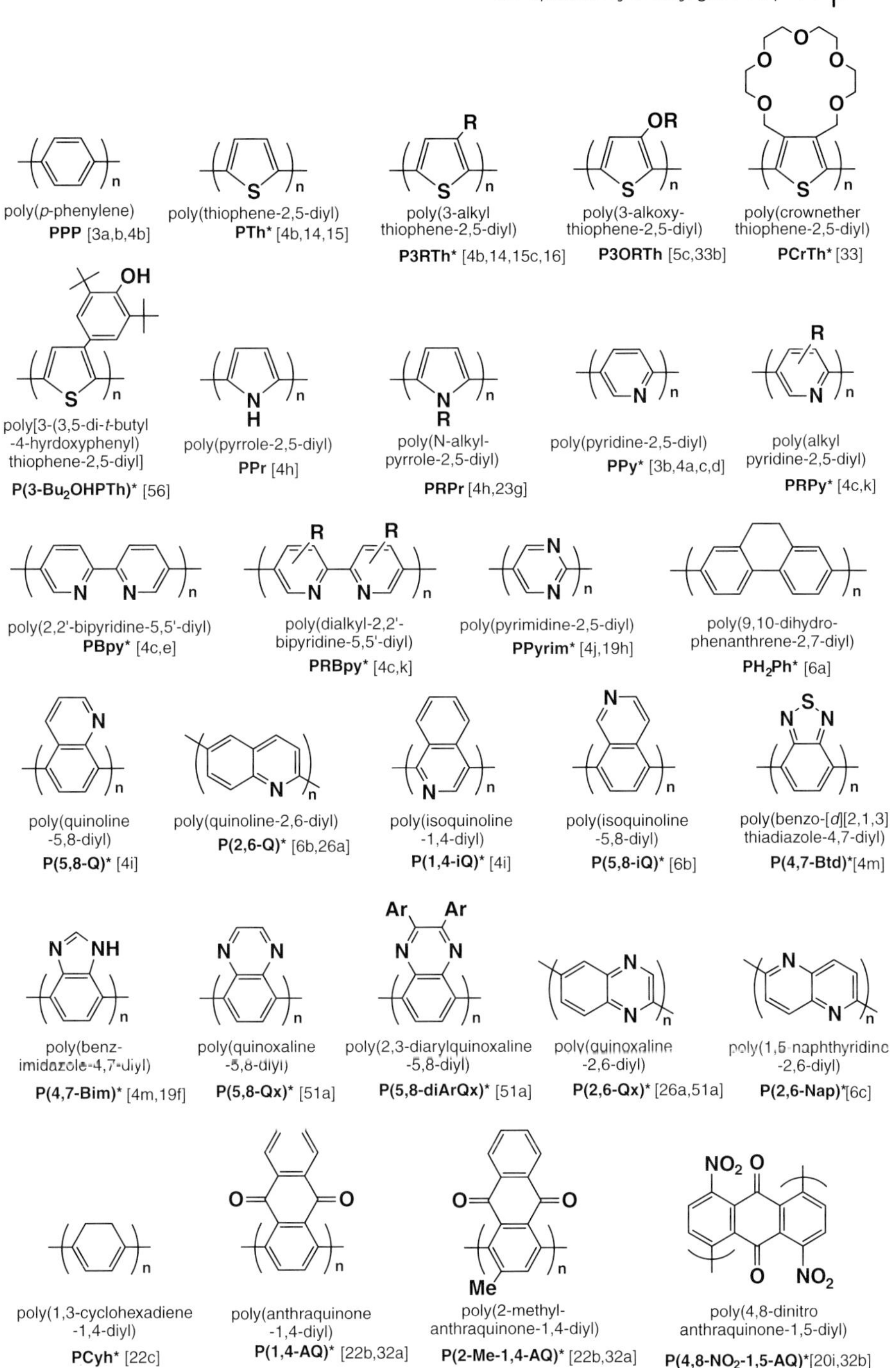

Figure 1.3 π-Conjugated polymers prepared by organometallic processes in our group. Polymers marked with an asterix were reported for the first time from the author's group.

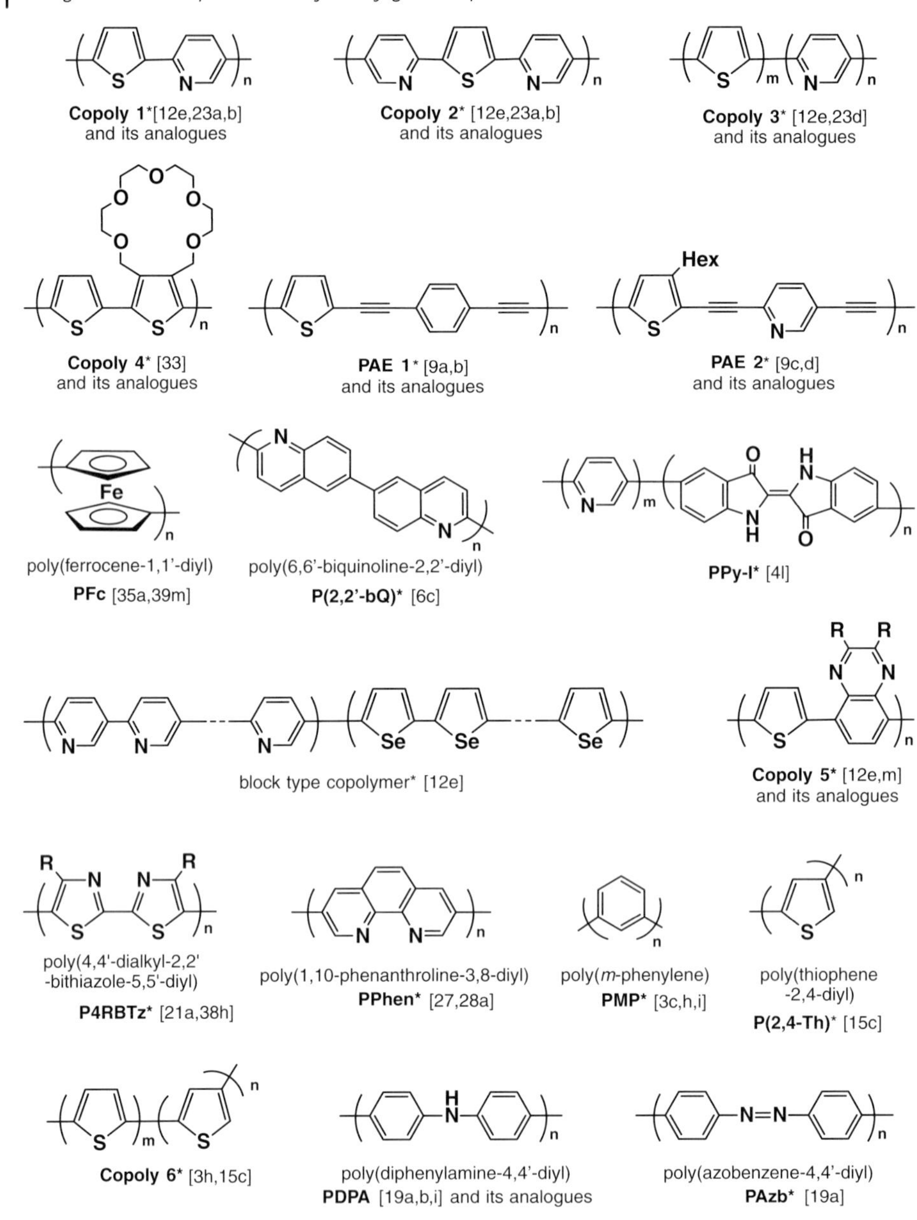

Figure 1.3 (*Continued*)

π-conjugated poly(arylene)s prepared by the organometallic polycondensation in our group. The polymers shown in Figure 1.3 are based on our previous review paper [3o], and examples of recently reported polymers will be shown later.

Some of the π-conjugated polymers shown in Figure 1.3 were patented as materials (e.g., poly(thiophene-2,5-diyl) PTh [14a] and poly(3-alkylthiophene-2,5-diyl)

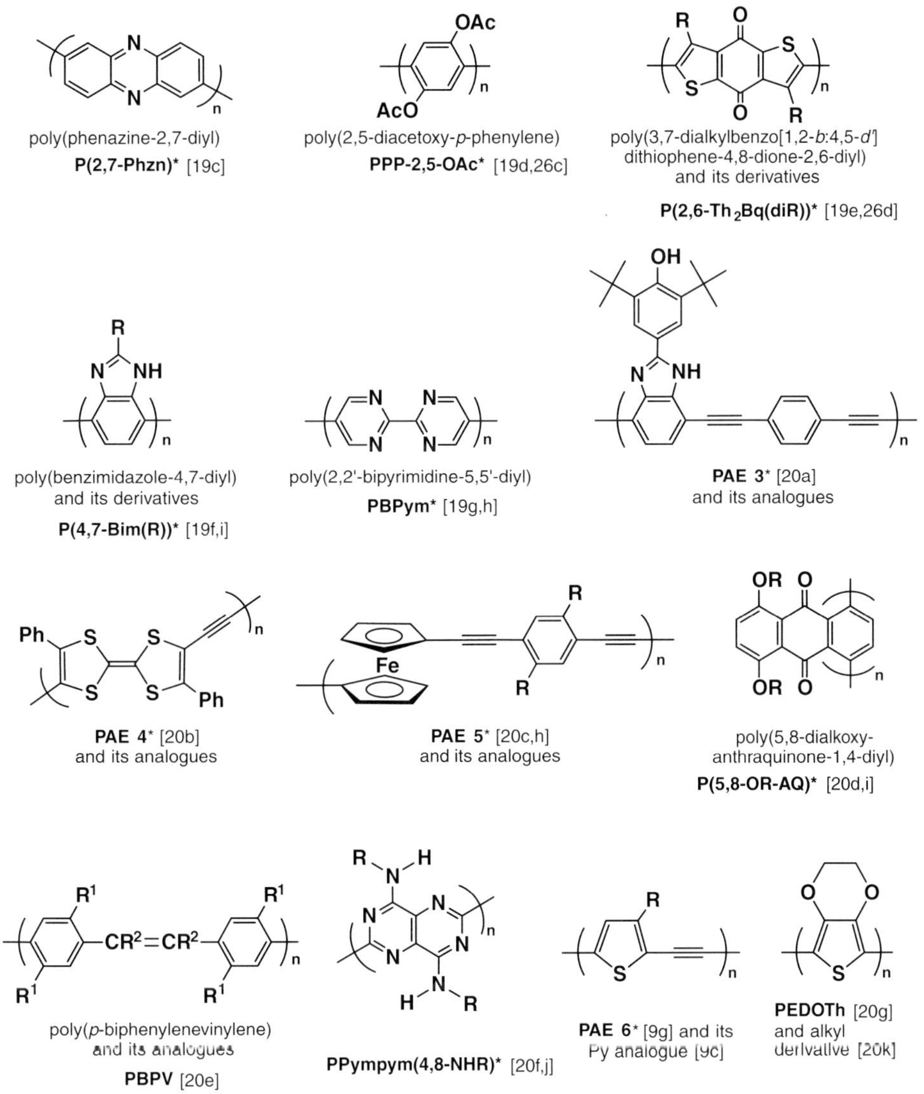

poly(phenazine-2,7-diyl)
P(2,7-Phzn)* [19c]

poly(2,5-diacetoxy-*p*-phenylene)
PPP-2,5-OAc* [19d,26c]

poly(3,7-dialkylbenzo[1,2-*b*:4,5-*d*′]dithiophene-4,8-dione-2,6-diyl) and its derivatives
P(2,6-Th$_2$Bq(diR))* [19e,26d]

poly(benzimidazole-4,7-diyl) and its derivatives
P(4,7-Bim(R))* [19f,i]

poly(2,2′-bipyrimidine-5,5′-diyl)
PBPym* [19g,h]

PAE 3* [20a] and its analogues

PAE 4* [20b] and its analogues

PAE 5* [20c,h] and its analogues

poly(5,8-dialkoxy-anthraquinone-1,4-diyl)
P(5,8-OR-AQ)* [20d,i]

poly(*p*-biphenylenevinylene) and its analogues
PBPV [20e]

PPympym(4,8-NHR)* [20f,j]

PAE 6* [9g] and its Py analogue [9c]

PEDOTh [20g] and alkyl derivative [20k]

Figure 1.3 (*Continued*)

P3RTh [14b,c]) under the name of our institute. PPP [3a,b] and PFc [35a,39m] with high crystallinity and well-defined bonding between the monomer units were also first prepared by the polycondensation.

PTh was designed as the first well-characterized and stable π-conjugated conducting polymer composed of a five-membered ring [15]. It was reported in 1982 that introduction of the alkyl group to PTh led to enhancement of solubility without losing the essential π-conjugation system [14b,c,16a]. Due to the increase in solubility, NMR analysis of the microstructure of P3RTh became possible [14c,15c,17]. For example,

s-*trans* (head-to-tail) s-*trans* (head-to-head) s-*cis* (head-to-tail)

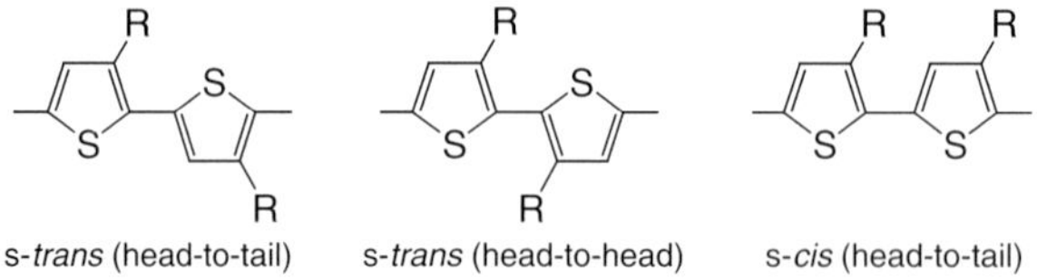

Figure 1.4 Microstructures of P3RTh.

the microstructure of P3RTh (R = CH_3) was discussed in term of head-to-tail and head-to-head joints (Figure 1.4) [14c,15c,17].

It has become possible to prepare head-to-tail type HT-P3RThs, which possess a highly controlled regioregularity, by organometallic polymerization methods [17a–f]. McCullough and his coworkers introduced Grignard reagent selectively to the 5-position of the RTh ring and polymerized it with Ni-catalyst [17a].

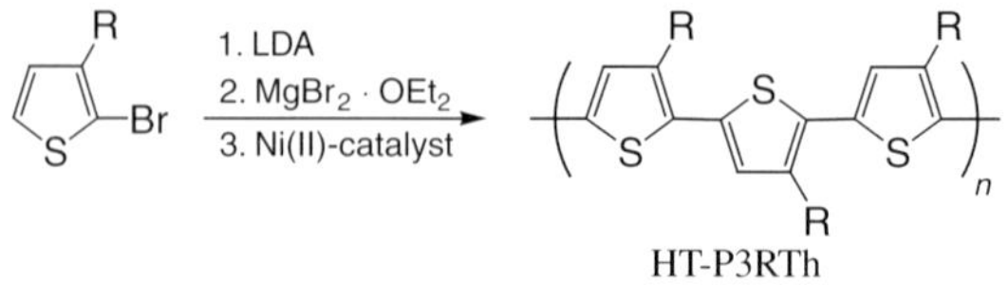

1. LDA
2. $MgBr_2 \cdot OEt_2$
3. Ni(II)-catalyst

HT-P3RTh (1.12)

On the other hand, Rieke and his coworkers reported that regio-controlled organozinc reagent was obtained with highly activated zinc (Rieke zinc) and that the organozinc reagent afforded regioregular P3RTh, in which the amount of head-to-tail product was as high as 98.5% [17b].

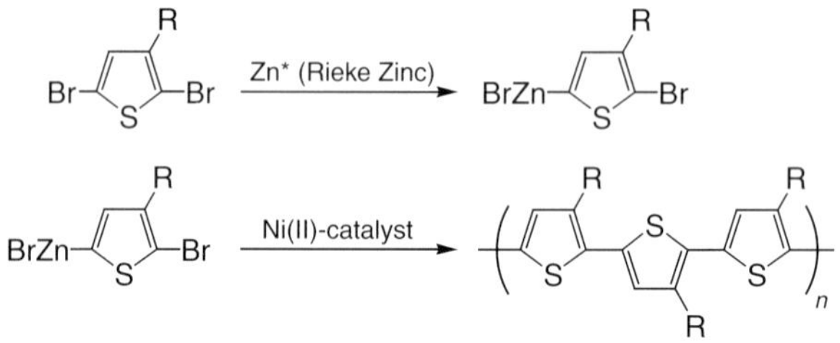

Zn* (Rieke Zinc)

Ni(II)-catalyst (1.13)

These regioregular HT-P3RThs exhibit higher crystallinity and higher electrical conductivity than the regio-irregular P3RThs, and form a π-stacked structure, as discussed later. Preparation of regio-regular HT-type polythiophene, with a *p*-alkylphenyl group at the 3-position, by oxidative polymerization has also been reported [17g]. Head-to-head type polythiophene with acetylenic –C≡CR side chains shows a strong tendency to form a π-stacked solid structure [17h–k], as discussed later. Syntheses of regioregular PAE-type polymers (e.g., PAE-6 in Figure 1.3) are also possible [9c,g].

$$\mathrm{X{-}Ar{-}C \equiv CH} \xrightarrow{\text{Pd-}Cu} \mathrm{{-}(Ar{-}C{\equiv}C)}_n{-} \tag{1.14}$$

For some of the π-conjugated polymers (e.g., RTh and P3RTh) depicted in Figure 1.3, other preparative methods (e.g., oxdative polymerization) [7a–g,17g] have also been developed.

Use of 4-chlorobenzyl chloride in the polycondensation using Mg (Eq. (1.15)) also seems to give a regio-controlled polymer due to a large difference in the reactivity between the two C–Cl bonds.

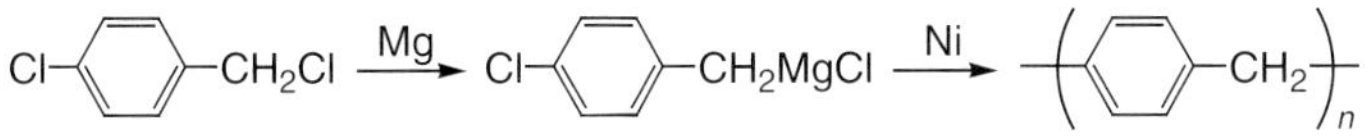

$$Cl-C_6H_4-CH_2Cl \xrightarrow{Mg} Cl-C_6H_4-CH_2MgCl \xrightarrow{Ni} -(C_6H_4-CH_2)_n- \quad (1.15)$$

The polymer gives rise to sharp X-ray diffraction peaks, supporting its crystalline structure [3b,4n]. The polycondensation proceeds well, even on addition of the transition metal complex prior to the formation of the Grignard reagent (Eq. (1.2)) [3b]; in this case, Mg may serve as a reducing reagent for the catalyst, similar to Zn in Eq. (1.4). For catalysis of the polycondensation with Mg, Ni-compounds are usually most effective, and conditions and catalysts have been examined for synthesis of PPP, PTh, and related copolymers [3h]. However, other transition metal (e.g., Pd and Fe) compounds sometimes exhibit catalytic activity [3b].

The molecular weight of the π-conjugated poly(arylene)s prepared by the organometallic polycondensation seems to depend on the solubility and crystallinity of the polymers. The polymerization is considered to proceed even in slurrys of oligomeric or polymeric propagating species deposited from the solvent [4b,c]. There seems to be a trend that crystalline polymers have a lower molecular weight whereas less crystalline and/or soluble propagating species (especially those with an alkyl chain) give a higher molecular weight polymer. For example the poly(arylene)s prepared by using $Ni(0)L_m$ have the following molecular weights:

PPy: $M_w = 4300$ [4c], 6300 [18a] with [η] of 2.29 dL g^{-1}
PBpy: 3200 [4c]
PRPy: 120 00–27 000 (R = CH_3), 36 000 (R = 2-hexyl)
PRBpy: 21 000 (R = 2-hexyl)
P3RTh: 190 000 (R = hexyl)
P(2-Me-1,4-AQ): 190 000
Copoly 3: about 5×10^4–5×10^6.

For PPy, its preparation [18a] on a larger (50 g) scale gave the polymer with a higher molecular weight ($M_w = 6300$, determined by light scattering method) than the preparation on a 1 g scale [4c], which afforded a polymer with $M_w = 4300$. PPy with $M_w = 6300$ showed an intrinsic viscosity [η] of as large as 2.29 dL g^{-1} [18a]. Data from elemental analyses agreed with the structure of the polymers. Worked-up PPy prepared by $Ni(0)L_m$ (Eq. (1.3)) contained 13 ppm Ni [18a] and negligible halogen. Sometimes Ni was not detected in ICP analysis of polymers prepared using the Ni(0) complex. The polymers prepared by $Ni(0)L_m$ often possess an H-terminated end group [4c,26c]. which is considered to be formed from Ni-terminated aryl groups [1g,2e] during the work-up, including treatment with acids such as HCl;

$$\text{----polymer-NiL}_m + H^+ \rightarrow \text{---polymer-H}$$

All of the poly(arylene)s shown in Figures 1.3 and 1.5, except for PPr [4h] and PCyh [22c], are stable in air. For example, PTh and P3RTh which can be stored long years [14–16] in an open atmosphere underwent virtually no change. Vacuum-deposited PTh film showed some electrical conductivity under air [22a]. PPr receives chemical redox reactions [4h] and PCyh is air-sensitive (*vide infra*). Many of the π-conjugated polymers shown in Figure 1.3 are soluble. However, PPP, PTh, and PCyh are insoluble.

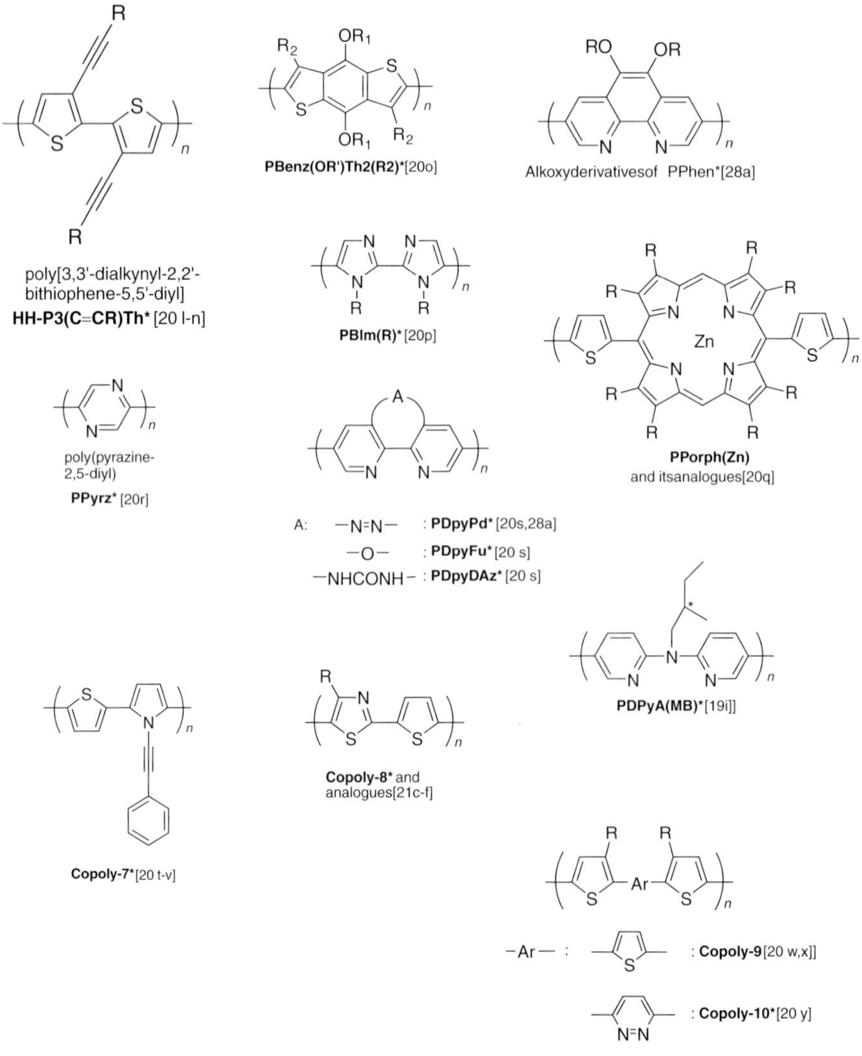

poly[3,3'-dialkynyl-2,2'-bithiophene-5,5'-diyl]
HH-P3(C≡CR)Th* [20 l-n]

PBenz(OR')Th2(R2)*[20o]

Alkoxyderivativesof PPhen*[28a]

PBIm(R)* [20p]

poly(pyrazine-2,5-diyl)
PPyrz* [20r]

PPorph(Zn)
and itsanalogues[20q]

A: —N=N— : **PDpyPd*** [20s,28a]
—O— : **PDpyFu*** [20 s]
—NHCONH— : **PDpyDAz*** [20 s]

PDPyA(MB)*[19i]]

Copoly-7*[20 t-v]

Copoly-8* and analogues[21c-f]

—Ar— : **Copoly-9**[20 w,x]]
Copoly-10*[20 y]

Figure 1.5 Examples of recently prepared π-conjugated polymers.

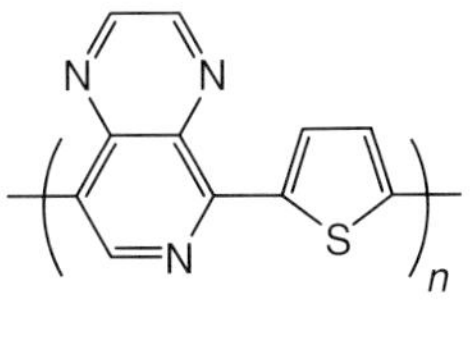

Copoly 7* [23 e]

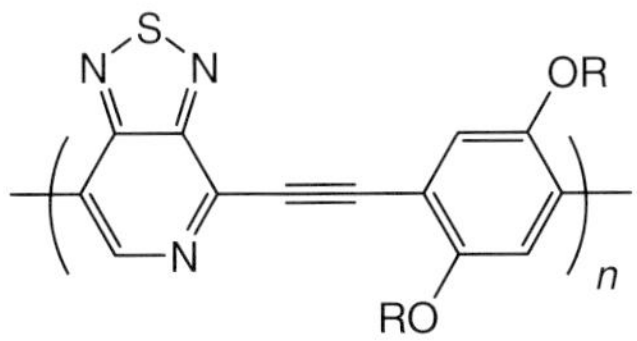

PAE 7* [23 h]

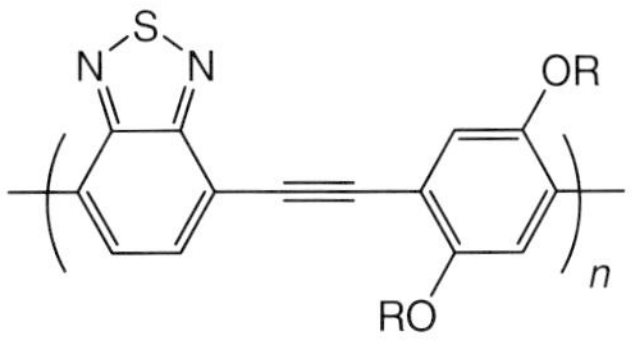

PAE 8* [23 i-k]

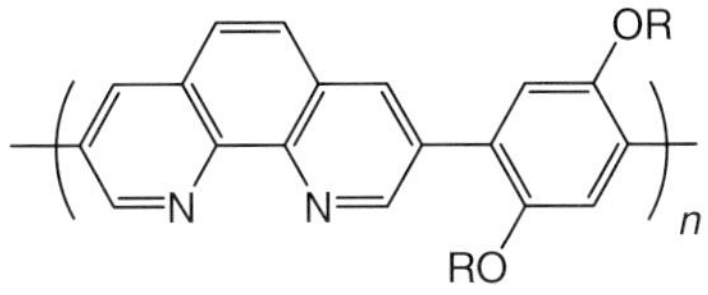

Copoly 8* [23 l,m]

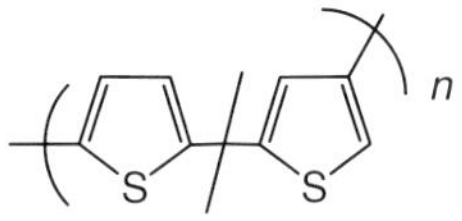

Copoly 6* and phenylene analogues [23 n,q]

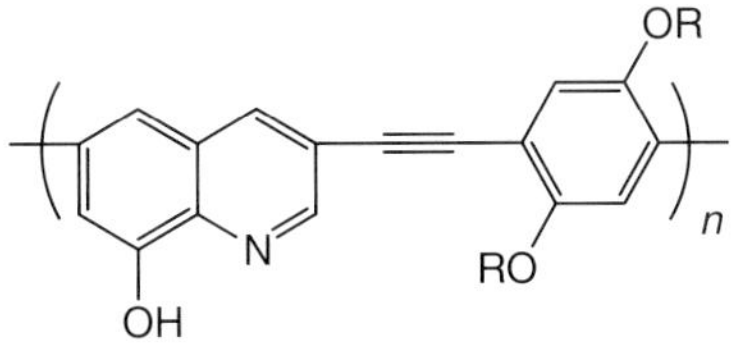

PAE 9* and analogues [23 r,s]

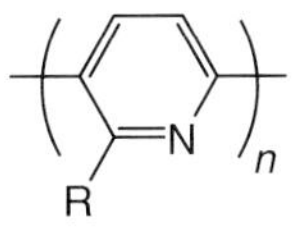

HT-P6RPy* [23 t]

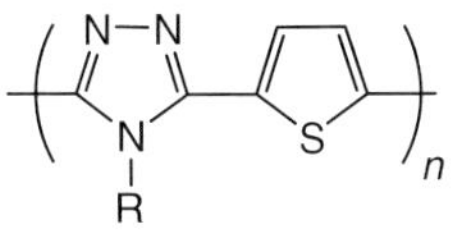

Copoly 9* [23 u]

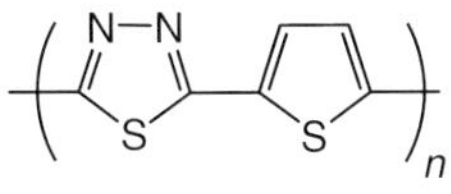

Copoly 10* [23 u]

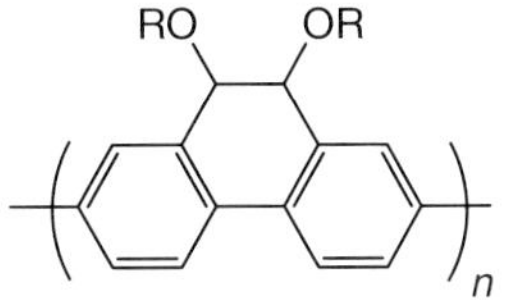

PH_2Ph(9,10-OR)* and its analogues [23 b-f]

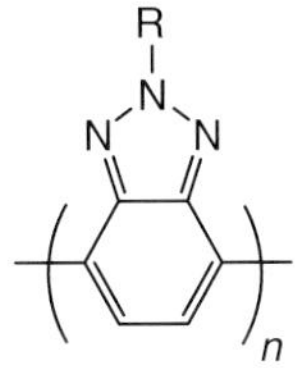

P[BTzR] and related polymers [23 v,w]

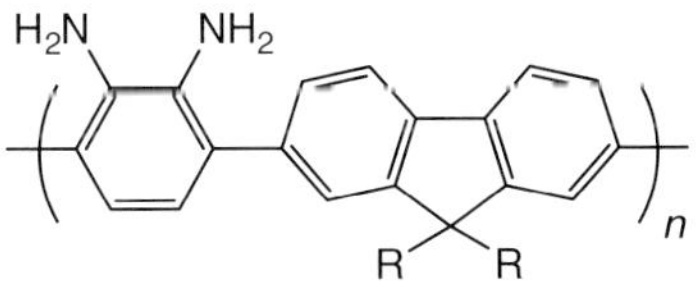

Copoly 11* and boron derivatives [23 x]

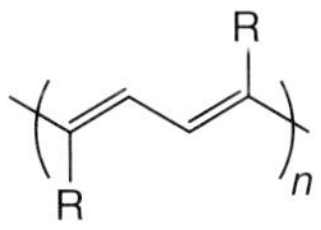

HH-P(CH=CR) [23 y]

Figure 1.5 (*Continued*)

Examples of π-conjugated polymers prepared recently in our group are shown in Figure 1.5. Poly(9,10-disubstituted-9,10-dihydrophenanthrene-2,7-diyl)s (Scheme 1.1) such as $PH_2Ph(9,10\text{-}OSiBu_3)$ were light emitting and showed strong circular dichroism [28b–e].

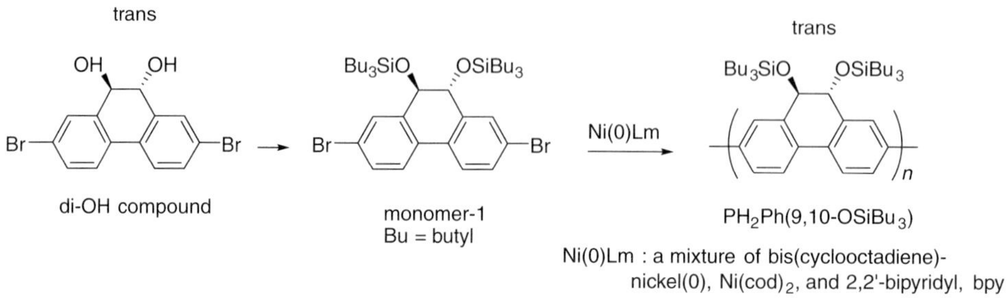

trans
OH OH
Br Br
di-OH compound
Bu_3SiO $OSiBu_3$
Br Br
monomer-1
Bu = butyl
Ni(0)Lm
trans
Bu_3SiO $OSiBu_3$
n
$PH_2Ph(9,10\text{-}OSiBu_3)$
Ni(0)Lm : a mixture of bis(cyclooctadiene)-nickel(0), $Ni(cod)_2$, and 2,2'-bipyridyl, bpy

Scheme 1.1 Preparation of poly(9,10-disubstituted-9,10-dihydrophenanthrene-2,7-diyl).

1.3 Optical Properties

1.3.1 UV–Vis Data

Because of the expansion of the π-conjugation system, π–π* absorption bands of the poly(arylene)s show red shifts from the bands of their corresponding monomeric compounds. The degree of the red shift reflects steric hindrance around the bond connecting the monomeric units. Thus PTh, P3RTh ($R = CH_3$), and P4RBTz ($R = CH_3$), which all inherently have minor steric hindrances in their intramonomer bonds, show large red shifts (Figure 1.6) [3d,3f,3o,18b]:

Degree of red shift:
Thiophene-PTh: about 21 000 cm^{-1}
3-Methylthiophene-P3RTh ($R = CH_3$): 19 700 cm^{-1}
4-Methylthiazole-P4RBTz ($R = CH_3$): 21 600 cm^{-1}.

On the other hand benzene (255 nm)–PPP (375 nm) [22a] and pyridine (248 nm)–PPy (373 nm) [4c] couples give somewhat smaller red shifts of about 13 000 cm^{-1}, partly due to a larger π-conjugation system of the basic unit and to the steric repulsion caused by the *o*-CH group. For an anthraquinone–P(1,4-AQ) couple, the red shift becomes much smaller (3500 cm^{-1}) [22b] for analogous reasons.

In the case of copolymers composed of electron-donating arylene and electron-withdrawing arylene units (see below), the copolymers are considered to have an intramolecular charge transfer structure. Copoly 1 gives rise to an absorption band at

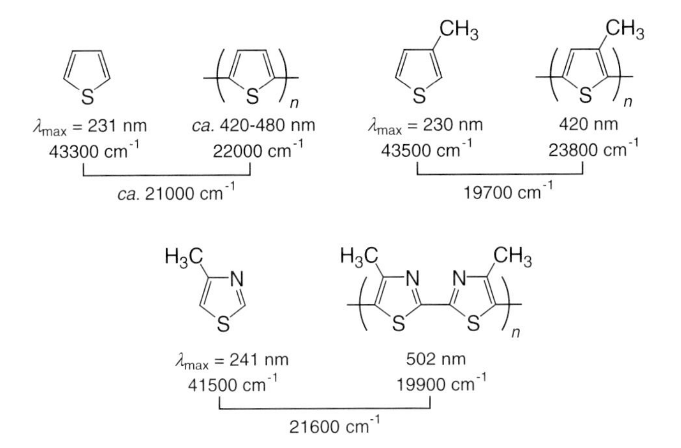

Figure 1.6 Shift of π–π^* transition energy by forming a polymer system.

wavelengths longer than the λ_{max} of the corresponding homopolymers PPy and PTh [12e,20y,21,23a–d].

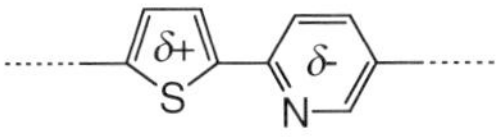

δ+ S δ- N

Copoly 1

Copoly 1 : λ_{max}=490 nm (PThPy)
PPy : λ_{max}=370 nm

and Copoly 5 (R=Ph) : λ_{max}=603 nm
Homopolymer : P(5,8-diArQx) : λ_{max}=444 nm
: PTh : λ_{max}=420-480 nm

Preparation of similar CT type copolymers, which also give λ_{max} at a longer wavelength, have been reported [21b,24].

Selenophene has electron-donating properties similar to those of thiophene, and the following copolymerization affords block-type copolymers (Scheme 1.2).

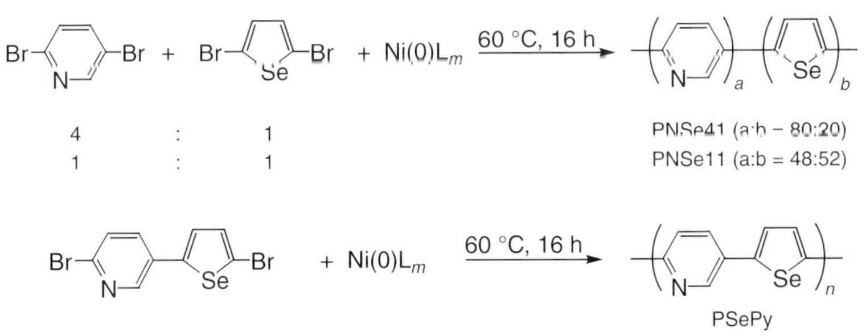

Br–(N)–Br + Br–(Se)–Br + Ni(0)L$_m$ —60 °C, 16 h→ (N)$_a$(Se)$_b$

4 : 1 PNSe41 (a:b = 80:20)
1 : 1 PNSe11 (a:b = 48:52)

Br–(N)–(Se)–Br + Ni(0)L$_m$ —60 °C, 16 h→ (N–Se)$_n$

PSePy

Scheme 1.2 Synthesis of block copolymer and alternating copolymer.

The block-type copolymers give rise to absorption bands assigned to the PPy block, the PSe block, and the CT unit formed between the two blocks. On the other hand, the alternating copolymer PSePy gives only the CT absorption band [12e]. Transfer of photo-energy in the block copolymer of pyridine and selenophene takes place very rapidly [29f].

For poly(naphthylene)-type polymers, a difference in the degree of the red shift has been noted between poly(naphthalene-2,6-diyl)-type and poly(naphthalene-1,4-diyl)-type polymers [3o,6b,6c,25,26a]:

$$\text{Quinoline-P(2,6-Q)}: \text{ red shift} = \text{about } 7500\ \text{cm}^{-1}$$
$$\text{Quinoline-P(5,8-Q)}: \text{ red shift} = 2500\ \text{cm}^{-1}$$

It was reported that *o*-substitution of PPP caused a shift of the π–π* absorption band to a shorter wavelength [6f,7a,26b,c]. PPympym(4,8-NHR) has no *o*-CH group, which gives the steric repulsion and may cause some twisting out of the main chain, and shows the π–π* absorption band at a longer wavelength than that of P(2,6-Q) [20f,j]:

$$\lambda_{max}: \begin{array}{c}\text{PPympym (4,8-NHOct)}\\ \text{Oct} = \text{octyl}\\ 452\ \text{nm}\end{array} > \begin{array}{c}\text{P(2,6-Q)}\\ \\ 403\ \text{nm (both for film)}\end{array}$$

In the case of PCyh, its color (black) indicates the formation of an effective π-conjugation system along the polymer chain, presumably due to the coplanarity of the polymer chain in the following *s-cis* conformation [22c]. In this conformation, the =C–H hydrogen of the diene unit can get between the $-CH_2$ hydrogens to form the coplanar chain. PCyh is highly reactive to oxygen in air [22c], similar to polyacetylene [22e,f], which may be due to the presence of the coplanar expanded π–conjugation system without the aromatic stabilization.

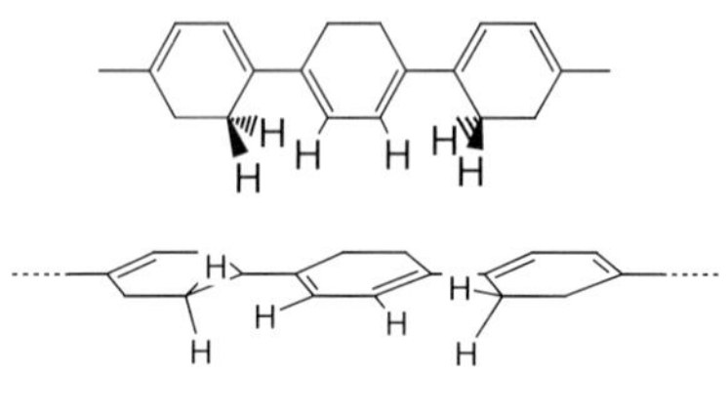

1.3.2 Photoluminescence

Most of the polymers shown in Figures 1.3 and 1.5 exhibit photoluminescence with an emission peak appearing at the onset of π–π* absorption. Linear rod-like polymers such as PPy [4c], PPhen [28] and PAE-2 [9d] often show excimer-like emission in films and solutions with high concentrations. Among PAE-1-type polymers, those containing an anthracene unit show especially strong fluorescence (Table 1.1) [9a].

In several cases, especially those concerned with PPy [12e,29a] energy transfer from a photoactivated π-conjugated unit to an energy-accepting π-conjugated unit

Table 1.1 Visible features of poly(aryleneethynylene) [9a].

Polymer	Color	Fluorescence
n	Yellow	Bluish purple
S *n*	Yellow	Bluish purple
n	Red	Green
N *n*	Light yellow	Purple
C=O C=O *n*	White	None
H_2C H_2C *n*	Yellow	Purple

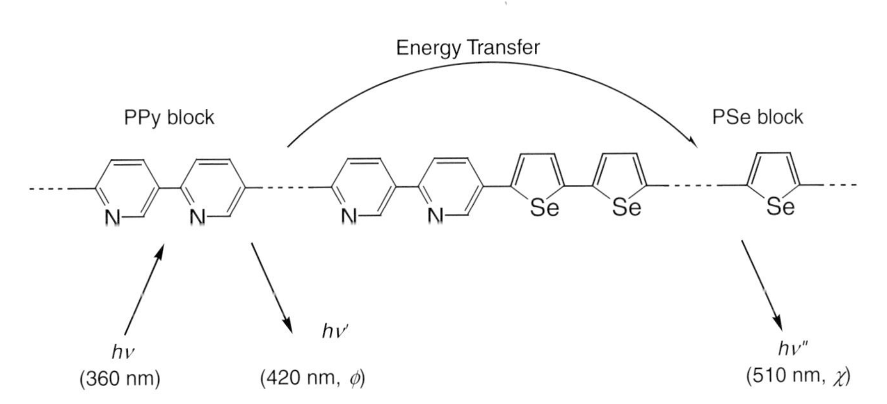

Scheme 1.3 Energy transfer from excited PPy block to PSe block.

has been observed during fluorescence (Scheme 1.3). The block structure of the above copolymer (Py-Se copolymer; Scheme 1.2) is confirmed by its UV–visible spectrum and solubility, and the copolymer undergoes such energy transfer.

Transfer of photoenergy accepted by the monomeric unit of P(5,8-diArQx) to the main chain π-conjugated system also take place [29b,c,f].

When PBpy forms a Ru complex (*vide infra*), the photoenergy accepted by the PBpy main chain is transferred into the Ru complex, and the photoemission occurs from the Ru complex [29a]. Similar energy transfers have been reported [29d,e].

1.3.3
Other Optical Properties

An interesting finding with Copolys 1–3 [12e,23b], Copoly-5 [12d,e,m], PAE-2 [9c, d,39], HT-P3RTh, and HH-P3(C≡CR)Th [20n] is that they give large $\chi(3)$ (third-order non-linear optical susceptibility) values of $3–5 \times 10^{-11}$ esu.

π-Conjugated polymers are considered to possess energy band structures similar to that of inorganic semiconductors such as TiO_2 (*vide infra*). Photogeneration of carriers is therefore expected to be possible in most kinds of π-conjugated polymers. Indeed, P(5,8-diArQx) (Ar = *p*-tolyl) gives rise to a photocurrent, and it was proposed that photocarriers are generated by the dissociation of excitons [31]. Chiral (*R,R*)-$PH_2Ph(9,10\text{-}OSiBu_3)$ shows strong circular dichroism (CD) when it forms a colloidal solution and film (for film the CD effect is about 5°/μm thickness of the film [28e].

1.4
Redox Behavior and Electrical Conductivity

π-Conjugated polymers are generally electrochemically active, and their cyclic voltammograms (e.g., that of poly(pyridine-2,5-diyl) PPy film on a Pt electrode [4c,d]) show p-doping (or oxidation) and n-doping (or reduction) peaks when the polymers are composed of electron-excessive units (e.g., pyrrole and thiophene) and electron-deficient units (e.g., pyridine and quinoxaline), respectively. Electrochemical reduction (n-doping) of PPy attains a peak cathodic potential E_{pc} of −2.43 V *vs.* Ag^+/Ag. Changing the scanning direction leads to oxidation (n-undoping) of the reduced PPy, giving rise to a peak anodic potential E_{pa} of −1.90 V. The n-doping and n-undoping are accompanied by a color change, as shown in Eq. (1.16); the doping level x in Eq. (1.16) is about 0.3.

$$(\text{pyridine-2,5-diyl})_n + nx\,NBu_4^+ + nx\,e \underset{-1.9\text{ V}}{\overset{-2.43\text{ V}}{\rightleftharpoons}} (\text{pyridine-2,5-diyl}^{-x}\ x\,NBu_4^+)_n \quad (1.16)$$

yellow — *vs.* Ag/Ag⁺ — reddish purple

Cyclic voltammographic (CV) data of P(2-Me-1,4-AQ) and its corresponding low molecular weight quinine have been recorded; the CV of P(2-Me-1,4-AQ) has been discussed based on a mixed oxidized state (Eq. (1.17)) [22b,32a].

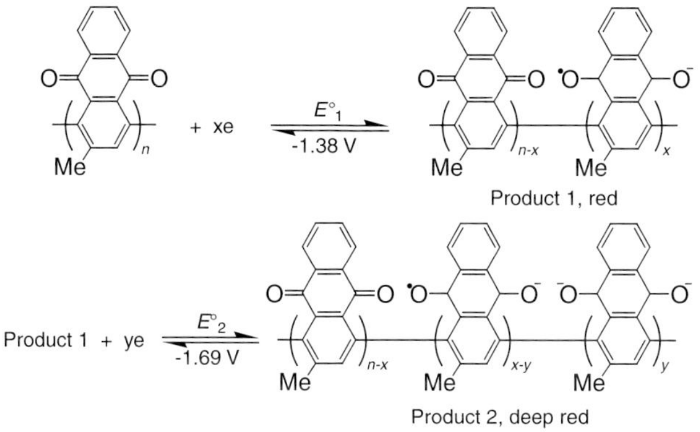

(1.17)

On the other hand, P(4,8-NO_2-1,5-AQ), bearing a strongly electron-withdrawing NO_2 group, shows an extremely low reduction potential ($E_1^0 = -0.74$ *vs.* Ag/Ag^+) [32b].

Poly(*p*-benzoquinone) PPBQ prepared from PPP-2,5-OAc undergoes electrochemical reduction (or n-doping) at a lower reduction potential (–0.5 V *vs.* Ag/Ag^+) due to the direct bonding of the *p*-benzoquinone unit in the π-conjugation system [19d,26c,32d].

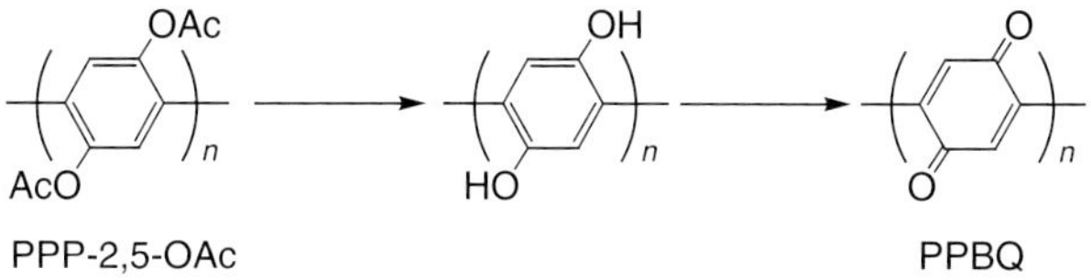

(1.18)

To our knowledge, this reduction potential is the lowest among those so far reported for π-conjugated poly(arylene)s.

Interestingly, P(4,8-NO2 1,5-AQ) gives rise to some electrical conductivity ($\sigma = 1.4 \times 10^{-6}\ S\ cm^{-1}$ at room temperature) even in the non-doped state [32b,c]. Some π-conjugated polymeric compounds (e.g., PBpy-transition metal complex [4c] and poly(arylene)-*N*-oxides) show similar electrical conductivity even in the non-doped state. The electrical conducting properties are considered to originate from generation of carrier by MLCT or from participation of resonance structures.

The ease of the electrochemical reduction of π-conjugated polymers simply reflects the electron-accepting ability of the monomeric repeating units [25]. A linear correlation holds between the reduction potential, E_{red}, of the polymer and the electron affinity E_a of the corresponding monomeric compound HArH for a wide range of poly(*p*-phenylene), poly(naphthalene-1,4-diyl), and poly(naphtalene-2,6-diyl) type polymers [19c,d,25,26].

$$E_{red} \text{ of } -(\text{Ar})_n- = a + \varrho \times E_a \text{ of X–Ar–X} \tag{1.19}$$

ϱ values of 0.75–0.8 have been obtained. A similar linear correlation is observed between the oxidation (or p-doping) potential E_{ox} of PPP, PTh, and PPr (*cf.* Figure 1.3) and the ionization potential IP of the corresponding monomeric compounds, although such a correlation is not observed between the E_{ox} of bulk metal in water and the IP of atomic metal due to a strong salvation effect [25]. A linear correlation between E_{red} of various poly(anthraquinone)s and the calculated E_a of the corresponding monomeric compounds has also been observed with a ϱ value of about 0.8 [20d].

PCrTh readily undergoes n-doping, in contrast to the other PTh analogues. The n-doped state is stabilized (even under air to some extent) due to a strong interaction of cations with ethereal oxygen, and n-undoping takes place at a potential considerably different from that of n-doping (Eq. (1.20)) [33].

(1.20)

The CT-type copolymers (e.g., Copolys 1–3) show great differences in p-doping and p-undoping potential due to an EC mechanism [12e,23a–e]. Due to an electron-withdrawing effect of the –C≡C– unit, the PAE type polymers are susceptible to electrochemical reduction [9c,d].

The π-conjugated polymers can also be chemically induced to undergo redox reactions (e.g., oxidation (or p-doping) by I_2 and $FeCl_3$ [4b,7h–j,14,16,17a,b,23f,g] and reduction (or n-doping) by Na [4a,c,I,7h–j,26a,51a]). The electrochemically and chemically oxidized or reduced polymers show electrical conductivity in the range 10^{-3} to 10^3 S cm^{-1} range, presumably due to cationic (positive) or anionic (negative) carriers formed along the π-conjugated system. Spectroscopic and XRD data support that I_2 and $FeCl_3$ are converted into I_5^- and $FeCl_4^-$ counter anions, respectively [4b,23g], in the chemical p-doping to form an ionic pair with the cationic center generated in the polymer chain. Oxidation of PFc with donor acceptors like TCNQ also gives electrically conducting materials [35a,39m].

The concept of electric conduction in π-conjugated polymers has been used to explain the "polaron", "bipolaron", "soliton", and "band model" (e.g., *cf.* Scheme 1.4) [7, 34]. The UPS spectrum of K-doped PBpy showed a peak assigned to the polaron state [34a].

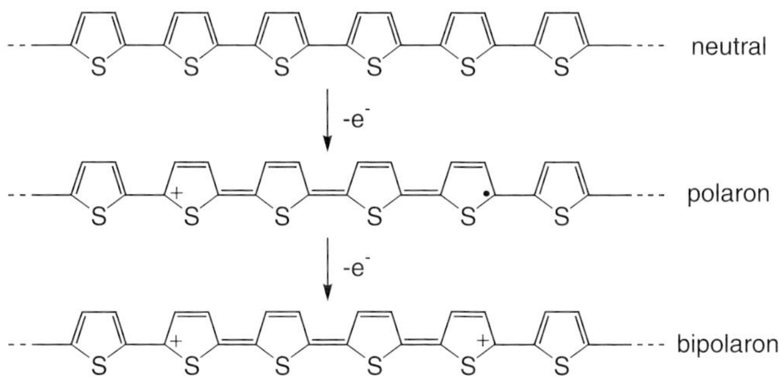

Scheme 1.4 Schematic model of polaron and bipolaron.

The iodine-doping of crystalline PTh prepared by the organometallic method gives another crystalline material with an I_n^- (n: presumably 5) counter ion [15d]; the iodine-doped crystalline PTh has electrical conductivity (σ) 30 S cm^{-1} whereas the introduction of thiophene-2,4-diyl unit (Copoly 6) causes a large decrease in the σ value. Derivatives of PTh [7,35b,c] and PPr [35d] are now used as conducting materials (e.g., as electrodes of a capacitor [35d])) industrially. In partially oxidized PFc, exchange of electrons between Fe(II) and Fe(III) species, which is related to the electrical conductivity, takes place on the Mössbauer time scale (10^{-6} s) [35a].

Due to the stabilization of the n-doped state of PCrTh (Eq. (1.20)), Na-doped (n-doped) PCrTh exhibits stability under air, as indicated by small changes in conductivity and by the IR spectrum of the Na-doped PCrTh under air. The n-doped PPy and PCrTh have electrical conductivities of 1.1×10^{-1} and 2×10^{-4} S cm^{-1}, respectively.

1.5
Linear Structure and Alignment on the Surface of Substrates

Because organometallic polycondensations give a π-conjugated poly(arylene) system with a well-defined bonding between the recurring arylene units, the polymers are considered to assume interesting structures such as rigid linear and helical structures. Furthermore, they often take an assembled structure. The rigid linear structure has been confirmed for several poly(arylene)s, based on the following observations.

1) The light-scattering analysis of PPy, PBpy and P(2,6-Q) yields a very large degree of depolarization ($\varrho_v = 0.2$–0.33) [4a,c,26a]. For example PPy gives a theoretically limiting ϱ_v value of 0.33 when irradiated with Ar laser light, indicating that it adopts an ideally linear structure with a very large anisotropy of polarizability [4a,c].

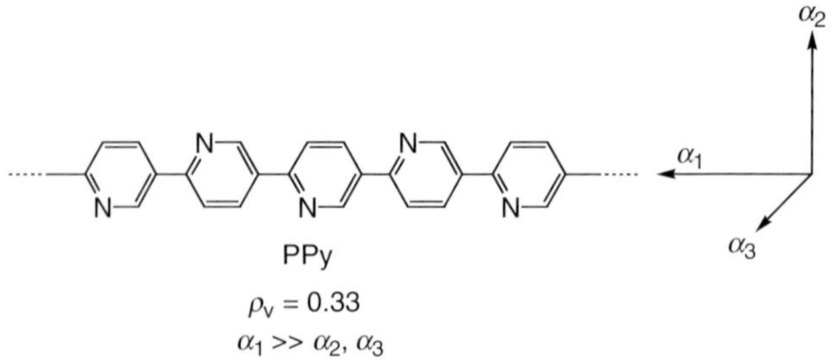

PPy

$\rho_V = 0.33$

$\alpha_1 >> \alpha_2, \alpha_3$

2) Vacuum evaporation of PPP, PTh and PBpy on carbon and metal substrates gives thin films in which the poly(arylene) molecules are aligned perpendicularly to the surface of the substrates (Figure 1.7) [3d,4b,c,22a,36].

   The alignment has been analyzed by a clear electron diffraction pattern. The report of the perpendicular alignment of PPP and PTh [36b] has been followed by many papers [7g,37] which report that oligomers (e.g., the hexamer) of thiophene are also arranged perpendicularly or somewhat tilted to the surface of the substrates. The ease of the perpendicular alignment of PPP increases in the following order on changing the substrate:

   Au, Ag $<$ Al $<$ C.

   The order is considered to reflect the magnitude of the interface energy between PPP and the substrate [22a,36c], which originates from a known metal–π–aromatic interaction for the Group 11 metals (Au and Ag).
3) Many of the π-conjugated poly(arylene)s exhibit excimer-like emissions in films and highly concentrated solutions, which can be attributed to a strong interaction between the linear rod-like molecules [4c,g,9c,d,28].
4) PBpy molecules can be aligned in parallel with the surface of a glass substrate due to coordination of 2,2′-bipyridyl with Si–O–H hydrogens on the surface of the substrate [3d,4c].
5) PPy, PBpy, and similar linear polymers give excellent polarizing films when included in stretched polymer (e.g., poly(vinyl alcohol)) films [4c,12e].

Control of the alignment of the poly(arylene) molecule may be crucial for the preparation of effective electronic and optical devices.

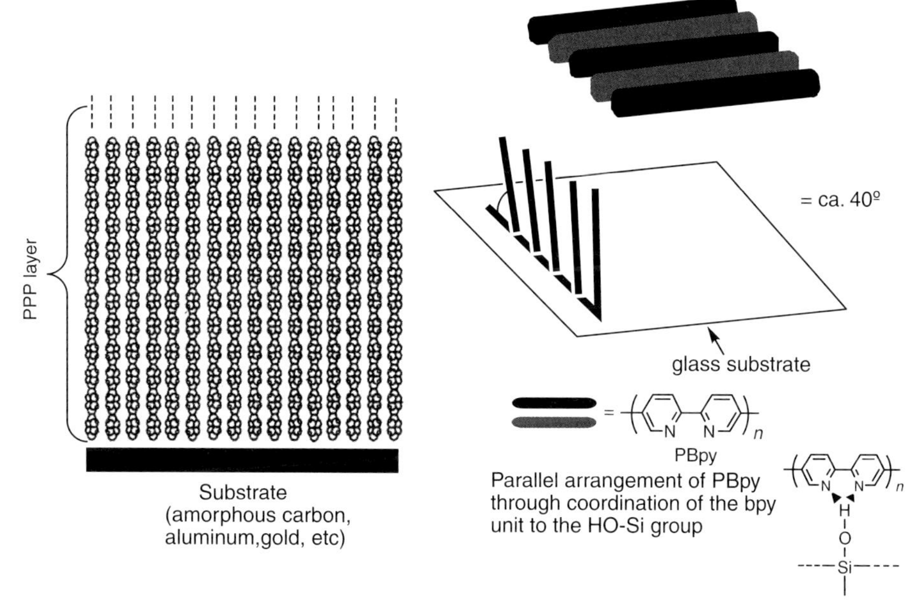

Figure 1.7 Perpendicular orientation of PPP on substrate and parallel arrangement of PBpy on a glass substrate.

1.6
Stacking in the Solid and Colloid

Regioregular head-to-tail HT-P3RTh (Eqs. (1.12) and (1.13)) [17, 38], head-to-head HH-P4RBTz [21], PAE-type polymers with CT structures (e.g., [23h–k]), and Copoly 10 [23u] with long R form stacked structures both in the solid and in colloidal solutions [17, 38]. They show sharp XRD (X-ray diffraction) peaks in a low angle region. HH-P4RBTz [38], P(2,6-Th_2Bq(diR)) [19e,26c], P(4,7-Bim(R)) [19f], P(5,8-OR-1,4-AQ) [29d], PBPV [20e], and HH-P3(C≡CR)Th (*cf.* Figures 1.3 and 1.5) with long side chains also give a sharp diffraction peak in a low angle region ($2\theta = 2$–$8°$ for CuK_α). The XRD peak in the low angle region is assigned to a distance between the core main chains separated by the long side chain (Figure 1.8).

The number density along the polymer main chain seems to determine the packing mode (the end-to-end or interdigitation packing). Plots of the *d* value *versus* number of carbons in the R group of the above described polymers give straight lines. When the slope is larger than the height of the CH_2 group (1.25 Å/C) [38h], the polymer does not have the interdigitation packing mode and is considered to take the end-to-end packing mode. On the other hand, the number density of the R group in P (2,6-Th_2Bq(diR)) is smaller than that in other polymers, and P(2,6-Th_2Bq(diR)) affords a linear line with a slope of about 1.2 Å/C, which corresponds to the interdigitation packing mode.

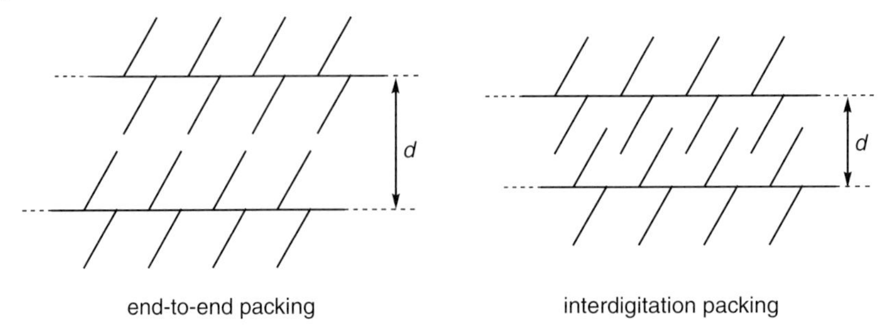

Figure 1.8 Packing modes of π-conjugated poly(arylene) with long side chains. One layer is depicted, and layers are considered to form the stacked structure.

The π-conjugated linear polymers adopt the packing mode, not only in the solid (film) but also in colloidal solutions, whose light scattering analyses sometimes reveal assembly of 10^3 polymer molecules [38h]. Revealing the driving force for the π-stacking of π-conjugated polymers is expected to give basic information on the controlling force for similar π-stacking observed with various molecules including graphite and DNA.

1.7
Chemical Reactivity and Catalysis

1.7.1
Metal Complexes and Modification of Nitrogen

The N-containing poly(arylene)s, especially those with chelating units like a bpy or 2,2′-bipyridmidine unit (PBpy and PBpym), form metal complexes (Figure 1.9) [4c,e, f,19g,29a].

PBpy forms metal complexes [4e,f], and various π-conjugated chelating polymers and their metal complexes have been synthesized and their electric and optical properties investigated [4c,12i,21,28,39a–g]. Other types of transition metal complexes have also been synthesized using PPP and its analogues [39h–k]. Ferrocene polymers have also been reported [20c,35,39k–m].

In the case of N-containing poly(arylene)s, chemical modification of imine nitrogen is also possible. For instance, quaternization with RX [40], *N*-oxidation with H_2O_2 [41], and *N*-ylidation with tetracyanoethylene oxide [41b] have been carried out for these polymers.

These polymer complexes and *N*-modified polymers show interesting electrochemical behavior. For example, a cyclic voltammogram of a film of a Ru-PBpy complex indicates an electron-exchange between Ru species, presumably *via* the π-conjugated system [4c], and RX (CH_3I, $(CH_3O)_2SO_2$, etc.) adducts of poly(quinoline)s show viologen-like redox behavior including electrochromism (Eq. (1.21)) [40].

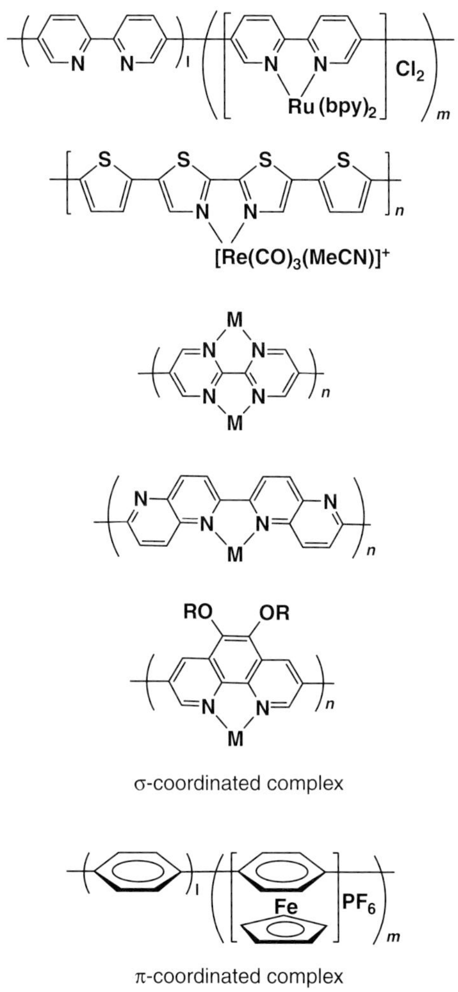

N N l N N Cl_2 Ru(bpy)$_2$ m

S S S S N N n

$[Re(CO)_3(MeCN)]^+$

M N N N N n M

N N N N n M

RO OR N N n M

σ-coordinated complex

l Fe PF_6 m

π-coordinated complex

Figure 1.9 Polymer complexes using π-conjugated polymers as ligands.

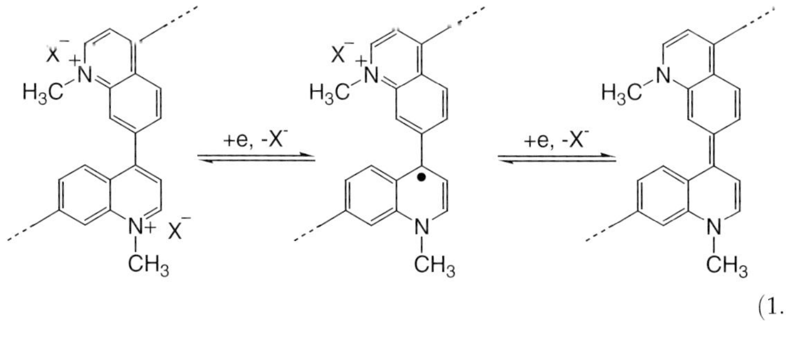

X^- + N H_3C N+ X^- CH_3 +e, -X^- X^- + N H_3C N CH_3 +e, -X^- H_3C N N CH_3

(1.21)

As described above, these *N*-modified polymers and the nitrated polymers give rise to some electrical conductivity even in the non-doped state.

From the viewpoint that π-conjugated polymers can be regarded as organic semiconductors, numerous attempts have been made to derive functions similar to those of inorganic semiconductors (e.g., TiO_2) from π-conjugated polymers. For example, photoinduced enhancement of electrical conductivity is observed for P(5,8-diArQx) (Ar = *p*-tolyl) [31] (*vide supra*). PPP and PPy prepared by organometallic polycondensation have been utilized as photocatalysts by Yanagida [42]. On the other hand, PBpy serves as a highly efficient photocatalyst for hydrogen evolution from aqueous media [43a]. This catalytic efficiency is superior to those of the other π-conjugated poly(arylene)s such as PPP and PPy [42], and this is attributed to the chelating ability and high hydrophilicity of PBpy. Electronic structures of PPP, PPy, and PBpy are compared with that of TiO_2 [22a,34,43a].

The metal complexes of the π-conjugated polymers sometimes show a highly active and stable catalytic effect (e.g., for oxidation reactions) which cannot be attained with low-molecular-weight metal complexes and metal complexes of non π-conjugated polymers [43b–d].

1.8
Electronic and Optical Devices (ECD, Battery, EL, Diode, Transistor, Nonlinear Optical Device, etc.)

1.8.1
Redox Functions

All π-conjugated poly(arylene)s change color when electrochemically oxidized or reduced (*cf.* Eqs. (1.16) and (1.17)). This phenomenon is called "electrochromism" and much effort has been made to derive practical benefits from it. Poly(vinyl alcohol) having a PTh pendant group serves as an excellent polymer electrolyte which shows electrochromism [44a,b]; poly(vinyl alcohol) serves as an excellent matrix polymer for polymer electrolytes [44c,d]. Changes in UV–visible spectra on electrochemical doping of π-conjugated polymers (e.g., n-doping of P(5,8-diArQx), Ar = C_6H_5) [45a]) have been reported.

The electrochemical redox behavior of π-conjugated polymers has been applied to batteries by many research groups [46f, 47]. Polymer batteries were reported by two groups in 1981 [46a,b]. PTh serves as a positive electrode material for Li and Zn batteries [46d–f]. The Li/LiI/PTh-I_2 solid electrolyte cell was also fabricated, and gave a high utilization of iodine [46f,g]. For a "storehouse of charge" not only π-conjugated polymers but also graphite can be used. Sanechika, who carried out his doctorate work in our group and learned the concepts of the polymer battery participated in the development of a graphite-based Li battery, which is now widely used as a lithium ion cell [48]. PDPA and its analogues have been tested as sensors for a lead battery [19b, 49a]. The N-containing polymers like PPy can transport H^+ in electrochemical processes [49b], and protonated polymers exhibit electrochemical activity similar to that of NAD and serve as an active material for a battery operating in an acidic medium [49c].

1.8.2
Electronic and Optical Devices

Recently, the utility of π-conjugated polymers as the material for an electroluminescence (EL) device or light emitting diode (LED) has been demonstrated [50]. PTh, P(5,8-diArQx), and P4RBTz serve as materials for the EL device [22a, 38c, 45b, 51]. PTh works as an excellent material for the hole-transporting layer (HTL) in the EL device [51b]. Many papers have been published on the utilization of poly(9,9-dialkylfluorene- 2,7-diyl)s and their related polymers in EL devices [50f–j], and the polymers are usually prepared by organometallic polycondensation. An Au/PTh/Al electric junction behaves as a rectifying diode [22a].

Thin layer transition (TLT) based on π-conjugated polymers or organic molecules is another target device of the π-conjugated polymers. The TFT behavior of PTh [22a, 52], Copoly-8, and Copoly-11 prepared by organometallic polycondensation has been examined [21d, 22a, 23u, 52].

Because the copolymers, Copoly 1–3 [12d, 23], PAE-2 and its analogues [9c,d, 30], and HH-P3(C≡CR)Th [20n] show large $\chi^{(3)}$ values, preparation of nonlinear optical devices using the materials is expected. A wave-guide using PAE-2 has been prepared [30b].

1.9
Conclusions

Organometallic polycondensation, based on organometallic chemistry, has contributed much to the preparation of π-conjugated polymers such as PTh and PRTh. The π-conjugated polymers prepared by organometallic polycondensation have a well-characterized structure. They have contributed to our understanding of basic chemical and physical properties as well as to elucidation of the structure of π-conjugated polymers. In addition, they are useful materials for electric and optical devices and are expected to be invaluable in their future application.

Acknowledgments

The author is grateful to the Emeritus Professors the late S. Kambara, the late S. Ikeda, and A. Yamamoto of our Institute for their guidance, discussion, and collaboration. Especially he thanks Emeritus Professor the late S. Kambara for encouraging him for a long period and Emeritus Professor A. Yamamoto for his organometallic research. Thanks are also due to Professors K. Osakada, T.-A.

Koizumi, and H. Fukumoto and Dr. Maruyama (now at Yokohama Rubber Co. Ltd.) of our Institute, Professor T. Kanbara of the University of Tsukuba, Professor I. Yamaguchi of Shimane University, and all other coworkers.

References

1 (a) Yamamoto, T., Yamamoto, A., and Ikeda, S. (1971) *J. Am. Chem. Soc.*, **93**, 3350 and 3360; (b) Kohara, T., Yamamoto, T., and Yamamoto, A. (1980) *J. Organomet. Chem.*, **192**, 265; (c) Tatsumi, K., Nakamura, A., Komiya, S., Yamamoto, T., and Yamamoto, A. (1984) *J. Am. Chem. Soc.*, **106**, 8181; (d) Komiya, S., Abe, Y., Yamamoto, A., and Yamamoto, T. (1983) *Organometallics*, **2**, 1466; (e) Åermark, B. and Ljungovist, A. (1978) *J. Organomet. Chem.*, **149**, 97; (f) Åermark, B., Johnasen, H., Ross, B., and Wahlgren, U. (1979) *J. Am. Chem. Soc.*, **101**, 5876; (g) Kim, Y.-J., Sato, R., Maruyama, T., Osakada, K., and Yamamoto, T. (1994) *J. Chem. Soc., Dalton Trans.*, 943; (h) Uchino, M., Asagi, K., Yamamoto, A., and Ikeda, S. (1975) *J. Organomet. Chem.*, **84**, 93; (i) Yamamoto, T. and Yamamoto, A. (1973) *J. Organomet. Chem.*, **57**, 127; (j) Binger, P. and Doyle, M. (1978) *J. Organomet. Chem.*, **162**, 195; (k) Yamamoto, T. and Abla, M. (1997) *J. Organometal. Chem.*, **535**, 209; (l) Murakami, Y. and Yamamoto, T. (1997) *Inorg. Chem.*, **36**, 5682; (m) Kochi, J.K. (1978) *Organometallic Mechanisms and Catalysis*, Academic Press, New York; (n) Collmann, J.P., Hegedus, L.S., Norton, J.R., and Finke, R.G. (1987) *Principles and Applications of Organometallic Chemistry*, University Science, Mill Valley, Carfornia, pp. 322 and 326; (o) Yamamoto, A. (1986) *Organotransition Metal Chemistry*, John Wiley, New York; (p) Crabtree, R.H. (1988) *The Organometallic Chemistry of the Transition Metals*, John Wiley, New York; (q) Dewar, M.J.S. (1951) *Bull. Soc. Chim. Fr.*, **18**, C71; (r) Chatt, J. and Duncanson, L.A. (1953) *J. Chem. Soc.*, 2939; (s) Yamamoto, T., Alba, M., and Murakami, Y. (2002) *Bull. Chem. Soc. Jpn.*, **75**, 1997; (t)Yamamoto, T. (2003) *Synlett*, 425; (u) Saruyama, T., Yamamoto, T., and Yamamoto, A. (1976) *Bull. Chem. Soc. Jpn.*, **49**, 546; (v) Yamamoto, T., Ishizu, J., Kohara, T., Komiya, S., and Yamamoto, A. (1980) *J. Am. Chem. Soc.*, **102**, 3758.

2 (a) Tamao, K., Sumitani, K., Kiso, Y., Zembayashi, M., Fujioka, A., Kodama, K., Nakajima, I., Minato, A., and Kumada, M. (1976) *Bull. Chem. Soc. Jpn.*, **49**, 1958 and references therein; (b) Kende, A.S., Lieberskind, L.S., and Braitsch, D. (1975) *Tetrahedron Lett.*, **16**, 3375; (c) Zembayashi, M., Tamao, K., Yoshida, J., and Kumada, M. (1977) *Tetrahedron Lett.*, **18**, 4089; (d) Semmelhack, M.F. and Ryono, L.S. (1975) *J. Am. Chem. Soc.*, **97**, 3873; (e) Zhou, Z.-H. and Yamamoto, T. (1991) *J. Organomet. Chem.*, **414**, 119; (f) Yamamoto, T., Wakabayashi, S., and Osakada, K. (1992) *J. Organomet. Chem.*, **428**, 223.

3 (a) Yamamoto, T. and Yamamoto, A. (1977) *Chem. Lett.*, 353; (b) Yamamoto, T., Hayashi, Y., and Yamamoto, A. (1978) *Bull. Chem. Soc. Jpn.*, **51**, 2091; (c) Yamamoto, T. and Yamamoto, A. (1976) *Polym. Prepr. Jpn.*, **25**, G2B07; (d) Yamamoto, T. (1992) *Prog. Polym. Sci.*, **17**, 1153; (e) Yamamoto, T. (1995) *J. Synth. Org. Chem. Jpn.*, **53**, 999; (f) Yamamoto, T. and Hayashida, H. (1998) *React. Functional Polym.*, **37**, 1; (g) Yamamoto, T., Taguchi, T., Sanechika, K., Hayashi, Y., and Yamamoto, A. (1983) *Macromolecules*, **16**, 1555; (h) Yamamoto, T., Abe, M., Wu, B., Choi, B.-K., Harada, Y., Takahashi, Y., Kawata, K., Sasaki, S., and Kubota, K. (2007) *Macromolecules*, **40**, 5504; (i) Kobayashi, S., Sasaki, S., Abe, M., Watanabe, S., Fukumoto, H., and Yamamoto, T. (2004) *Macromolecules*, **37**, 7986; (j) Yamamoto, T. (2002) *Organomet. Chem.*, **653**, 195; (k) Yamamoto, T. and Yamaguchi, I. (2003) in *Electronic and Optical Properties of Conjugated Molecular Systems in Condensed Phases* (ed. S. Hotta,),

Research Signopost, Kerala, p. 1; (l) Yamamoto, T., Yamaguchi, I., and Yasuda, T. (2005) *Adv. Polym. Sci.*, **177**, 181; (m) Yamamoto, T. and Koizumi, T.-A. (2007) *Polymer,* **48**, 5449; (n) Yamamoto, T. and Sato, T. (2009) in *Bottom-up Nanofabrication* (eds K. Ariga, and H.S. Nalwa,), American Scientific Publishers, Valencia, California, p. 229; (o) Yamamoto, T. (1999) *Bull. Chem. Soc. Jpn.*, **72**, 621.

4 (a) Yamamoto, T., Ito, T., and Kubota, K. (1988) *Chem. Lett.*, 153; (b) Yamamoto, T., Morita, A., Miyazaki, Y., Maruyama, T., Wakayama, H., Zhou, Z.-H., Nakamura, Y., Kanbara, T., Sasaki, S., and Kubota, K. (1992) *Macromolecules,* **25**, 1214; (c) Yamamoto, T., Maruyama, T., Zhou, Z.-H., Ito, T., Fukuda, T., Yoneda, Y., Begum, F., Ikeda, T., Sasaki, S., Takezoe, H., Fukuda, A., and Kubota, K. (1994) *J. Am. Chem. Soc.*, **116**, 4832; (d) Yamamoto, T., Ito, T., Sanechika, K., and Hishinuma, M. (1988) *Synth. Met.*, **25**, 103; (e) Yamamoto, T., Zhou, Z.-H., Maruyama, T., Yoneda, Y., and Kanbara, T. (1990) *Chem. Lett.*, 223; (f) Yamamoto, T., Yoneda, Y., and Maruyama, T. (1992) *J. Chem. Soc., Chem. Commun.*, 1652; (g) Yamamoto, T., Maruyama, T., Ikeda, T., and Sisido, M. (1990) *J. Chem. Soc., Chem. Commun.*, 1306; (h) Yamamoto, T., Zhou, Z.-H., Ando, I., and Kikuchi, M. (1993) *Makromol. Chem. Rapid Commun.*, **14**, 833; (i) Kanbara, T., Saito, N., and Yamamoto, T. (1991) *Macromolecules,* **24**, 5883; (j) Kanbara, T., Kushida, T., Saito, N., Kuwajima, I., Kubota, K., and Yamamoto, T. (1998) *Chem. Lett.*, 583; (k) Maruyama, T., Kubota, K., and Yamamoto, T. (1993) *Macromolecules,* **26**, 4055; (l) Yamamoto, T. and Kizu, K. (1995) *J. Phys. Chem.*, **99**, 8; (m) Kanbara, T. and Yamamoto, T. (1993) *Chem. Lett.*, 419; (n) Yamamoto, T., Kim, S.-B., and Choi, B.K. (1999) *J. Polym. Sci.: Part B: Polym. Phys.*, **37**, 2544.

5 (a) Yamamoto, T., Osakada, K., Wakabayashi, T., and Yamamoto, A. (1985) *Makromol. Chem. Rapid Commun.*, **6**, 671; (b) Yamamoto, T., (1986) Japan Kokai Patent, 1986-233014; (c) Yamamoto, T., Kashiwazaki, A., and Kato, K. (1989) *Makromol. Chem.*, **190**, 1649; (d) Ueda, M. and Ichikawa, F. (1990) *Macromolecules,* **23**, 926; (e) Grab, M.G., Fering, A.E., Auman, B.C., Percec, V., Zhao, M., and Hill, D.H. (1996) *Macromolecules,* **29**, 7284.

6 (a) Saito, N., Kanbara, T., Sato, T., and Yamamoto, T. (1993) *Polym. Bull.*, **30**, 285; (b) Saito, N., Kanbara, T., Nakamura, Y., Yamamoto, T., and Kubota, K. (1994) *Macromolecules,* **27**, 756; (c) Saito, N. and Yamamoto, T. (1995) *Macromolecules,* **28**, 4260; (d) Yamamoto, T. and Saito, N. (1996) *Macromol. Chem. Phys.*, **197**, 165; (e) Fauvarque, J.-F., Digua, A., Petit, M.-A., and Savard, J. (1985) *Makromol. Chem.*, **186**, 2415; (f) Yamamoto, T., Hayashida, N., and Maruyama, T. (1997) *Macromol. Chem. Phys.*, **198**, 341; (g) Kitada, K. and Ozaki, S. (1995) *Polym. J.*, **27**, 1161.

7 (a) Skotheim, T.A. and Reynolds, J.R. (eds) (2007) *Handbook of Conducting Polymers,* 3rd edn, CRC Press, Boca Raton, FL; (b) Salaneck, W.R., Clark, D.L., and Samuelsen, E.J.E. (1990) *Science and Applications of Conducting Polymers,* Adam Hilger, New York; (c) MacDiarmid, A.G. and Heeger, A.J. (1981) NRL Memo, Rep., AD-A05816208; (d) Nalwa, H.S. (ed.) (1997) *Handbook of Organic Conductive Molecules and Polymers,* John Wiley, Chickester; (e) Roncali, J. (1992) *Chem. Rev.*, **92**, 711; (f) McCullough, R.D. (1998) *Adv. Mater.*, **10**, 93; (g) Fichou, D. (1999) *Handbook of Oligo- and Polythiophenes,* Wiley-VCH, Weinheim; (h) Shirakawa, H. (2001) *Angew. Chem. Int. Ed.*, **40**, 2575; (i) MacDiarmid, A.G. (2001) *Angew. Chem. Int. Ed.*, **40**, 2581; (j) Heeger, A.J. (2001) *Angew. Chem. Int. Ed.*, **40**, 2591.

8 (a) Ozawa, F., Kurihara, K., Yamamoto, T., and Yamamoto, A. (1958) *Bull. Chem. Soc. Jpn.*, **58**, 399; (b) Osakada, K., Sakata, R., and Yamamoto, T. (1997) *Organometallics,* **16**, 5354; (c) Osakada, K., Sakata, R., and Yamamoto, T. (1997) *J. Chem. Soc., Dalton Trans.*, 1265; (d) Ozawa, F., Fujimori, M., Yamamoto, T., and Yamamoto, A. (1986) *Organometallics,* **5**, 2144; (e) Choudary, B.M., Mahdi, S.M., kantam, M.L., Sreedhar, B., and Iwasawa, Y. (2004) *J. Am. Chem. Soc.*, **126**, 2292;

(f) Osakada, K., Onodera, H., and Nishihara, Y. (2005) *Organometallics*, **24**, 190; (g) Koizumi, T.-A., Yamazaki, A., and Yamamoto, T. (2008) *Dalton Trans.*, 3949.

9 (a) Sanechika, K., Yamamoto, T., and Yamamoto, A. (1984) *Bull. Chem. Soc. Jpn.*, **57**, 752; (b) Sanechika, K., Yamamoto, T., and Yamamoto, A. (1981) *Polym. Prepr. Jpn.*, **30**, 160; (c) Yamamoto, T., Takagi, M., Kizu, K., Maruyama, T., Kubota, K., Kanbara, H., Kurihara, T., and Kaino, T. (1983) *J. Chem. Soc., Chem. Commun.*, 797; (d) Yamamoto, T., Yamada, W., Takagi, M., Kizu, K., Maruyama, T., Ooba, N., Tomaru, S., Kurihara, T., Kaino, T., and Kubota, K. (1994) *Macromolecules*, **27**, 6620; (e) Yamamoto, T., Honda, K., Ooba, N., and Tomaru, S. (1998) *Macromolecules*, **31**, 7; (f) Trumbo, D.L. and Marvel, C.S. (1986) *J. Polym. Sci., Part A., Polym. Chem.*, **24**, 2311; (g) Hayashi, H. and Yamamoto, T. (1997) *Macromolecules*, **30**, 330; (h) Li, J. and Pang, Y. (1997) *Macromolecules*, **30**, 7487.

10 (a) Dieck, H.A. and Heck, R.F. (1975) *J. Organomet. Chem.*, **93**, 259; (b) Sonogashira, K., Tohda, Y., and Hagihara, N. (1975) *Tetrahedron Lett.*, **16**, 4467; (c) Soga, K., Ohgisawa, M., and Shiono, T. (1989) *Makromol. Chem., Rapid Commun.*, **10**, 503.

11 (a) Stille, J.K. (1986) *Angew. Chem., Int. Ed. Engl.*, **25**, 508; (b) Suzuki, A. (1982) *Acc. Chem. Res.*, **15**, 178; (c) Suzuki, A. (1985) *Pure Appl. Chem.*, **57**, 1749 (d) Kosugi, M., Koshiba, M., Sano, H., and Migita, T. (1985) *Bull. Chem. Soc. Jpn.*, **58**, 1075; (e) Mizoroki, T., Mori, K., and Ozaki, A. (1971) *Bull. Chem. Soc. Jpn.*, **44**, 581; (f) Heck, R.F. and Nolley, J.P. (1972) *J. Org. Chem.*, **37**, 2320.

12 (a) Remmers, M., Schize, M., and Wegner, G. (1996) *Macromol. Rapid Commun.*, **17**, 239; (b) Hu, Q.-S., Vitharana, D., Liu, G., Jain, V., and Pu, L. (1996) *Macromolecules*, **29**, 5075; (c) Bochmann, M. and Lu, J. (1994) *J. Polym. Sci., Part A: Polym. Chem.*, **32**, 2493; (d) Kanbara, T., Miyazaki, Y., and Yamamoto, T. (1995) *J. Polym. Sci., Part A: Polym. Chem.*, **33**, 999; (e) Yamamoto, T., Zhou, Z.-H., Kanbara, T., Shimura, M., Kizu, K., Maruyama, T., Nakamura, Y., Fukuda, T., Lee, B.-L., Ooba, N., Tomaru, S., Kurihara, T., Kaino, T., Kubota, K., and Sasaki, S. (1996) *J. Am. Chem. Soc.*, **118**, 10389; (f) Tamao, K., Yamaguchi, S., Ito, Y., Matsuzaki, Y., Yamabe, T., Fukushima, M., and Mori, S. (1995) *Macromolecules*, **28**, 8668; (g) Nishide, H., Kaneko, T., Torii, S., Kuzumaki, Y., and Tsuchida, E. (1996) *Bull. Chem. Soc. Jpn.*, **69**, 499; (h) Chan, W.-K. and Yu, L. (1995) *Macromolecules*, **28**, 6410; (i) Peng, Z., Gharavi, A.R., and Yu, L. (1997) *J. Am. Chem. Soc.*, **119**, 4622; (j) Schluter, A.D. and Wegner, G. (1993) *Acta Polym.*, **44**, 59; (k) Bochmann, M. and Kelly, K. (1989) *J. Chem. Soc., Chem. Commun.*, 532; (l) Bao, Z., Chan, W.K., and Lu, L. (1995) *J. Am. Chem. Soc.*, **117**, 12426; (m) Yamamoto, T., Lee, B.-L., Kokubo, H., Kishida, H., Hirota, K., Wakabayashi, T., and Okamoto, H. (2003) *Macromol. Rapid Commun.*, **24**, 440.

13 (a) Yamamoto, T., Ishizu, J., Kohara, T., Komiya, S., and Yamamoto, A. (1980) *J. Am. Chem. Soc.*, **102**, 3758; (b) Yamamoto, T., Ishizu, J., and Yamamoto, A. (1981) *J. Am. Chem. Soc.*, **103**, 6863; (c) Yamamoto, T., Miyashita, S., Komiya, S., Ito, T., and Yamamoto, A. (1982) *Organometallics*, **1**, 808; (d) Hidai, M., Kashiwagi, T., Ikeuchi, T., and Uchida, Y. (1971) *J. Organomet. Chem.*, **30**, 279; (e) Fahey, D.R. and Mahan, J.E. (1979) *J. Am. Chem. Soc.*, **36**, 269; (f) Parshall, G.W. (1974) *J. Am. Chem. Soc.*, **96**, 2360; (g) Yamamoto, T., Kohara, T., and Yamamoto, A. (1981) *Bull. Chem. Soc. Jpn.*, **54**, 1720.

14 (a) Yamamoto, T., Sanechika, K., and Yamamoto, A.,Japan Kokai Pat. 1981-47421. Jpn. Pat. 120926 (1984) and 1236581 (1984); (b) Yamamoto, T., Sanechika, K., and Yamamoto, A.,Jpn. Kokai Pat. 1983-147426 (applied: 26 February 1981); Jpn. Pat. 1258016 (1985) and 1262404 (1985); (c) US Pat. 4521589 (1985)

15 (a) Yamamoto, T., Sanechika, K., and Yamamoto, A. (1980) *J. Polym. Sci., Polym. Lett. Ed.*, **18**, 9, received April 20, 1979; (b) Sanechika, K., Yamamoto, T., and Yamamoto, A. (1979) *Polym. Prepr. Jpn.*, **27**, 966; (c) Yamamoto, T., Sanechika, K., and Yamamoto, A. (1983) *Bull. Chem. Soc.*

Jpn., **56**, 1497 and 1503; (d) Yamamoto, T., Morita, A., Miyazaki, Y., Maruyama, T., Wakayama, H., Zhou, Z.-H., Nakamura, Y., Kanbara, T., Sasaki, S., and Kubota, K. (1992) *Macromolecules*, **25**, 1214.

16 (a) Yamamoto, T. and Sanechika, K. (1982) *Chem. Ind. (London)*, 301; (b) Miyazaki, Y. and Yamamoto, T. (1994) *Synth. Met.*, **64**, 69.

17 (a) McCullough, R.D., Tristam-Nagle, S., Williams, S.P., Lowe, R.D., and Jayaraman, M. (1993) *J. Am. Chem. Soc.*, **115**, 4910; (b) Chen, T., Wu, X., and Rieke, R.D. (1995) *J. Am. Chem. Soc.*, **117**, 233; (c) McCullough, R.D. and Lowe, R.D. (1992) *J. Chem. Soc., Chem. Commun.*, 70; (d) McCullough, R.D., Ewband, P.C., and Loewe, R.S. (1997) *J. Am. Chem. Soc.*, **119**, 633; (e) McCullough, R.D., Lowe, R.D., Jayaraman, M., and Anderson, D.L. (1993) *J. Org. Chem.*, **58**, 904; (f) Chen, T.-A., and Rieke, R.D. (1992) *J. Am. Chem. Soc.*, **114**, 10087; (g) Andersson, M.R., Selse, D., Berggren, M., Järvinen, H., Hjertberg, T., Inganäs, O., Wennerström, O., and Österholm, J.-E. (1994) *Macromolecules*, **27**, 6503; (h) Sato, T., Cai, Z., Shiono, T., and Yamamoto, T. (2006) *Polymer*, **47**, 37; (i) Sato, T., Kishida, H., Nakamura, A., Fukuda, T., and Yamamoto, T. (2007) *Synth. Met.*, **157**, 318; (j) Yamamoto, T. and Sato, T. (2006) *Jpn. J. Appl. Phys.*, **45**, L301; (k) Yamamoto, T., Sato, T., IIjima, T., Abe, M., Fukumoto, H., Koizumi, T.-A., Usui, M., Nakamura, Y., Yagi, T., Tajima, H., Okada, T., Sasaki, S., Kishida, H., Nakamura, A., Fukuda, T., Emoto, A., Ushijima, H., Kurosaki, C., and Hirota, H., (2009) *Bull. Chem. Soc. Jpn.*, **82**, 896.

18 (a) Yamamoto, T., Takeuchi, M., and Kubota, K. (2000) *J. Polym. Sci.: B, Polym. Phys.*, **38**, 1348; (b) Yamamoto, T. (1997) in *Handbook of Organic Conductive Molecules and Polymers*, vol. 2 (ed. H.S. Nalwa), John Wiley, Chichester, p. 171.

19 (a) Yamamoto, T., Kim, S.-B., and Yamamoto, T. (1996) *Chem. Lett.*, 413; (b) Kim, S.-B., Harada, K., and Yamamoto, T. (1998) *Macromolecules*, **31**, 988; (c) Yamamoto, T. and Okuda, T. (1999) *J. Electroanal. Chem.*, **490**, 242; (d) Yamamoto, T. and Kimura, T. (1998) *Macromolecules*, **31**, 2683; (e) Yamamoto, T. and Shiraishi, K. (1998) *Chem. Lett.*, 895; (f) Yamamoto, T., Sugiyama, K., Kanbara, T., Hayashi, H., and Etori, H. (1998) *Macromol. Chem. Phys.*, **199**, 1807; (g) Yamamoto, T., Hayashida, N., Maruyama, T., and Kubota, K. (1998) *Chem. Lett.*, 1125; (h) Hayashida, N. and Yamamoto, T. (1999) *Bull. Chem. Soc. Jpn.*, **72**, 1153; (i) Horie, M., Yamaguchi, I., and Yamamoto, T. (2006) *Macromolecules*, **39**, 7493; (j) Tanimoto, A., Shiraishi, K., and Yamamoto, T. (2004) *Bull. Chem. Soc. Jpn.*, **77**, 597.

20 (a) Hayashi, H. and Yamamoto, T. (1998) *Macromolecules*, **31**, 6063; (b) Yamamoto, T. and Shimizu, T. (1997) *J. Mater. Chem.*, **7**, 1967; (c) Yamamoto, T., Morikita, T., Maruyama, T., Kubota, K., and Katada, M. (1997) *Macromolecules*, **30**, 5390; (d) Muramatsu, Y. and Yamamoto, T. (1998) *Denki Kagaku*, **66**, 223; (e) Yamamoto, T., Xu, Y., and Koinuma, H. (1998) *Chem. Lett.*, 613; (f) Yamamoto, T. and Lee, B.-L. (1998) *Chem. Lett.*, 1663; (g) Yamamoto, T. and Abla, M. (1999) *Synth. Met.*, **100**, 237; (h) Morikita, T. and Yamamoto, T. (2001) *J. Organomet. Chem.*, **637–639**, 809; (i) Yamamoto, T., Muramatsu, Y., Lee, B.-L., Kokubo, H., Sasaki, S., Hasegawa, M., Yagi, T., and Kubota, K. (2003) *Chem. Mater.*, **15**, 4384; (j) Yamamoto, T. and Lee, B.-L. (2002) *Macromolecules*, **35**, 2993; (k) Yamamoto, T., Shiraishi, K., Alba, M., Yamaguchi, I., and Groenendaal, L.B. (2002) *Polymer*, **43**, 711; (l) Sato, T., Cai, Z., Shiono, T., and Yamamoto, T. (2006) *Polymer*, **47**, 37; (m) Yamamoto, T. and Sato, T. (2006) *Jpn. J. Appl. Phys.*, **45**, 301; (n) Yamamoto, T., Sato, T., Iijima, T., Abe, M., Fukumoto, H., Koizumi, T.-A., Usui, M., Nakamura, Y., Yagi, T., Tajima, H., Okada, T., Sasaki, S., Kishida, H., Nakamura, A., Fukuda, T., Emoto, A., Ushijima, H., Kurosaki, C., and Hirota, H. (2009) *Bull. Chem. Soc. Jpn.*, **82**, 896; (o) Shiraishi, K. and Yamamoto, T. (2002) *Synth. Met.*, **130**, 139; (p) Yamamoto, T., Uemura, T., Tanimoto, A., and Sasaki, S. (2003) *Macromolecules*, **36**, 1047; (q) Yamamoto, T., Fukushima, N., Nakajima, H., Maruyama, T., and Yamaguchi, T. (2000) *Macromolecules*, **33**,

5988; (r) Yamamoto, T., Fujiwara, Y., Fukumoto, H., Nakamura, Y., Koshihara, S., and Ishikawa, T. (2003) *Polymer*, **44**, 4487; (s) Choi, B.-K. and Yamamoto, T. (2003) *Electrochem. Commun.*, **5**, 566; (t)Yamamoto, T., Yoshizawa, M., Mahmut, A., Abe, M., Kuroda, S.-I., Imase, T., and Sasaki, S. (2005) *J. Polym. Sci. Part A: Polym. Chem.*, **43**, 6223; (v) Yamashita, R., Koizumi, T.-A., Sasaki, S., and Yamamoto, T. (2007) *Polym. J.*, **39**, 1202; (v) Yamamoto, T. and Yamashita, R. (2008) *Polym. J.*, **40**, 775; (w)Kokubo, H. and Yamamoto, T. (2001) *Macromol. Chem. Phys.*, **202**, 1031; (x) Kokubo, H., Sato, T., and Yamamoto, T. (2006) *Macromolecules*, **39**, 3959; (y) Yasuda, T., Sakai, Y., Aramaki, S., and Yamamoto, T. (2004) *Chem. Mater.*, **16**, 4616.

21 (a) Yamamoto, T., Suganuma, H., Maruyama, T., and Kubota, K. (1995) *J. Chem. Soc., Chem. Commun.*, 1613; (b) Nanos, J.L., Kampf, J.W., Curtis, M.D., Gonzalez, L., and Martin, D.C. (1995) *Chem. Mater.*, **7**, 2332; (c) Yamamoto, T., Arai, M., kokubo, H., and Sasaki, S. (2003) *Macromolecules*, **36**, 7986; (d) Yamamoto, T., kokubo, H., Kobashi, M., and Sakai, Y. (2004) *Chem. Mater.*, **16**, 4616; (e) Yamamoto, T., Otsuka, S., Namekawa, K., Fukumoto, H., Yamaguchi, I., Fukuda, T., Asakawa, N., Yamanobe, T., Shiono, T., and Cai, Z. (2006) *Polymer*, **47**, 6038; (f) Otsuka, S., Fukumoto, H., and Yamamoto, T. (2008) *Bull. Chem. Soc. Jpn.*, **81**, 536.

22 (a) Yamamoto, T., Kanbara, T., Mori, C., Wakayama, H., Fukuda, T., Inoue, T., and Sasaki, S. (1996) *J. Phys. Chem.*, **30**, 12631; (b) Yamamoto, T. and Etori, H. (1995) *Macromolecules*, **28**, 3371; (c) Yamamoto, T., Saito, H., and Osakada, K. (1992) *Polym. Bull.*, **29**, 597, the reflection spectrum of PCyh indicates that it has an absorption peak at about 600nm (T. Yamamoto and T. Fukuda, unpublished data). The peak position locates near that of *cis*-polyacetylene (λ_{max} = 217 695 nm.) [22d]. The degree of the red shift (cf. Figure 1.6) from 1,3-butadiene (λ_{max} = 217 nm) is estimated at 29400 cm^{-1}; (d) Ito, T., Shirakawa, H., and Ikeda, S. (1974) *J. Polym. Sci., Polym. Chem. Ed.*, **12**, 11 and (1975) **13**, 1943; (e) Hatano, M., Kanbara, S., and Okamoto, S. (1961) *J. Polym. Sci.*, **51**, S26; (f) Gibson, H.W. and Pockan, J.M. (1982) *Macromolecules*, **15**, 242; (g) Yamamoto, T., Wakayama, H., Fukuda, T., and Kanbara, T. (1992) *J. Phys. Chem.*, **96**, 8677.

23 (a) Zhou, Z.-H., Maruyama, T., Kanbara, T., Ikeda, T., Ichimura, K., Yamamoto, T., and Tokuda, K. (1991) *J. Chem. Soc., Chem. Commun.*, 1210; (b) Kurihara, T., Kaino, T., Zhou, Z.-H., Kanbara, T., and Yamamoto, T. (1992) *Electronics Lett.*, **28**, 681; (c) Kizu, K., Maruyama, T., and Yamamoto, T. (1995) *Polym. J.*, **27**, 205; (d) Yamamoto, T., Shimura, M., Osakada, K., and Kubota, K. (1992) *Chem. Lett.*, 1003; (e) Lee, B.-L. and Yamamoto, T. (1999) *Macromolecules*, **32**, 1375; (f) Komarudin, D., Morita, A., Osakada, K., and Yamamoto, T. (1998) *Polym. J.*, **30**, 860; (g) Yamamoto, T., Sanechika, K., and Sakai, H. (1990) *J. Macromol. Sci. Chem.*, **A27**, 1147; (h) Fukumoto, H. and Yamamoto, T. (2008) *J. Polym. Sci. Part A: Polym. Chem.*, **46**, 2975; (i) Yamamoto, T., Kokubo, H., and Morikita, T. (2001) *J. Polym. Sci. Part B: Polym. Phys.*, **39**, 1713; (j) Morikita, T., Yamaguchi, I., and Yamamoto, T. (2001) *Adv. Mater.*, **13**, 1862; (k) Yamamoto, T., Fang, Q., and Morikita, T. (2003) *Macromolecules*, **36**, 4262; (l) Yasuda, T., Yamaguchi, I., and Yamamoto, T. (2003) *Adv. Mater.*, **15**, 293; (m) Yasuda, T. and Yamamoto, T. (2003) *Macromolecules*, **36**, 7513; (n) Yamamoto, T., Abe, M., Wu, B., Choi, B.-K., Harada, Y., Takahashi, Y., Kawata, K., Kokubo, H., Sasaki, S., and kubota, K. (2007) *Macromolecules*, **40**, 5504; (o) Abe, M. and Yamamoto, T. (2006) *Synth. Met.*, **156**, 1118; (p) Abe, M., Karim, S.M., Yamaguchi, I., Kuroda, S., Kubota, K., and Yamamoto, T. (2005) *Bull. Chem. Soc. Jpn.*, **78**, 534;(q) Abe, M., (2007) Dissertation for Ph.D., Tokyo Institute of Technology; (r) Iijima, T. and Yamamoto, T. (2004) *Macromol. Rapid Commun.*, **25**, 669; (s) Iijima, T., Kuroda, S.-I., and Yamamoto, T. (2008) *Macromolecules*, **41**, 1654; (t) Fukumoto, H., Kimura, R., Sasaki, S., Kubota, K., and Yamamoto, T. (2005) *J. Polym. Sci. Part B: Polym. Phys.*, **43**, 215; (u) Yasuda, T., Imase, T., Sasaki, S., and

Yamamoto, T. (2005) *Macromolecuels*, **38**, 1500; (v) Tanimoto, A. and Yamamoto, T. (2004) *Adv. Synth. Catal.*, **346**, 1818; (w) Tanimoto, A. and Yamamoto, T. (2006) *Macromolecules*, **39**, 3546; (x) Yamaguchi, I., Choi, B.-J., Koizumi, T.-A., Kubota, K., and Yamamoto, T. (2007) *Macromolecules*, **40**, 438; (y) Yamamoto, T., Ohya, K., Kobayashi, K., Okamoto, K., Koie, S., Fukumoto, H., Koizumi, T.-A., and Yamaguchi, I. (2009) *Macromolecules*, **42**, 3207.

24 (a) Karikomi, M., Kitamura, C., Tanaka, S., and Yamashita, Y. (1995) *J. Am. Chem. Soc.*, **117**, 6791; (b) Zhang, Q.T. and Tour, J.M. (1998) *J. Am. Chem. Soc.*, **120**, 5355.

25 (a) Yamamoto, T. (1996) *J. Polym. Sci., Part A: Polym. Chem.*, **34**, 997; (b) Yamamoto, T. and Kokubo, H. (2005) *Electrochim. Acta*, **50**, 1453.

26 (a) Saito, N., Kanbara, T., Kushida, T., Kubota, K., and Yamamoto, T. (1993) *Chem. Lett.*, 1775; (b) Vahlenkamp, T. and Wegner, G. (1994) *Macromol. Chem. Phys.*, **195**, 1933; (c) Yamamoto, T., Kimura, T., and Shiraishi, K. (1999) *Macromolecules*, **32**, 8886; (d) Shiraishi, K. and Yamamoto, T. (2002) *Polym. J.*, **34**, 727.

27 Saitoh, Y. and Yamamoto, T. (1995) *Chem. Lett.*, 785.

28 (a) Yamamoto, T., Saitoh, Y., Anzai, K., Fukumoto, H., Yasuda, T., Fujiwara, Y., Choi, B.-K., Kubota, K., and Miyamae, T. (2003) *Macromolecules*, **36**, 6722; (b) Yamamoto, T., Asao, T., and Fukumoto, H. (2004) *Polymer*, **45**, 8085; (c) Iijima, T. and Yamamoto, T. (2005) *Chem. Lett.*, 1672; (d) Yamamoto, T., Iijima, T., Ozawa, Y., Toriumi, K., Kubota, K., and Sasaki, S. (2007) *J. Polym. Sci. Part A: Polym. Chem.*, **45**, 548; (e) Yamamoto, T., Iijima, T., Ozawa, Y., Toriumi, K., Kubota, K., and Sasaki, S. (2007) *J. Polym. Sci. part A: Polym. Chem.*, **45**, 548;(f) Tokimitsu, R. (2009) C Dissertation for M. Eng. (Tokyo Institute of Technology).

29 (a) Yamamoto, T., Yoneda, Y., and Kizu, K. (1995) *Makromol. Rapid Commun.*, **16**, 549; (b) Yamamoto, T. and Lee, B.-L. (1996) *Chem. Lett.*, 65; (c) Yamamoto, T. and Lee, B.-L. (1996) *Chem. Lett.*, 679; (d) Davadoss, C., Bharathi, P., and Moore, J.S. (1996) *J. Am. Chem. Soc.*, **118**, 9635; (e) Aktin, M.R., Stemp, E.D.A., Turro, C., Turro, N.J., and Barton, J.K. (1996) *J. Am. Chem. Soc.*, **118**, 2267; (f) Yamamoto, T., Lee, B.-L., Nurulla, I., Yasuda, T., Yamaguchi, I., Wada, A., Hirose, C., Tasumi, M., Sakamoto, A., and Kobayashi, E. (2005) *J Phys. Chem. B*, **109**, 10605.

30 (a) Ooba, N., Tomaru, S., Kurihara, T., Kaino, T., Yamada, W., Takagi, M., and Yamamoto, T. (1995) *Jpn. J. Appl. Phys.*, **34**, 3139; (b) Ooba, N., Asobe, M., Tomaru, S., Kaino, T., Yamada, W., Takagi, M., and Yamamoto, T. (1996) *Nonlinear Optics*, **15**, 481.

31 Suruga, K., Fukuda, T., Ishikawa, K., Takezoe, H., Fukuda, A., Kanbara, T., and Yamamoto, T. (1996) *Synth. Met.*, **79**, 149.

32 (a) Etori, H., Kanbara, T., and Yamamoto, T. (1994) *Chem. Lett.*, 461; (b) Muramatsu, Y. and Yamamoto, T. (1997) *Chem. Lett.*, 581; (c) Yamamoto, T., Muramatsu, Y., Shimizu, T., and Yamada, W. (1998) *Macromol. Rapid Commun.*, **19**, 263; (d) Yamamoto, K., Asada, T., Nishide, H., and Tsuchida, E. (1990) *Bull. Chem. Soc. Jpn.*, **63**, 1211.

33 (a) Miyazaki, Y. and Yamamoto, T. (1994) *Chem. Lett.*, 41; (b) Yamamoto, T., Omote, M., Miyazaki, Y., Kashiwazaki, A., Lee, B.-L., Kanbara, T., Osakada, K., Inoue, T., and Kubota, K. (1997) *Macromolecules*, **30**, 7158.

34 (a) Miyamae, T., Yoshimura, D., Ishii, H., Ouchi, Y., Miyazaki, T., Koike, T., Yamamoto, T., and Seki, K. (1996) *J. Electron. Spectrosc.*, **78**, 399; (b) Miyamae, T., Aoki, M., Etori, H., Muramatsu, Y., Saito, Y., Yamamoto, T., Sakurai, Y., Seki, K., and Ueno, N. (1998) *J. Electron Spectrosc. Relat. Phenom.*, **89–91**, 905.

35 (a) Yamamoto, T., Sanechika, K., Yamamoto, A., Katada, M., Motoyama, I., and Sano, H. (1983) *Inorg. Chim. Acta*, **73f**, 75; (b) Jonas, F. and Heywang, G. (1994) *Electrochim. Acta*, **39**, 1345; (c) Patil, A.O., Ikenoue, Y., Basescu, N., Chen, J., Wudl, F., and Heeger, A.J. (1989) *Synth. Met.*, **29**, 151; (d) Kudoh, Y., Fukuyama, M., and Yoshimura, S. (1994) *Synth. Met.*, **66**, 157.

36 (a) Yamamoto, T., Kanbara, T., and Mori, C. (1990) *Chem. Lett.*, 1211; (b) Yamamoto,

T., Kanbara, T., and Mori, C. (1990) *Synth. Met.*, **38**, 399; (c) Kanbara, T., Mori, C., Wakayama, H., Fukuda, T., Zhou, Z.-H., Maruyama, T., Osakada, K., and Yamamoto, T. (1992) *Solid State Commun.*, **83**, 771.

37 (a) Ostoja, P., Guerri, S., Rossini, S., Servidori, M., Taliani, C., and Zamboni, R. (1993) *Synth. Met.*, **54**, 447; (b) Porzio, W., Destri, S., Mascherpa, M., Rossini, S., and Brükner, S. (1993) *Synth. Met.*, **55–57**, 408; (c) Pozzio, W., Destri, S., Mascherpa, M., and Brükner, S. (1993) *Acta Polym.*, **44**, 266; (d) Xervet, B., Ries, S., Trotel, M., Alnot, P., Horowitz, G., and Garnier, F. (1993) *Adv. Mater.*, **5**, 461; (e) Taliani, C. and Blinov, L.M. (1996) *Adv. Mater.*, **8**, 353; (f) Lovinger, A.J., Davis, D.D., Dodabalapur, A., and Katz, H.E. (1996) *Chem. Mater.*, **8**, 2836.

38 (a) Yamamoto, T. (1996) *Chem. Lett.*, 703; (b) Yamamoto, T., Maruyama, T., Suganuma, H., Arai, M., Komarudin, D., and Sasaki, S. (1997) *Chem. Lett.*, 139; (c) Yamamoto, T., Suganuma, H., Maruyama, T., Inoue, T., Muramatsu, Y., Arai, M., Komarudin, D., Ooba, N., Tomaru, S., Sasaki, S., and Kubota, K. (1997) *Chem. Mater.*, **9**, 1217; (d) Gonzalez Londa, L. and Martin, D.C. (1997) *Macromolecules*, **30**, 1524; (e) Yang, C., Orfino, F.P., and Holdcroft, S. (1996) *Macromolecules*, **29**, 6510; (f) Winokur, M.J., Wamsley, P., Monlton, J., Smith, P., and Heeger, A.J. (1991) *Macromolecules*, **24**, 3812;
(g) Märdalen, J., Samuelsen, E.J., Gautun, O.R., and Carlsen, P.H. (1992) *Synth. Met.*, **48**, 363; (h) Yamamoto, T., Komarudin, D., Arai, M., Lee, B.-L., Suganuma, H., Asakawa, N., Inoue, Y., Kubota, K., Sasaki, S., Fukuda, T., and Matsuda, H. (1998) *J. Am. Chem. Soc*, **120**, 2047; (i) Yamamot, T., Lee, B.-L., Suganuma, H., and Sasaki, S. (1998) *Polym. J.*, **30**, 83.

39 (a) Wolf, M.O. and Wrighton, M.S. (1994) *Chem. Mater.*, **6**, 1526; (b) Maruyama, T. and Yamamoto, T. (1995) *Inorg. Chim. Acta*, **238**, 9; (c) Zhu, S.S. and Swager, T.M. (1997) *J. Am. Chem. Soc.*, **119**, 12568; (d) Ley, K.D., Whittle, C.E., Barberger, M.D., and Schanze, K.S. (1997) *J. Am. Chem. Soc.*, **119**, 3423; (e) Wang, B. and Wasielewski, M.R. (1997) *J. Am. Chem. Soc.*, **119**, 12; (f) Zhu, S.S. and Swager, T.M. (1996) *Adv. Mater.*, **8**, 49;
(g) Wärnmark, K., Baxter, P.N.W., and Lehn, J.-M. (1998) *J. Chem. Soc., Chem. Commun.*, 993; (h) Yaniger, S.I., Rose, D.J., Mckenna, W.P., and Eyring, E.M. (1984) *Macromolecules*, **17**, 2579; (i) Matsuda, J., Aramaki, K., and Nishihara, H. (1995) *J. Chem. Soc., FaradayTrans.*, **91**, 1477;
(j) Funaki, H., Aramaki, K., and Nishihara, H. (1995) *Synth. Met.*, **74**, 59;
(k) Nishihara, H. (1997) *J. Org. Synth., Jpn.*, **55**, 410; (l) Spilner, I.J. and Pellegrini, P. Jr. (1965) *J. Chem. Soc.*, 3800;
(m) Sanechika, K., Yamamoto, T., and Yamamoto, A. (1981) *Polym J.*, **13**, 255.

40 (a) Kanbara, T. and Yamamoto, T. (1993) *Macromolecules*, **26**, 1975; (b) Yamamoto, T. and Kanbara, T. (1994) US Pat. 5310829.

41 (a) Yamamoto, T. and Saito, N. (1996) *Chem. Lett.*, 127; (b) Yamamoto, T., Lee, B.-L., Hayashi, H., Saito, N., and Maruyama, T. (1997) *Polymer*, **38**, 4233.

42 (a) Yanagida, S., Kabumoto, A., Mizumoto, K., Pac, C., and Yoshino, K. (1985) *J. Chem. Soc., Chem. Commun.*, 474; (b) Matsuoka, S., Kohzuki, T., Kuwana, Y., Nakamura, A., and Yanagida, S. (1992) *J. Chem. Soc., Perkin Trans. 2*, 679.

43 (a) Maruyama, T. and Yamamoto, T. (1997) *J. Phys. Chem. B*, **101**, 3806; (b) Sato, Y. and Yamamoto, T. (2000) Jpn. Pat. Kokai 2000-5605; (c) Sato, Y., Kagatoni, M., Yamamoto, T., and Souma, Y. (1999) *Appl. Cat A: General*, **185**, 219; (d) Sato, Y., Yamamoto, T., and Souma, Y. (2000) *J. Cat.*, **65**, 123.

44 (a) Wakabayashi, M., Miyazaki, Y., Kanbara, T., Osakada, K., Yamamoto, T., and Ishibashi, A. (1993) *Synth. Met.*, **55/57**, 3632; (b) Yamamoto, T. and Wakabayashi, M.,Eur. Pat. Appl. EP 597, 234 (1994) and US Pat. 5561206 (1996); (c) Yamamoto, T., Inami, M., and Kanbara, T. (1994) *Chem. Mater.*, **6**, 44; (d) Kanbara, T., Inami, M., and Yamamoto, T. (1991) *J. Power Sources*, **36**, 87.

45 (a) Kanbara, T. and Yamamoto, T. (1993) *Chem. Lett.*, 1459; (b) Yamamoto, T., Inoue, T., and Kanbara, T. (1994) *Jpn. J. Appl. Phys.*, **33**, L250.

46 (a) Yamamoto, T. (1981) *J. Chem. Soc., Chem. Commun.*, 187; (b) MacInnes, D. Jr., Dury, M.A., Nigrey, P.J., Nairns, D.P., MacDiarmid, A.G., and Heeger, A.J. (1981) *J. Chem. Soc., Chem. Commun.*, 317; (c) Schacklette, L.W., Elsenbaumer, R.L., Chance, R.R., Sowa, J.M., Ivory, D.M., Miller, G.G., and Baughmann, R.H. (1982) *J. Chem. Soc., Chem. Commun.*, 361; (d) Yamamoto, T., Zama, M., Hishinuma, M., and Yamamoto, A. (1987) *J. Appl. Electrochem.*, **17**, 607; (e) Yamamoto, T., Zama, M., and Yamamoto, A. (1985) *Chem. Lett.*, 563; (f) Hishinuma, M., Zama, M., Osakada, K., and Yamamoto, T. (1987) *Inorg. Chim. Acta*, **128**, 185; (g) Yamamoto, T., Zama, M., and Yamamoto, A. (1984) *Chem. Lett.*, 1577.

47 (a) MacDiarmid, A.G., Yang, L.S., Haung, W.S., and Humphry, B.D. (1987) *Synth. Met.*, **18**, 393; (b) Chiang, J.C. and MacDiarmid, A.G. (1986) *Synth. Met.*, **13**, 193; (c) Daifuku, E. (1988) *Kino Zairyo*, **8**, 17; (d) Yamamoto, T. and Matsunaga, T. (1990) Polymer Battery (in Japanese), Kyoritsu, Tokyo.

48 For example: Sanechika, K. and Yoshino, A. (1985) Jpn. Kokai Tokkyo Koho, 1985-253157 (Chem. Abstr. 104, P152474m (1985)).

49 (a) Yamamoto, T. and Sano, S. (1993) US. Pat. 5273841; (b) Yamamoto, T., Matsuzaki, T., Minetomo, A., Kawazu, Y., and Ohashi, O. (1996) *Bull. Chem. Soc. Jpn.*, **69**, 3461; (c) Yamamoto, T., Nishiyama, T., Harada, G., and Takeuchi, M. (1999) *J. Power Sources*, **79**, 281.

50 (a) Burroughes, J.H., Bradley, D.D.C., Brown, A.R., Marks, R.N., Mackay, K., Friend, R.H., Burn, P.L., and Holmes, A.B. (1990) *Nature*, **347**, 539; (b) Brown, D., Heeger, A.J., and Kroemer, H. (1991) *J. Electron Mater.*, **20**, 945; (c) Bradley, D.D.C. (1993) *Synth. Met.*, **54**, 401; (d) Ueda, M., Ohmori, Y., Noguchi, T., and Ohnishi, T. (1993) *Jpn. J. Appl. Phys.*, **32**, L921; (e) Adachi, C., Tsutsui, T., and Saito, S. (1990) *Appl. Phys. Lett.*, **57**, 531; (f) Bernius, M.T., Inbasekaran, M., O'Brien, J., and Wu, W. (2000) *Adv. Mater.*, **12**, 1737; (g) Morteani, A.C., Dhoot, A.S., Kim, J.-S., Silva, C., Greenham, N.C., Murphy, C., Moons, E., Cina, S., Burroughes, H., and Friend, R.H. (2003) *Adv. Mater.*, **15**, 1708; (h) Gomg, X., Ostrowski, J.C., Bazan, G.C., Moses, D., Heeger, A.J., Liu, M.S., and Jen, A.K.-Y. (2003) *Adv. Mater.*, **15**, 45; (i) List, E., Guentner, R., Scanducci de Freitas, P., and Scherf, U. (2002) *Adv. Mater.*, **14**, 374; (j) Scherf, U. and List, E. (2002) *Adv. Mater.*, **14**, 477.

51 (a) Yamamoto, T., Sugiyama, K., Kushida, T., Inoue, T., and Kanbara, T. (1996) *J. Am. Chem. Soc.*, **118**, 3930; (b) Eur. Pat. 98105161.8-2102 (1998).

52 Yamamoto, T., Wakayama, H., Kanbara, T., Sasaki, K., Tsutsui, K., and Furukawa, J. (1994) *Denki Kagaku*, **62**, 84.

2
Catalyst-Transfer Condensation Polymerization for Precision Synthesis of π-Conjugated Polymers

Tsutomu Yokozawa

2.1
Introduction

π-Conjugated polymers containing aromatic rings in the backbone are an attractive class of materials owing to their potential as organic electronic materials and devices such as field effect transistors (FETs), organic light-emitting diodes (OLEDs), and photovoltaic cells. These polymers have generally been synthesized by condensation polymerization methods, such as electrochemical polymerization [1] and metal-mediated polycondensation reactions [2–4]. Therefore, the molecular weight of those polymers is thought to be difficult to control within a narrow molecular weight distribution. However, uncontrolled molecular weight and broad molecular weight distribution do not stem inherently from the condensation polymerization reaction, that is, condensation steps with elimination of a small molecule species, but from a polymerization mechanism for step-growth polymerization, in which all the end groups of monomers and oligomers in the reaction mixture react equally with each other. If the mechanism of condensation polymerization could be converted from step-growth to chain-growth, π-conjugated polymers with defined molecular weight and narrow molecular weight distribution would be obtained.

This change of mechanism in condensation polymerization is evidently not impossible, because nature already uses a chain-growth condensation polymerization process to synthesize perfectly monodisperse biopolymers such as polypeptides [5, 6], DNA [7], and RNA [8, 9]. For example, in the biosynthesis of polypeptides, the amino group of an aminoacyl-tRNA, monomer reacts selectively with the terminal ester moiety of polypeptidyl-tRNA in a ribosome to elongate the peptide chain. Furthermore, artificial chain-growth condensation polymerization has been recently developed and used to synthesize well-defined aromatic polyamides, polyesters, and polyethers [10, 11].

In this research, chain-growth coupling polymerization with a catalyst for the synthesis of π-conjugated polymers has been investigated and it was found that Kumada–Tamao coupling polymerization with a Ni catalyst and Suzuki–Miyaura coupling polymerization with a Pd catalyst, which had been believed to be a typical

Conjugated Polymer Synthesis. Edited by Yoshiki Chujo

ISBN: 978-3-527-32267-1

step-growth condensation polymerization, have a latent potential to involve a chain-growth polymerization mechanism. Surprisingly, the mechanism is similar to that of the biosynthesis mentioned above; the catalyst activates the polymer end group, followed by reaction with the monomer and the transfer of the catalyst to the elongated polymer end group. This chapter describes chain-growth Kumada–Tamao coupling polymerization and Suzuki–Miyaura coupling polymerization for the synthesis of precision π-conjugated polymers.

2.2 Kumada–Tamao Coupling Polymerization with Ni Catalyst

2.2.1 Polythiophene

2.2.1.1 Discovery and Mechanism of Catalyst-Transfer Condensation Polymerization

Poly(alkylthiophene) has received much attention in recent years because of its small band gap, high electrical conductivity [12, 13], and interesting properties such as light emitting ability [14] and high field effect mobility [15, 16]. The polymerization of Grignard thiophene monomer **1a** with Ni(dppp)Cl_2 (dppp = 1,3-bis(diphenylphosphino)propane) was well known as a regioregulated synthetic method for poly(alkylthiophene)s developed by McCullough *et al.*, but the polymers obtained possessed broad molecular weight distribution [17–20]. However, Yokozawa and coworkers found that the M_n values of the polymers increased in proportion to the monomer conversion, with narrow polydispersities being retained, controlled by the amount of the Ni catalyst; the M_n values were proportional to the feed ratio of $[\mathbf{1a}]_0/[\text{Ni catalyst}]_0$ when the polymerization was carried out at room temperature, with care to use the exact amount of isopropyl magnesium chloride for generation of monomer **1a** from the corresponding bromoiodothiophene (Scheme 2.1) [21]. Furthermore, the M_w/M_n ratios were around 1.1 up to M_n of 28 700 when the polymerization of **1a** was quenched with hydrochloric acid [22]. McCullough and coworkers also reported that a similar zinc monomer [23] and **1a** from the corresponding dibromothiophene showed the same polymerization behavior [24].

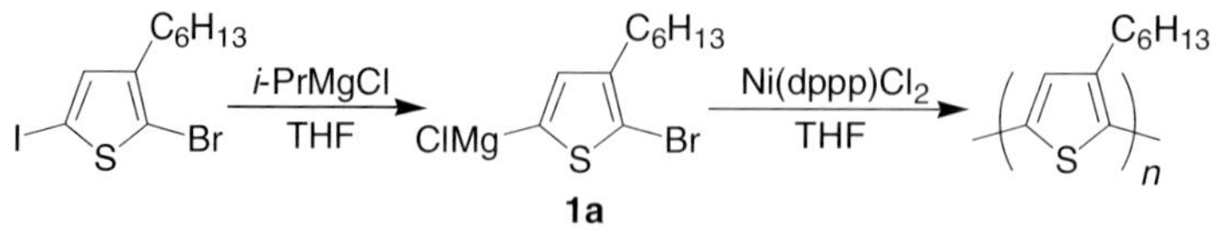

Scheme 2.1

After a detailed study of the polymerization of **1a**, four important points were clarified: (i) the polymer end groups are uniform among molecules; one end group is Br and the other is H; (ii) the propagating end group is a polymer–Ni–Br complex; (iii) one Ni molecule forms one polymer chain; and (iv) the chain initiator is a dimer of **1a** formed *in situ*. On the basis of these results, a catalyst-transfer condensation polymerization mechanism has been proposed (Scheme 2.2). Thus, $Ni(dppp)Cl_2$ reacts with 2 equiv of **1a**, and the coupling reaction occurs with concomitant generation of a zero-valent Ni complex. The Ni(0) complex does not diffuse to the reaction mixture but is inserted into the intramolecular C–Br bond. Another **1a** reacts with this Ni, followed by the coupling reaction and transfer of the Ni catalyst to the next C–Br bond. Growth continues in such a way that the Ni catalyst moves to the polymer end group [25].

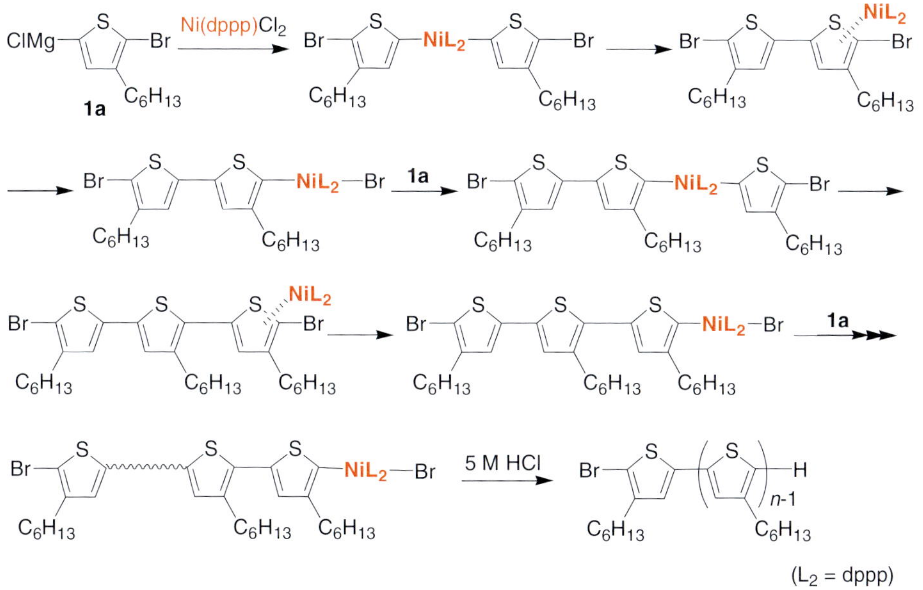

Scheme 2.2

Several other reactions involving similar intramolecular transfer of metal catalysts have been reported [26–34]. Van der Boom and coworkers demonstrated that the reaction of $Ni(PEt_3)_4$ with a brominated vinylarene results in selective η2-C=C coordination, followed by intramolecular "ring-walking" of the metal center and aryl-bromide oxidative addition, even in the presence of arylI-containing substrates (Scheme 2.3) [33]. Nakamura and coworkers studied the Ni-catalyzed cross-coupling

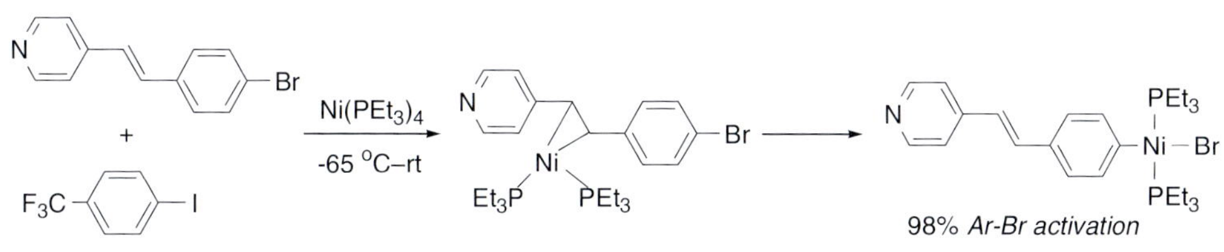

Scheme 2.3

reaction by analysis of kinetic isotope effects and theoretical calculations, and indicated that the first irreversible step of the reaction is the π-complexation of the Ni catalyst on the π-face of the haloarene. In other words, once a Ni/haloarene π-complex forms through ligand change, it does not dissociate but proceeds quickly to the oxidative addition step in an intramolecular manner.

The influence of the phosphine ligand of the Ni catalyst on the catalyst-transfer condensation polymerization was investigated [35, 36]. The M_n value and the M_w/M_n ratio of polymer were strongly affected by the ligands of the Ni catalyst: $Ni(dppe)Cl_2$ (dppe = 1,2-bis(diphenylphosphino)ethane) and $Ni(PPh_3)_4$ gave a polymer with a slightly lower M_n and a slightly broad molecular weight distribution, whereas Ni $(PPh_3)_2Cl_2$, $Ni(dppb)Cl_2$ (dppb = 1,4-bis(diphenylphosphino)butane), and Ni(dppf) Cl_2 (dppf = 1,1′-bis(diphenylphosphino)ferrocene) gave polymers with low M_ns and broad molecular weight distributions. Finally, $Ni(dppp)Cl_2$ resulted in an M_n value close to the theoretical value based on the feed ratio of monomer to catalyst and the narrowest M_w/M_n ratio. The influence of the molecular weight and molecular weight distribution of poly(3-hexylthiophene) on the characteristics of field effect transistor (FET) [37, 38] and photovoltaic cells [39] was investigated.

Taking advantage of the polythiophene end group containing the Ni complex, McCullough and coworkers introduced functional groups on one or both ends of the polymer by using Grignard reagents. Allyl, ethynyl, and vinyl Grignard reagents afford monofunctionalized polythiophenes, whereas aryl and alkyl Grignard reagents yield difunctionalized polythiophenes. By utilizing the proper protecting groups, hydroxy, formyl, and amino groups can also be incorporated onto the polymer chain ends [40]. Kiriy and coworkers used $(PPh_3)_2Ni(Ph)Br$ as an initiator and synthesized Ph-terminated poly(3-hexylthiophene) [41, 42]. This method was applied to growing conductive polymer brushes of poly(3-hexylthiophene) via surface-initiated chain-growth condensation polymerization of **1a**. Thus, the Ni(II) macroinitiator was prepared by the reaction of $Ni(PPh_3)_4$ with photo-cross-linked poly(4-bromostyrene) films. Exposure of the initiator layers to the monomer solution led to polymerization from the surface, resulting in poly(3-hexylthiophene) brushes (Scheme 2.4). Recently, Locklin and coworkers prepared unsubstituted polythiophene brushes from gold surfaces functionalized with self-assembled monolayers of 2-bromothiophene moi-

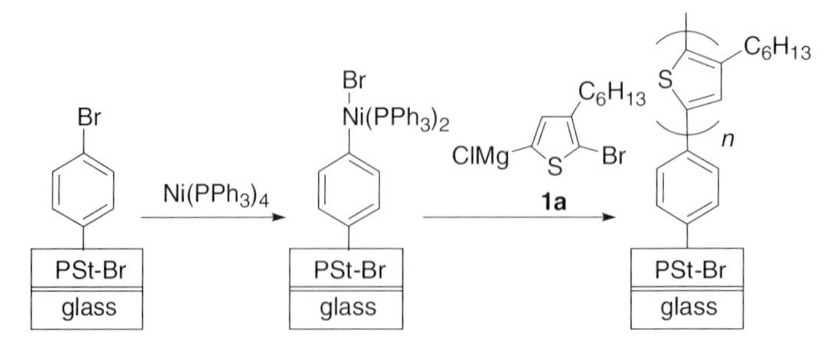

Scheme 2.4

ety, which underwent oxidative addition with $Ni(COD)(PPh_3)_2$ (COD = 1,5-cyclooocatadiene) complex, which is more reactive than $Ni(PPh_3)_4$ [43].

2.2.1.2 A Variety of Monomers

The chain-growth polymerization of other substituted thiophene monomers instead of **1a** with the hexyl group was investigated (Scheme 2.5). The polymerization of butylthiophene monomer **1b** with $Ni(dppp)Cl_2$ afforded polymer with low polydispersity ($M_w/M_n = 1.33$–1.43), although the M_n value is less than 5500 due to the low solubility of poly**1b** [44]. Aryl-substituted monomer **1c**, the polymer of which may have a stabilized π-conjugated main chain system by virtue of the pendent aromatic group, gave a polymer with $M_w/M_n = 2.15$, probably due to the low solubility of the conjugated poly**1c** in the reaction solvent [45]. The polymers from alkoxy-substituted monomers **1d** possessed M_w/M_n of 1.5–1.7 [46, 47]. The polymerization of alkoxymethyl-substituted monomer **1e** with $Ni(dppp)Cl_2$ gave a polymer with M_w/M_n of 1.42, whereas the polymerization with $Ni(dppe)Cl_2$ instead of $Ni(dppp)Cl_2$ resulted in a decrease of M_w/M_n to 1.15 [48]. The polymerization of a thiophene monomer **1f** containing an ester moiety also showed chain-growth polymerization behavior to afford polymers with M_w/M_n of 1.25–1.5 [49]. The ω-Bromohexyl group is tolerable when the corresponding dibromothiophene is converted to Grignard type monomer **1g**, and the polymerization with $Ni(dppp)Cl_2$ gave a polymer with M_w/M_n of 1.30. The bromohexyl group in the polymer is converted to the 5-hexenyl group, which is utilized for cross-linking upon thermal treatment [50].

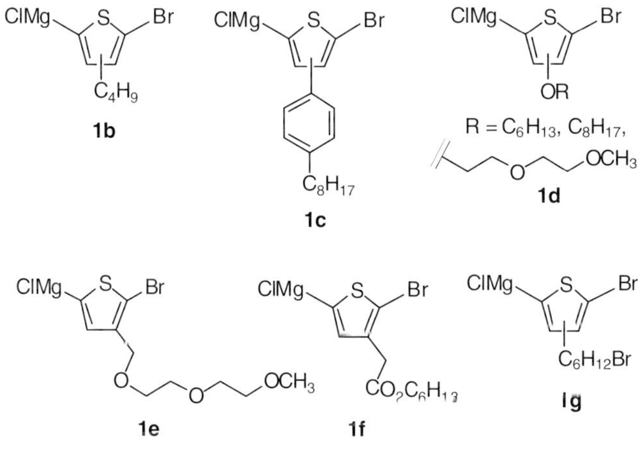

Scheme 2.5

Kiriy and coworkers studied the polymerization of thiophene dimer **2** and trimer **3** with $(PPh_3)_2Ni(Ph)Br$ as an initiator to see how far intramolecular catalyst transfer takes place (Scheme 2.6, Table 2.1) [51]. The polymerization of **2a** and **3a** led predominantly to Ph-terminated polymers, indicating the chain-growth polymerization mechanism, although an increase in the monomer molecular length some-

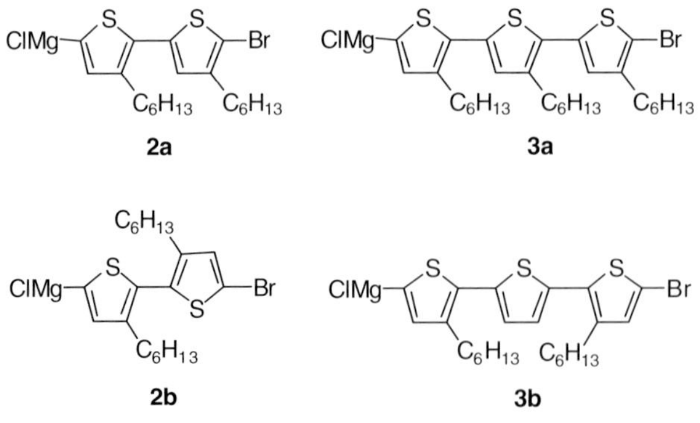

Scheme 2.6

Table 2.1 Ph–Ni$(PPh_3)_2$–Br initiated polymerization.

Monomer	M_n [a]	M_w/M_n [a]	Ph/H and Ph/Br, % [b]
2a	5300	1.67	87
3a	5700	1.74	69
2b	3300	1.91	45
3b	4200	2.0	31

a) GPC data.
b) ^{1}H NMR data.

what decreased the fraction of the Ph-terminated products. However, the polymerization of **2b** and **3b** gave less than half the fraction of the Ph-terminated products. Accordingly, the chain-growth performance turns out to be sensitive to the substitution pattern of the polymerized monomers. Monomers having alkyl substituents in an ortho (with respect to the growing site) position gave better results, possibly due to higher stability of the intermediate ortho-substituted aryl-nickel complexes.

Venkataraman and coworkers tried the polymerization of another thiophene dimer **4** with Ni(dppp)Cl_2 (Scheme 2.7) [52]. Polymerization at ambient temperature afforded very little of the desired polymer, but polymerization at 70 °C yielded the polymer bearing alternating protected butynyl and hexyl side chains. The M_n value was 15 300 and M_w/M_n was 1.28–1.35. The obtained polymer was functionalized by azide bearing a naphthalimide moiety under "click chemistry" conditions.

2.2.1.3 **Block Polythiophenes**

Since several substituted thiophene monomers undergo chain-growth polymerization in a living polymerization fashion, a lot of block copolythiophenes **5** [24, 52–59] have been synthesized by successive polymerization in one pot (Scheme 2.8). Of these block copolymers, **5a** was first synthesized by McCullough and coworkers [24]. Yokozawa *et al.* synthesized block copolymer **5b** having both hydrophobic and

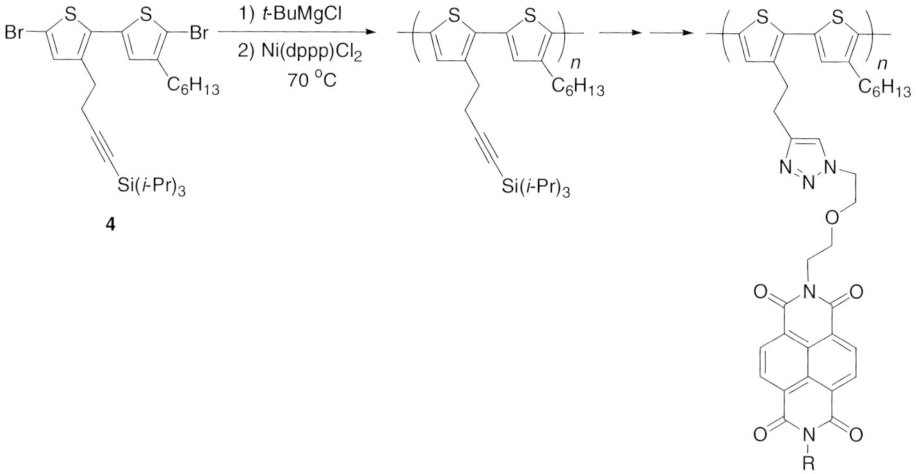

Scheme 2.7

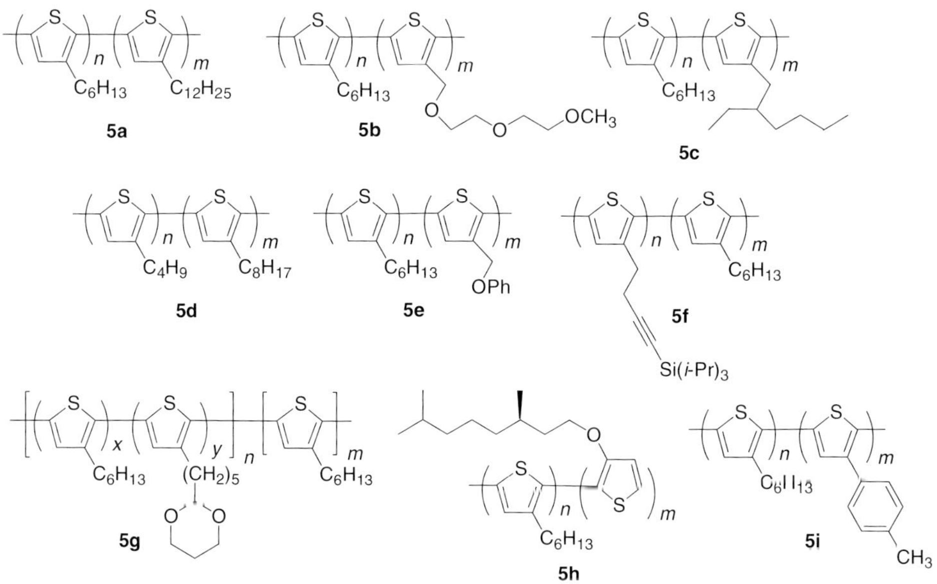

Scheme 2.8

hydrophilic side chains in each segment [53]. The thin films of **5c** [54], **5e** [55], and **5i** [58] showed nanofiber structures after annealing, probably because of microphase separation of the crystalline poly(hexylthiophene) segment and the other amorphous segment. The block copolymer **5d** is a crystalline–crystalline diblock copolymer, self-assembled into crystalline nanowires in solution. In melt-phase assembly, a microphase-separated lamellar structure with two crystalline domains characteristic of the two different side chains was observed [59]. The side chain of block copolymer **5f** [52] was converted to the naphthalimide moiety, as shown in Scheme 2.7. The diblock copolymer **5g** [56] consists of a block made from a random copolymerization of hexylthiophene monomer and (1,3-dioxa-2-octyl)thiophene monomer and the pure poly(hexylthiophene) block. The 1,3-dioxone moiety was converted to the formyl group, followed by introduction of fullerene C_{60} with the aid of *N*-methylglycine. In the block copolymer **5h**, the chiral segment influences the supramolecular organization of the poly(hexylthiophene) segment [57].

2.2.1.4 **Block Copolymers of Polythiophene and Other Polymers**

Block copolymers composed of polythiophene and conventional polymer have also been synthesized. McCullough and coworkers synthesized block copolymers of poly (3-hexylthiophene) and polystyrene or poly(methyl acrylate) by ATRP of the vinyl monomer from a polythiophene macroinitiator, which was prepared in several steps [60]. After the development of catalyst-transfer condensation polymerization of polythiophene, the block copolymer of polythiophene and poly(alkyl acrylate) was prepared more easily. As mentioned above, the vinyl-terminated polythiophene was first prepared. The vinyl group was converted to the 2-hydroxyethyl group by hydroboration, followed by esterification with 2-bromopropionyl bromide to give a macroinitiator for ATRP (Scheme 2.9) [61]. The allyl-terminated polythiophene was also converted to a macroinitiator for ATRP, which led to block copolymers of polythiophene and poly(alkyl methacrylate) [62] or poly(acrylic acid) [63]. This ally-terminated polythiophene has a bromine atom at the other end, which has an adverse effect on the purity of block copolymers prepared by ATRP. Hawker, Kim, and coworkers reported that replacement of the bromine with a phenyl group, followed by functionalization of the allyl group for the ATRP initiator unit, allowed access to narrower molecular weight distribution diblock copolymers of polythiophene and ATRP-derived vinyl block [64]. Recently, reversible addition fragmentation chain transfer (RAFT) polymerization or nitroxide-mediated polymerization (NMP) was used instead of ATRP for the synthesis of the block copolymer of polythiophene and polyisoprene or polystyrene, because ATRP generates materials containing traces of metals from the catalyst [65]. A block copolymer of poly(hexylthiophene) and poly(perylene bisimide acrylate) was also recently synthesized by NMP from a polythiophene macroinitiator (Scheme 2.10)

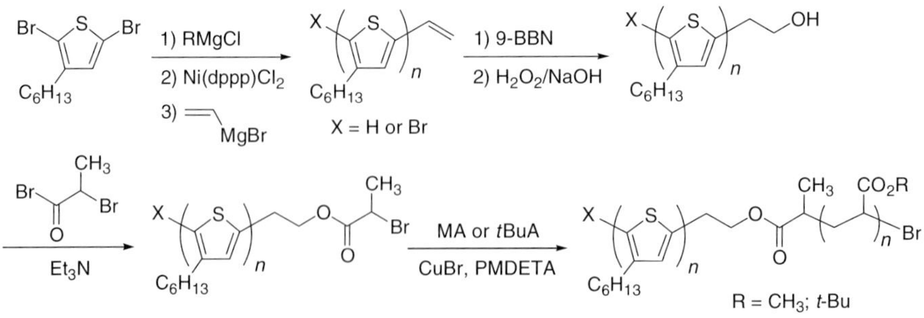

1) RMgCl 2) Ni(dppp)Cl_2 3) =\MgBr; X = H or Br; 1) 9-BBN 2) H_2O_2/NaOH; Et_3N; MA or *t*BuA, CuBr, PMDETA; R = CH_3; *t*-Bu

Scheme 2.9

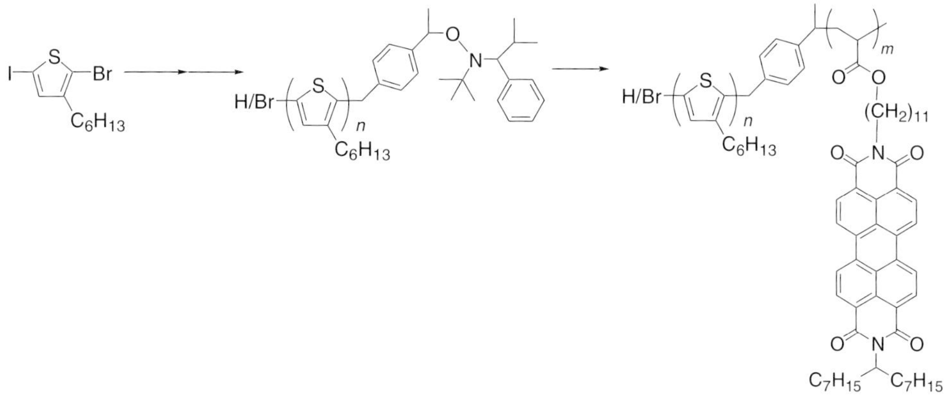

Scheme 2.10

[66, 67]. This block copolymer is a crystalline–crystalline donor–acceptor block copolymer and shows microphase separation, implying efficient photovoltaic applications. A similar polythiophene macroinitiator for NMP was used for the synthesis of fullerene-grafted rod–coil block copolymers [68].

Frisbie and Hillmyer used the hydroxyethyl-terminated polythiophene in Scheme 2.9 as a macroinitiator for the ring-opening polymerization of D,L-lactide (Scheme 2.11) [69]. The hydroxy-terminated polythiophene was converted to the corresponding aluminum alkoxide macroinitiator with triethylaluminum, followed by ring-opening polymerization of the lactide to yield a block copolymer of polythiophene and polylactide. In thin films of the block copolymers, microphase separated domains were formed. Upon chemical etching of the polylactide block, nanopitted film, where the crystallinity of the polythiophene phase remained, was observed.

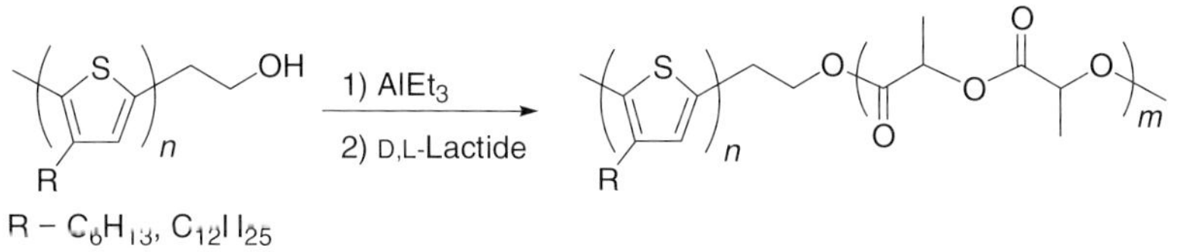

Scheme 2.11

Dai, Su, and coworkers synthesized block copolymers of polythiophene and poly(vinylpyridine) by means of living anionic polymerization of 4-vinylpyridine from the vinyl-terminated polythiophene macroinitiator (Scheme 2.12) [70] These block copolymers were able to undergo microphase separation and self-assembly into nanostructures of sphere, cylinder, lamellae, and nanofiber structure with increasing polythiophene segment ratios.

Meijer and coworkers used the allyl-terminated polythiophene to synthesize a block copolymer of poly(3-hexylthiophene) and polyethylene as a crystalline–

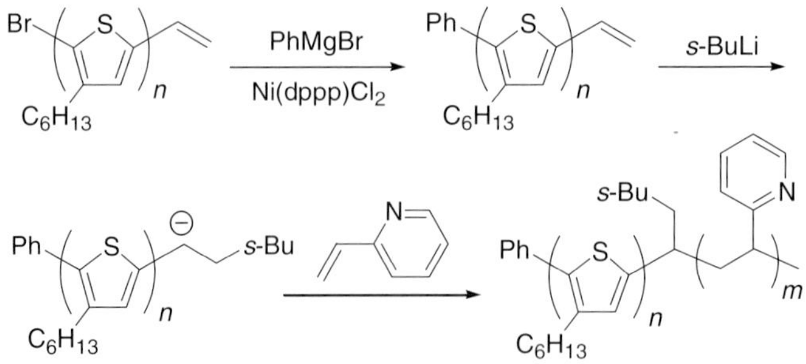

Scheme 2.12

crystalline block copolymer: the ring-opening metathesis polymerization of cyclooctene in the presence of the allyl-terminated polythiophene was followed by hydrogenation (Scheme 2.13) [71].

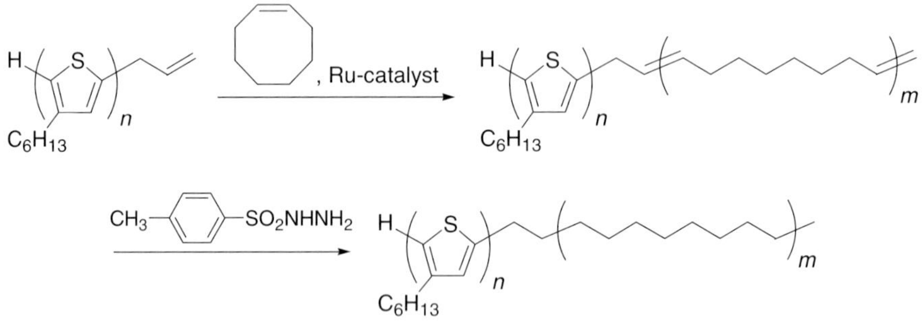

Scheme 2.13

Fréchet and coworkers reported the synthesis of a dendron-modified polythiophene, in which one terminus of poly(3-hexylthiophene) is linked to the focal point of a polyester-type dendron (Scheme 2.14) [72]. Introduction of the benzylamine moiety to the end group of polythiophene was carried out by the Suzuki–Miyaura

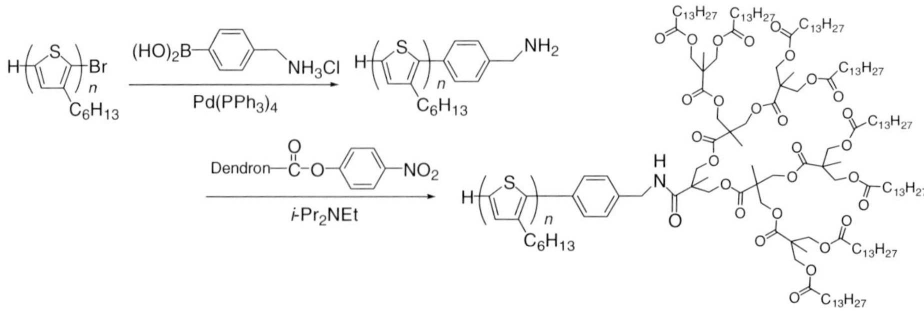

Scheme 2.14

coupling reaction of the polythiophene with the H/Br end groups and the corresponding boronic acid hydrochloric acid. The amino-functionalized polythiophene was then treated with dendron active ester in the presence of base. Despite the insulating nature of the polyester dendron, a device made with this dendron-modified polythiophene functioned as a transistor, and showed moderate field effect mobilities.

Triblock copolymers of polystyrene-*b*-polythiophene-*b*-polystyrene were independently synthesized by two groups. Cloutet and coworkers conducted a "click reaction" of terminal azide-functionalized polystyrenes, which were prepared by ATRP, and alkyne-functionalized poly(3-hexylthiophene) (Scheme 2.15A) [73]. The point of this reaction is the pentynyl group as a terminal alkynyl group of polythiophene; the "click" reaction of ω-ethynyl-polythiophene did not work. They surmised that conjugation of the ethynyl group with polythiophene can hinder "click" reactions. Ueda and coworkers introduced 1,1-diphenylethylene moieties on the α,ω-ends of polythiophene *in situ* and conducted a linking reaction with a living anionic polymer of styrene (Scheme 2.15B) [74].

Scherf and coworkers reported all-conjugated donor–acceptor–donor-type triblock copolymer by using the polythiophene with Br/H end groups. Cyano-substituted poly(phenylene vinylene), oligo(benzothiadiazole), and oligo(dicyanostilbene) prepolymers with two bromoaryl terminals as the central acceptor building block were first generated with $Ni(COD)_2$ under Yamamoto-type aryl–aryl coupling conditions, and were then decorated with two polythiophenes with the Br/H end groups under the same conditions [75, 76]. Two examples are shown in Scheme 2.16. In this method, however, the products need to be fractionated by Soxhlet extraction with several solvents to remove the homopolymer of polythiophene.

In a similar manner, polyfluorene, obtained by Suzuki–Miyaura coupling polymerization, was coupled with the bromo-terminated poly[3-(6-bromohexyl)thiophene] to yield a diblock copolymer of polythiophene and polyfluorene (Scheme 2.17) [77]. The separation of the target block copolymer and homopolymeric side products was accomplished by several solvent extraction steps. The bromohexyl group on the polythiophene block was converted to a phosphonate [77] or ammonium [76] moiety to produce amphiphilic, all-conjugated rod–rod diblock copolymers. The aggregation behavior of the former block copolymer at the air/water interface was investigated with a LB technique [78]. The AFM showed the initial formation of vesicular species which undergo fusion to lamellar monolayers at higher surface pressures, followed by irreversible destruction of the monolayers and, finally, extrusion/reformation of the vesicles.

2.2.1.5 **Graft Copolymers**

Chochos and coworkers synthesized a thiophene-grafted copolymer by using polythiophene with Br/H end groups. Thus, Suzuki–Miyaura coupling was performed between the bromo-terminated polythiophene and 4-vinylphenylboronic acid, followed by radical copolymerization with *N,N*-dimethylacrylamide to afford the graft copolymer (Scheme 2.18) [79]. They studied the interaction of the

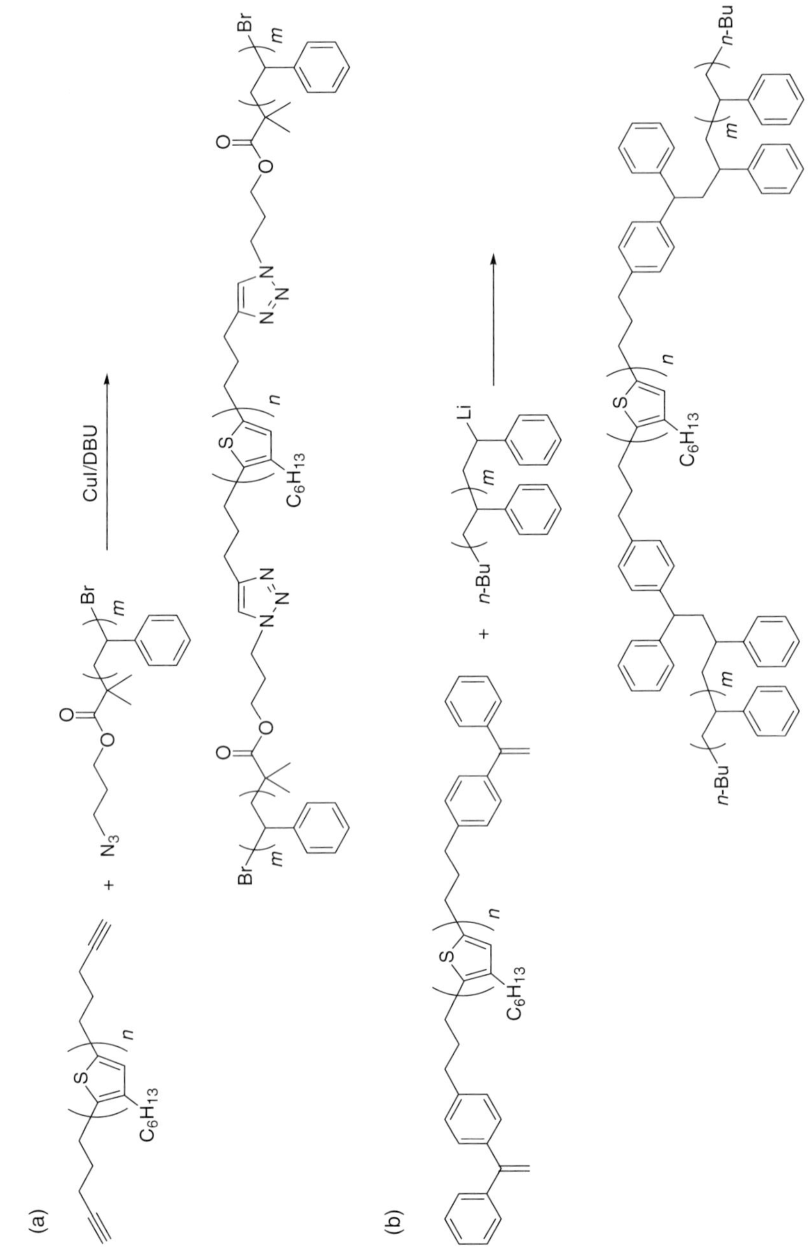

Scheme 2.15

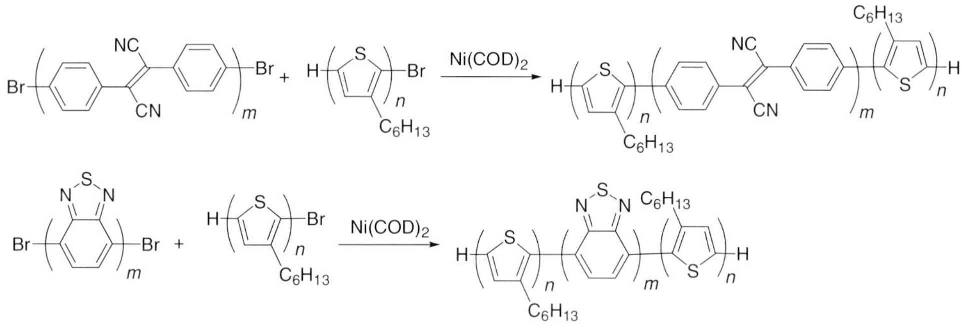

Br NC CN Br m + H S Br n C_6H_{13} $Ni(COD)_2$ H S n C_6H_{13} NC CN m C_6H_{13} S n H

N S N Br Br m + H S Br n C_6H_{13} $Ni(COD)_2$ H S n C_6H_{13} N S N C_6H_{13} m S n H

Scheme 2.16

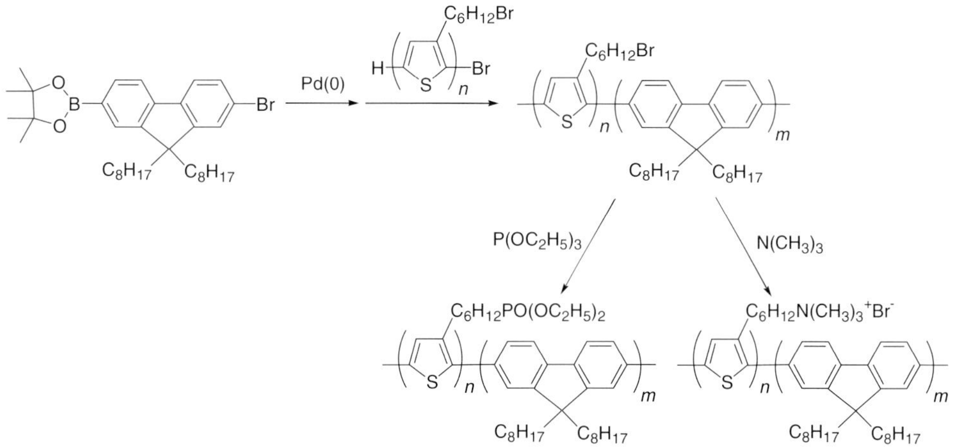

O B O Br C_8H_{17} C_8H_{17} Pd(0) $C_6H_{12}Br$ H S Br n $C_6H_{12}Br$ S n m C_8H_{17} C_8H_{17}

$P(OC_2H_5)_3$ $N(CH_3)_3$

$C_6H_{12}PO(OC_2H_5)_2$ S n m C_8H_{17} C_8H_{17} $C_6H_{12}N(CH_3)_3{}^+Br^-$ S n m C_8H_{17} C_8H_{17}

Scheme 2.17

H S Br n C_6H_{13} + B HO OH $Pd(PPh_3)_4$ 2M Na_2CO_3 H S n C_6H_{13}

n C_6H_{13} S H + O=C NMe_2 AIBN x y O=C NMe_2 n C_6H_{13} S H

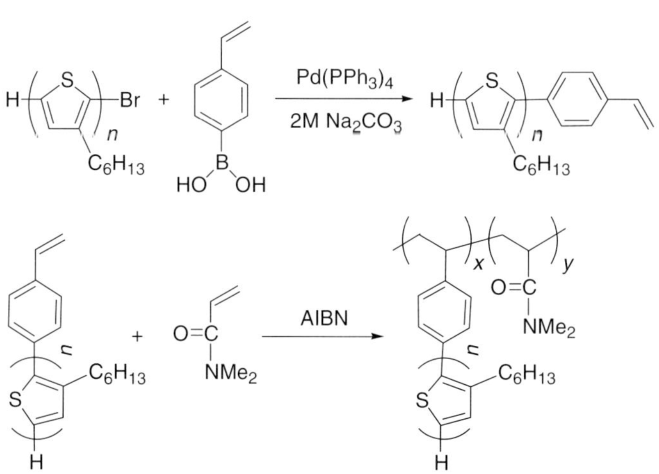

Scheme 2.18

polythiophene chains on the copolymer in water and several organic solvents. It was demonstrated that the polythiophene chains adopt a coil conformation in solvents such as THF and chloroform. However, the polythiophene chains were organized in a single chain packing form (intrachain interactions) in polar solvents such as ethanol and methanol. When water was used as solvent, the polythiophene chains self-assembled into a stack-like structure due to the increased interchain interactions.

Kiriy and coworkers conducted grafting of poly(3-hexylthiophene) from poly(4-vinylpyridine)-*b*-poly(4-iodostyrene) immobilized on silica particles. This block copolymer adheres strongly to a variety of polar substrates including silicon wafers, glasses, or a metal oxide surface by a polar poly(4-vinylpyridine) block, forming polymer brushes of poly(4-iodostyrene) chains, which reacted with $Ni(PPh_3)_4$ to generate the Ni initiator for the catalyst-transfer polymerization of **1a**. Unfortunately, the polythiophene grafts of the products are relatively short (~10 nm) (Scheme 2.19) [80].

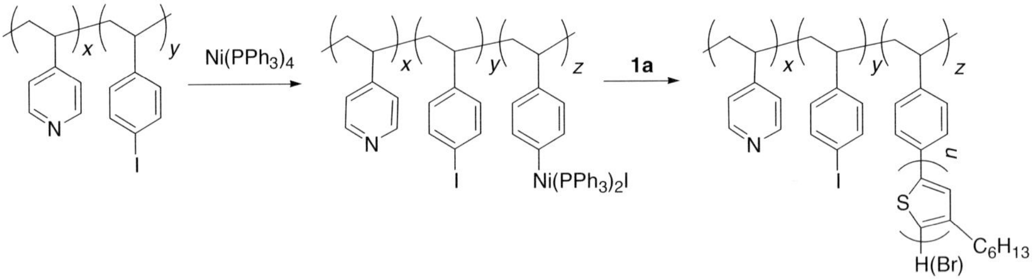

Scheme 2.19

2.2.2 Polyphenylenes

It is important to clarify whether catalyst-transfer condensation polymerization is specific to polythiophene, or whether it is generally applicable to the synthesis of well-defined π-conjugated polymers. Yokozawa and coworkers investigated the synthesis of poly(*p*-phenylene), to see whether a monomer **6** containing no heteroatom in the aromatic ring would undergo catalyst-transfer polymerization. However, all polymers obtained in the polymerization with $Ni(dppp)Cl_2$, $Ni(dppe)Cl_2$ or $Ni(dppf)Cl_2$ possessed low molecular weights and broad polydispersities. Nevertheless, they found that LiCl was necessary to optimize the chain-growth condensation polymerization leading to poly(*p*-phenylene) with low polydispersity, and that the molecular weight was controlled by the feed ratio of **6** to the Ni catalyst (Scheme 2.20) [81].

Furthermore, block copolymers of polythiophene and poly(*p*-phenylene) by the successive catalyst-transfer condensation polymerization of monomers **1a** or **1e** and **6**

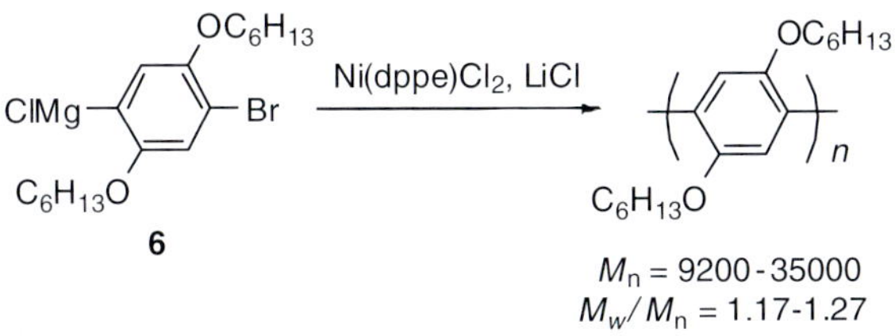

OC_6H_{13} ; ClMg ; Br ; $C_6H_{13}O$; **6** ; $Ni(dppe)Cl_2$, LiCl ; OC_6H_{13} ; *n* ; $C_6H_{13}O$

M_n = 9200-35000
M_w/M_n = 1.17-1.27

Scheme 2.20

with a Ni catalyst were synthesized [82]. This is the first example of the successive catalyst-transfer condensation polymerization for the synthesis of block copolymers consisting of a different type of π-conjugated polymers The polymerization of **1** and then **6** with a Ni catalyst yielded polymers with broad molecular weight distribution, whereas the reverse order of polymerization resulted in well-defined block copolymers of poly(*p*-phenylene) and polythiophene (Scheme 2.21). Successful block copolymerization of **6** and then **1** may be accounted for by the π-donor ability of

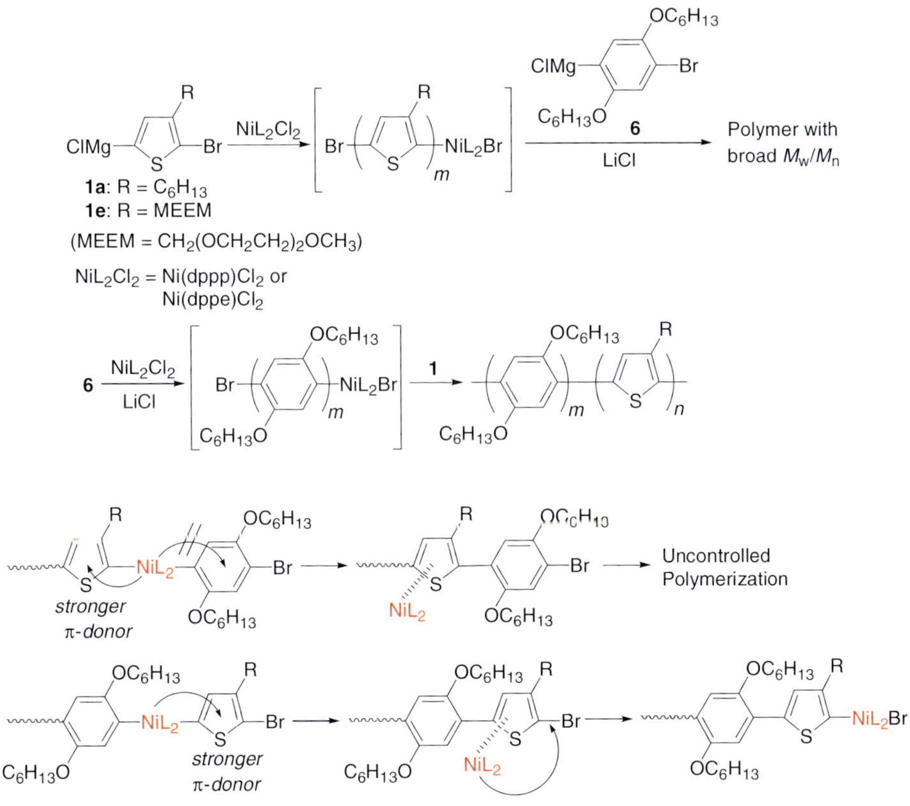

ClMg ; R ; S ; Br ; NiL_2Cl_2 ; Br ; R ; S ; *m* ; NiL_2Br ; OC_6H_{13} ; ClMg ; Br ; $C_6H_{13}O$; **6** ; LiCl ; Polymer with broad M_w/M_n

1a: R = C_6H_{13}
1e: R = MEEM
(MEEM = $CH_2(OCH_2CH_2)_2OCH_3$)
NiL_2Cl_2 = $Ni(dppp)Cl_2$ or $Ni(dppe)Cl_2$

6 ; NiL_2Cl_2 ; LiCl ; Br ; OC_6H_{13} ; $C_6H_{13}O$; *m* ; NiL_2Br ; **1** ; OC_6H_{13} ; R ; $C_6H_{13}O$; *m* ; S ; *n*

R ; S ; NiL_2 ; stronger π-donor ; OC_6H_{13} ; Br ; OC_6H_{13} ; R ; S ; NiL_2 ; OC_6H_{13} ; Br ; OC_6H_{13} ; Uncontrolled Polymerization

OC_6H_{13} ; NiL_2 ; R ; S ; Br ; stronger π-donor ; $C_6H_{13}O$; OC_6H_{13} ; R ; S ; Br ; NiL_2 ; $C_6H_{13}O$; OC_6H_{13} ; R ; S ; NiL_2Br ; OC_6H_{13}

Scheme 2.21

polythiophene and poly(*p*-phenylene) because the π-electrons of the polymers are considered to assist the transfer of the Ni catalyst in catalyst-transfer polymerization. When **6** is added to the reaction mixture of polythiophene as a prepolymer, the Ni catalyst is difficult to move to the terminal C–Br bond of the phenylene ring of the elongated **6** unit, because the thiophene ring has stronger π-donor ability than the phenylene ring. On the other hand, when **1** is added to poly(*p*-phenylene) as a prepolymer, the Ni catalyst moves smoothly to the C–Br bond of the thiophene ring with stronger π-donor ability (Scheme 2.21).

2.2.3
Polypyrroles

Yokozawa and coworkers also investigated the condensation polymerization of Grignard-type *N*-hexylpyrrole monomer **7**, which was generated by the reaction of the corresponding dibromopyrrole with *i*-PrMgCl, with a Ni catalyst (Scheme 2.22). When Ni(dppp)Cl_2 was used as a catalyst in a similar manner to the polymerization of hexylthiophene monomer **1a**, a polymer with M_w/M_n of 1.26 was obtained, accompanied by low-molecular-weight oligomers. On the other hand, polymerization with Ni(dppe)Cl_2 afforded a polymer with a narrower polydispersity ($M_w/M_n = 1.19$), though oligomeric by-products were still formed. To suppress the formation of the oligomeric by-products, they examined the effect of several additives, and found that polymerization of **7** with Ni(dppe)Cl_2 in the presence of additional dppe, equimolar to the catalyst, at 0 °C gave the polypyrrole with a narrow molecular weight distribution ($M_w/M_n = 1.11$) without formation of by-products. The conversion-number average molecular weight (M_n) and feed ratio–M_n relationships indicated that this polymerization proceeded in a catalyst-transfer polymerization manner [83]. McCullough and coworkers used *i*-PrMgCl·LiCl for generation of Grignard-type *N*-dodecylpyrrole monomer and polymerized with Ni(dppp)Cl_2 to afford a polypyrrole with $M_n = 11\,900$ and $M_w/M_n = 1.4$ [84].

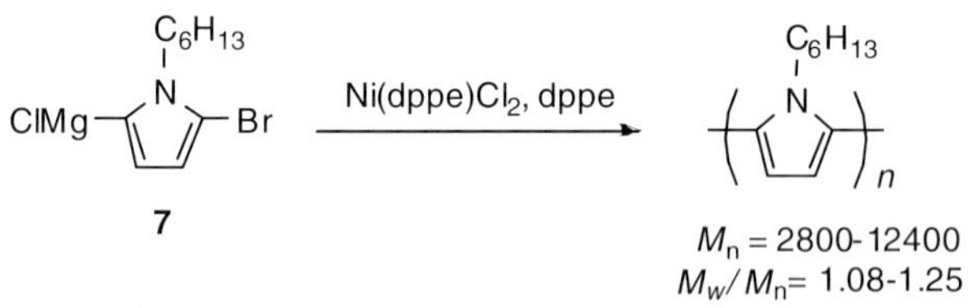

C_6H_{13} ... ClMg ... Br — **7** — Ni(dppe)Cl_2, dppe → C_6H_{13} ... *n*

M_n = 2800-12400
M_w/M_n = 1.08-1.25

Scheme 2.22

Yokozawa and coworkers synthesized a block copolymer of poly(*p*-phenylene) and polypyrrole by their method. The *p*-phenylene monomer **6** was polymerized first with Ni(dppe)Cl_2 in the presence of LiCl, followed by addition of **7** and dppe to yield the desired block copolymer with a narrow polydispersity ($M_w/M_n = 1.16$) (Scheme 2.23).

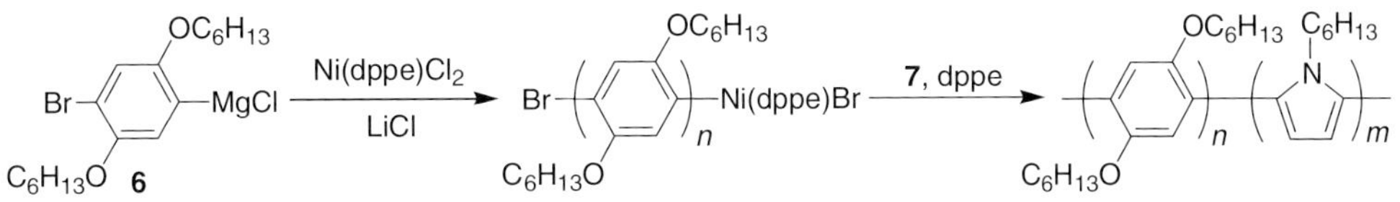

Scheme 2.23

The block copolymerization in reverse order afforded polymer with a somewhat broader polydispersity ($M_w/M_n = 1.38$).

2.2.4
Polyfluorenes and Polycarbazoles

The synthesis of well-defined polyfluorene was first conducted by catalyst transfer Suzuki–Miyaura coupling polymerization (see Section 2.3.1). After this report, Geng and coworkers reported the polymerization of Grignard-type fluorene monomer **8** with Ni(dppp)Cl_2 [85]. The monomer **8** was generated from the corresponding bromoiodofluorene with *i*-PrMgCl·LiCl. The polymerization proceeded very rapidly at 0 °C and was almost over in 10 min. High molecular weight polyfluorene ($M_n = 18$ 800–86 000) was obtained in the very beginning and the M_n remained almost constant for different conversions of monomer **8**. The M_w/M_n of the resulting polymers was in the range 1.49–1.77. The behavior of this polymerization resembles that of conventional radical polymerization that proceeds in a chain-growth polymerization manner accompanying a chain transfer reaction. The ^{1}H NMR spectrum of the polymer indicated that a noticeable amount of polymers were the Br/Br ended polyfluorene, which would be formed by reductive elimination followed by intermolecular transfer of the Ni(0) catalyst that can initiate polymerization of another chain (Scheme 2.24). McCullough and coworkers conducted a similar polymerization of **8**, generated from the corresponding dibromofluorene with *t*-BuMgCl·LiCl·15-crown-5 [84].

Carter and coworkers conducted the polymerization of lithiated fluorene monomer **9**, generated by the reaction of the corresponding dibromofluorene with *t*-BuLi *in situ*, with Ni(dppp)Cl_2 (Scheme 2.25) [86]. The ratio of *t*-BuLi to the dibromofluorene had a direct effect on the molecular weight; use of excess *t*-BuLi decreased the molecular weight. Furthermore, the observed molecular weight distributions were around 2. Accordingly, they proposed a step-growth polymerization mechanism.

The synthesis of polycarbazole was attempted by polymerization of a Grignard monomer. However, the magnesium halogen exchange of 2,7-dibromo-*N*-octylcarbazole with *i*-PrMgCl·LiCl was significantly slower compared to both the above fluorene and pyrrole monomers. In contrast, lithium tributylmagnesate (*n*-Bu_3MgLi) was effective for the magnesium halogen exchange. The polymerization of this Grignard type monomer was carried out with Ni(dppp)Cl_2 to afford a polycarbazole with $M_n = 2600$ and $M_w/M_n = 1.23$ (Scheme 2.26) [84].

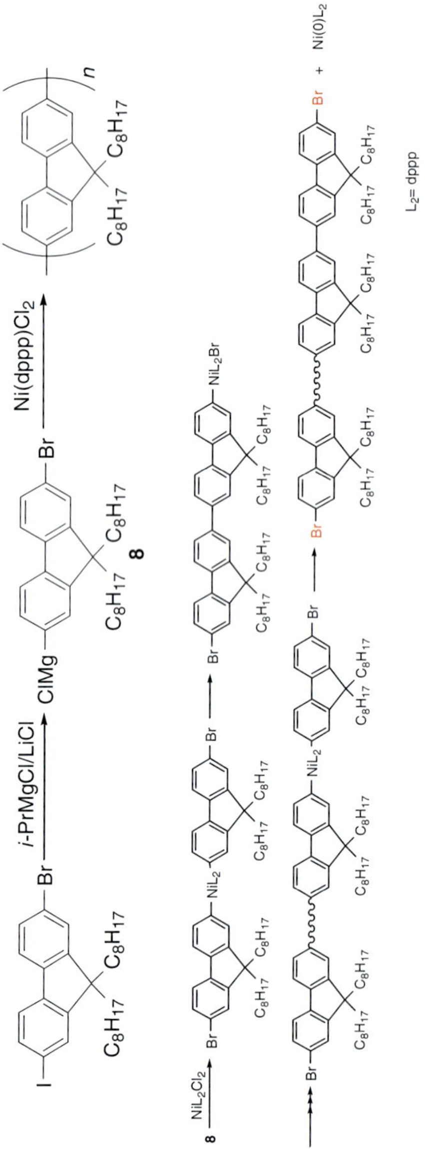

Scheme 2.24

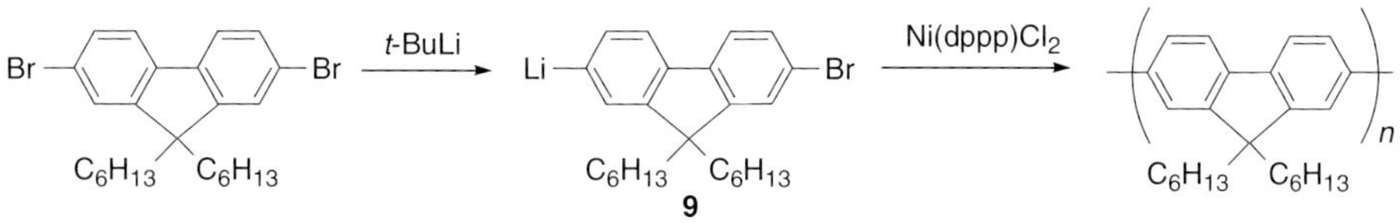

Scheme 2.25

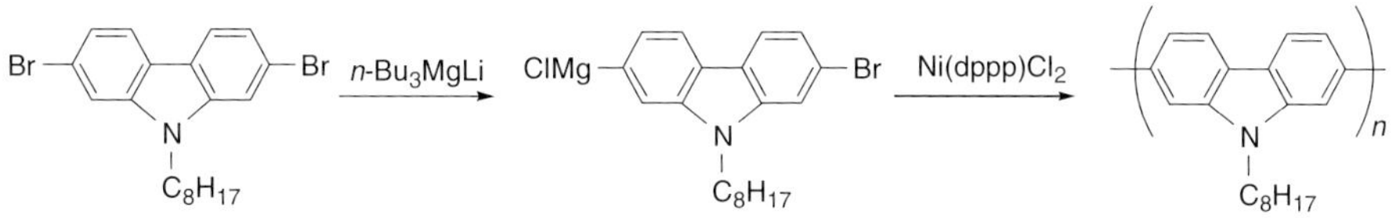

Scheme 2.26

2.3
Suzuki–Miyaura Coupling Polymerization with Pd Catalyst

2.3.1
Polyfluorenes

Suzuki–Miyaura coupling polymerization is one of the most efficient methods for synthesis of π-conjugated polymers, but the polymerizations reported so far have been step-growth polymerization [87]. Yokozawa and coworkers investigated chain-growth Suzuki–Miyaura coupling polymerization. In this polymerization, stable arylpalladium(II) halide complex can be used as an externally added initiator, and the aryl group of the complex served as an initiator unit for the polymer. The polymerization of a fluorene monomer **10** was carried out in the presence of $^{t}Bu_3PPd(Ph)Br$ as catalyst to yield polyfluorene with a narrow polydispersity (Scheme 2.27). The molecular weight of the obtained polymer increased linearly in proportion to the conversion of monomer with low polydispersity throughout the polymerization, and also increased linearly in proportion to the feed ratio of **10** to the initiator up to 17 700 with low polydispersity, indicating that this Suzuki–Miyaura coupling condensation polymerization proceeded through a chain-growth polymerization mechanism [88], as shown in the model reactions [28–30]. The MALDI-TOF mass spectrum of the

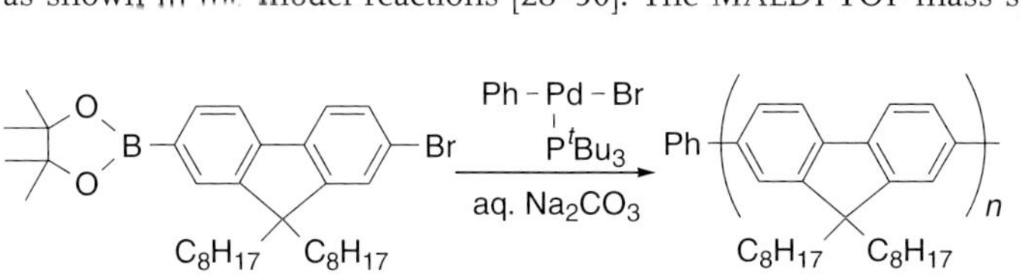

Scheme 2.27

obtained polyfluorene showed that all the polymers bore the phenyl group at one end. This observation strongly supported the view that ^{t}Bu$_3$PPd(Ph)Br served as an initiator.

Kiriy and coworkers applied this chemistry to surface-initiated polymerization. The bromophenyl group of cross-linked poly(4-bromostyrene) deposited on a silicon wafer was treated with Pd(t-Bu$_3$P)$_2$ at 70 °C to generate the initiator site. The polymerization of a similar fluorene monomer having a bis(2-ethylhexyl) group proceeded from the surface initiation site, and resulted in grafted polyfluorene films with thickness up to 100 nm [89].

Reynolds and coworkers recently reported one-pot Suzuki–Miyaura coupling polymerization from 2,7-dibromo-9,9-dioctylfluorene and bis(pinacolato)diboran in the presence of $Pd_2(dba)_3/HPCy_3BF_4$ and CsF. The molecular weight distribution of the obtained polymer was about 2, and the polymerization may proceed in a step-growth polymerization manner (Scheme 2.28) [90].

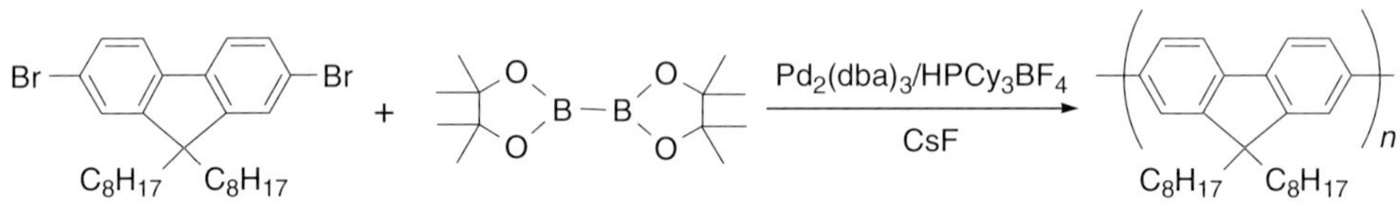

Br, Br, C_8H_{17}, C_8H_{17} + O, B–B, O; $Pd_2(dba)_3/HPCy_3BF_4$, CsF → C_8H_{17}, C_8H_{17}, n

Scheme 2.28

2.3.2 Polyphenylenes

The polymerization system using ^{t}Bu$_3$PPd(Ph)Br as an externally added initiator, which was effective for the catalyst-transfer polymerization of fluorene monomer **10**, was applied to the polymerization of phenylene monomer **11**. When the polymerization was carried out under the same conditions using aqueous Na_2CO_3 at room temperature, polymer with a broad molecular weight distribution was obtained. However, the polymerization of **11** with CsF at 0 °C yielded polymer with a relatively narrow molecular weight distribution ($M_w/M_n < 1.5$). Under these polymerization conditions, the molecular weight of the obtained polymer increased linearly in proportion to the conversion of monomer with low polydispersity throughout the polymerization, and also increased linearly in proportion to the feed ratio of **11** to the initiator up to 21 000 with low polydispersity (Scheme 2.29). Furthermore, the

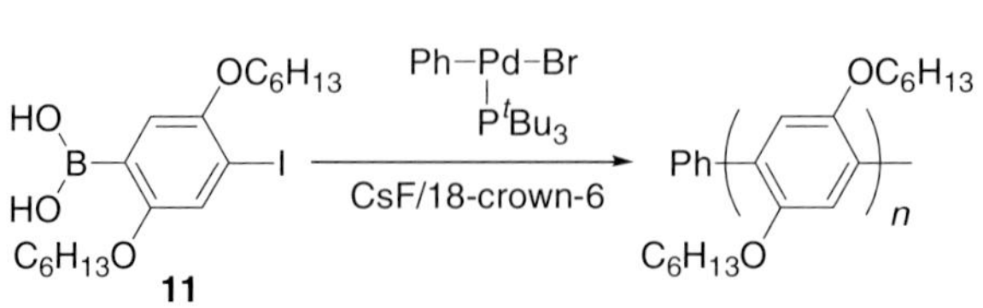

HO, B, HO, OC_6H_{13}, I, $C_6H_{13}O$, **11**; Ph–Pd–Br, P^{t}Bu$_3$, CsF/18-crown-6 → Ph, OC_6H_{13}, $C_6H_{13}O$, n

Scheme 2.29

polymerization of **10** under these condition and then the successive polymerization of **11** were carried out to yield a block copolymer of polyfluorene and polyphenylene (Scheme 2.30) [91].

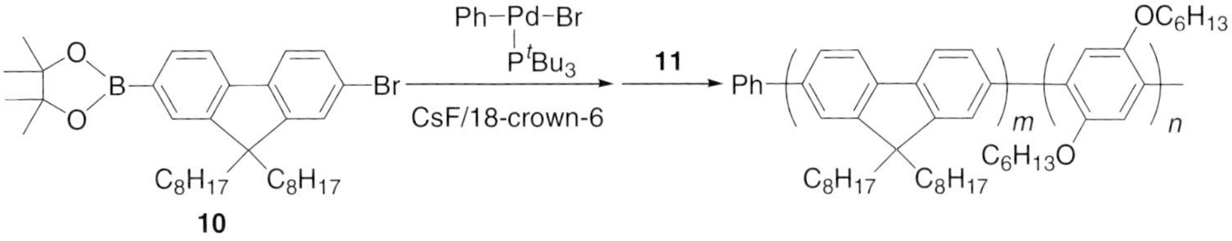

Scheme 2.30

2.4 Conclusion

Only in the last five years has it been revealed that π-conjugated polymers can be synthesized in a controlled fashion by catalyst-transfer condensation polymerization, which was discovered in Kumada–Tamao coupling polymerization with a Ni catalyst for the synthesis of polythiophene. This polymerization method is applicable not only to Kumada–Tamao coupling polymerization for the synthesis of polyphenylene and polypyrrole but also to Suzuki–Miyaura coupling polymerization leading to polyfluorene and polyphenylene. However, the controlled π-conjugated polymers have been limited so far to π-donor polymers, except for polyfluorene. This is understandable because the π-electrons of the polymers are considered to assist the transfer of the catalyst in catalyst-transfer polymerization. Hereafter, polymerization of well-defined π-acceptor polymers should be investigated by studying other catalysts and other coupling polymerizations. Successive catalyst-transfer polymerization of π-donor monomer and π-acceptor monomer will enable us to synthesize a variety of block copolymers of donor and acceptor π-conjugated polymers, which are promising materials for light-emitting diodes, photovoltaic cells, and so on.

References

1 Skotheim, T.A., Elsenbaumer, R.L., and Reynolds, J.R. (eds) (1997) *Handbook of Conducting Polymers*, 2nd edn Revised and Expanded, Marcel Dekker, Inc., New York.

2 Yamamoto, T. (2002) *Macromol. Rapid Commun.*, **23**, 583–606.

3 Babudri, F., Farinola, G.M., and Naso, F. (2004) *J. Mater. Chem.*, **14**, 11–34.

4 Barbarella, G., Melucci, M., and Sotgiu, G. (2005) *Adv. Mater.*, **17**, 1581–1593.

5 Weissbach, H. and Pestka, S. (eds) (1977) *Molecular Mechanisms of Protein Biosynthesis*, Academic Press, New York.

6 Bermek, E. (ed.) (1985) Mechanisms of protein synthesis: structure-function relations, control mechanisms, and evolutionary aspects. Proceedings of the

Symposium on Molecular Mechanisms in Protein Synthesis Held at Beyaz Koesk, Emirgan, Bosphorus, Istanbul.

7 Kornberg, A. (1960) *Science*, **131**, 1503–1508.

8 Travers, A. (1976) *Nature*, **263**, 641–646.

9 Chamberlin, M.J. (1982) *Enzymes*, Academic Press, New York, 3rd edn, vol. 15, pp. 61–86.

10 Yokozawa, T. and Yokoyama, A. (2007) *Prog. Polym. Sci.*, **32**, 147–172.

11 Yokoyama, A. and Yokozawa, T. (2007) *Macromolecules*, **40**, 4093–4101.

12 Roncali, J. (1997) *Chem. Rev.*, **97**, 173–205.

13 McCullough, R.D. (1998) *Adv. Mater.*, **10**, 93–116.

14 Chen, F., Mehta, P.G., Takiff, L., and McCullough, R.D. (1996) *J. Mater. Chem.*, **6**, 1763–1766.

15 Sirringhaus, H., Tessler, N., and Friend, R.H. (1998) *Science*, **280**, 1741–1744.

16 Sirringhaus, H., Brown, P.J., Friend, R.H., Nielsen, M.M., Bechgaard, K., Langeveld-Voss, B.M.W., Spiering, A.J.H., Janssen, R.A.J., Meijer, E.W., Herwig, P., and de Leeuw, D.M. (1999) *Nature*, **401**, 685–688.

17 McCullough, R.D. and Lowe, R.D. (1992) *J. Chem. Soc., Chem. Commun.*, 70–72.

18 McCullough, R.D., Lowe, R.D., Jayaraman, M., and Anderson, D.L. (1993) *J. Org. Chem.*, **58**, 904–912.

19 Loewe, R.S., Khersonsky, S.M., and McCullough, R.D. (1999) *Adv. Mater.*, **11**, 250–253.

20 Loewe, R.S., Ewbank, P.C., Liu, J.S., Zhai, L., and McCullough, R.D. (2001) *Macromolecules*, **34**, 4324–4333.

21 Yokoyama, A., Miyakoshi, R., and Yokozawa, T. (2004) *Macromolecules*, **37**, 1169–1171.

22 Miyakoshi, R., Yokoyama, A., and Yokozawa, T. (2004) *Macromol. Rapid Commun.*, **25**, 1663–1666.

23 Sheina, E.E., Liu, J.S., Iovu, M.C., Laird, D.W., and McCullough, R.D. (2004) *Macromolecules*, **37**, 3526–3528.

24 Iovu, M.C., Sheina, E.E., Gil, R.R., and McCullough, R.D. (2005) *Macromolecules*, **38**, 8649–8656.

25 Miyakoshi, R., Yokoyama, A., and Yokozawa, T. (2005) *J. Am. Chem. Soc.*, **127**, 17542–17547.

26 Nomura, N., Tsurugi, K., and Okada, M. (2001) *Angew. Chem. Int. Ed.*, **40**, 1932–1935.

27 Nomura, N., Tsurugi, K., RajanBabu, T.V., and Kondo, T. (2004) *J. Am. Chem. Soc.*, **126**, 5354–5355.

28 Sinclair, D.J. and Sherburn, M.S. (2005) *J. Org. Chem.*, **70**, 3730–3733.

29 Dong, C.G. and Hu, Q.S. (2005) *J. Am. Chem. Soc.*, **127**, 10006–10007.

30 Weber, S.K., Galbrecht, F., and Scherf, U. (2006) *Org. Lett.*, **8**, 4039–4041.

31 Strawser, D., Karton, A., Zenkina, E.V., Iron, M.A., Shimon, L.J.W., Martin, J.M.L., and van der Boom, M.E. (2005) *J. Am. Chem. Soc.*, **127**, 9322–9323.

32 Zenkina, O., Altman, M., Leitus, G., Shimon, L.J.W., Cohen, R., and van der Boom, M.E. (2007) *Organometallics*, **26**, 4528–4534.

33 Zenkina, O.V., Karton, A., Freeman, D., Shimon, L.J.W., Martin, J.M.L., and van der Boom, M.E. (2008) *Inorg. Chem.*, **47**, 5114–5121.

34 Yoshikai, N., Matsuda, H., and Nakamura, E. (2008) *J. Am. Chem. Soc.*, **130**, 15258–15259.

35 Miyakoshi, R., Yokoyama, A., and Yokozawa, T. (2008) *J. Polym. Sci., Part A: Polym. Chem.*, **46**, 753–765.

36 Mao, Y.X., Wang, Y., and Lucht, B.L. (2004) *J. Polym. Sci., Part A: Polym. Chem.*, **42**, 5538–5547.

37 Zhang, R., Li, B., Iovu, M.C., Jeffries-El, M., Sauve, G., Cooper, J., Jia, S.J., Tristram-Nagle, S., Smilgies, D.M., Lambeth, D.N., McCullough, R.D., and Kowalewski, T. (2006) *J. Am. Chem. Soc.*, **128**, 3480–3481.

38 Verilhac, J.M., LeBlevennec, G., Djurado, D., Rieutord, F., Chouiki, M., Travers, J.P., and Pron, A. (2006) *Synth. Met.*, **156**, 815–823.

39 Hiorns, R.C., de Bettignies, R., Leroy, J., Bailly, S., Firon, M., Sentein, C., Preud'homme, H., and Dagron-Lartigau, C. (2006) *Eur. Phys. J.: Appl. Phys.*, **36**, 295–300.

40 Jeffries-El, M., Sauve, G., and McCullough, R.D. (2005) *Macromolecules*, **38**, 10346–10352.

41 Senkovskyy, V., Khanduyeva, N., Komber, H., Oertel, U., Stamm, M., Kuckling, D.,

and Kiriy, A. (2007) *J. Am. Chem. Soc.*, **129**, 6626–6632.
42 Khanduyeva, N., Senkovskyy, V., Beryozkina, T., Bocharova, V., Simon, F., Nitschke, M., Stamm, M., Grotzschel, R., and Kiriy, A. (2008) *Macromolecules*, **41**, 7383–7389.
43 Sontag, S.K., Marshall, N., and Locklin, J. (2009) *Chem. Commun.*, 3354–3356.
44 Hiorns, R.C., Khoukh, A., Gourdet, B., and Dagron-Lartigau, C. (2006) *Polym. Int.*, **55**, 608–620.
45 Ouhib, F., Hiorns, R.C., Bailly, S., de Bettignies, R., Khoukh, A., Preud'homme, H., Desbrieres, J., and Dagron-Lartigau, C. (2007) *Eur. Phys. J.: Appl. Phys.*, **37**, 343–346.
46 Sheina, E.E., Khersonsky, S.M., Jones, E.G., and McCullough, R.D. (2005) *Chem. Mater.*, **17**, 3317–3319.
47 Koeckelberghs, G., Vangheluwe, M., Van Doorsselaere, K., Robijns, E., Persoons, A., and Verbiest, T. (2006) *Macromol. Rapid Commun.*, **27**, 1920–1925.
48 Adachi, I., Miyakoshi, R., Yokoyama, A., and Yokozawa, T. (2006) *Macromolecules*, **39**, 7793–7795.
49 Vallat, P., Lamps, J.P., Schosseler, F., Rawiso, M., and Catala, J.M. (2007) *Macromolecules*, **40**, 2600–2602.
50 Miyanishi, S., Tajima, K., and Hashimoto, K. (2009) *Macromolecules*, **42**, 1610–1618.
51 Beryozkina, T., Senkovskyy, V., Kaul, E., and Kiriy, A. (2008) *Macromolecules*, **41**, 7817–7823.
52 Benanti, T.L., Kalaydjian, A., and Venkataraman, D. (2008) *Macromolecules*, **41**, 8312–8315.
53 Yokozawa, T., Adachi, I., Miyakoshi, R., and Yokoyama, A. (2007) *High Perform. Polym.*, **19**, 684–699.
54 Zhang, Y., Tajima, K., Hirota, K., and Hashimoto, K. (2008) *J. Am. Chem. Soc.*, **130**, 7812–7813.
55 Ohshimizu, K. and Ueda, M. (2008) *Macromolecules*, **41**, 5289–5294.
56 Ouhib, F., Khoukh, A., Ledeuil, J.B., Martinez, H., Desbrieres, J., and Dagron-Lartigau, C. (2008) *Macromolecules*, **41**, 9736–9743.
57 Van den Ber, K., Huybrechts, J., Verbiest, T., and Koeckelberghs, G. (2008) *Chem. Eur. J.*, **14**, 9122–9125.
58 Ouhib, F., Hiorns, R.C., de Bettignies, R., Bailly, S., Desbrieres, J., and Dagron-Lartigau, C. (2008) *Thin Solid Films*, **516**, 7199–7204.
59 Wu, P.-T., Ren, G., Li, C., Mezzenga, R., and Jenekhe, S.A. (2009) *Macromolecules*, **42**, 2317–2320.
60 Liu, J., Sheina, E., Kowalewski, T., and McCullough, R.D. (2002) *Angew. Chem. Int. Ed.*, **41**, 329–332.
61 Iovu, M.C., Jeffries-El, M., Sheina, E.E., Cooper, J.R., and McCullough, R.D. (2005) *Polymer*, **46**, 8582–8586.
62 Iovu, M.C., Zhang, R., Cooper, J.R., Smilgies, D.M., Javier, A.E., Sheina, E.E., Kowalewski, T., and McCullough, R.D. (2007) *Macromol. Rapid Commun.*, **28**, 1816–1824.
63 Craley, C.R., Zhang, R., Kowalewski, T., McCullough, R.D., and Stefan, M.C. (2009) *Macromol. Rapid Commun.*, **30**, 11–16.
64 Lee, Y., Fukukawa, K.I., Bang, J., Hawker, C.J., and Kim, J.K. (2008) *J. Polym. Sci., Part A: Polym. Chem.*, **46**, 8200–8205.
65 Iovu, M.C., Craley, C.R., Jeffries-El, M., Krankowski, A.B., Zhang, R., Kowalewski, T., and McCullough, R.D. (2007) *Macromolecules*, **40**, 4733–4735.
66 Sommer, M., Lang, A.S., and Thelakkat, M. (2008) *Angew. Chem. Int. Ed.*, **47**, 7901–7904.
67 Zhang, Q., Cirpan, A., Russell, T.P., and Emrick, T. (2009) *Macromolecules*, **42**, 1079–1082.
68 Richard, F., Brochon, C., Leclerc, N., Eckhardt, D., Heiser, T., and Hadziioannou, G. (2008) *Macromol. Rapid Commun.*, **29**, 885–891.
69 Boudouris, B.W., Frisbie, C.D., and Hillmyer, M.A. (2008) *Macromolecules*, **41**, 67–75.
70 Dai, C.A., Yen, W.C., Lee, Y.H., Ho, C.C., and Su, W.F. (2007) *J. Am. Chem. Soc.*, **129**, 11036–11038.
71 Radano, C.P., Scherman, O.A., Stingelin-Stutzmann, N., Muller, C., Breiby, D.W., Smith, P., Janssen, R.A.J., and Meijer, E.W. (2005) *J. Am. Chem. Soc.*, **127**, 12502–12503.
72 Watanabe, N., Mauldin, C., and Frechet, J.M.J. (2007) *Macromolecules*, **40**, 6793–6795.

73 Urien, M., Erothu, H., Cloutet, E., Hiorns, R.C., Vignau, L., and Cramail, H. (2008) *Macromolecules*, **41**, 7033–7040.

74 Higashihara, T., Ohshimizu, K., Hirao, A., and Ueday, M. (2008) *Macromolecules*, **41**, 9505–9507.

75 Tu, G.L., Li, H.B., Forster, M., Heiderhoff, R., Balk, L.J., and Scherf, U. (2006) *Macromolecules*, **39**, 4327–4331.

76 Scherf, U., Gutacker, A., and Koenen, N. (2008) *Acc. Chem. Res.*, **41**, 1086–1097.

77 Tu, G.L., Li, H.B., Forster, M., Heiderhoff, R., Balk, L.J., Sigel, R., and Scherf, U. (2007) *Small*, **3**, 1001–1006.

78 Park, J.Y., Koenen, N., Forster, M., Ponnapati, R., Scherf, U., and Advincula, R. (2008) *Macromolecules*, **41**, 6169–6175.

79 Stefopoulos, A.A., Chochos, C.L., Bokias, G., and Kallitsis, J.K. (2008) *Langmuir*, **24**, 11103–11110.

80 Khanduyeva, N., Senkovskyy, V., Beryozkina, T., Horecha, M., Stamm, M., Uhrich, C., Riede, M., Leo, K., and Kiriy, A. (2009) *J. Am. Chem. Soc.*, **131**, 153–161.

81 Miyakoshi, R., Shimono, K., Yokoyama, A., and Yokozawa, T. (2006) *J. Am. Chem. Soc.*, **128**, 16012–16013.

82 Miyakoshi, R., Yokoyama, A., and Yokozawa, T. (2008) *Chem. Lett.*, **37**, 1022–1023.

83 Yokoyama, A., Kato, A., Miyakoshi, R., and Yokozawa, T. (2008) *Macromolecules*, **41**, 7271–7273.

84 Stefan, M.C., Javier, A.E., Osaka, I., and McCullough, R.D. (2009) *Macromolecules*, **42**, 30–32.

85 Huang, L., Wu, S.P., Qu, Y., Geng, Y.H., and Wang, F.S. (2008) *Macromolecules*, **41**, 8944–8947.

86 Jhaveri, S.B., Peterson, J.J., and Carter, K.R. (2008) *Macromolecules*, **41**, 8977–8979.

87 Sakamoto, J., Rehahn, M., Wegner, G., and Schlüter, A.D. (2009) *Macromol. Rapid Commun.*, **30**, 653–687.

88 Yokoyama, A., Suzuki, H., Kubota, Y., Ohuchi, K., Higashimura, H., and Yokozawa, T. (2007) *J. Am. Chem. Soc.*, **129**, 7236–7237.

89 Beryozkina, T., Boyko, K., Khanduyeva, N., Senkovskyy, V., Horecha, M., Oertel, U., Simon, F., Stamm, M., and Kiriy, A. (2009) *Angew. Chem. Int. Ed.*, **48**, 2695–2698.

90 Walczak, R.M., Brookins, R.N., Savage, A.M., van der Aa, E.M., and Reynolds, J.R. (2009) *Macromolecules*, **42**, 1445–1447.

91 Yokozawa, T., Kohno, H, Ohta, Y., and Yokoyama, A. Macromolecules, in press.

3
Regioregular and Regiosymmetric Polythiophenes

Itaru Osaka and Richard D. McCullough

3.1
Introduction

π-Conjugated polymers are an important class of conducting or semiconducting materials for the next generation of electronic and optical devices [1, 2]. One of the most interesting properties of π-conjugated polymers is that they possess excellent processability: the polymers can be deposited by simple printing techniques. Among the number of π-conjugated polymers, polythiophenes have become leading materials in organic electronics due to their desirable physical properties [3–8]. Polythiophene-based devices such as organic field-effect transistors [9–11] and organic photovoltaics [12–14] are state-of-the-art technology in the field.

The key to achieving high performances in such devices is to design and control intrachain and interchain π overlap. It is obvious that synthesis can determine π architectures, and synthesis is the first and most important step toward molecular engineering of materials having reasonable properties. The most striking example of the importance of synthesis can be understood through the example of regioregular head-to-tail coupled polythiophenes [15, 16]. This familiar material has demonstrated that design and strategic synthesis can lead to dramatically improved properties in π-conjugated polymers. While great effort has been directed to polythiophene applications, the chemistry of polythiophenes has also greatly expanded in the past decade. This chapter addresses the recent issues in design and synthesis of the regioregular polythiophene family.

3.2
Synthesis of Polythiophene and Regioirregular Polythiophenes

Synthesis of polythiophenes can be classified into two methods: chemical and electrochemical. Most polythiophenes reported today are, however, synthesized

Conjugated Polymer Synthesis. Edited by Yoshiki Chujo

ISBN: 978-3-527-32267-1

through chemical methods, and we will focus on chemical synthesis in this chapter. Generally, polythiophenes can be chemically synthesized by two routes, as depicted in Scheme 3.1: (i) transition metal-catalyzed polymerization and (ii) oxidative polymerization. These polymerization methods were developed in the study of unsubstituted polythiophene and have been used widely for synthesis of 3- or 3,4-substituted polythiophenes and their copolymers.

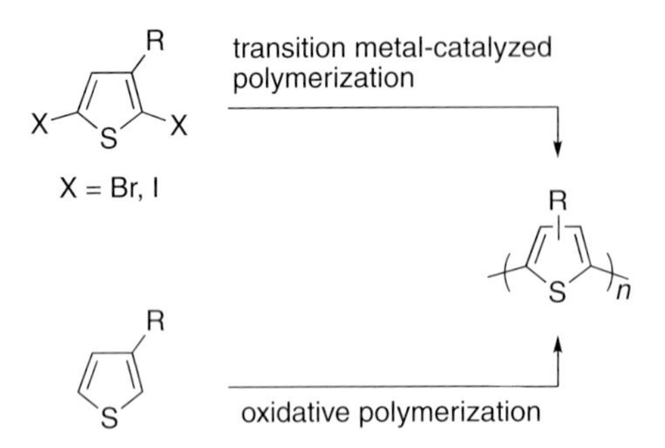

Scheme 3.1 General synthetic routes for the synthesis of polythiophene and 3-substituted polythiophenes.

Transition metal-catalyzed polymerizations commonly start with 2,5-dihalogenated thiophenes. The first step for a classical metal-catalyzed route is preparation of a mono-Grignard monomer by treatment of the dihalogenothiophene with magnesium. Polymerization is then employed by the addition of a transition metal catalyst such as Ni(II) or Pd(II) to give a 2,5-coupled polythiophene [17–19], an extension of the Kumada coupling reaction. Another familiar metal-catalyzed route is also a polymerization of a 2,5-dihalogenothiophene, using a Ni(0) reagent, for example, $Ni(cod)_2$ (cod = 1,5-cyclooctadiene) with an appropriate ligand, for example, 2,2′-bipyridine, known as Yamamoto polymerization [20].

For many years oxidative polymerization using $FeCl_3$ had been the synthetic method of choice for the preparation of polythiophenes [21]. In this method, thiophene or 3- (or 3,4-) substituted thiophenes are simply polymerized in the presence of $FeCl_3$, generating the corresponding polymers. However, these methods are not controlled polymerizations and allow the possibility of the 2,4-linkage of thiophenes in polymer chains, limiting the π-electron system, whereas transition-metal catalyzed polymerizations lead primarily to 2,5-coupled polythiophenes.

It is important to point out that in these methods, when polymerizing 3-substituted thiophenes, the coupling occurs with no regiochemical control and hence produces structurally irregular polymers [4, 5, 8]. These irregular polymers are denoted as regio*ir*regular polythiophenes or, in the case of poly(3-alkylthiophene)s, regio*ir*regular poly(3-alkylthiophene)s (irP3ATs). They are distinct from regioregular polythiophenes, which will be described later. Since the irregular structure causes the loss of π-conjugation,and thereby leads to severely limited electrical properties, control of regioregularity in the polymer backbone is crucial for maximizing polymer performance.

3.3 Head-to-Tail Coupled Regioregular Poly(3-Alkylthiophene)s

3.3.1 Design and Synthesis of rrP3ATs

Due to the asymmetric nature of the 3-alkylthiophene molecule, three relative orientations are available (Figure 3.1) when two 3-alkylthiophene rings are coupled between the 2- and 5- positions. The first orientation is 2,5′ or head-to-tail (HT) coupling, the second is 2,2′ or head-to-head (HH) coupling, and the third is 5,5′ or tail-to-tail (TT) coupling. The orientation becomes even more complicated when coupling three 3-alkylthiophene rings, enabling the creation of four possible chemically distinct triad regioisomers (Figure 3.1) [22]. The loss of regioregularity, that is to say, contamination of HH couplings, causes a sterically twisted structure in the polymer backbone again, giving rise to the loss of π-conjugation (Figure 3.2) [15, 23]. This is primarily due to the increased repulsive interaction between the alkyl substituent on the 3-position of the thiophene ring and the sp^2 lone pair on sulfur, or perhaps between two alkyl substituents on the alternate thiophene rings. These steric interactions lead to greater bandgaps, with concomitant destruction of high conductivity. Regioregular HT coupled poly(3-alkylthiophene)s (rrP3ATs) consist of almost perfect HT arrangements. Thus they can easily access planar polymer back-

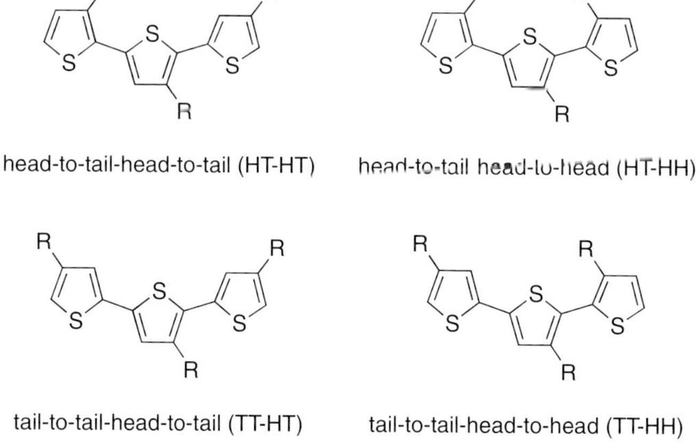

Figure 3.1 Coupling regiochemistry of 3-alkylthiophenes.

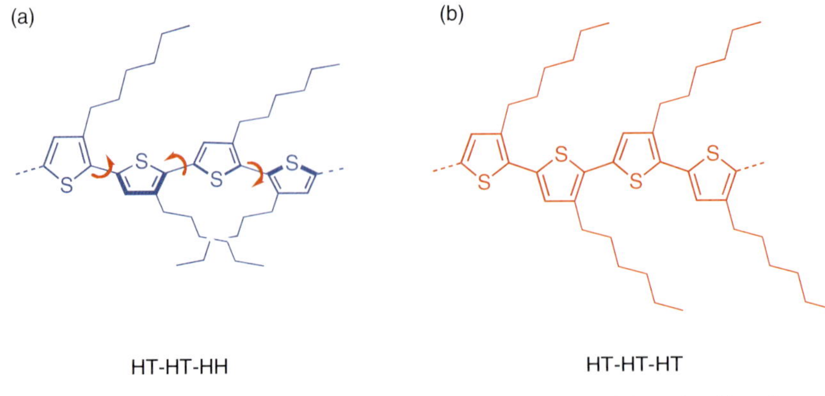

Figure 3.2 Regioirregular P3AT in nonplanar (a) and HT regioregular P3AT (rrP3AT) in planar structure (b).

bones and self-assemble to form well-defined, organized three-dimensional polycrystalline structures [24]. These structures provide efficient intrachain and interchain charge carrier pathways, leading to high conductivity and other desirable properties for polythiophenes.

rrP3ATs are most commonly synthesized by one of the following three methods: McCullough [15], Rieke [16], or Grignard metathesis (GRIM) [25] (Scheme 3.2). All these methods use nickel-catalyzed cross-coupling and produce comparable rrP3ATs. Palladium-catalyzed cross-coupling reactions, Stille [26] and Suzuki [27], can also be used for the synthesis of rrP3ATs (Scheme 3.2). The key synthetic strategy feature is the regioselective metallation of the corresponding monomers, which generates regiospecific intermediates 2-bromo-5-metalo-3-alkylthiophenes, whereas the classic synthesis of irP3ATs using magnesium gives a mixture of 2,5-exchanged intermediates. Although a small portion of the undesirable 2,5-exchanged intermediate is generated in the case of the three nickel-catalyzed methods, the desired intermediate reacts preferentially to produce fully HT arranged rrP3AT. All these synthetic methods for the production of rrP3ATs yield comparable materials that are not spectroscopically distinct. The three nickel-catalyzed methods, however, offer an advantage over the two palladium-catalyzed methods in terms of accessibility to rrP3ATs; this is because the nickel methods can produce rrP3ATs in one pot while the palladium methods require isolation and purification of the intermediates.

3.3.1.1 McCullough Method

The first synthesis of rrP3ATs was reported by McCullough early in 1992 [15]. This method regiospecifically generates 2-bromo-5-bromomagnesio-3-alkylthiophene **2**, which is achieved by treating 2-bromo-3-alkylthiophene **1** with LDA (lithium diisopropylamide) at $-40\,^{\circ}C$ followed by the addition of $MgBr_2{\cdot}Et_2O$ (recrystallized from Et_2O in a dry box). Quenching studies performed on intermediate **2** indicate that

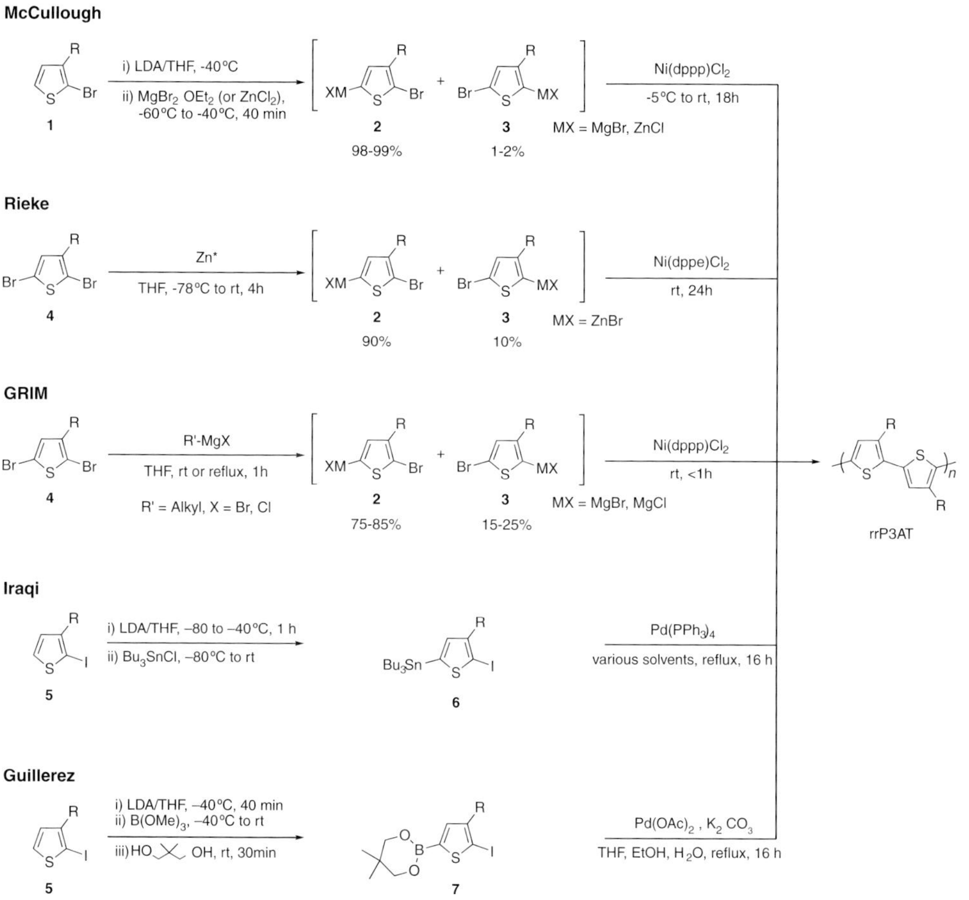

McCullough

1 → i) LDA/THF, -40°C; ii) $MgBr_2\ OEt_2$ (or $ZnCl_2$), -60°C to -40°C, 40 min → [2 (98-99%) + 3 (1-2%)], MX = MgBr, ZnCl → $Ni(dppp)Cl_2$, -5°C to rt, 18h

Rieke

4 → Zn*, THF, -78°C to rt, 4h → [2 (90%) + 3 (10%)], MX = ZnBr → $Ni(dppe)Cl_2$, rt, 24h

GRIM

4 → R'-MgX, THF, rt or reflux, 1h, R' = Alkyl, X = Br, Cl → [2 (75-85%) + 3 (15-25%)], MX = MgBr, MgCl → $Ni(dppp)Cl_2$, rt, <1h → rrP3AT

Iraqi

5 → i) LDA/THF, –80 to –40°C, 1 h; ii) Bu_3SnCl, –80°C to rt → 6 → $Pd(PPh_3)_4$, various solvents, reflux, 16 h

Guillerez

5 → i) LDA/THF, –40°C, 40 min; ii) $B(OMe)_3$, –40°C to rt; iii) HO⨯OH, rt, 30min → 7 → $Pd(OAc)_2$, K_2CO_3, THF, EtOH, H_2O, reflux, 16 h

Scheme 3.2 Synthetic methodologies of the production of rrP3ATs.

98–99% of the desired monomer and less than 1–2% of 2,5-exchanged intermediate **3** are produced [23]. The polymerization is then employed *in situ* by a cross coupling reaction using a catalytic amount of $Ni(dppp)Cl_2$ (dppp = 1,3-diphenylphosphino-propane) using the Kumada cross-coupling reaction [28], affording rrP3AT (~70% yield). The cross-coupling polymerization of these intermediates occurs without any scrambling. $MgBr_2{\cdot}Et_2O$ can be replaced with $ZnCl_2$, generating intermediate 2-bromo-5-chlorozincio-3-alkylthiophene **2**, which is thus regarded as the Negishi cross-coupling reaction [29]. This allows greater solubility of the reactive intermediate at cryogenic temperatures. The resulting polymer is precipitated in MeOH (or MeOH with a small amount of aqueous HCl), washed (fractionated) with sequential MeOH and hexane Soxhlet extractions, and then recovered by Soxhlet extraction with chloroform. Purified rrP3ATs afford HT–HT regioregularity of 98–100% as seen by ^{1}H NMR, and the number averaged molecular weights (M_n) are typically 20 K–40 K with polydispersities (PDI) of around 1.4 [15, 23, 24, 30, 31].

3.3.1.2 Rieke Method

The second approach to rrP3ATs was reported by Rieke and coworkers soon after the report of the McCullough method [16, 32–34]. In this method, treating 2,5-dibromo-3-alkylthiophenes **4** with highly reactive "Rieke Zinc" (Zn^*) [35, 36] yields a mixture of isomeric intermediates **2** and **3** in a ratio of 90:10. The ratio between these regioisomers is dependent on reaction temperature and, to a much lesser extent, the steric influence of the alkyl substituent. *In situ* addition of Ni(dppe)Cl_2 (dppe = 1,3-diphenylphosphinoethane) gives rrP3ATs with ~75% yield after purification. Use of a palladium catalyst, $Pd(PPh_3)_4$, yields completely regioirregular P3ATs. As an alternative approach, 2-bromo-5-iodo-3-alkylthiophene reacts with Zn^* to form only the desired intermediate **2**. This species then reacts in an identical fashion to form rrP3AT when the Ni catalyst is added. Molecular weights for rrP3ATs prepared by this method are typically $M_n = 24–34$ K (PDI = 1.4).

3.3.1.3 GRIM Method

An economical new synthesis for rrP3ATs, known as the Grignard metathesis (GRIM) method, was reported in 1999, as a modification of the McCullough method [25, 37]. This method, discovered by the McCullough group, does not require the use of either $MgBr_2{\cdot}Et_2O$ or $ZnCl_2$, which are both expensive and air sensitive. One strong advantage of this method is that the reaction can be carried out at room temperature and consequently it offers quick and easy preparation of rrP3ATs and enables the production of large-scale high-molecular weight rrP3ATs. In this method, 2,5-dibromo-3-alkylthiophene **4** is treated with 1 equiv of any Grignard reagent (R′MgX) to form a mixture of intermediates **2** and **3** in a ratio of 85 : 15 to 75 : 25 [37]. This ratio appears to be independent of reaction time, temperature, or Grignard reagent used. Although the ratio of desirable to undesirable isomers is relatively higher (when compared to either the McCullough or Rieke method), this method still affords rrP3ATs with high regioregularity of >99% HT couplings. Use of 2-bromo-5-iodo-3-alkylthiophene as the starting material can generate only desirable intermediate **2**. The typical M_n of rrP3ATs synthesized through this method is 20–35 K with very low PDI of 1.2–1.4.

3.3.1.4 Palladium-Catalyzed Polymerization Methods

Stille [38] and Suzuki [39] palladium-catalyzed cross-coupling reactions are alternative methods to access rrP3ATs. These methods require cryogenic conditions for preparing the corresponding organometallic monomer **6** or **7**, which then must be isolated and purified. Iraqi *et al.* investigated the synthesis of rrP3ATs through the Stille reaction, using 3-hexyl-2-iodo-5-(tri-n-butylstannyl)thiophene **6**, with a variety of solvents [26]. In all cases rrP3ATs with greater than 96% of HT couplings were obtained. Molecular weights of rrP3ATs prepared by this method are $M_n = 10$ K–16 K with PDI of 1.2–1.4 after purification. Suzuki reaction using 3-octyl-2-iodo-5-boronatothiophene **7** has been employed by Guillerez to give rrP3ATs with 96–97% HT couplings and a weight averaged molecular weight (M_w) of 27 K in 51% yield [27].

3.3.2
Mechanism of the Nickel-Catalyzed Polymerization

The synthesis of rrP3ATs is based on transition-metal-catalyzed cross-coupling reactions [40], in which the mechanism consists of a catalytic cycle of three consecutive steps: oxidative addition, transmetalation, and reductive elimination. Since the nickel-catalyzed polymerization is formally a polycondensation reaction, it was generally accepted that it proceeds via a step-growth mechanism [41]. However, it has been proposed by Yokozawa *et al.* that the nickel-catalyzed cross-coupling polymerization (McCullough and GRIM method) proceeds via a chain-growth mechanism [42]. Conversion vs. M_n and conversion vs. M_w/M_n plots of the polymerization of the monomer in the case of the GRIM method have been reported. M_n increased proportionally as monomer conversion increased and M_w/M_n ratios were 1.30–1.39 throughout the polymerization. This new insight is of particular interest because chain-growth polymerization affords well-defined rrP3ATs with very narrow polydispersities [43, 44].

At the same time as the Yokozawa report, the McCullough group also reported that the degree of rrP3AT polymerization increased with monomer conversion and could be predicted by the molar ratio of monomer to nickel initiator [45]. In addition, they further proposed that this polymerization system is not only a chain growth system but also essentially a living system [46]. This allows control of the molecular weight of the polymer as a function of reaction time and amount of Ni catalyst. The proposed mechanism for the regioregular polymerization of 3-alkylthiophene is outlined in Scheme 3.3 [3, 8, 45, 46]. The first step is the reaction of two equivalents of intermediate **2** with Ni(dppp)Cl_2 affording the organonickel compound **8**, and the reductive elimination that immediately occurs to form an associated pair of the 2,2′-dibromo-5,5′-bithiophene (tail-to-tail coupling) and Ni(0) [**9**·**10**]. Dimer **9** undergoes fast oxidative addition to the nickel center, generating the new organonickel compound **11**. Transmetalation with another **2**, which forms **12**, and reductive elimination gives an associated pair of terthiophene and Ni(0) [**13**·**10**]. Growth of the polymer chain occurs by an insertion of one monomer at a time, as shown in the reaction cycle (**14**–**15**–[**16**·**10**]–**14**), where the Ni(dppp) moiety is always incorporated into the polymer chain as an end group via the formation of a π-complex. In this fashion, Ni(dppp)Cl_2 is believed to act as an initiator rather than a catalyst, and therefore limits polymerization to one end of the polymer chain. The prediction of the mechanism as living is supported by two experimental results: first, the degree of rrP3AT polymerization has been found to increase with monomer conversion and can be predicted by the molar ratio of the monomer to the nickel initiator; second, addition of various Grignard reagents (R′MgX) at the end of polymerization results in end-capping of rrP3AT with the R′ end group [47, 48]. This living nature offers the synthesis of molecular weight-controlled and very pure rrP3ATs, which allow one to conduct systematic studies of structure–property relationships [49]. It also enables the synthesis of multiblock copolymers that may be useful for electronic devices [46, 50–52]. As a consequence, this synthetic

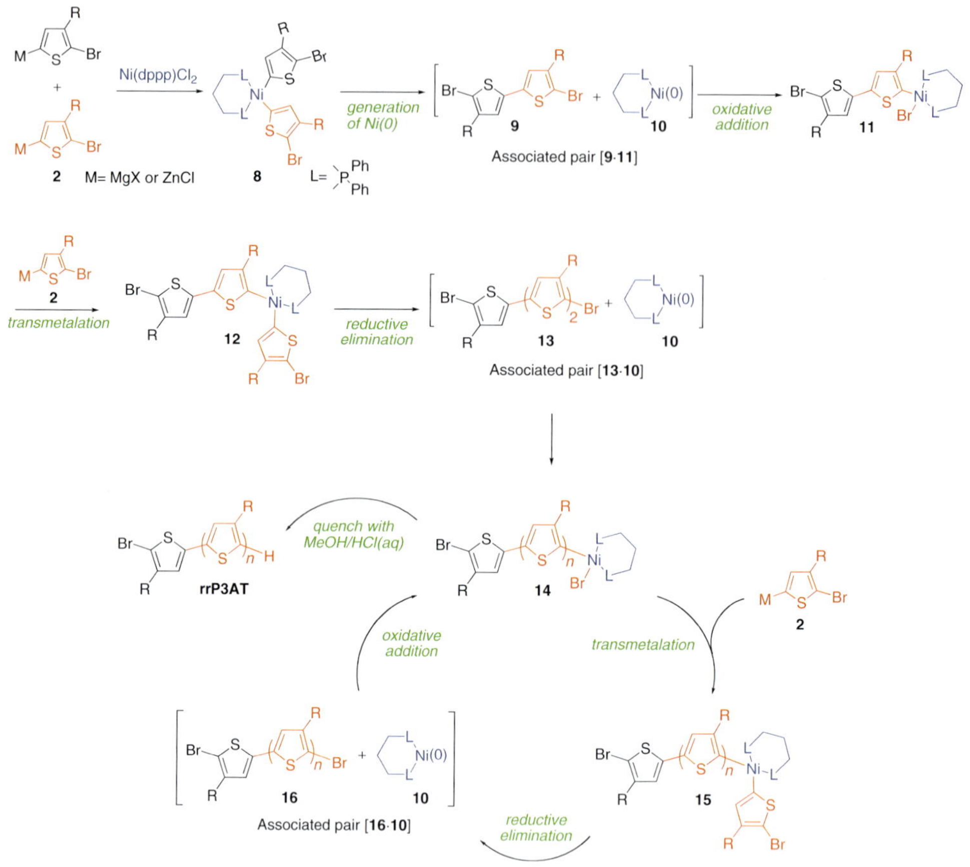

Scheme 3.3 Proposed mechanism of the nickel-initiated cross-coupling polymerization for the synthesis of rrP3ATs.

methodology has expanded the research field and, hence, has led to a better understanding of polythiophene chemistry as well as its physics.

3.4 Side Chain Functionalized HT Regioregular Polythiophenes

Early work has shown that the introduction of alkyl side chains to polythiophene backbones improved solubility and processability. The introduction of other functionalized side chains can alter the material's electrical and optical properties as well as solubility. A large number of side chain functionalized HT regioregular poly(3-substituted thiophene)s (rrPTs) have been synthesized to date (Table 3.1). Interestingly, some rrPTs with functional side chains have been successfully synthesized via oxidative coupling methods that usually give regioirregular backbones. rrPTs with

Table 3.1 Summary of side chain functionalized HT regioregular polythiophenes.

Entry	Side chain		Method	Ref.
1	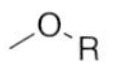 O R	R 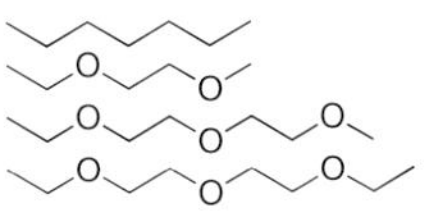 =	GRIM	[53]
2	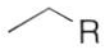 R	R = O, O O, O O O, S 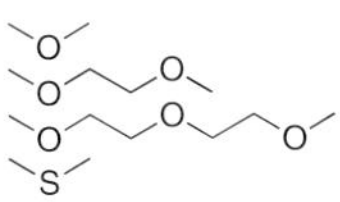	McCullough	[54]
3	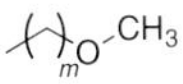 $(\ )_m$ O CH_3	$m = 5, 6, 10$	GRIM	[55]
4	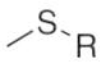 S R	R 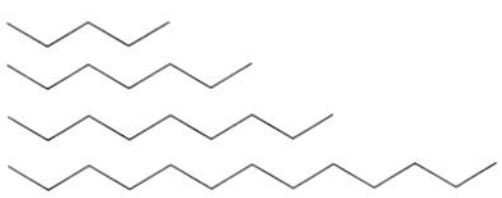 =	Rieke	[56]
5	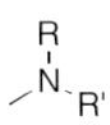 R N R'	R = -H R' 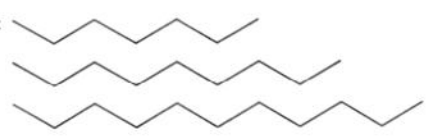 =	$FeCl_3$	[57]
		R = $-CH_3$ R' 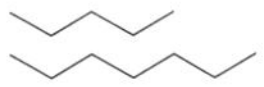 =	$NOBF_4$	[57]
6	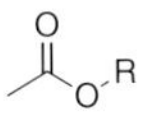 O O R	R 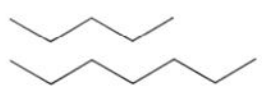 =	GRIM	[58]
7	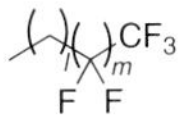 $(\ )_l(\ )_m CF_3$ F F	(a) $l = 11$, $m = 3$	GRIM	[59]
		(b) $l = 0$, $m = 7$	McCullough	[60]
8	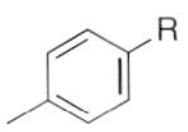 R	R = 	$FeCl_3$	[61]
9	R^*	(a) R^* = O, O 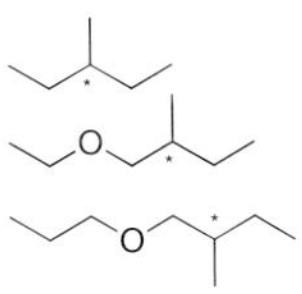	McCullough	[62, 63]
		(b) R^* 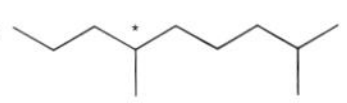 =	McCullough	[64]
		(c) R^* 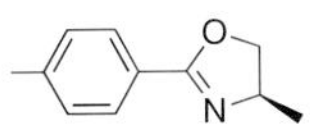 = O N	McCullough	[65]

(*Continued*)

Table 3.1 (Continued)

Entry	Side chain			Method	Ref.
10	$(\ )_m$R	(a) $m = 2, 6$	R = (tetrahydropyranyloxy)	McCullough	[66]
		(b) R = $-SiMe_3$		GRIM	[67]
		(c) $m = 2$	R = (4,4-dimethyloxazoline)	Stille	[68]
		(d) $m = 6$	R = $-P(O)(OEt)_2$	Stille	[69]
		(e) $m = 6$	R = $-Br$	McCullough, GRIM	[70, 71]

more reactive functional side chains could be synthesized by postpolymerization functionalization following the regioregular polymerization of thiophene monomers with protected groups.

3.4.1
Heteroatom-Containing Groups

Introducing electron-donating groups, such as alkoxy or alkylthio groups, with the heteroatom directly connected to the ring in the 3-position of thiophene is expected to decrease the band gap. This decrease occurs by raising their HOMO levels, which leads to low oxidation potentials and a highly stable conducting state. rrPTs with alkoxy substituents have been synthesized by the GRIM or McCullough method (Table 3.1 entry 1–3). UV–Vis absorption spectra of poly(3-(2-(2-methoxyethoxy) ethoxy)thiophene) (PMEET) showed that the λ_{max} was more than 100 nm red shifted from rrP3ATs, indicating that the introduction of the electron-donating group in the side chain lowers the band gap of PTs (entry 1) [53]. PMEET showed a high conductivity of 650 S cm^{-1} upon iodine doping, and the conductivity remained high, at 150 S cm^{-1}, after 2 months. A series of rrPTs with alkoxy groups, where the oxygen is not directly connected to the ring, was synthesized by the McCullough method (entry 2) [54]. While the shorter side chain polymer only gave low molecular weight due to low solubility, the longer side chain polymer gave high molecular weight. These polymers, after iodine doping, exhibited very high conductivities, with averages of 500–1000 S cm^{-1}. The polymer also exhibited ion-binding properties to Li^+, Pb^{2+}, and Hg^{2+}. rrPTs with a methoxy group on the end of the alkyl chains were also synthesized by the GRIM method (entry 3) [55].

rrPTs with alkylthio groups were synthesized by the Rieke method (entry 4) [56]. They showed good solubility in carbon disulfide, whereas solubility in common organic solvents such as chloroform, THF, and xylene was fairly low. The iodine doped polymer films showed high conductivity of 100 S cm^{-1}. UV–Vis absorption spectroscopy revealed that the band gap was smaller when compared to rrP3ATs.

Alkylamino and dialkylamino substituted rrPTs were reported by Rasmussen *et al.* (entry 5) [57]. Corresponding alkylamino and dialkylamino thiophenes were polymerized by the $FeCl_3$ and $NOBF_4$ methods, to give ~88% and ~90% HT couplings, respectively. These regioregularities show higher HT preference as compared to most polyalkylthiophenes prepared by oxidative polymerization ($FeCl_3$ method: ~70–80% HT couplings).

The ester group, which is an electron-withdrawing functional group, was also introduced into rrPT (entry 6) [58]. The polymers synthesized by the GRIM method showed a λ_{max} slightly blue shifted from that of rrP3ATs, which may be due to the electron-withdrawing nature of the carbonyl group, thus giving rise to a wider band gap.

PTs bearing partially fluorinated alkyl side chains are unique materials that provide unusual properties, including hydrophobicity, chemical and oxidative resistance, and self-organization of fluoroalkyl chains. Collard *et al.* have reported the synthesis of rrPTs with partially fluorinated alkyl side chains using the GRIM method (entry 7a) [59]. This polymer shows liquid crystalline behavior and thus forms a highly ordered solid-state structure. Synthesis of rrPTs with perfluoroalkyl side chains has also been reported using the McCullough method (entry 7b), giving polymers with ~86% HT couplings [60].

3.4.2
Aromatic-Containing Group

Poly(3-(4-octylphenyl)thiophene) is another example of regioregular HT polythiophenes that was synthesized via the oxidative coupling method (entry 8) [61]. In this polymerization, $FeCl_3$ was slowly added to the reaction mixture, allowing ~94% HT couplings. The rationale for the success of this method was reported to be that the slow addition of $FeCl_3$ keeps the Fe^{3+}/Fe^{2+} ratio low and, hence, lowers the oxidation potential providing a more controlled polymerization.

3.4.3
Chiral Groups

rrPTs with chiral side chains are fascinating materials that induce a helical packing of the polymer backbone into an aggregated chiral superstructure, leading to optical activity (entry 9) [62–65]. The first synthesis of chiral rrPTs, for which the McCullough method was used, was reported by Meijer, (entry 9a) [62]. The polymer solution showed a strong CD (circular dichroism) signal at its π–π^* transition at low temperature or upon addition of poor solvents, indicating the formation of a helical conformation of the backbone in the aggregate; in contrast, the regioirregular polymer showed only a weak CD signal. rrPTs with chiral oxazoline group showed similar behavior (entry 9c) [65].

3.4.4
γ-Functionalized Groups

Functionalization at the end of rrPT side chains is desirable in terms of tuning polymer properties. One reason why end of side chain functionalization is interesting is that using protective groups at the γ-end of the side chains, enables the incorporation of functional groups that can be reactive to polymerization conditions. Tetrahydropyranyl (entry 10a) [66] and trimethylsilyl (entry 10b) [67] γ functionalized rrPTs were synthesized through the McCullough and GRIM methods, respectively. The oxazolinyl γ-functionalized rrPT (entry 10c) was synthesized by a CuO-modified Stille coupling reaction, after which the polymer was modified by a post-polymerization hydrolysis reaction to give a carboxylic acid group on the side chain; this proved suitable for chromatic chemosensing [68]. The rrPT bearing phosphonic ester group (entry 10d) was synthesized by the Stille method [69]. Subsequent deprotection of the ester led to the phosphonic acid-functionalized rrPT, forming a supramolecular assembly when a tetraalkylammonium hydroxide salt was added to the polymer solution. Another approach to functionalizing the side chain of rrPTs incorporated reactive groups at the γ-end of the side chains, which are stable to polymerization conditions. rrPT with bromohexyl side chain (entry 10e) was synthesized by Iraqi using the McCullough method, and then reacted with 2-carboxyanthraquinone to give a highly redox active rrPT [70]. The bromohexyl-substituted rrPT was also synthesized by the GRIM method, and the γ-end of the side chain was subsequently functionalized to various groups, such as carboxylic acid, amine or thiol [71].

3.5
End Group Functionalized HT Regioregular Polythiophenes

While side chain functionalization has been demonstrated to be an effective way to tune the physical and electronic properties of rrP3ATs, end group functionalization is another approach to alter those properties. End group functionalization is expected to lead to a number of new uses for polythiophenes, including end group driven self-assembly onto surfaces and into conducting polymer assembled networks, as well as the synthesis of reactive end groups that can be used as building blocks for the synthesis of block copolymers. Two approaches have been investigated to alter end group composition: postpolymerization and *in situ* methods.

3.5.1
Postpolymerization End Group Functionalization

rrP3ATs synthesized through either the McCullough, Rieke, or GRIM route contain a large majority of one end group– a proton at one terminus and a bromine on the other

(H/Br) [46, 72]. This bromine end can be converted to H by treating the polymer with an excess of Grignard reagents and subsequent aqueous work-up, yielding an H/H type polymer **17**, as shown in Scheme 3.4 [72, 73]. Further, both hydrogen end groups can be converted to aldehyde **18** by the Vilsmeier reaction, and subsequently the aldehyde groups can be reduced to hydroxymethyl groups on both ends (**19**) [73].

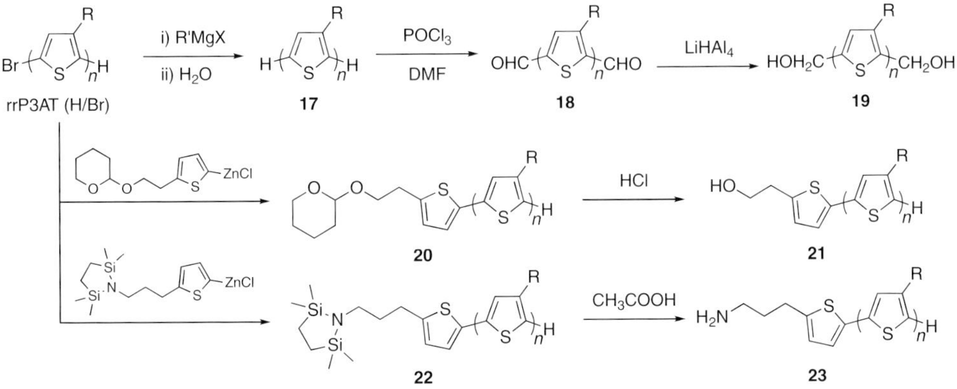

Scheme 3.4 Postpolymerization end group functionalization of rrP3ATs.

On the other hand, in one study the bromine end of the H/Br polymer was successfully replaced with a thiophene with tetrahydropyranyl protected hydroxy group (**20**) by a Negishi coupling reaction [74]. This end group was then deprotected to yield rrP3HT with a hydroxy end group (**21**). Similarly, a thiophene bearing STABASE [75]-protected amino group was substituted with a bromine end group (**22**), and subsequently deprotected to yield rrP3HT with an amino end group (**23**). Furthermore, these end groups can be converted to polyacrylates or polystyrenes, leading to rrP3AT–based rod-coil block copolymers.

3.5.2
In Situ End Group Functionalization

The *in situ* method offers an advantage over the postpolymerization method since an end group(s) can be modified in one pot. The first attempt toward *in situ* end group functionalization was reported by Janssen using the McCullough method, in which 2-thienylmagnesium bromide or 5-trimethylsilyl-2-thienylmagnesium bromide was added to the reaction mixture with an additional Ni catalyst; this yielded a mixture of H/H, mono- and di-capped polymer chains [76].

We have also reported a simple and versatile method to achieve *in situ* functionalization of rrP3HT using the GRIM method [47, 48]. Since the GRIM method follows a living mechanism, rrP3AT is still bound to the nickel catalyst at the end of the reaction. Therefore, a simple addition of another Grignard reagent effectively terminates the reaction and "end caps" the polymer chains (Scheme 3.5). In this

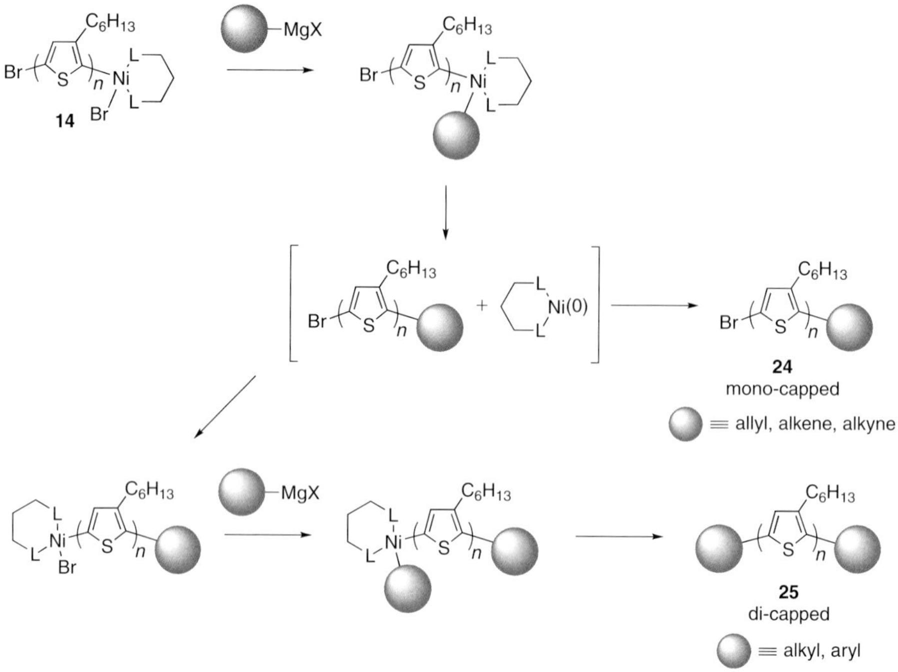

Scheme 3.5 *In situ* end capping of rrP3HT.

system, a variety of different types of Grignard reagents (alkyl, allyl, vinyl, aryl, etc.) successfully achieved end-capping, giving both mono-capped (**24**) and di-capped (**25**) rrP3AT (Table 3.2). When the end group is allyl, ethynyl, or vinyl groups, the nickel catalyst is postulated to be bound to the end group through a nickel-π complex to yield mono-capped polymers, eliminating the possibility of di-capping. However when the end groups are alkyl or aryl, the polymers react further to yield di-capped polymers. Aminophenyl (hydrolyzed from bis(trimethylsilyl)aminophenyl during work-up) is an exception to the aryl end groups found to mono-cap the rrP3HT chain. This simple procedure allows the synthesis of a library of end-capped polymers that can be used to better understand the effect of end-group composition on polymer morphology and self-assembly. Moreover, these *in situ* end-capped polymers have led to the library synthesis of a large number of rrP3AT block copolymers.

3.6
Block Copolymers Derived from HT Regioregular Polythiophenes

Block copolymers composed of two different polymeric segments are fascinating materials that generally self-organize and microphase separate [77]. If one or more of the blocks is a conjugated polymer, phase separation could lead to unique advanced

Table 3.2 Summary of end groups mono- and di-capped on rrP3HT.

End groups	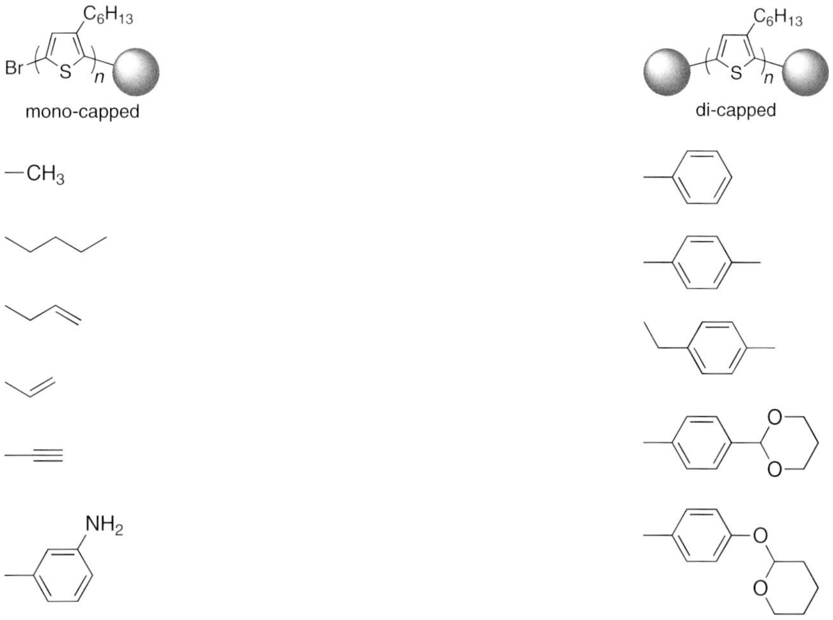
C_6H_{13}, Br, S, n — mono-capped	C_6H_{13}, S, n — di-capped
$-CH_3$	
	O, O
NH_2	O, O

new materials that could be used as components in nano-electronic devices. Herein, we describe the synthesis of rrP3AT-based block copolymers comprised of all-conjugated segments and of conjugated and non-conjugated segments.

3.6.1 All-Conjugated Block Copolymers

Block copolymers consisting of different π-conjugated segments are particularly interesting since different electrical or optical characters from each segment can be wrapped into one polymer chain; this may lead to very unique properties, especially when self-assembled to form nanostructured morphologies [78–81]. To realize highly ordered and nanostructured films, high molecular weights of the individual blocks and narrow polydispersities are preferred. As you will see, the living nature of the synthetic methodology of rrP3ATs has allowed the straightforward access to those block copolymers by simple chain extension reaction in one pot.

The first synthesis of all-conjugated rrP3AT-based block copolymers was reported by Iovu and McCullough *et al.* [46]. In their study, a diblock copolymer **26** with R = hexyl, R′ = dodecyl (P3HT-*b*-P3DDT, entry 1, Table 3.3) and a triblock copolymer

Table 3.3 Reported all-conjugated rrP3AT-based diblock copolymers.

Entry	R	R′	Ref.
1	C_6H_{13}	$C_{12}H_{25}$	[46]
2[a),b]	C_6H_{13}	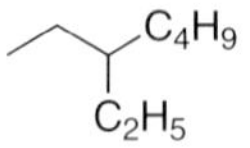 C_4H_9, C_2H_5	[82]
3[b]	C_8H_{17}	C_4H_9	[83]
4[a),b]	C_6H_{13}	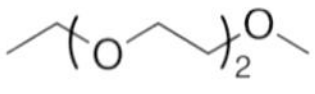 $(O)_2$, O	[84]
5[a),b]	C_6H_{13}	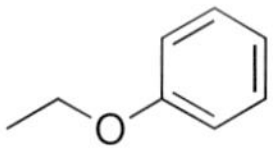 O	[86]
6	 TiPS	C_6H_{13}	[87]

a) 2-bromo-5-iodo-3-alkylthiophenes were used as the monomers.
b) was used instead of ^{t}BuMgCl.

27 with R = dodecyl, R′ = hexyl (P3DDT-*b*-P3HT-*b*-P3DDT) were synthesized by sequential addition of dibrominated monomers of 3-hexylthiophene and 3-dodecylthiophene (Scheme 3.6). The molecular weights of P3HT-*b*-P3DDT were as high as $M_n = 21$ K, which, of course, can be controlled by reaction time and catalyst concentration, given that the synthesis has a living nature. The polymer's PDI of 1.44 was slightly higher than that of the homopolymer, which was 1.2; this indicates the formation of some dead or inactive chains during the chain extension process. The chain length of P3DDT-*b*-P3HT-*b*-P3DDT was carefully controlled due to its low solubility, an issue that could cause polymer precipitation during the reaction process. Molecular weights of the triblock copolymer were as high as $M_n = 9.8$ K with relatively wide PDI of 1.53. rrP3AT-based diblock copolymers with R = hexyl, R′ = 2-ethylhexyl (entry 2, Table 3.3) [82] and R = octyl, R′ = butyl (entry 3, Table 3.3) [83] were also synthesized by essentially the same procedures.

Yokozawa *et al.* synthesized rrP3AT-based diblock copolymers of hydrophobic 3-hexylthiophene and hydrophilic 3-[2-(2-methoxyethoxy)ethoxy]methylthiophene (entry 4, Table 3.3) using this methodology [84]. It was found that the PDI for these polymers was dependent on catalyst choice due to reactivity differences between the two monomers. In addition to ascertaining that the synthesis of poly(*p*-phenylene) (PPP) via the GRIM method is a chain-growth system, as will be described in the next section, they also reported on the synthesis of interesting diblock copolymers consisting of an rrP3AT segment and a PPP segment (**28**, **29**, Scheme 3.7) [85]. While high molecular weight ($M_n = 19.4$ K) with narrow PDI (1.24) block copolymer **28** was successfully obtained when the polymerization was initiated with PPP followed by a chain extension by rrP3ATs, a low molecular weight ($M_n = 5.6$ K), broad

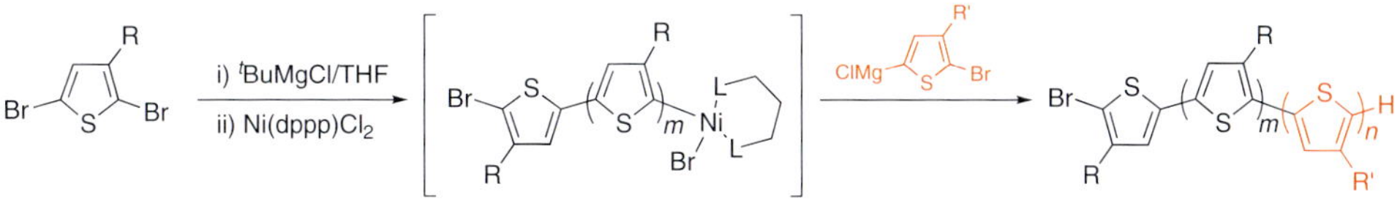

Scheme 3.6 Synthesis of all-conjugated diblock copopymers and triblock copolymers by chain extension through sequential monomer addition.

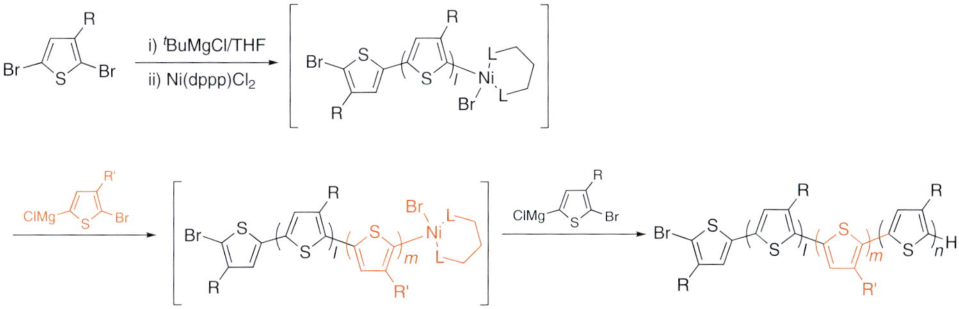

PDI (2.36) block copolymer **29** was obtained when the polymerization was performed in the reverse order. They proposed that, in the latter case, the Ni catalyst does not smoothly transfer to the phenylene rings since thiophene and Ni form a relatively strong π-complex, and thus do not allow efficient postpolymerization of the PPP segment. In addition, poly(3-hexylthiophene)-*b*-poly(3-phenoxymethylthiophene) (entry 5) [86] and poly(3-TIPS-butynylthiophene-*b*-poly(3-hexylthiophene) (P3TBT-*b*-P3HT, entry 6) [87] were also reported. The side chain of the first block in P3TBT-*b*-

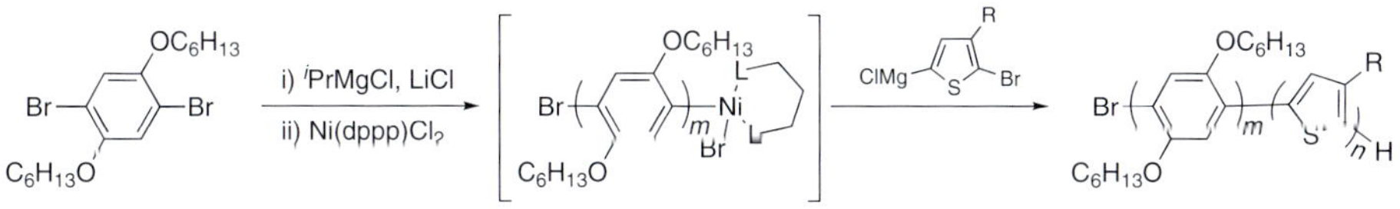

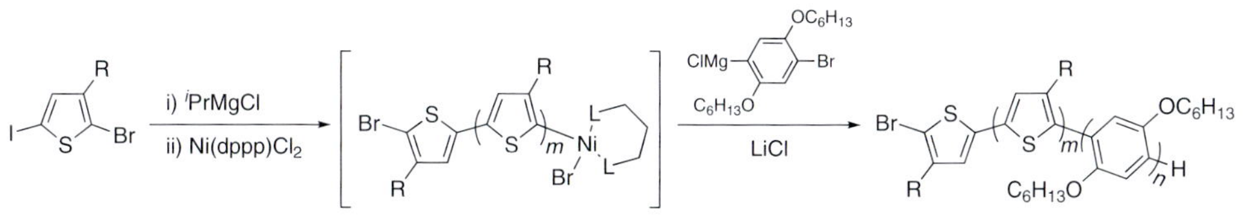

Scheme 3.7 Synthesis of rrP3AT–PPP diblock copopymers using the GRIM method.

P3HT can be γ-functionalized to an azide linked with naphthalene dicarboximide moieties via "click chemistry".

A different synthetic methodology was used to prepare a triblock copolymer comprising rrP3HT and polyfluorene (PF) (**31**, Scheme 3.8). Scherf *et al.* first synthesized an end-functionalized PF segment (**30**), and it was then reacted with a GRIM type reagent to form an rrP3HT segment at its terminal position [88]. The length of the rrP3HT block was limited to about six to seven repeating thiophene units. A triblock copolymer P3HT-*b*-CN-PPV-*b*-rrP3HT (CN-PPV = cyano-substituted poly-*p*-phenylenevinylene) **33** was also synthesized by the same group using a different methodology [89]. Using the Yamamoto method a cyano-substituted dibromophenylenevinylene monomer was polymerized; subsequently the polymerization was terminated by adding the pre-prepared rrP3HT with monobromo end group, giving a triblock copolymer with molecular weight of $M_n \approx 53$ K (PDI ≈ 1.5).

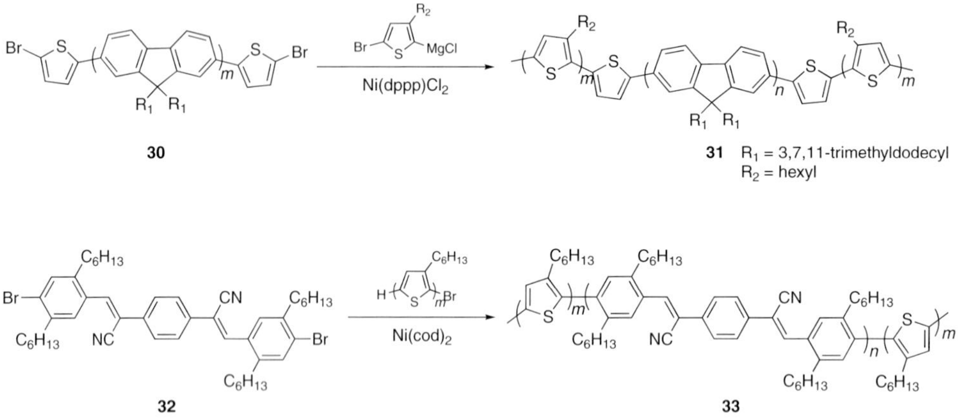

Ni(dppp)Cl_2

30

31 R_1 = 3,7,11-trimethyldodecyl
R_2 = hexyl

Ni(cod)$_2$

32

33

Scheme 3.8 Synthesis of all–conjugated triblock copolymers, P3HT-*b*-PF-*b*-P3HT **31** and P3HT-*b*-CN-PPV-*b*-P3HT **33**.

Several successful methodologies for the synthesis of all-π-conjugated block copolymers have been shown. As some block copolymers showed nanosized mesostructures, these copolymers seem to be favorable for use in organic electronic devices such as bulk hetero-junction solar cells.

3.6.2
Conjugated–Non-Conjugated Block Copolymers

Conjugated–non-conjugated block copolymers, namely rod–coil block copolymers, can produce numerous phase separated nano- or microstructures that may be of use in various applications [90]. Incorporation of flexible coil-like segments may improve the mechanical properties and processability of rigid rod rrP3ATs, in which high crystallinity may deteriorate the reproducibility of the self-assembled structures in films and thus impede device performance. The McCullough group reported the first

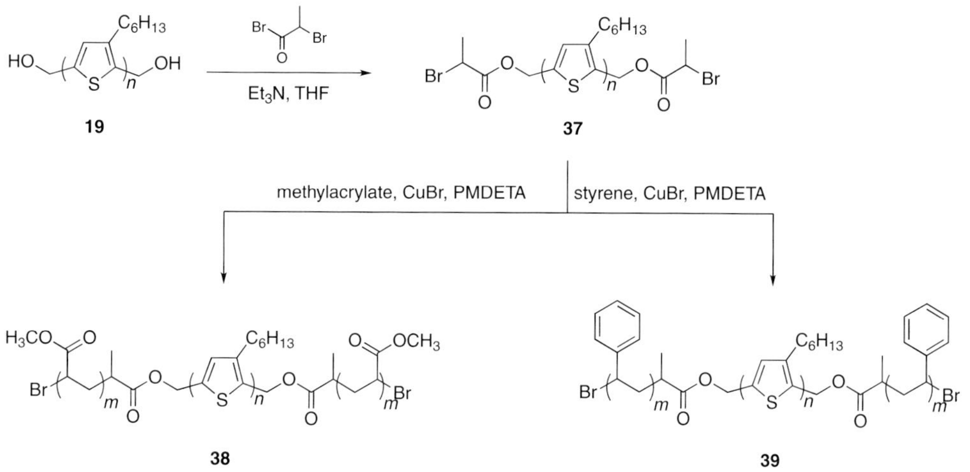

Scheme 3.9 Conjugated–non-conjugated rrP3HT diblock copolymers synthesized by ATRP (atom transfer radical polymerization) following postpolymerization end group functionalization.

conjugated–non-conjugated rrP3AT-based block copolymers [73]. In this report, end groups of rrP3HTs synthesized via the McCullough method were modified by postpolymerization functionalization, giving mono-capped polymer **20** and di-capped polymer **19**, as shown in Scheme 3.4 (Section 3.5.1). Atom transfer radical polymerization (ATRP) [91, 92] was then employed on a rrP3HT macroinitiator (**34**, **37**) to introduce polymethylacrylate (PMA) (**35**, **38**) and polystyrene (PS) segments (**36**, **39**), giving di- and triblock copolymers, respectively (Schemes 3.9 and 3.10). The weight percentage of the PMA and PS blocks was completely controlled by the feed

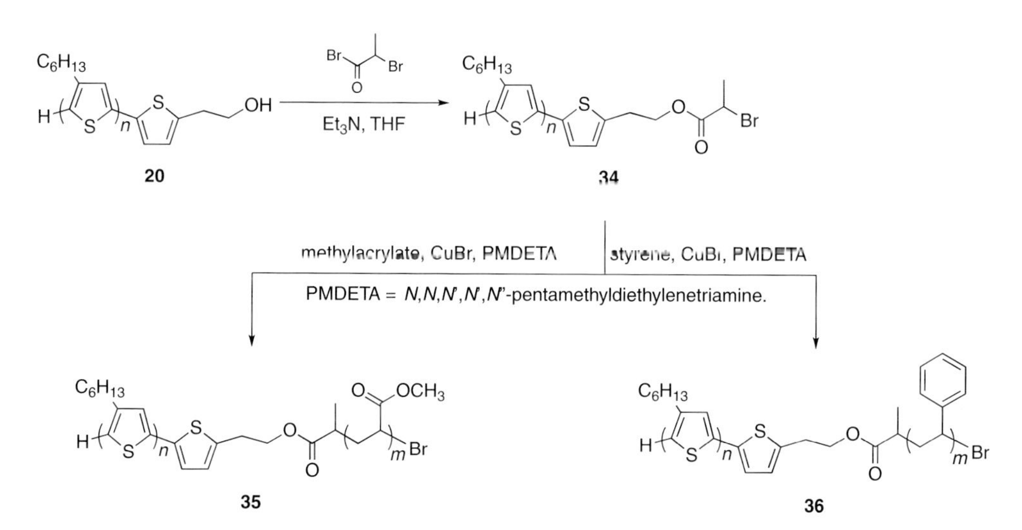

Scheme 3.10 Conjugated–non-conjugated rrP3HT triblock copolymers synthesized by ATRP following postpolymerization end group functionalization.

ratio of the monomers. The conductivities of the block copolymers decreased as the ratio of insulating blocks (PS and PMA) increased, as expected.

McCullough and coworkers have also reported on a series of rrP3HT-based block copolymers, in which non-conjugated coil blocks of PS, polyacrylates, polymethacrylate, and isoprene were introduced (Scheme 3.11) [50–52, 93, 94]. Importantly, all of these diblock copolymers were synthesized strategically via *in situ* allyl-terminated rrP3HT **43**. Allyl-terminated rrP3HT **43** was converted to its corresponding macroinitiators (**41**, **44**, **50**, **52**), and then copolymerization was employed to give the series of block copolymers. The rrP3HT–PS diblock copolymers were synthesized by anionic polymerization (**42**) [50], ATRP (**45**) [50], and reversible addition fragmentation chain transfer polymerization (RAFT) (**53**) [93]. AFM study of thin films of

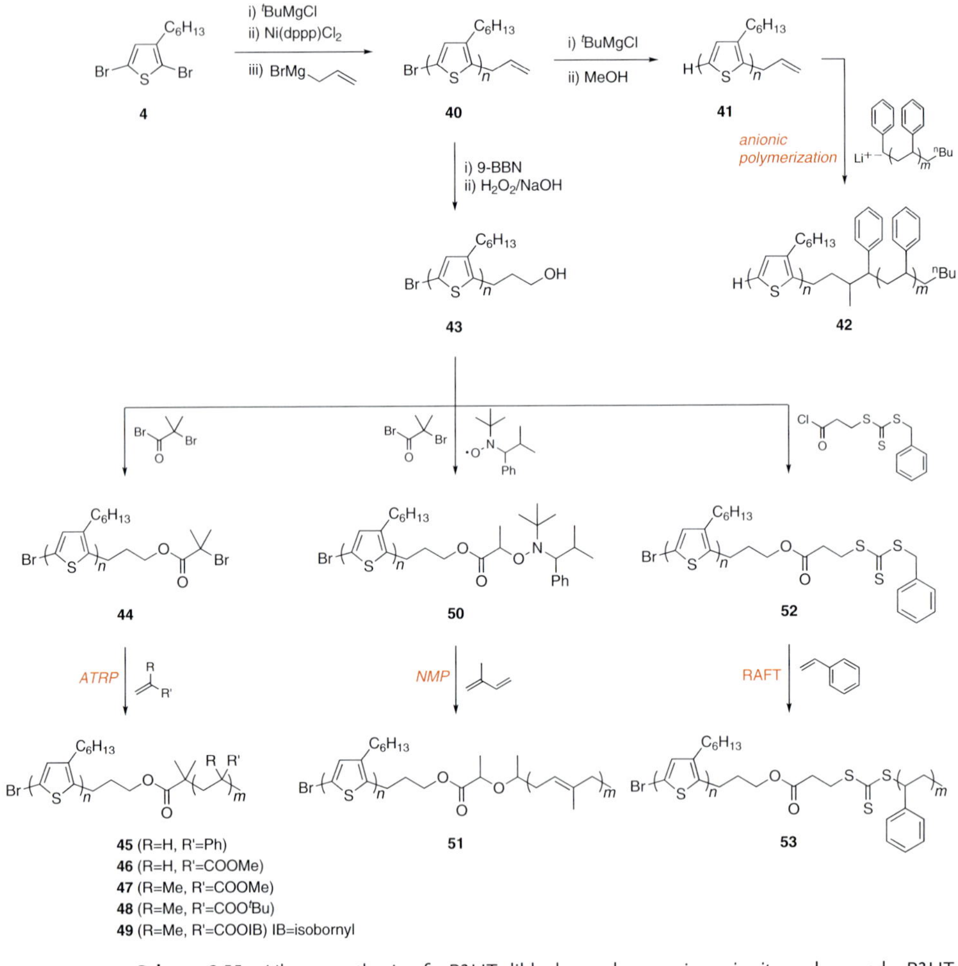

45 (R=H, R'=Ph)
46 (R=H, R'=COOMe)
47 (R=Me, R'=COOMe)
48 (R=Me, R'=COO^{t}Bu)
49 (R=Me, R'=COOIB) IB=isobornyl

Scheme 3.11 Library synthesis of rrP3HT diblock copolymers via an *in-situ* end-capped rrP3HT using anionic polymerization, ATRP, NMP (nitroxide mediated polymerization), and RAFT (reversible addition fragmentation transfer polymerization).

diblock copolymers **42** and **45** showed the formation of sparsely packed nanowire morphologies, relative to those observed in rrP3HT; these may be due to the presence of the PS coil blocks. rrP3HT diblock copolymers integrated with polyacrylate (**46**) and polymethacrylate (**47–49**) segments were also synthesized by ATRP [51, 52, 94], as shown in Scheme 3.11. It was found that the nature and the molar ratio of coil segments had a profound impact on nanofibrillar width, length, and distribution, as well as interfibrillar ordering. Interestingly, block copolymer **46** exhibited high field-effect mobilities, even at high polymethylacrylate (PMA) content (rrP3HT: PMA = 43:57) [94].

A few other rrP3HT-based block copolymers have been reported (Figure 3.3). Recently, "click" chemistry has been used to prepare rrP3HT–PS di- (**54**) and tri-block

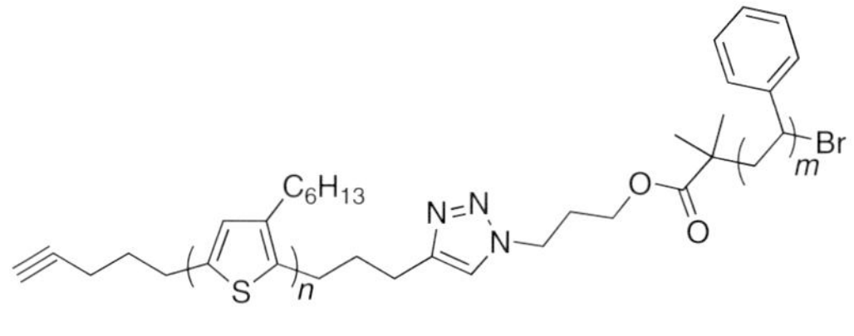

54

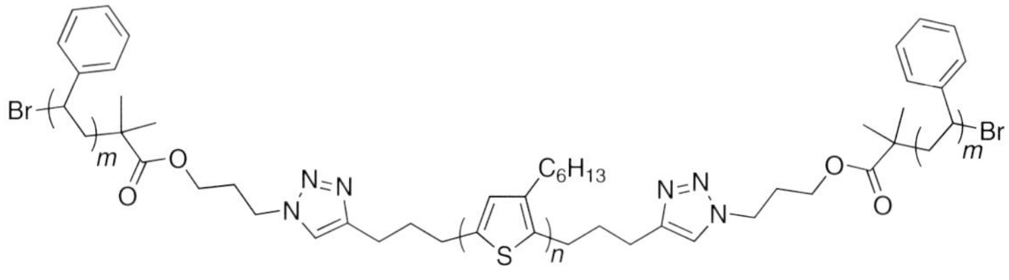

55

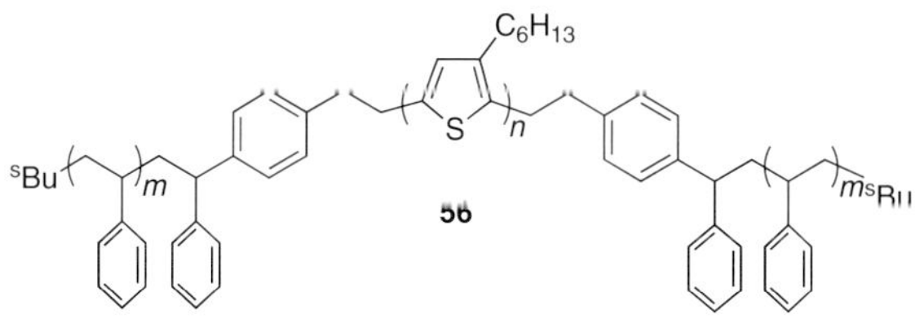

56

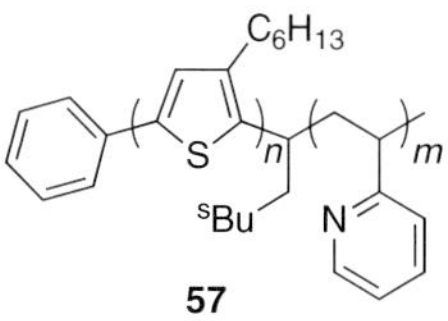

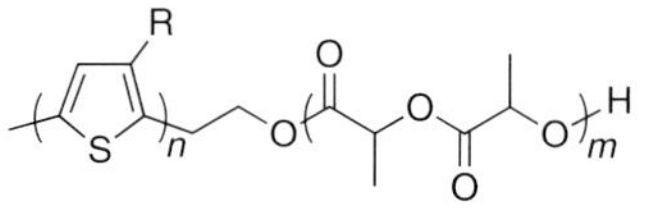

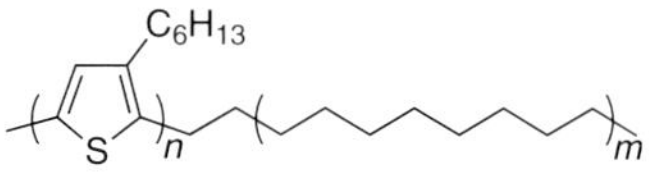

57

58 (R = $-C_6H_{13}$, $-C_{12}H_{25}$)

59

Figure 3.3 rrP3AT-based block copolymers synthesized via *in situ* end group functionalization.

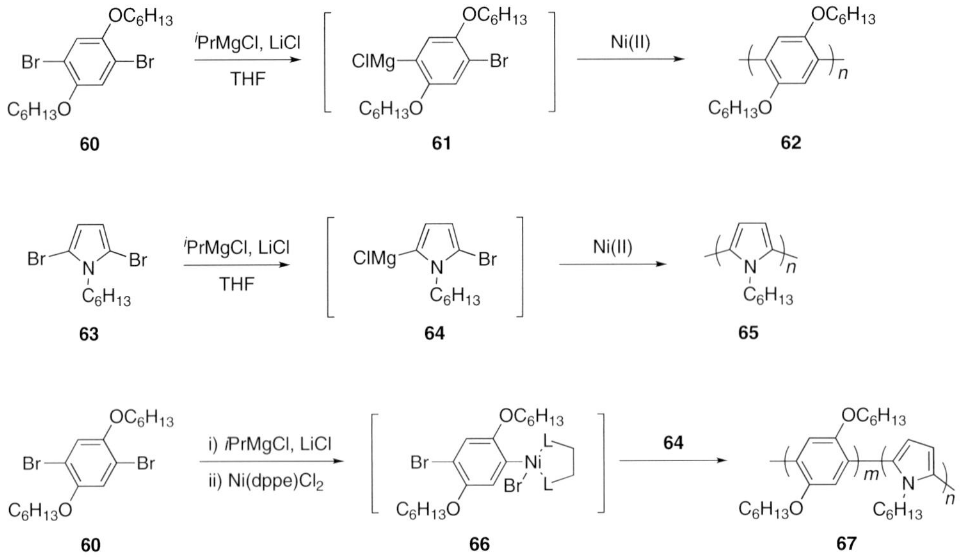

Scheme 3.12 Synthesis of poly-*p*-phenylene, polypyrrole and their diblock copolymer via the GRIM method as reported by the Yokozawa group.

(**55**) copolymers [95]. A triblock copolymer consisting of rrP3HT and polystyrene (**56**) was also prepared by anionic polymerization following an *in situ* end-capping of rrP3HT [96]. Poly-2-vinylpyridine was incorporated by anionic polymerization with different composition ratios (**57**) [97]. A polylactide (PLA) coil block was integrated into rrP3AT (**58**) using ring-opening polymerization [98]. Polyethylene segments were introduced via ruthenium-catalyzed ring-opening metathesis polymerization, giving crystalline–crystalline rrP3HT diblock copolymer (**59**) [99]. It is important to note that these copolymers were also synthesized via the *in situ* end-functionalized rrP3ATs described in the previous section.

A number of rrP3AT-based block copolymers with various kinds of insulating blocks have been synthesized and characterized. Thin films of block copolymers have indeed shown phase separated nano- or microstructures. Their morphologies have changed as a function of insulating block size, which will possibly lead to better control of electronic device performance.

3.7
Universal Use of the GRIM Method

As has been described in this chapter, nickel-mediated cross-coupling polymerizations for the synthesis of rrP3ATs, and particularly the GRIM (Grignard metathesis) method, are versatile synthetic methodologies that can create a wide variety of regioregular polythiophenes, including side chain and end group functionalized

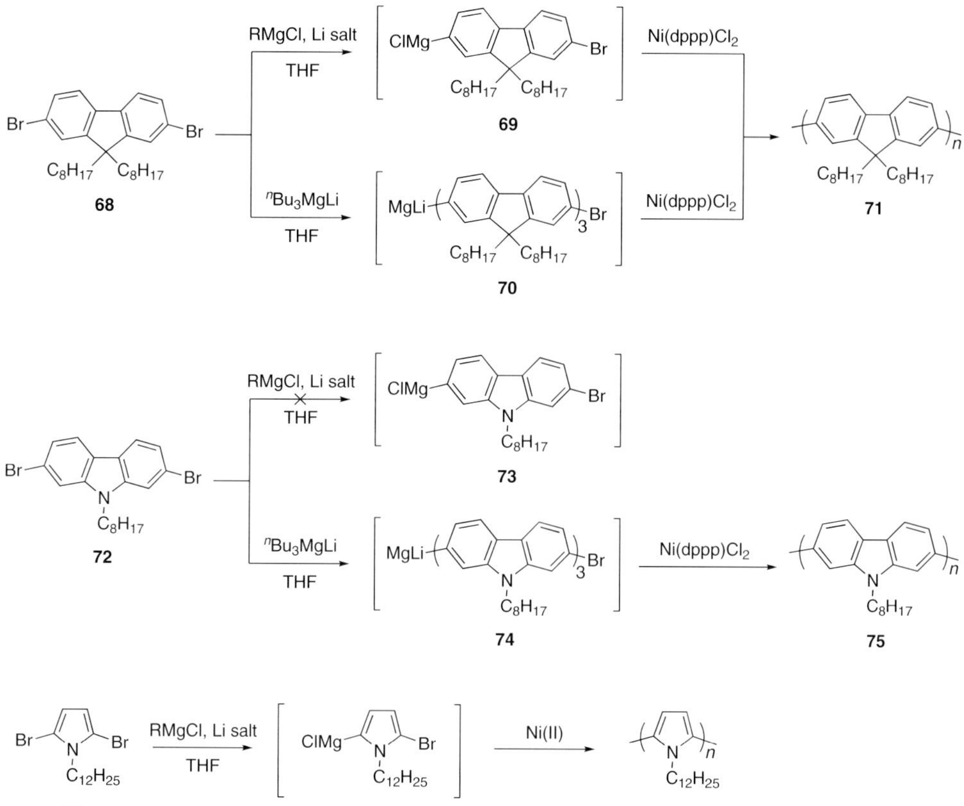

Scheme 3.13 Universal use of the GRIM method for the synthesis of π-conjugated polymers as reported by the McCullough group.

polymers and block copolymers. The GRIM method can also be applied to the synthesis of other aryl or heteroaryl π-conjugated polymers. In general, the syntheses of those polymers are performed through metal-catalyzed polycondensation reactions based on Yamamoto, Suzuki, or Stille coupling reactions. Polymerization using any of these methodologies usually requires one or more days and high temperatures ($> 100\,^{\circ}C$). However, employing the GRIM method can shorten polymerization times and lower reaction temperatures, facilitating the preparation of these aryl and heteroaryl polymers.

Yokozawa *et al.* reported on the successful polymerization of 1,4-dibromo-2,5-dihexyloxybenzene **60** via the GRIM method, giving well-defined poly(*p*-phenylene) (PPP) (Scheme 3.12) [100]. First, they performed a standard procedure to obtain PPP, where ^{i}PrMgCl was used to generate intermediate **61**, followed by an *in situ* addition of a Ni catalyst. Although polymerization proceeded smoothly at room temperature, the obtained PPP had low molecular weights and broad polydispersities. After careful analysis of the GPC profile, they speculated that chain-growth polymerization might

Table 3.4 Summary of rsPATs.

Polymer	Structure	R	Method	Ref.
88	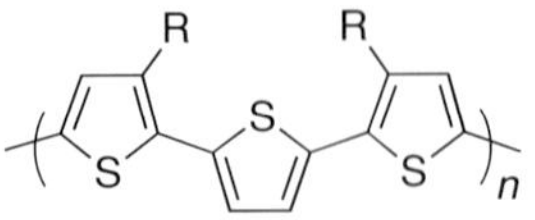	(a) -C_6H_{13}	A, B	[105, 107]
		(b) -C_8H_{17}	A, C	[106, 108]
89	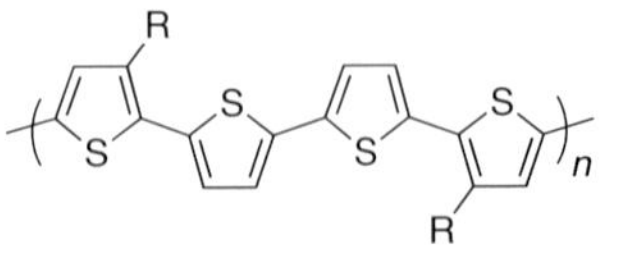	-$C_{12}H_{25}$	A	[109]
90	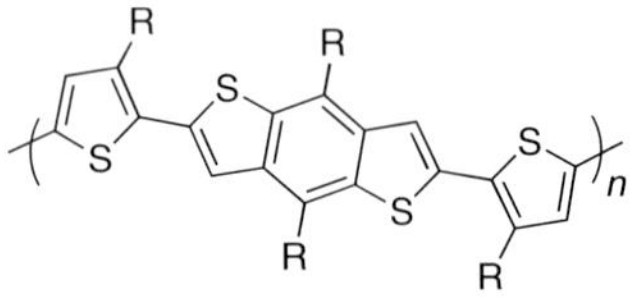	-C_6H_{13}	A	[110]
91	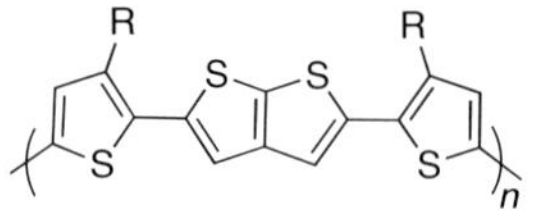	-C_8H_{17}, $C_{10}H_{21}$, $C_{12}H_{25}$	C	[112]
92	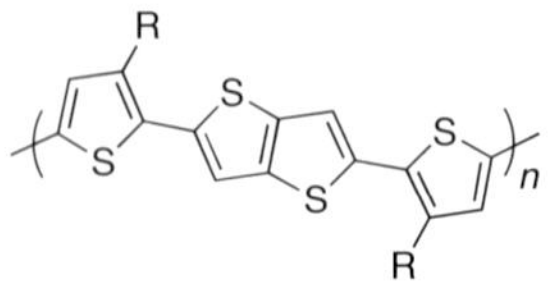	-$C_{10}H_{21}$, $C_{12}H_{25}$, $C_{14}H_{29}$	C	[113]
93	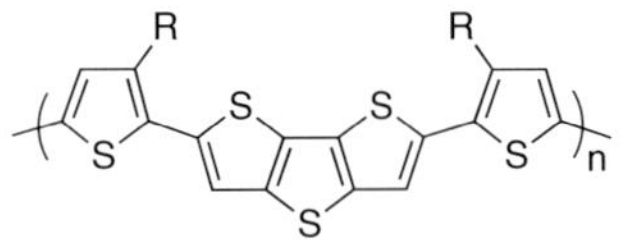	-$C_{12}H_{25}$	C	[115]
94	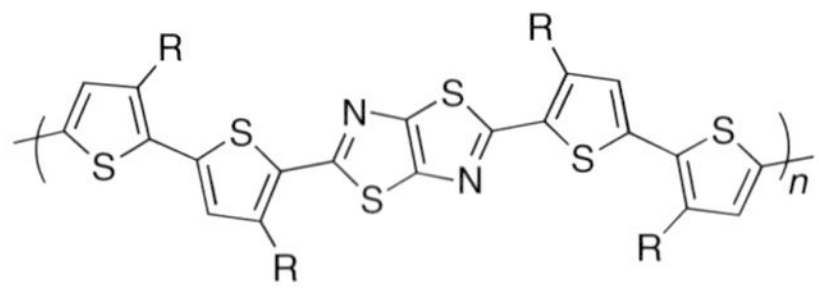	-C_6H_{13}, $C_{12}H_{25}$, $C_{14}H_{29}$	C	[114, 117]
95	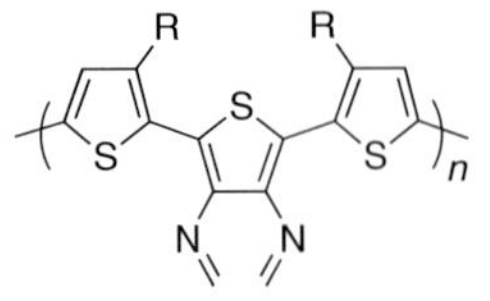	-$C_{12}H_{25}$	D	[116]

Table 3.4 (*Continued*)

Polymer	Structure	R	Method	Ref.
96		-C_6H_{13}, $C_{12}H_{25}$, $C_{14}H_{29}$	E	[114]

be able to proceed once the aggregation of Grignard monomer **61** was broken. Accordingly, polymerization of **60** in the presence of LiCl, which is known to break the aggregation of Grignard reagents and enhance the metal–halogen exchange, gave PPP in shorter times than without LiCl; additionally, this modification allowed high molecular weights and narrow polydispersities. Furthermore, the M_n values increased in proportion to both monomer conversion and feed ratio. These features, together with the fact that the addition of a fresh feed of **61** to a prepolymer afforded a higher molecular weight polymer, imply that the polymerization of PPP via the GRIM is a chain-growth system and is also essentially a living system. They have also reported on the synthesis of polypyrrole (PPy, **65**) and its diblock copolymer with PPP (**67**) [101]. Polymerization of *N*-hexylpyrrole **63** was more effective in the presence of dppe (1,2-bis(diphenylphophino)ethane), because it suppressed the formation of byproducts. Sequential addition of **60** and **63** gave well-defined PPP and PPy diblock copolymer **67** with high molecular weight ($M_n \approx 16$ K) and narrow PDI (1.1–1.2), while the addition of **60** and **63** in reverse order gave **67** with slightly narrower PDI (1.2–1.3).

McCullough and coworkers have also investigated a universal polymerization method they call "universal GRIM", which can be applied to the synthesis of aryl and heteroaryl monomers. Polyfluorene (PF, **71**), polycarbazole (PCz, **75**), and PPy (**78**) have been successfully synthesized using the universal GRIM (Scheme 3.13) [102]. A Li salt was added to a reaction mixture to break the polymeric aggregates of Grignard reagents, which in turn led to the formation of a "turbo-Grignard" and enhanced the reactivity of the Grignard reagent. Note that this result is similar to that described previously in Yokozawa's report. R_3MgLi can be an alternative to RMgCl with a Li salt system. 9,9-Dioctyl-2,7-dibromofluorene **69** was systematically tested with these reagents (Table 3.4). The magnesium–halogen exchange of **69** with either ^{i}PrMgCl or ^{t}BuMgCl was unsuccessful. Use of ^{i}PrMgCl complex with a Li salt (LiCl or LiO^{t}Bu) or use of $^{n}Bu_3MgLi$ as the Grignard reagent successfully resulted in the formation of a mono-Grignard reagent. ^{t}BuMgCl worked well with an addition of LiCl and crown ether 15-c-5. *In situ* addition of Ni(dppp)Cl_2, where polymerization was carried out at room temperature and in 20 min, afforded PF **70** with molecular weights of $M_n \approx 29$ K with PDI of < 1.7. The M_n values for GRIM type PF were comparable to those for conventional Suzuki or Yamamoto type PF, and the PDI

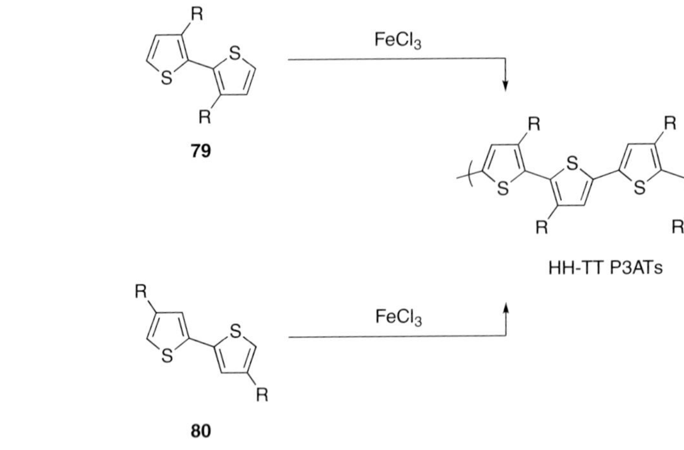

Scheme 3.14 Synthesis of HH-TT P3ATs.

values were narrower, indicating that the universal GRIM method has advantages over conventional methods for the synthesis of PF.

Magnesium–halogen exchange of a carbazole monomer (**72**) was only successful when ^{n}Bu$_3$MgLi was used. Polymerization yield was 50%, and a polymer with $M_n = 2.6$ K and PDI = 1.23 was obtained. The universal GRIM method worked well for the synthesis of PPy **78**, with the use of ^{i}PrMgCl · LiCl as the Grignard reagent for

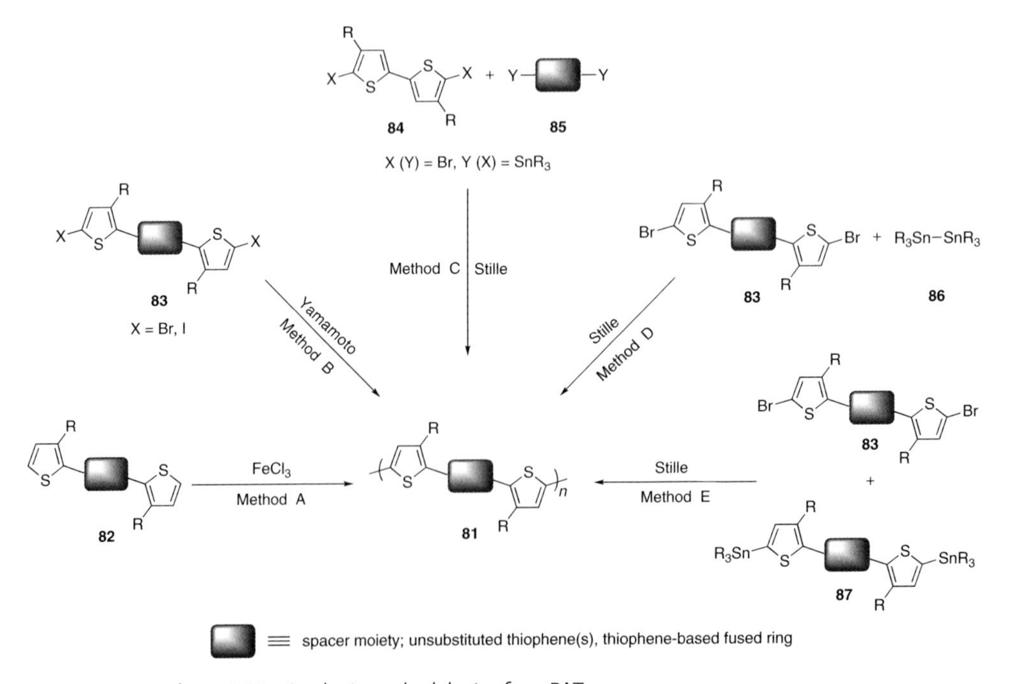

Scheme 3.15 Synthetic methodologies for rsPATs.

magnesium–halogen exchange. A polymer with $M_n = 12$ K and PDI $= 1.40$ was obtained upon addition of Ni(dppp)Cl_2.

Universal use of the GRIM method has been successful and has proven to have an advantage over the conventional method for the synthesis of aryl or heteroaryl π-conjugated polymers. This is because the methodology affords polymers with sufficient molecular weight in shorter reaction times and with narrower polydispersities. This method also paves the way toward the synthesis of block copolymers, and opens the door for further development of this field.

3.8 Regiosymmetric Polythiophenes

Regiocontrolled polymerization of asymmetric 3-alkylthiophene, giving HT-coupled rrP3AT, has been a major approach for the synthesis of regioregular (regiochemically defined) polyalkylthiophenes (PATs). Another approach to the preparation of a regioregular PAT is to polymerize symmetric thiophene monomer(s), yielding a "regiosymmetric" poly(alkylthiophene) (rsPAT) [3, 4]. Earlier works by Wudl [103] and Krische [104] demonstrated the polymerization of symmetric HH (**79**) and TT (**80**) dimers of 3-alkylthiophene, respectively, affording essentially the same rsPAT (Scheme 3.14), namely HH-TT PAT. However, HH-TT PAT possesses unfavorable HH arrangements that destroy the main chain coplanarity and exhibit limited π-conjugations and low conductivities. To improve the electrical properties of such rsPATs, the steric impact of the HH arrangement must be reduced. One efficient design strategy for reducing the steric impact is to incorporate unsubstituted thiophene ring(s) or a thiophene-based fused-ring between HH coupled 3-alkylthiophenes as a spacer moiety (Scheme 3.15). Thiophene-based fused-rings that are to be incorporated in the polymer chain can be substituted with alkyl groups as long as the alkyl moieties do not interfere with coplanarity of the polymer chain. Synthetic methodologies of rsPAT **81**, one of the primary structures of rsPATs, are summarized in Scheme 3.15. This structure is typically synthesized by one of the following methods: $FeCl_3$-mediated polymerization (Method A), Yamamoto polymerization using Ni(cod)$_2$ (Method B), and Stille coupling reaction (Methods C–E). Three different routes have been developed to synthesize rsPATs using the Stille reaction. In Method C, a dibromo- (or distannyl-) bithiophene (**84**) and a distannyl (or dibromo-) symmetric monomer (**85**) are coupled to yield **81**. In Method D, **84** is reacted with hexaalkylditin **86** to give **81**. In Method E, dibromo (**84**) and distannyl (**87**) symmetric monomers consisting of the same building unit are coupled. Table 3.4 summarizes the representative polymers synthesized by the above methods.

Poly(3,3″-dialkyl-α-terthiophene) **88** is the first rsPAT of this type, where the spacer is thiophene, synthesized via Method A by Gallazi *et al.* [105]. Electrical conductivity of an iodine-doped pressed pellet was 100 S cm^{-1}, which is comparable to that of rrP3ATs. The same polymer, including the polymer with different alkyl chain, was also synthesized by Method A [106], B [107], and C [108]. rsPAT with bithiophene (**89**), known as PQT [109], and benzodithiophene (**90**) [110] as the spacer was synthesized using Method A. PQT was also synthesized through Method C [111]. Using Method C, thieno[2,3-*b*]thiophene **91** [112] and thieno[3,2-*b*]thiophene **92**, known as PBTTT [113], were introduced into this structure, in which a microwave was used to shorten reaction times. Thiazolothiazole-containing polymers with additional bithiophene unit (**93**, PTzQT) [114] or dithienothiophene (**94**) [115]-containing polymers, were also synthesized via Method C. Thienopyrazine **95** [116] and thiazolothiazole **96** [114] polymers were synthesized using Method D and E, respectively, and represent some of the limited examples of this type.

Typical design and synthetic strategies have been described here. However, the number of reports on rsPATs is simply too large to be completely described. Among the number of reported rsPATs, some of the polymers noted above represent a very important class of materials in organic electronics due to their striking device performance. For instance, PQT **89**, reported by the Xerox group, is the first polymer semiconductor to exceed rrP3HT in field-effect mobility [109]. PBTTTs **92**, developed by the Merck group, have demonstrated the highest mobility among the polymer semiconductors reported today [113]. We have reported on other unique polymers, PTzQTs **93**, that show very high mobilities despite having low molecular weight and an amorphous-like nature [117]. Although the methods described here generate rsPATs with large polydispersities (typically > 2), these can introduce various kinds of fused-rings into regioregular polythiophenes, a potential advantage over the synthetic methodologies of rrP3ATs.

3.9 Summary

The discovery of the synthesis of rrP3ATs, regarded as one of the most important breakthroughs in the field of organic electronics, has opened the door to highly conductive processable materials as well as to the rapid and broad development of the field. Their ease of synthesis, controllable molecular weights, and narrow polydispersities, due to the living nature of the polymerization, have allowed systematic studies of reigioregular polythiophene and their structure–property relationships. This advanced synthesis of rrP3ATs has also been successful in producing a wide variety of functionalized polythiophenes – functionalized not only at the 3-position of the thiophene rings but also at the end position of the polymer chains. End group functionalization has expanded rrP3ATs into well-defined block copolymers with conjugated or non-conjugated systems. Furthermore, the methodology for the synthesis of rrP3ATs has proven to be universally applicable to other aromatic π-conjugated polymers. In addition, design and synthetic strategies of rsPATs are of particular interest because they allow the generation of new functional polymers with various heteroaromatic rings that can be feasible for organic electronic devices. Accordingly, the polythiophene family will remain one of the most versatile plastic materials and continue to lead the way in organic electronics.

References

1 Skotheim, T.A., Elsenbaumer, R.L., and Reynolds, J.R. (1998) *Handbook of Conducting Polymers*, 2nd edn Revised and Expanded, Marcel Dekker, Inc., New York.

2 *Handbook of Conducting Polymers*, (2007) 3rd edn (eds T.A. Skotheim and J.R. Reynolds), CRC Press, Boca Raton, FL.

3 Osaka, I. and McCullough, R.D. (2008) *Acc. Chem. Res.*, **41**, 1202.

4 McCullough, R.D. (1998) *Adv. Mater.*, **10**, 93.

5 Ewbank, P.C. and McCullough, R.D. (1998) in *Handbook of Conducting Polymers*, 2nd edn (eds T.A. Skotheim, R.L. Elsenbaumer, and J.R. Reynolds), Marcel Dekker, New York, p. 225.

6 McCullough, R.D. (1999) *Handbook of Oligo- Polythiophenes* (ed. D. Fichou), Wiley-VCH Verlag GmbH, Weinheim, Germany, p. 1.

7 Blanchard, P., Leriche, P., Frere, P., and Roncali, J. (2007) in *Handbook of Conducting Polymers*, 3rd edn, vol. 1 (eds T.A. Skotheim and J.R. Reynolds), CRC Press, Boca Raton, FL, p. 13/1.

8 Jeffries-El, M. and McCullough, R.D. (2007) in *Handbook of Conducting Polymers*, 3rd edn, vol. 1 (eds T.A. Skotheim and J.R. Reynolds), CRC Press, Boca Raton, FL, p. 9/1.

9 Bao, Z., Dodabalapur, A., and Lovinger, A.J. (1996) *Appl. Phys. Lett.*, **69**, 4108.

10 Sirringhaus, H., Brown, P.J., Friend, R.H., Nielsen, M.M., Bechgaard, K., Langeveld-Voss, B.M.W., Spiering, A.J.H., Janssen, R.A.J., Meijer, E.W., Herwig, P., and De Leeuw, D.M. (1999) *Nature*, **401**, 685.

11 Bao, Z. and Locklin, J. (2007) *Organic Field-Effect Transistors*, CRC Press, Boca Raton, FL.

12 Thompson, B.C. and Frechet, J.M.J. (2008) *Angew. Chem., Int. Ed.*, **47**, 58.

13 Ewbank, P.C., Laird, D., and McCullough, R.D. (2008) in *Organic Photovoltaics* (eds C. Brabec, V. Dyakonov, and U. Scherf), Wiley-VCH Verlag GmbH, Weinheim, Germany, p. 3.

14 Guenes, S., Neugebauer, H., and Sariciftci, N.S. (2007) *Chem. Rev.*, **107**, 1324.

15 McCullough, R.D. and Lowe, R.D. (1992) *J. Chem. Soc., Chem. Commun.*, 70.

16 Chen, T.A. and Rieke, R.D. (1992) *J. Am. Chem. Soc.*, **114**, 10087.

17 Lin, J.W.P. and Dudek, L.P. (1980) *J. Polym. Sci., Polym. Chem. Ed.*, **18**, 2869.

18 Yamamoto, T., Sanechika, K., and Yamamoto, A. (1980) *J. Polym. Sci., Polym. Lett. Ed.*, **18**, 9.

19 Jen, K.Y., Oboodi, R., and Elsenbaumer, R.L. (1985) *Polym. Mater. Sci. Eng.*, **53**, 79.

20 Yamamoto, T., Morita, A., Miyazaki, Y., Maruyama, T., Wakayama, H., Zhou, Z.H., Nakamura, Y., Kanbara, T., Sasaki, S., and Kubota, K. (1992) *Macromolecules*, **25**, 1214.

21 Yoshino, K., Hayashi, S., and Sugimoto, R. (1984) *Jpn. J. Appl. Phys., Part 2*, **23**, 899.

22 Sato, M. and Morii, H. (1991) *Macromolecules*, **24**, 1196.

23 McCullough, R.D., Lowe, R.D., Jayaraman, M., and Anderson, D.L. (1993) *J. Org. Chem.*, **58**, 904.

24 McCullough, R.D., Tristram-Nagle, S., Williams, S.P., Lowe, R.D., and Jayaraman, M. (1993) *J. Am. Chem. Soc.*, **115**, 4910.

25 Loewe, R.S., Khersonsky, S.M., and McCullough, R.D. (1999) *Adv. Mater.*, **11**, 250.

26 Iraqi, A. and Barker, G.W. (1998) *J. Mater. Chem.*, **8**, 25.

27 Guillerez, S. and Bidan, G. (1998) *Synth. Met.*, **93**, 123.

28 Tamao, K., Sumitani, K., and Kumada, M. (1972) *J. Am. Chem. Soc.*, **94**, 4374.

29 Negishi, E. (1982) *Acc. Chem. Res.*, **15**, 340.

30 McCullough, R.D., Williams, S.P., Tristram-Nagle, S., Jayaraman, M., Ewbank, P.C., and Miller, L. (1995) *Synth. Met.*, **69**, 279.

31 McCullough, R.D., Lowe, R.D., Jayaraman, M., Ewbank, P.C., Anderson, D.L., and Tristram-Nagle, S. (1993) *Synth. Met.*, **55**, 1198.

32 Chen, T.-A., Wu, X., and Rieke, R.D. (1995) *J. Am. Chem. Soc.*, **117**, 233.
33 Chen, T.-A., and Rieke, R.D. (1993) *Synth. Met.*, **60**, 175.
34 Chen, T.A., O'Brien, R.A., and Rieke, R.D. (1993) *Macromolecules*, **26**, 3462.
35 Rieke, R.D. (1977) *Acc. Chem. Res.*, **10**, 301.
36 Rieke, R.D., Li, P.T.-J., Burns, T.P., and Uhm, S.T. (1981) *J. Org. Chem.*, **46**, 4323.
37 Loewe, R.S., Ewbank, P.C., Liu, J., Zhai, L., and McCullough, R.D. (2001) *Macromolecules*, **34**, 4324.
38 Stille, J.K. (1986) *Angew. Chem.*, **98**, 504.
39 Miyaura, N. and Suzuki, A. (1995) *Chem. Rev.*, **95**, 2457.
40 de Meijere, A., and Diederich, F. (eds) (2004) *Metal-Catalyzed Cross-Coupling Reactions*, Second Completely Revised and Enlarged Edition, Wiley-VCH Verlag GmbH & Co. KGaA, Weinheim, Germany.
41 Odian, G. (2004) *Principles of Polymerization*, 4th edn, Wiley-Interscience, New York.
42 Yokoyama, A., Miyakoshi, R., and Yokozawa, T. (2004) *Macromolecules*, **37**, 1169.
43 Miyakoshi, R., Yokoyama, A., and Yokozawa, T. (2004) *Macromol. Rapid Commun.*, **25**, 1663.
44 Miyakoshi, R., Yokoyama, A., and Yokozawa, T. (2005) *J. Am. Chem. Soc.*, **127**, 17542.
45 Sheina, E.E., Liu, J., Iovu, M.C., Laird, D.W., and McCullough, R.D. (2004) *Macromolecules*, **37**, 3526.
46 Iovu, M.C., Sheina, E.E., Gil, R.R., and McCullough, R.D. (2005) *Macromolecules*, **38**, 8649.
47 Jeffries-El, M., Sauve, G., and McCullough, R.D. (2004) *Adv. Mater.*, **16**, 1017.
48 Jeffries-El, M., Sauve, G., and McCullough, R.D. (2005) *Macromolecules*, **38**, 10346.
49 Zhang, R., Li, B., Iovu, M.C., Jeffries-El, M., Sauve, G., Cooper, J., Jia, S., Tristram-Nagle, S., Smilgies, D.M., Lambeth, D.N., McCullough, R.D., and Kowalewski, T. (2006) *J. Am. Chem. Soc.*, **128**, 3480.
50 Iovu, M.C., Jeffries-El, M., Zhang, R., Kowalewski, T., and McCullough, R.D. (2006) *J. Macromol. Sci., Part A: Pure Appl. Chem.*, **43**, 1991.
51 Iovu, M.C., Jeffries-El, M., Sheina, E.E., Cooper, J.R., and McCullough, R.D. (2005) *Polymer*, **46**, 8582.
52 Iovu, M.C., Zhang, R., Cooper, J.R., Smilgies, D.M., Javier, A.E., Sheina, E.E., Kowalewski, T., and McCullough, R.D. (2007) *Macromol. Rapid Commun.*, **28**, 1816.
53 Sheina, E.E., Khersonsky, S.M., Jones, E.G., and McCullough, R.D. (2005) *Chem. Mater.*, **17**, 3317.
54 McCullough, R.D. and Williams, S.P. (1993) *J. Am. Chem. Soc.*, **115**, 11608.
55 Bolognesi, A., Porzio, W., Bajo, G., Zannoni, G., and Fannig, L. (1999) *Acta Polym.*, **50**, 151.
56 Wu, X., Chen, T.-A., and Rieke, R.D. (1996) *Macromolecules*, **29**, 7671.
57 Ogawa, K., Stafford, J.A., Rothstein, S.D., Tallman, D.E., and Rasmussen, S.C. (2005) *Synth. Met.*, **152**, 137.
58 Amarasekara, A.S., and Pomerantz, M. (2003) *Synthesis*, 2255.
59 Hong, X.M. and Collard, D.M. (2000) *Macromolecules*, **33**, 6916.
60 Li, L., Counts, K.E., Kurosawa, S., Teja, A.S., and Collard, D.M. (2004) *Adv. Mater.*, **16**, 477.
61 Andersson, M.R., Selse, D., Berggren, M., Jaervinen, H., Hjertberg, T., Inganaes, O., Wennerstroem, O., and Oesterholm, J.E. (1994) *Macromolecules*, **27**, 6503.
62 Bouman, M.M., Havinga, E.E., Janssen, R.A.J., and Meijer, E.W. (1994) *Mol. Cryst. Liq. Cryst. Sci. Technol., Sect. A*, **256**, 439.
63 Bidan, G., Guillerez, S., and Sorokin, V. (1996) *Adv. Mater. (Weinheim, Ger.)*, **8**, 157.
64 Langeveld-Voss, B.M.W., Christiaans, M.P.T., Janssen, R.A.J., and Meijer, E.W. (1998) *Macromolecules*, **31**, 6702.
65 Goto, H., Okamoto, Y., and Yashima, E. (2002) *Macromolecules*, **35**, 4590.
66 Yu, J. and Holdcroft, S. (2000) *Macromolecules*, **33**, 5073.
67 Lanzi, M., Costa-Bizzarri, P., Della-Casa, C., Paganin, L., and Fraleoni, A. (2002) *Polymer*, **44**, 535.

68 McCullough, R.D., Ewbank, P.C., and Loewe, R.S. (1997) *J. Am. Chem. Soc.*, **119**, 633.

69 Stokes, K.K., Heuze, K., and McCullough, R.D. (2003) *Macromolecules*, **36**, 7114.

70 Iraqi, A., Crayston, J.A., and Walton, J.C. (1998) *J. Mater. Chem.*, **8**, 31.

71 Zhai, L., Pilston, R.L., Zaiger, K.L., Stokes, K.K., and McCullough, R.D. (2003) *Macromolecules*, **36**, 61.

72 Liu, J., Loewe, R.S., and McCullough, R.D. (1999) *Macromolecules*, **32**, 5777.

73 Liu, J., Sheina, E., Kowalewski, T., and McCullough, R.D. (2002) *Angew. Chem., Int. Ed.*, **41**, 329.

74 Liu, J. and McCullough, R.D. (2002) *Macromolecules*, **35**, 9882.

75 Djuric, S., Venit, J., and Magnus, P. (1981) *Tetrahedron Lett.*, **22**, 1787.

76 Langeveld-Voss, B.M.W., Janssen, R.A.J., Spiering, A.J.H., van Dongen, J.L.J., Vonk, E.C., and Claessens, H.A. (2000) *Chem. Commun. (Cambridge)*, 81.

77 Massimo Lazzari, G.L. and Lecommandoux, S. (2006) *Block Copolymers in Nanoscience*, Wiley-VCH Verlag GmbH & Co. KGaA, Weinheim, Germany.

78 Schmitt, C., Nothofer, H.-G., Falcou, A., and Scherf, U. (2001) *Macromol. Rapid Commun.*, **22**, 624.

79 Sun, S., Fan, Z., Wang, Y., Haliburton, J., Taft, C., Maaref, S., Seo, K., and Bonner, C.E. (2003) *Synth. Met.*, **137**, 883.

80 Sun, S.-S. (2003) *Sol. Energy Mater. Sol. Cells*, **79**, 257.

81 Scherf, U., Gutacker, A., and Koenen, N. (2008) *Acc. Chem. Res.*, **41**, 1086.

82 Zhang, Y., Tajima, K., Hirota, K., and Hashimoto, K. (2008) *J. Am. Chem. Soc.*, **130**, 7812.

83 Wu, P.-T., Ren, G., Li, C., Mezzenga, R., and Jenekhe, S.A. (2009) *Macromolecules*, **42**, 4.

84 Yokozawa, T., Adachi, I., Miyakoshi, R., and Yokoyama, A. (2007) *High Perform. Polym.*, **19**, 684.

85 Miyakoshi, R., Yokoyama, A., and Yokozawa, T. (2008) *Chem. Lett.*, **37**, 1022.

86 Ohshimizu, K. and Ueda, M. (2008) *Macromolecules*, **41**, 5289.

87 Benanti, T.L., Kalaydjian, A., and Venkataraman, D. (2008) *Macromolecules*, **41**, 8312.

88 Asawapirom, U., Guentner, R., Forster, M., and Scherf, U. (2005) *Thin Solid Films*, **477**, 48.

89 Tu, G., Li, H., Forster, M., Heiderhoff, R., Balk, L.J., and Scherf, U. (2006) *Macromolecules*, **39**, 4327.

90 Jenekhe, S.A. and Chen, X.L. (1998) *Science*, **279**, 1903.

91 Matyjaszewski, K. (2000) *Controlled/Living Radical Polymerization: Progress in ATRP, NMP and RAFT*, vol. **768**, American Chemical Society, Washington DC.

92 Matyjaszewski, K. and Xia, J. (2001) *Chem. Rev.*, **101**, 2921.

93 Iovu, M.C., Craley, C.R., Jeffries-El, M., Krankowski, A.B., Zhang, R., Kowalewski, T., and McCullough, R.D. (2007) *Macromolecules*, **40**, 4733.

94 Sauve, G. and McCullough, R.D. (2007) *Adv. Mater.*, **19**, 1822.

95 Urien, M., Erothu, H., Cloutet, E., Hiorns, R.C., Vignau, L., and Cramail, H. (2008) *Macromolecules*, **41**, 7033.

96 Higashihara, T., Ohshimizu, K., Hirao, A., and Ueda, M. (2008) *Macromolecules*, **41**, 9505.

97 Dai, C.-A., Yen, W.-C., Lee, Y.-H., Ho, C.-C., and Su, W.-F. (2007) *J. Am. Chem. Soc.*, **129**, 11036.

98 Boudouris, B.W., Frisbie, C.D., and Hillmyer, M.A. (2008) *Macromolecules*, **41**, 67.

99 Radano, C.P., Scherman, O.A., Stingelin-Stutzmann, N., Mueller, C., Breiby, D.W., Smith, P., Janssen, R.A.J., and Meijer, E.W. (2005) *J. Am. Chem. Soc.*, **127**, 12502.

100 Miyakoshi, R., Shimono, K., Yokoyama, A., and Yokozawa, T. (2006) *J. Am. Chem. Soc.*, **128**, 16012.

101 Yokoyama, A., Kato, A., Miyakoshi, R., and Yokozawa, T. (2008) *Macromolecules*, **41**, 7271.

102 Stefan, M.C., Javier, A.E., Osaka, I., and McCullough, R.D. (2009) *Macromolecules*, **42**, 30.

103 Maior, R.M.S., Hinkelmann, K., Eckert, H., and Wudl, F. (1990) *Macromolecules*, **23**, 1268.

104 Zagorska, M. and Krische, B. (1990) *Polymer*, **31**, 1379.

105 Gallazzi, M.C., Castellani, L., Marin, R.A., and Zerbi, G. (1993) *J. Polym. Sci., Part A: Polym. Chem.*, **31**, 3339.

106 Wu, Y., Liu, P., Gardner, S., and Ong, B.S. (2005) *Chem. Mater.*, **17**, 221.

107 Kokubo, H. and Yamamoto, T. (2001) *Macromol. Chem. Phys.*, **202**, 1031.

108 McCulloch, I., Bailey, C., Giles, M., Heeney, M., Love, I., Shkunov, M., Sparrowe, D., and Tierney, S. (2005) *Chem. Mater.*, **17**, 1381.

109 Ong, B.S., Wu, Y., Liu, P., and Gardner, S. (2004) *J. Am. Chem. Soc.*, **126**, 3378.

110 Pan, H., Li, Y., Wu, Y., Liu, P., Ong, B.S., Zhu, S., and Xu, G. (2007) *J. Am. Chem. Soc.*, **129**, 4112.

111 Thompson, B.C., Kim, B.J., Kavulak, D.F., Sivula, K., Mauldin, C., and Frechet, J.M.J. (2007) *Macromolecules*, **40**, 7425.

112 Heeney, M., Bailey, C., Genevicius, K., Shkunov, M., Sparrowe, D., Tierney, S., and McCulloch, I. (2005) *J. Am. Chem. Soc.*, **127**, 1078.

113 McCulloch, I., Heeney, M., Bailey, C., Genevicius, K., MacDonald, I., Shkunov, M., Sparrowe, D., Tierney, S., Wagner, R., Zhang, W., Chabinyc, M.L., Kline, R.J., McGehee, M.D., and Toney, M.F. (2006) *Nat. Mater.*, **5**, 328.

114 Osaka, I., Sauve, G., Zhang, R., Kowalewski, T., and McCullough, R.D. (2007) *Adv. Mater.*, **19**, 4160.

115 Li, J., Qin, F., Li, C.M., Bao, Q., Chan-Park, M.B., Zhang, W., Qin, J., and Ong, B.S. (2008) *Chem. Mater.*, **20**, 2057.

116 Zhu, Y., Champion, R.D., and Jenekhe, S.A. (2006) *Macromolecules*, **39**, 8712.

117 Osaka, I., Zhang, R., Sauve, G., Smilgies, D.-M., Kowalewski, T., and McCullough, R.D. (2009) *J. Am. Chem. Soc.*, **131**, 2521.

4
Functional Hyperbranched Polymers Constructed from Acetylenic A_n-Type Building Blocks

Jianzhao Liu, Jacky W.Y. Lam, and Ben Zhong Tang

4.1
Introduction

Most commodity plastics used nowadays, such as polypropylene, polystyrene and poly(methyl methacrylate), are manufactured by chain polymerization of vinyl monomers. Thus conventional chain polymerization is mainly based on olefin chemistry. When vinyl monomers are polymerized, their double bonds are transformed to single bonds, giving saturated polymers with electrical and optical inactivity. Whereas an enormous amount of research has been done on olefin polymerization, acetylene chemistry has been much less utilized in polymer synthesis, in spite of the promise that acetylene polymerizations can generate unsaturated or conjugated polymers with electrical conductivity, optical activity, and photonic susceptibility [1–3].

A dozen research groups have done some pioneering work in the area of acetylene polymerizations [4–15]. The area is, however, still full of challenges. For example, most catalysts for acetylene polymerizations are labile transition-metal salts or sensitive organometallic complexes; their stable organic congeners or organocatalysts are virtually unknown. Only a few types of acetylene reactions been successfully developed into useful polymerization techniques, and it has been difficult to polymerize functionalized acetylene monomers because many of the traditional catalysts have little tolerance to polar groups. Many acetylene-based conjugated polymers are unstable, which has greatly limited the scope of their practical application. Almost all acetylene polymers with high molecular weight possess a linear structure, and few hyperbranched polymers have been derived from acetylene monomers.

Attracted by the great variety of acetylene derivatives and their fabulously rich reactivities [15–17], we have embarked upon a research program on acetylene polymerizations, with the aim of developing new polymerization techniques based on acetylene chemistry and cultivating acetylenic compounds into versatile building blocks for the construction of new conjugated macromolecules. We have previously focused our attention on the syntheses of linear polyacetylenes from acetylene

Conjugated Polymer Synthesis. Edited by Yoshiki Chujo

ISBN: 978-3-527-32267-1

monomers with single triple bonds, that is, monoynes (Scheme 4.1). Our effort in the exploration of functionality-tolerant catalysts and the optimization of polymerization processes has enabled us to successfully prepare a large number of functional polyacetylenes with various pendant groups. In addition to being stable and processable, the new polyacetylenes with appropriate skeleton-pendant combinations have shown an array of novel properties, such as luminescence, photoconductivity, photonic patternability, helical chirality, liquid crystallinity, optical nonlinearity, solvatochromism, cytocompatibility, and biological activity [18, 19]. The properties of the polyacetylenes can be tuned internally and manipulated externally: the former is accomplished by varying their molecular structures, especially their functional pendants, while the latter is achieved by applying thermal, mechanical, electrical, photonic, and chemical stimuli [18].

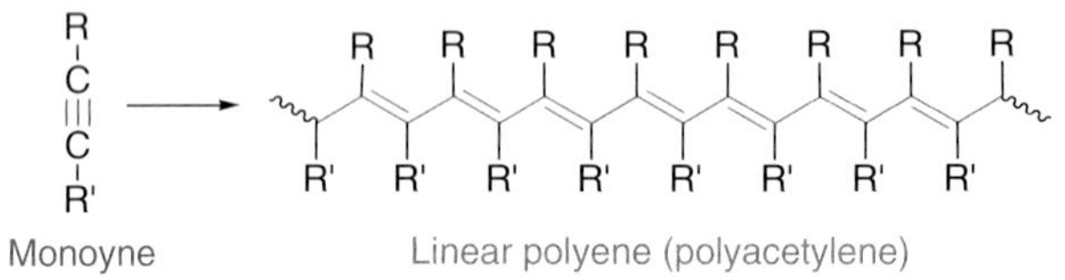

Monoyne

Linear polyene (polyacetylene)

Scheme 4.1 Synthesis of linear polyacetylenes from monoyne monomers.

We have recently extended our research effort from one-dimensional to three-dimensional systems and have worked on the syntheses of nonlinear hyperbranched polymers from acetylenic monomers having two and three triple bonds, that is, diynes and triynes (Scheme 4.2). In this chapter, we will briefly review our work in this area: we will summarize our results on the polymer syntheses, outline the functional properties of the new polymers, and illustrate their potential applications. We have successfully developed new catalysts for acetylene polymerizations, especially organocatalysts with great functionality-tolerance. We have established new synthetic routes to hyperbranched polymers and prepared new polyphenylenes, polydiynes and polytriazoles by polycyclotrimerization, polycoupling and polycycloaddition reactions of diyne and triyne monomers. Traditionally, hyperbranched polymers have often been prepared from polycondensations of AB_n ($n \geq 2$), especially AB_2, monomers (Scheme 4.3). The new A_n ($n = 2, 3$) protocol developed in this study thus provides polymer chemists with a new synthetic tool that is free of the complications suffered by the AB_2 system, such as the difficulties in monomer preparation, isolation and storage, because of the coexistence of two mutually reactive groups A and B in one monomer species [20–26]. The new hyperbranched polymers have been found to show unique properties including thermal curability, photonic susceptibility, pattern formation, fluorescent imaging, light refraction, optical nonlinearity, and metallic complexation. Furthermore, the metallized polymers can be readily pyrolyzed into soft ferromagnetic ceramics with high magnetizability and into nanoparticles with catalytic activity for carbon nanotube fabrication.

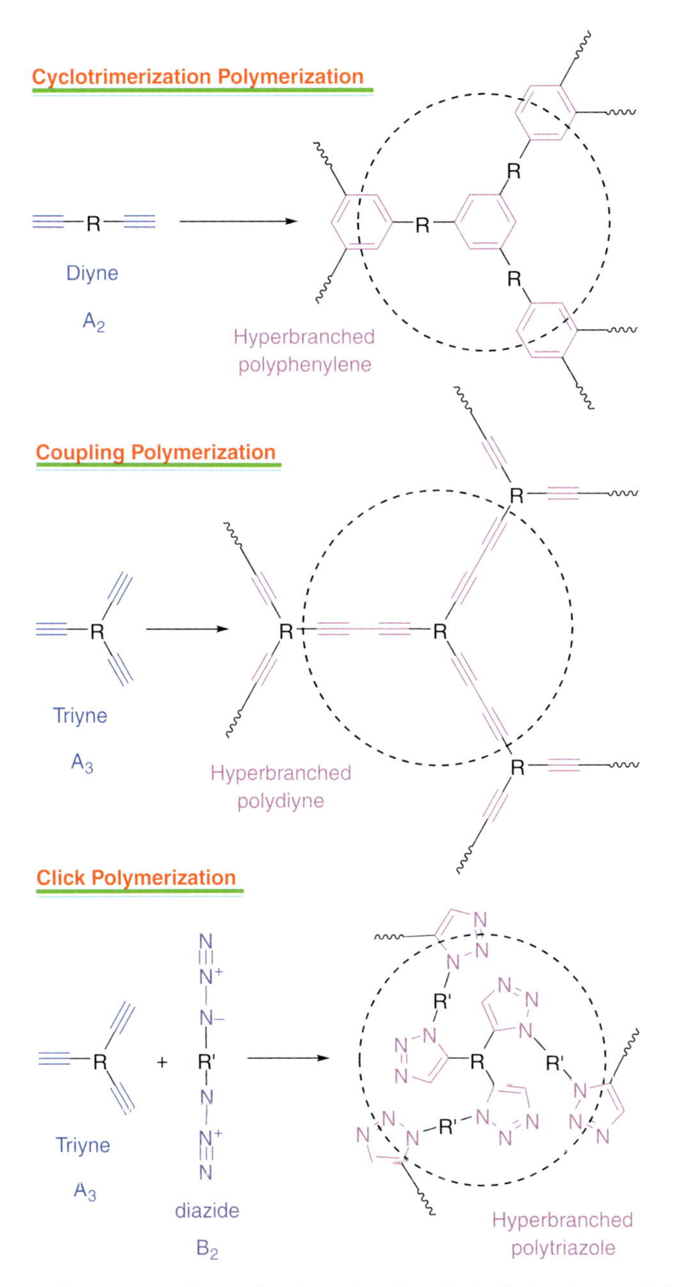

Scheme 4.2 Syntheses of nonlinear hyperbranched polyphenylenes, polydiynes and polytriazoles from polycyclotrimerization, polycoupling and polycycloaddition of diyne, triyne and diazide monomers.

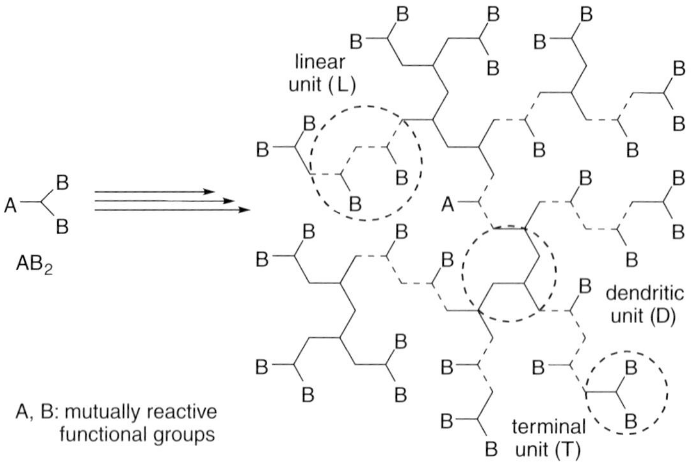

Scheme 4.3 A common approach to hyperbranched polymers, with structural illustration of the dendritic, linear and terminal units.

4.2
Hyperbranched Polymers Constructed from Acetylenic A_n-Type Building Blocks

4.2.1
Hyperbranched Poly(Alkylenephenylene)s (*hb*-PAPs)

Acetylene cyclotrimerization is a century-old reaction for the effective transformation of monoyne molecules to benzene rings. When this reaction is applied to diyne molecules, hyperbranched polymers may be generated. This A_2 polycyclotrimerization approach will circumvent the synthetic difficulties encountered by the AB_2 system and produce stable polymers consisting of robust benzene rings. With these anticipations in mind, we started our work on the synthesis of hyperbranched polymers by diyne polycyclotrimerization.

4.2.1.1 Synthesis

We first studied the homopolycyclotrimerization of aliphatic diynes with different lengths of alkyl spacers (Scheme 4.4) [27–31]. Binary mixtures of MtX_5 and Ph_4Sn were found to be effective catalysts for polycyclotrimerizations of terminal and internal diyne monomers. We investigated the effects on the diynepolycyclotrimerization of such parameters as monomer structure, reaction time, monomer/catalyst concentration, and so on. The correct combination of these factors ensured that the polymerization reactions proceeded smoothly. Under optimal conditions, completely soluble homopolymers with high molecular weights (M_w up to $\sim 1.4 \times 10^6$) and predominantly 1,2,4-benzenetriyl core structures are obtained in high yields (up to 93%). Internal diynes can be polymerized into hexasubstituted *hb*-PAPs in moderate yields and molecular weights, probably due to the steric hindrance of the terminal groups (R).

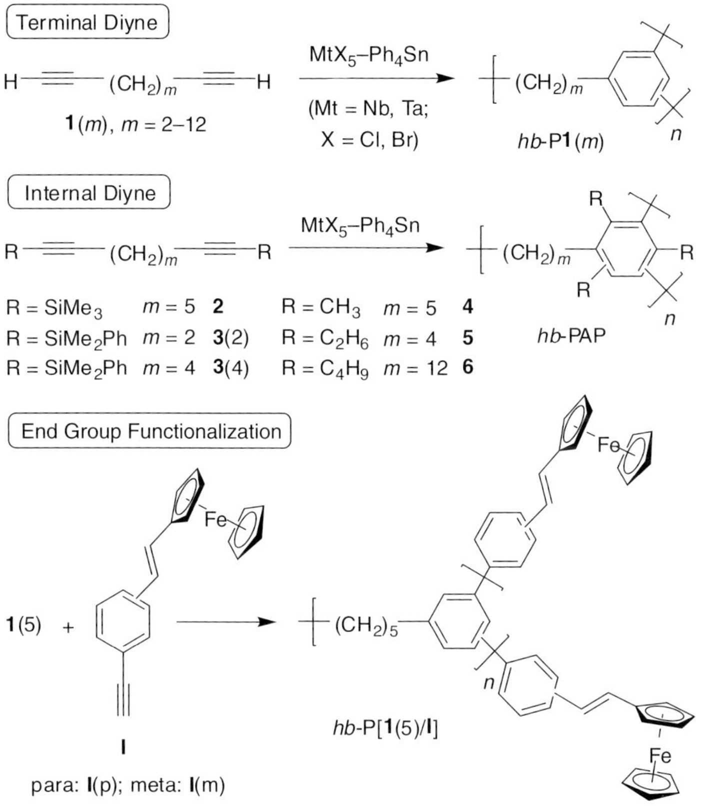

Terminal Diyne

H—≡—$(CH_2)_m$—≡—H → MtX_5–Ph_4Sn (Mt = Nb, Ta; X = Cl, Br) → *hb*-P**1**(*m*)

1(*m*), *m* = 2–12

Internal Diyne

R—≡—$(CH_2)_m$—≡—R → MtX_5–Ph_4Sn → *hb*-PAP

R = $SiMe_3$ *m* = 5 **2**; R = $SiMe_2Ph$ *m* = 2 **3**(2); R = $SiMe_2Ph$ *m* = 4 **3**(4)

R = CH_3 *m* = 5 **4**; R = C_2H_6 *m* = 4 **5**; R = C_4H_9 *m* = 12 **6**

End Group Functionalization

1(5) + **I** → *hb*-P[**1**(5)/**I**]

para: **I**(p); meta: **I**(m)

Scheme 4.4 Polycyclotrimerizations of terminal and internal diynes and syntheses of functional *hb*-PAPs end-capped by ferrocenyl groups.

Copolycyclotrimerization of the diynes with monoynes can be used to incorporate functional groups into the *hb*-PAP structure at the molecular level. This is demonstrated by the synthesis of *hb*-P[**1**(5)/**I**]: the *hb*-PAP is decorated by redox-active ferrocenyl units on its periphery [31]. We have recently found that single component catalysts, such as $TaBr_5$ and $NbBr_5$, can also initiate the polycyclotrimerization. The homopolycyclotrimerization of 1,7-octadiyne [**1**(4)] catalyzed by $TaBr_5$, for example, furnishes an *hb*-PAP with a high molecular weight (M_w: 5.5×10^4) in 81% yield [30].

Size exclusion chromatography (SEC) is nowadays a standard tool for the estimation of the molecular weights of polymers. The SEC system calibrated with linear polymer standards such as PS, however, often underestimates the molecular weights of hyperbranched polymers because of their nearly spherical topological structures [32]. Our *hb*-PAPs also display such a difference: their absolute M_w values are 2–7 times higher than the relative values estimated by an SEC system. The highest absolute M_w value can reach 3.8×10^6. Despite their high molecular weights, solutions of the *hb*-PAPs exhibit low intrinsic viscosities ($[\eta] < 0.2\,dL\,g^{-1}$). The $[\eta]$ value is insensitive to the molecular weight of the polymer. This is not surprising

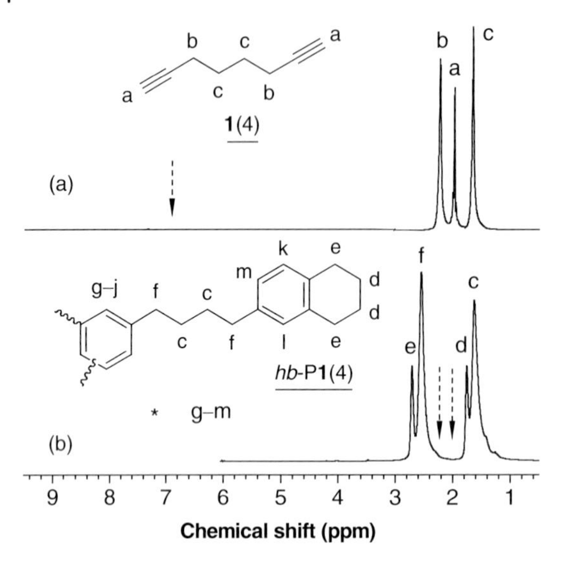

Figure 4.1 ^{1}H NMR spectra of chloroform-*d* solutions of (A) 1,7-octadiyne [**1**(4)] and (B) its hyperbranched polymer [*hb*-P**1**(4)]. The solvent peak is marked with an asterisk.

because there is practically no chain entanglement in the hyperbranched polymer systems [20–26].

4.2.1.2 **Structures**

The structures of the *hb*-PAPs were fully characterized by standard spectroscopic methods. Figure 4.1 shows examples of the ^{1}H NMR spectra of **1**(4) and its polymer *hb*-P**1**(4) [30]. The resonance peak of the ethynyl protons (a) of **1**(4) at δ 1.97 disappears after the polymerization, and new, broad peaks emerge at δ 6.8–7.0 (g–m), confirming the [2 + 2 + 2] polycyclization of the acetylenic triple bonds into benzene rings. The trialkylbenzene ring is formed in symmetrical 1,3,5- and asymmetrical 1,2,4-fashions with the latter being predominant ($F_{1,2,4} = 74\%$) [30]. The propargylic protons (b) of **1**(4) are transformed to benzylic protons (f) of *hb*-P**1**(4), whose resonance is accordingly downfield shifted from δ 2.23 to δ 2.56.

The appearance of two peaks at δ 2.71 (e) and δ 1.76 (d) associated with the back-biting reaction of the flexible diyne monomer confirms another key feature of the *hb*-PAP structure. The back-biting reaction plays an important role in the polycyclotrimerization of the aliphatic diynes. When this end-capping reaction is active, the propagating branches will be readily terminated, giving only oligomer products. On the other hand, if the back-biting reaction is too sluggish, the polymerization will become difficult to control, resulting in the formation of cross-linked gels. Fine tuning the back-biting reaction will help control the reaction output. For example, diynes with two [**1**(2)] and three short methylene spacers [**1**(3)] have a high tendency toward back-biting and form *hb*-PAPs with low molecular weights (Table 4.1, nos. 1 and 2). No terminal triple bonds but strong back-biting signals are observed in their ^{1}H NMR spectra. Diynes with long spacers [**1**(m), $m \geq 6$] show a low back-biting

Table 4.1 Analysis of the products obtained from the attempted polycyclotrimerizations of aliphatic diynes **1**(*m*) with different lengths (*m*) of methylene spacers[a].

No.	*m*	S[b]	Triple-bond signal[c]	Back-biting signal[c]	M_w[d]	PDI[d]
1	2	√	Not observed	Strong	~3000	2.3
2	3	√	Not observed	Very strong	~900	1.5
3	4	√	Observed	Medium	~60 000	4.9
4	5	√	Not observed	Medium	~40 000	5.0
5	6	√	Observed	Very weak	~600 000	23.0
6	8	×				
7	9	√	Observed	Weak	~200 000	7.9
8	10	×				

a) Polymerization reactions carried out in toluene at room temperature under nitrogen using $TaCl_5$–Ph_4Sn as catalyst.
b) Solubility (*S*) tested at room temperature in common organic solvents including toluene, benzene, chloroform, dichloromethane (DCM), and THF. Symbol: √ = soluble, × = insoluble.
c) In the ^1H NMR spectrum of *hb*-PAP.
d) [d]Polydispersity index (M_w/M_n) estimated by SEC in THF on the basis of a linear PS calibration.

activity, which is readily confirmed by the existence of their unterminated triple bonds. As a result, the polymerizations are difficult to control, giving polymers with very high M_ws and extremely broad PDIs (Table 4.1, nos. 5 and 7) or even totally insoluble gels (Table 4.1, nos. 6 and 8).

An elaborate comparison of the polymerization behavior of the aliphatic diynes with different spacer lengths reveals a unique even–odd effect on their tendencies toward back-biting reaction. In the diynes with an odd number of methylene spacers, their triple bonds are on the same side, which facilitates the back-biting reaction (Scheme 4.5). On the other hand, in the diynes with an even number of methylene units, their triple bonds are located on opposite sides. This unfavorable positioning frustrates the back-biting reaction. Consequently, *hb*-P**1**(4) possesses triple bond residues in its final structure but its congener *hb*-P**1**(5) does not.

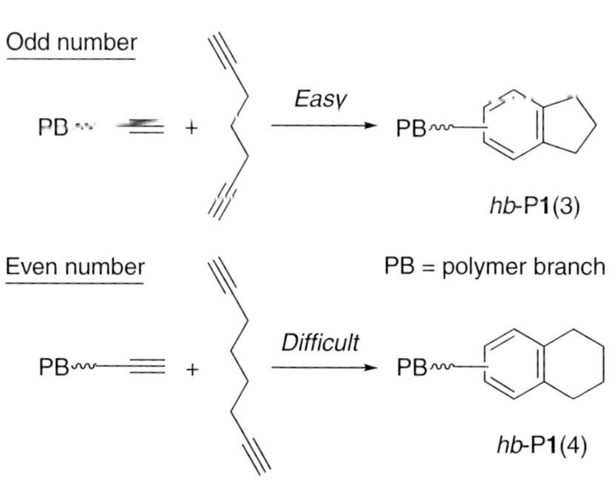

Scheme 4.5 Odd–even effect on the back-biting reaction in the diyne polycyclotrimerization.

The *hb*-PAPs contain numerous branching units, resulting from the cyclization propagation and back-biting termination. In addition to the branching structures, there exists a pseudo-"linear" structure in the polymers, formed by the reaction of the closely located triple bonds in a 1,2,4-substituted benzene ring (Scheme 4.6). The acetylene triple bonds in the 1 and 2 positions can form a new benzene ring through cyclotrimerization with another triple bond from a monomer or a polymer branch. Although this may also be considered as a "cross-linking" reaction, the structural motif is not detrimental to the solubility of *hb*-PAP due to its overall "linear" propagation mode. Combining all these three structural features, the *hb*-PAPs possess a molecular architecture resembling that of glycogen, a hyperbranched natural polymer [30].

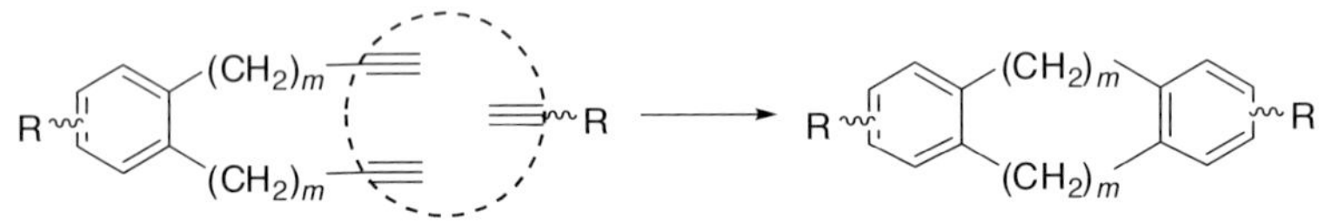

Scheme 4.6 Pseudo-"linear" propagation mode in the polycyclotrimerizations of aliphatic diynes.

4.2.1.3 Properties

The *hb*-PAPs are thermally very stable. As can be seen from the thermogravimetric analysis (TGA) curves shown in Figure 4.2, the polymers lose little of their weight

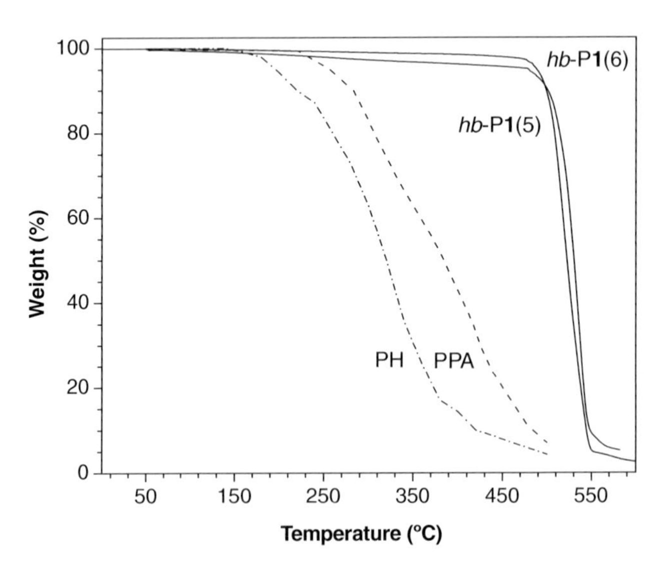

Figure 4.2 TGA thermograms of hyperbranched poly(alkylenephenylenes) *hb*-P**1**(5) and *hb*-P**1**(6). Data for poly(phenylacetylene) (PPA) and poly(1-hexyne) (PH) are shown for comparison.

when heated to a temperature as high as ~500 °C. Their thermal stabilities are much higher than those of PPA and PH, the linear polyacetylenes prepared by the metathesis polymerizations of phenylacetylene and 1-hexyne, which start to decompose at temperatures as low as ~220 and ~150 °C, respectively. Only glass transitions, but no melting transitions, are detected by differential scanning calorimetry (DSC) analyses of the *hb*-PAPs, indicative of an amorphous morphology associated with their irregular and highly branched molecular structure. The glass transition temperature (T_g) of *hb*-PAP decreases with increase in the spacer length, due to the internal plasticization effect of its flexible spacer between the benzene rings.

Thin solid films of the *hb*-PAPs are highly transparent and absorb almost no visible light. This is easy to understand, because the polymers are comprised of isolated benzene rings. The thin film of *hb*-P**1**(5), for example, displays an optical dispersion (D) as low as 0.009 in the visible spectral region, much superior to those of the commercially important "organic glasses" such as PMMA ($D = 0.0175$) and polycarbonate (PC; $D = 0.0297$) [29].

The *hb*-PAPs contain many benzyl units, which readily form radical species upon photoexcitation. Recombination of the radicals should result in curing or hardening of the polymers. Indeed, thin films of the *hb*-PAPs are readily cross-linked upon illumination with a UV lamp. Figure 4.3 shows the formation of an insoluble gel upon exposure of a thin film of *hb*-P**1**(4) to UV irradiation. After ~20 min exposure, almost the whole film is cross-linked with an F_g value of ~100%, indicative of a high photosensitivity of the polymer in spite of its irregular hyperbranched structure.

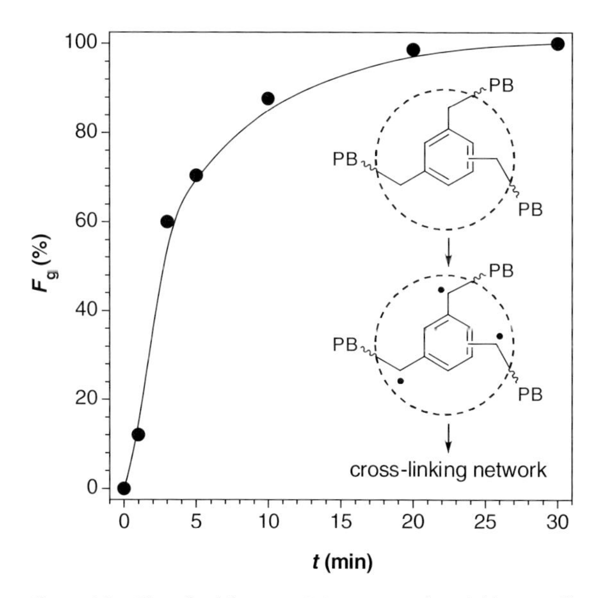

Figure 4.3 Plot of gel fraction (F_g) in UV-irradiated *hb*-P**1**(4) film versus exposure time (t). PB stands for a polymeric branch.

4.2.2
Hyperbranched Poly(Arylenephenylene)s (*hb*-PArPs)

The exciting results described above prompted us to forge ahead in this new area of research. We replaced the saturated methylene chains in the alkyl diynes with unsaturated aromatic rings in aryl diynes. Different from the isolated benzene rings in the *hb*-PAPs discussed above, the new benzene rings of the *hb*-PArPs formed by the polycyclotrimerizations of the arylenediynes are interconnected by the "old" aromatic units from the monomers. This should endow the *hb*-PArPs with an extended π-conjugation and hence interesting electronic and optical properties.

4.2.2.1 Synthesis

We designed and synthesized a wide variety of organic and organometallic arylenediynes as well as monoynes, examples of which are shown in Scheme 4.7 [33–45]. Most of the acetylenic monomers were prepared in high yields by the palladium-catalyzed cross-coupling of silylacetylenes with dihaloarenes followed by base-catalyzed desilylation. Despite their rigid molecular structures and lack of back-biting capability, the homopolycyclotrimerizations of metallolyldiynes (**26** and **27**), carbazolyldiynes (**28–31**) and silyldiynes (**36–40**) initiated by the Ta, Nb and Co catalysts proceeded smoothly, giving soluble *hb*-PArPs with high molecular weights in high yields [40–45]. The large free volumes and the irregular structures created by the nonplanar, nonlinear metallolyl, carbazolyl and silyl units may have conferred the excellent solubility on the homopolymers. Homopolycyclotrimerizations of other aromatic diynes including **7–25** and **32–35** all proceeded very rapidly, giving polymeric products with partial or total insolubility in common organic solvents, due to the involved cross-linking reactions.

In order to suppress the undesired cross-linking reactions in the polycyclotrimerization processes and to improve the macroscopic processability of the resultant polymers, copolycyclotrimerizations of the aromatic diynes with monoynes (**II–VIII**) were carried out. This approach worked very well: all the copolycyclotrimerization reactions proceeded smoothly with good controllability, producing completely soluble *hb*-PArPs with high molecular weights (M_w up to $\sim 1.8 \times 10^5$) in high yields (up to 99.7%). Among the monoynes, 1-alkynes (**II–V**) are generally better comonomers than their 1-arylacetylene congeners (**VI–XIII**) when the molecular weights of the copolymers are concerned. This is probably due to the following effects. First, the long alkyl chains may confer higher solubility on the propagating species, hence enabling their uninterrupted, continued growth into bigger polymers. The second may be an electronic effect. The electron-donating alkyl groups make the triple bonds of 1-alkynes electronically richer, thus makingthem likely to interact favorably and react with the electronically poorer aromatic diynes [46]. In other words, the donor–acceptor interactions have promoted the formation of higher molecular weight *hb*-PArPs from the monomer pairs of aliphatic monoynes and aromatic diynes.

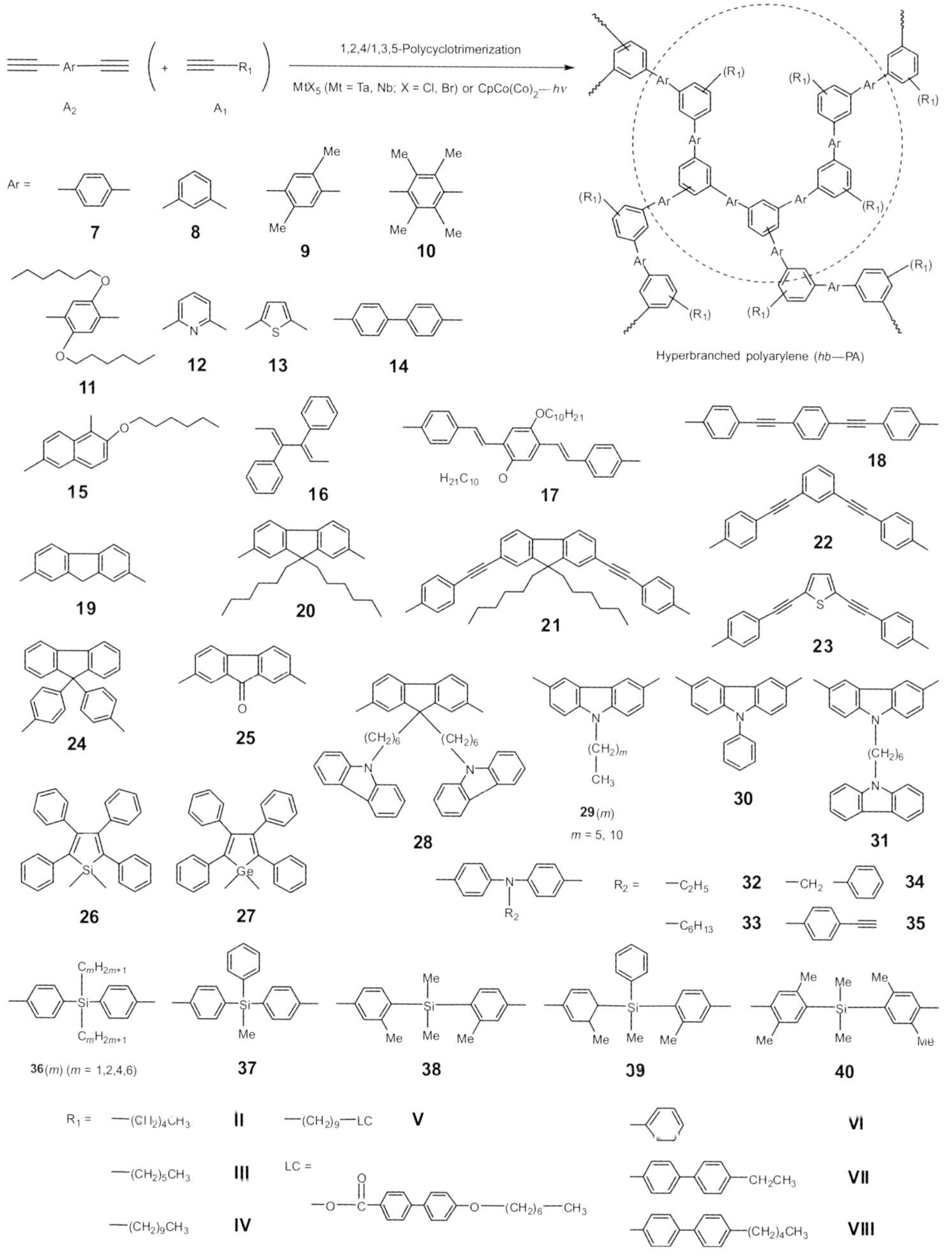

1,2,4/1,3,5-Polycyclotrimerization

MtX_5 (Mt = Ta, Nb; X = Cl, Br) or $CpCo(CO)_2$—$h\nu$

A_2 A_1

Hyperbranched polyarylene (*hb*—PA)

Ar = 7 8 9 10 11 12 13 14 15 16 17 18 19 20 21 22 23 24 25 28 29(*m*) *m* = 5, 10 30 31 26 27

R_2 = $—C_2H_5$ 32 $—CH_2$— 34 $—C_6H_{13}$ 33 35

36(*m*) (*m* = 1,2,4,6) 37 38 39 40

R_1 = $—(CH_2)_4CH_3$ II $—(CH_2)_9$—LC V VI

$—(CH_2)_5CH_3$ III LC = $—O—C(=O)—C_6H_4—C_6H_4—O—(CH_2)_6—CH_3$ $—C_6H_4—C_6H_4—CH_2CH_3$ VII

$—(CH_2)_9CH_3$ IV $—C_6H_4—C_6H_4—(CH_2)_4CH_3$ VIII

Scheme 4.7 (Co)polycyclotrimerizations of aromatic diynes (with monoynes).

4.2.2.2 Structures

Structural analyses of the homopolymers by spectroscopic methods confirm that the triple bonds of the arylenediynes have undergone [2 + 2 + 2] polycyclotrimerizations to form new benzene rings. The ratio of the 1,2,4- to 1,3,5-isomers of the trisubstituted benzene rings is estimated to be typically ~2.2: 1. Careful inspection of the ^{1}H NMR spectra shows that the number of the terminal acetylene triple bonds in the final *hb*-PArPs is much smaller than that in an "ideal" hyperbranched structure produced by the diyne polycyclotrimerization. This suggests that intrasphere ring formation has been involved in the polycyclotrimerization. Unfortunately, however, the signals from the benzene rings newly formed by the intrasphere cyclotrimerizations are indistinguishable from each other and also indistinct from those of the benzene rings formed by the "normal" polycyclotrimerization reactions in the ^{1}H NMR spectra, making it difficult to determine experimentally the probabilities of the intracyclotrimerization reactions.

To solve this problem, computational simulation was used [47]. Models of the polymers were built, and the probabilities of the growth modes were estimated according to the minimized energy of the structures. An example of the outputs of the computer simulations is shown in Figure 4.4. The overall two-dimensional structure of *hb*-P**36**(1) looks like a star-shaped polymer containing many small cyclic units. The total number of the triple bonds left in the hyperbranched structure and the total number of aromatic protons are in good agreement with the numbers estimated from the ^{1}H NMR spectral data. The computer simulation model is

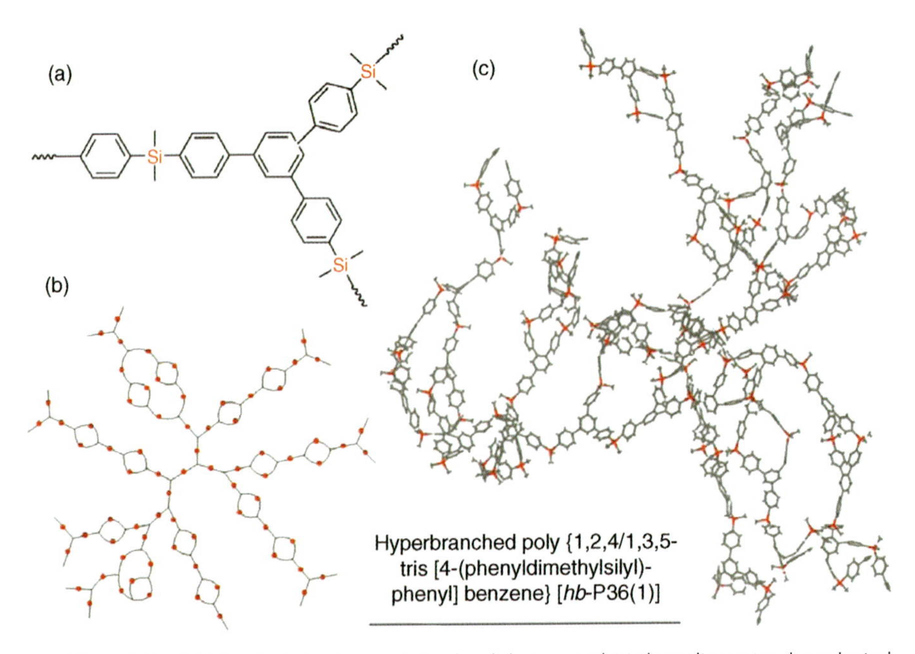

Figure 4.4 (a) Chemical structure and simulated (b) two- and (c) three-dimensional topological structures of *hb*-P**36**(1).

therefore consistent with the structure of the real polymer. Although the intracyclotrimerization occupies more than 1/3 (or 36%) of the probabilities of the propagation modes, most of the cycles are small in size, formed by only two monomer repeat units, mainly due to the close proximity of the 1 and 2 positions of the newly formed 1,2,4-trisubstituted benzenes. These small rings are strung together to form a willow twig-like structure, well accounting for the excellent solubility of the polymer.

Similarly, spectral characterizations of the copolymers substantiate their hyperbranched structures. Estimations of the ratios of the diynes to monoynes incorporated into the copolymer structures reveal that the monoynes function as growth-controlling agents, impeding the intracyclotrimerization reactions [44, 48, 49]. As an optimal ratio of diyne to monoyne, 1: 1.5 has often been found to work well for most diynes yielding completely soluble, high molecular weight *hb*-PArPs. The copolycyclotrimerization of aromatic triynes such as **35** is inherently much more difficult to control. For the aliphatic monoynes like 1-octyne (**III**), a monoyne/triyne ratio of 3 : 1 is needed to obtain a soluble *hb*-PArP. A larger monoyne to diyne ratio (4 : 1) is required for the aromatic monoynes such as phenylacetylene (**VI**) [48].

Even when such a large excess of growth-controlling agent of monoyne is used, the resultant *hb*-PArPs still contain internal cyclic structures [48]. According to the ratio of monoyne to triyne units found in *hb*-P(**35**/**III**), different propagation modes for the intrasphere cyclotrimerization reactions are proposed (Scheme 4.8). Taking the 1,2,4- and 1,3,5-isomeric structures of the trisubstituted benzenes into account, formation of "small"- and "medium"-sized cycles is highly possible. Similar to the silyldiyne homopolycyclotrimerization, the close intramolecular proximity of the two triple bonds originating from an ortho connection allows growth termination by the triple bond of 1-octyne, furnishing a second 1,2,4-trisubstituted benzene ring (Scheme 4.8a). This intrasphere "cross-linking" leads to the formation of a small cyclic structure, yet the polymer is still soluble due to the overall pseudo-"linear" propagation mode.

Another possible pathway is the ring closure of three triyne monomer units connected in a meta fashion with a monoyne via 1,2,4- or 1,3,5-cyclization. Such a reaction is likely to produce a medium-sized ring, serving as a core for the macrodentritic propagation with three growing arms (Scheme 4.8b). Para-substituted polymer branches resulting from 1,2,4-cyclotrimerization inherently cannot form any ring structures and may only be involved in the formation of oval-shaped "macrocycles". Such cyclic substructures are possibly formed via combined ortho- and meta-linkages (Scheme 4.8c), which has little effect on the solubility of the *hb*-PArPs, as again the overall structure is still propagating in a pseudo-"linear" mode.

4.2.2.3 **Properties**

The *hb*-PArPs are comprised of aromatic rings and are thus expected to be thermally stable. This is indeed the case, as proved by the TGA analyses. For example, *hb*-P**36**, an organosilicon hybrid polymer, loses merely 5% of its weight at a temperature as high as 595 °C (Figure 4.5). Other polymers show similar performance. Most polymers graphitize in >50 wt% yields upon pyrolysis at 800 °C, with *hb*-P(**14**/**II**)

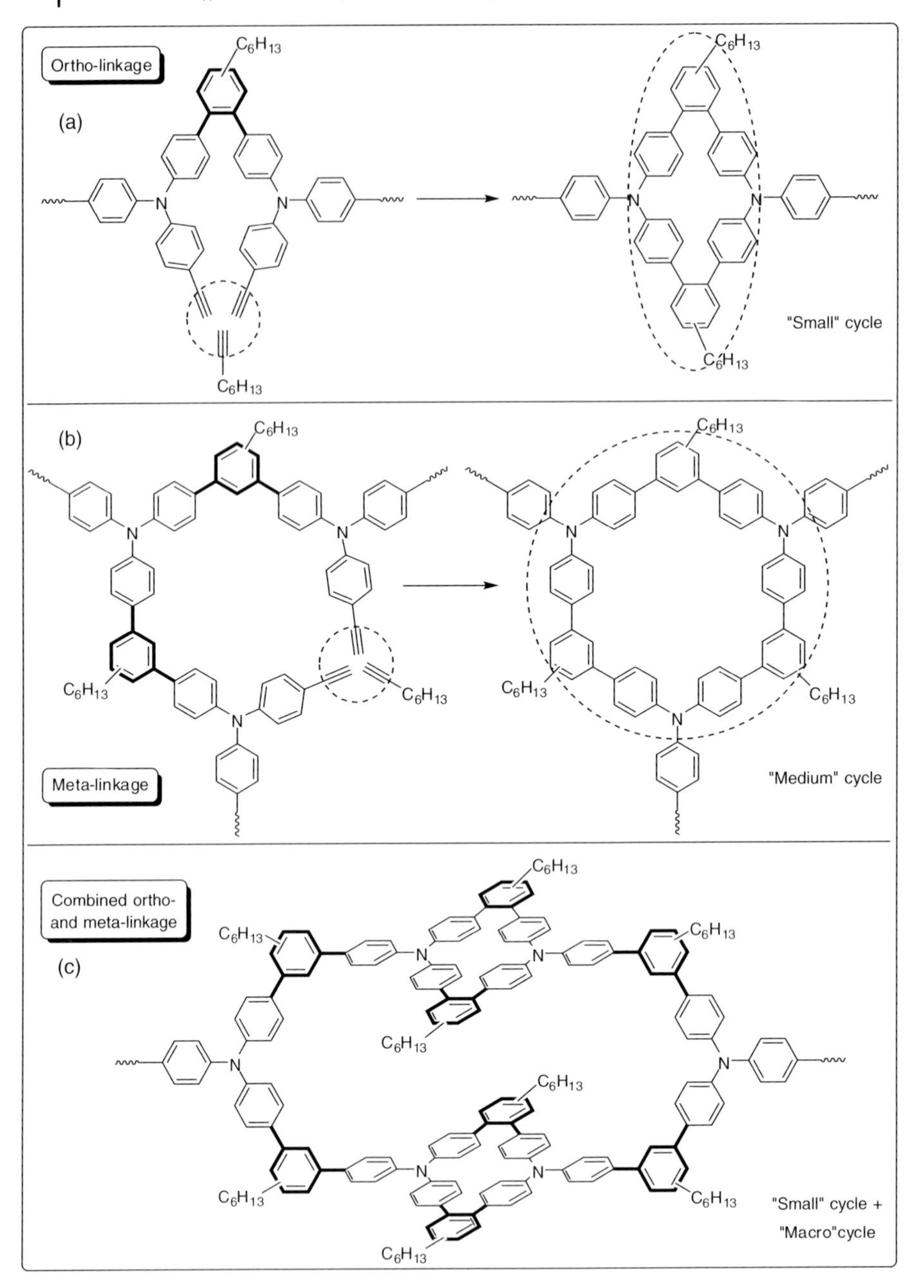

Scheme 4.8 Formation of intramolecular cyclic structures in triyne/monoyne copolycyclotrimerization.

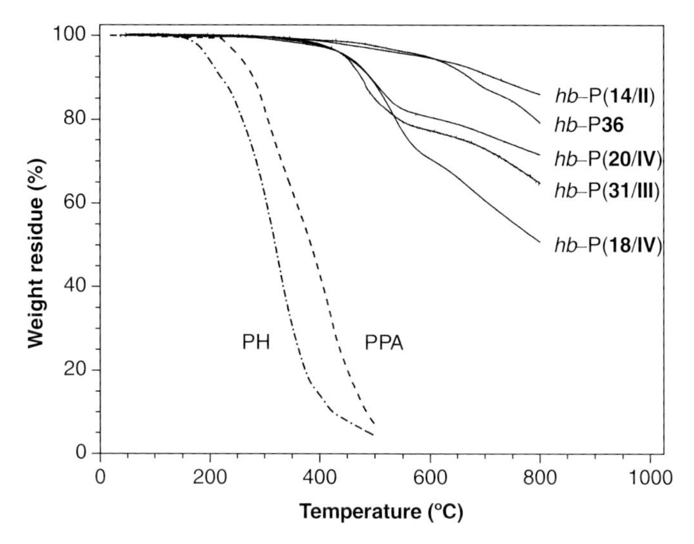

Figure 4.5 TGA thermograms of *hb*-PArPs; data for linear polyacetylenes of PPA and PH are shown for comparison.

carbonizing in a yield as high as 86 wt%. The thermal stabilities of the *hb*-PArPs are similar to that of poly(*p*-phenylene) (PPP; $T_d \sim 550\,^\circ$C) but much better than those of polyacetylenes such as PPA and PH ($T_d < 220\,^\circ$C). The difference in the thermal stability is mainly due to the difference in the molecular structure: PPP is made of thermally stable aromatic rings, whereas PPA and PH are comprised of labile polyacetylene chains. The outstanding thermal stabilities of the *hb*-PArPs thus verify their polyarylene structure.

The aromatic architecture of the *hb*-PArPs imparts not only outstanding thermal stabilities but also unique optical properties. Thanks to their π-conjugated structure, the polymers emit deep-blue to blue–green lights, whose intensities are higher than that of poly(1-phenyl-1-octyne), a highly luminescent polyacetylene [18]. Polymers *hb*-P(**7**/**II**), *hb*-P(**14**/**I**), *hb*-P(**17**/**II**), *hb*-P(**19**/**II**), *hb*-P(**19**/**III**) and *hb*-P(**28**/**II**) all give very high fluorescence quantum yields ($\Phi_F > 70\%$), with *hb*-P(**19**/**III**) showing a Φ_F value of almost unity (98%).

During our search for efficient light-emitting materials, we have discovered a group of molecules called silole **26** and germole **27**, which are nonemissive in solution but become strongly emissive when aggregated in poor solvents or fabricated into solid films. We coined a new term of "aggregation-induced emission" (AIE) for this phenomenon [50–52]. Through experimental and theoretical studies [53], the AIE mechanism is understood as follows. In dilute solution at room temperature, the active intramolecular rotations of the peripheral phenyl rings of silole around the axes of the single bonds linked to the central silole core nonradiatively annihilate the excitons, thereby making the silole molecules nonemissive. In the aggregates, the propeller shape of the silole molecules prevents them from forming excimeric species but the physical constraint in the solid state restricts their intramolecular rotations, blocking the nonradiative relaxation channels and populating the radiative decay, thus making the silole molecules luminescent.

We prepared *hb*-PArPs from the silole-containing diyne monomers in anticipation of getting highly emissive polymers in the solid state. The diyne 1,1-dihexyl-2,5-bis(4-ethynylphenyl)-3,4-diphenylsilole is AIE-active and thus nonemissive in solution, while its polymer (*hb*-PDHTPSP) is somewhat luminescent (Figure 4.6). The silole units in the hyperbranched polymer are knitted together by the rigid benzene rings and located within a stiffened polymer sphere, which restricts the intramolecular rotations of the peripheral benzene rings around the central silole core to some extent and thus makes the polymer somewhat luminescent in the solution state. The emission of the polymer is enhanced by aggregate formation: the intensity of its photoluminescence is progressively increased with gradual addition of water, a poor solvent, into its dilute THF solution. We called this effect "aggregation-induced emission enhancement (AIEE)". The AIEE system is thus operating with a mechanism similar to that in the AIE system discussed above.

The *hb*-PArPs are nonlinear optically active and strongly attenuate the optical power of harsh laser pulses. As shown in Figure 4.7, in the low energy region, the fluence transmitted from a solution of *hb*-P(**18**/**IV**) increases linearly with increase in the incident fluence. The transmitted fluence, however, starts to deviate from linearity at an incident fluence of $\sim$250 mJ cm^{-2} and reaches a saturation plateau of $\sim$140 mJ cm^{-2}. The optical limiting performance of *hb*-P(**18**/**IV**) is superior to that of C_{60}, a well-known optical limiter [54, 55]. Taking into account that C_{60} is a three-dimensionally π-conjugated buckyball, it may be concluded that the three-dimensionally π-conjugated electronic structure of the *hb*-PArPs is responsible for their optical limiting properties. Compared to *hb*-P(**18**/**IV**), *hb*-P(**21**/**IV**) is a better optical limiter (with a saturation plateau as low as $\sim$50 mJ cm^{-2}) but P(**19**/**III**) is a poorer one [33]. Clearly, the optical limiting power of the *hb*-PArP is sensitive to a change in its molecular structure, thus offering the opportunity to tune its performance through molecular engineering.

4.2.3
Hyperbranched Poly(Aroylphenylene)s (*hb*-PAkPs) and Poly(Aroxycarbonylphenylene)s (*hb*-PAePs)

The *hb*-PAPs and *hb*-PArPs discussed in Chapters 2 and 3 are both synthesized from transition metal-catalyzed alkyne polycyclotrimerizations. Due to the coordination nature of the metal-catalyzed polycyclotrimerization reactions, the resultant polymers possess regiorandom structures consisting of 1,3,5- and 1,2,4-trisubstituted benzene isomers. The structural irregularity may not be disadvantageous for some practical applications and is actually beneficial to the enhancement of solubility and hence processability of the polymers. It, however, makes it a challenging job to completely characterize the molecular structures of the polymers by spectroscopic methods. Furthermore, the Ta- and Nb-based catalysts are highly air- and moisture-sensitive and intolerant of polar functional groups. The residues of the metallic catalysts, which are difficult to completely remove from the polymerization products,

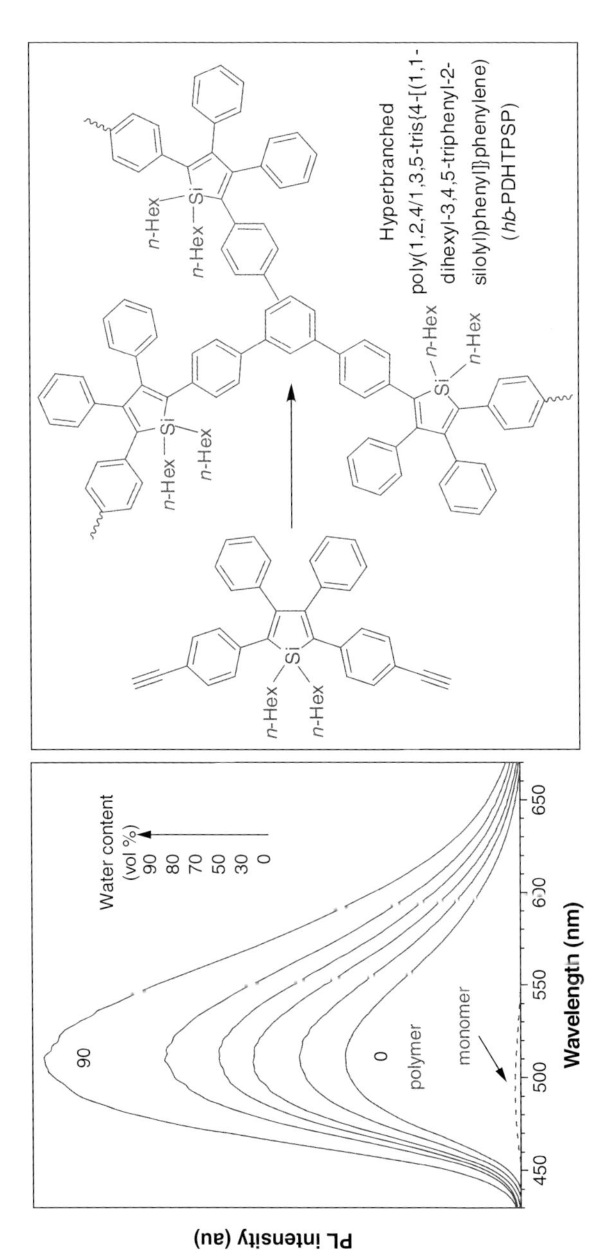

Figure 4.6 Photoluminescence spectra of *hb*-PDHTPSP in THF/water mixtures with different water contents; polymer concentration: 10 μM.

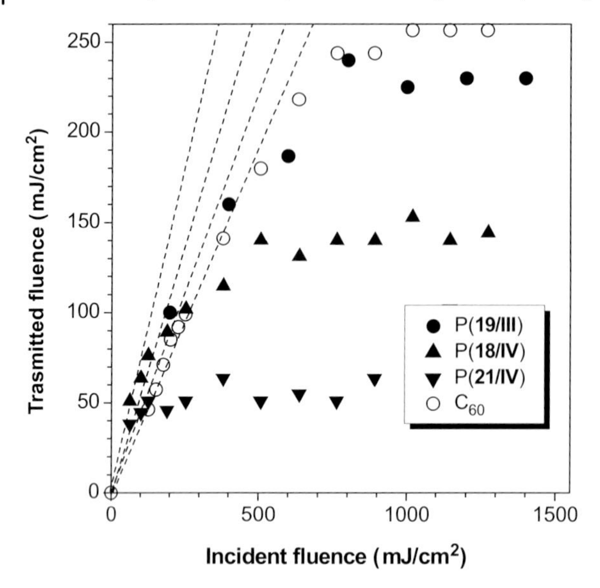

Figure 4.7 Optical limiting responses to 8 ns, 532 nm optical pulses, of DCM solutions (0.86 mg mL^{-1}) of *hb*-PArPs. Data for a toluene solution of C_{60} (0.16 mg mL^{-1}) is shown for comparison.

are detrimental to the properties, especially the optical and photonic activities, of the polymers. It would be nice if an acetylene polycyclotrimerization could proceed in a regioselective manner in the absence of metallic catalysts. We thus explored the possibility of developing metal-free, regioselective acetylene polycyclotrimerization reactions.

4.2.3.1 **Synthesis**

We noted that benzoylacetylene could be cyclotrimerized to form 1,3,5-tris(phenylcarbonyl)benzene in the presence of diethylamine or when refluxed in *N,N*-dimethylformamide (DMF) [56]. We tried to develop this reaction into a new polycyclotrimerization technique for the synthesis of 1,3,5-regioregular hyperbranched polymers from functionalized acetylenes. We prepared a series of bis(aroylacetylene)s with various linkers (R; Scheme 4.9) and studied their polymerization behavior. None of the catalysts, however, worked well for the polycyclotrimerizations of the bis(aroylacetylene)s: the reactions in the presence of diethylamine gave polymers in very low yields (down to 7%), while those carried out in refluxing DMF/tetralin mixtures were sluggish, taking as long as 3 days to get some satisfactory results.

We thus searched for new catalysts and eventually found that piperidine was a good organocatalyst. The reactions catalyzed by piperidine were completed in shorter times and produced polymers in much higher yields: for example, the polycyclotrimerizations of **41**(6) and **43**(4) initiated by piperidine produced their corresponding polymers, that is, *hb*-P**41**(6) and *hb*-P**43**(4), in virtually quantitative yields in 24 h [57]. Spectroscopic analyses prove that the aroyldiynes are regioselectively

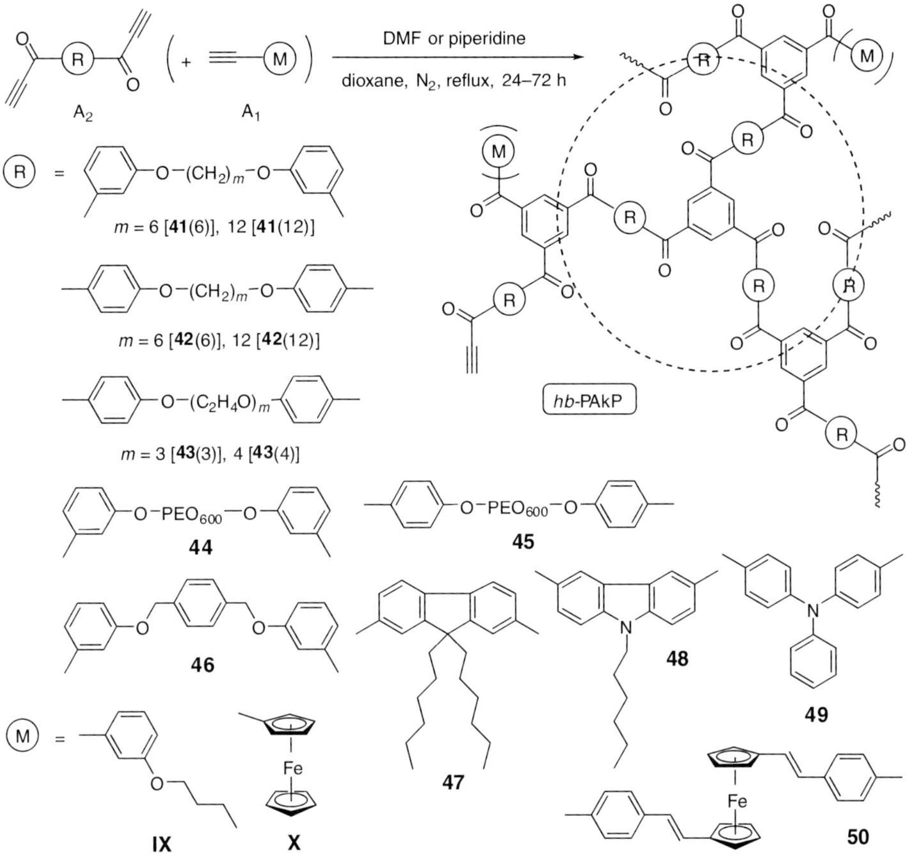

Scheme 4.9 Metal-free, 1,3,5-regioselective polycyclotrimerization of bis(aroylacetylene)s.

polycyclotrimerized by piperidine into *hb*-PAkPs with 1,3,5-conformation. This regioselectivity stems from an ionic, instead of a coordinative, mechanism of the base-catalyzed polycyclotrimerization where piperidine reacts with an aroylethynyl group in a Michael addition mode via the formation of ketoenamines [57].

The bis(aroylacetylene)s are, however, difficult to prepare, taking many reaction steps. If the carbonyl linkage between the triple bond and aromatic ring in the bis (aroylacetylene) can be replaced by an ester group, it will make the monomer synthesis much easier. Propiolic acid (or acetylenecarboxylic acid) is a commercially available compound and can be readily esterified with diols to form bipropiolates. We designed and prepared a group of arylene bipropiolates by the esterification of propiolic acid with arylene diols in the presence of *N,N′*-dicyclohexylcarbodiimide (DCC), 4-dimethylaminopyridine (DMAP) and *p*-toluenesulfonic acid (TsOH), examples of which are shown in Scheme 4.10.

To examine whether the arylene bipropiolates could be polycyclotrimerized, we conducted a model reaction using a monoyne of phenyl propiolate as model

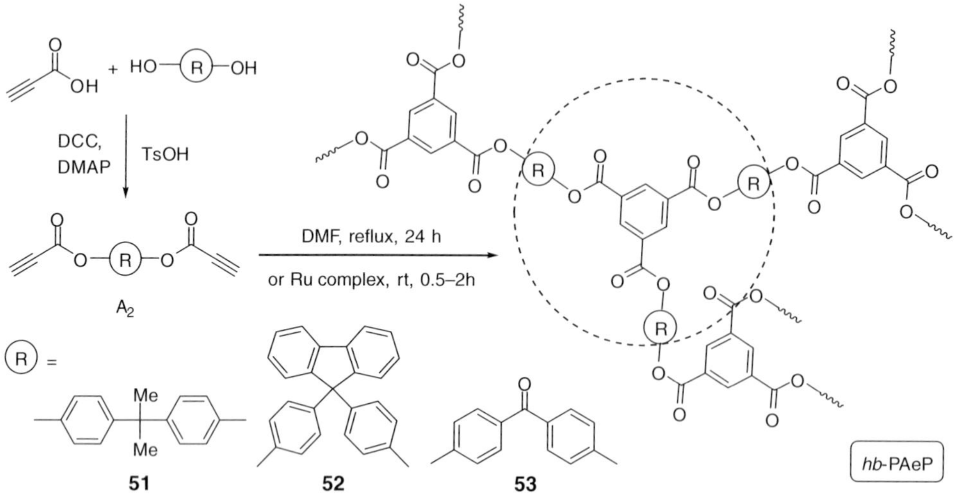

Scheme 4.10 1,3,5-Regioselective polycyclotrimerization of arylene bipropiolate monomers catalyzed by nonmetallic or metallic catalysts.

compound and found that it readily underwent 1,3,5-cyclotrimerization in refluxing DMF. Based on the result of the model reaction, we carried out polycyclotrimerization reactions of the arylene bipropiolates in refluxing DMF and succeeded in transformation of the monomers into their corresponding polymers *hb*-PAePs in a 1,3,5-regioselective fashion [58]. The polymerization can even be conducted in air: for example, heating **52** in refluxing DMF in air for 24 h gives an *hb*-P(**52**) with a molecular weight of 13 200 in 82% yield. We also found that the bipropiolate monomers could be readily polycyclotrimerized by organoruthenium complexes such as $Cp^*Ru(PPh_3)_2Cl$, $CpRu(PPh_3)_2Cl$, $Ru(OAc)_2(PPh_3)_2$ and $[RuCl_2(allyl)_2]_2$. The polycyclotrimerization of **51** catalyzed by $Cp^*Ru(PPh_3)_2Cl$ in THF at room temperature, for example, produced an *hb*-P**51** with a molecular weight of 12 100 in 78% yield in a reaction time as short as 2 h.

4.2.3.2 Structures

Hyperbranched structures of the *hb*-PAkPs and *hb*-PAePs are readily confirmed spectroscopically. Thanks to the 1,3,5-regioregularity of the polymer structure endowed by the regioselectivity of the diyne polycyclotrimerization, the NMR spectra of the polymers are much simpler than those of their *hb*-PArP congeners. The presence of three basic structures of an *hb*-PAkP, viz., *D*, *L* and *T* units, is readily verified by the numbers of unreacted triple bonds in the repeat branches (Scheme 4.11).

Due to the structural flexibility of the monomers, especially **41**–**45**, the polymer branches can also be terminated by an additional end-capping reaction of dimerization, in which two triple bonds of the same diyne molecule react with a ketoenamine to form a new benzene ring. The end group resulting from this back-biting reaction is called “a cyclophane terminal” (T_c). From their corresponding

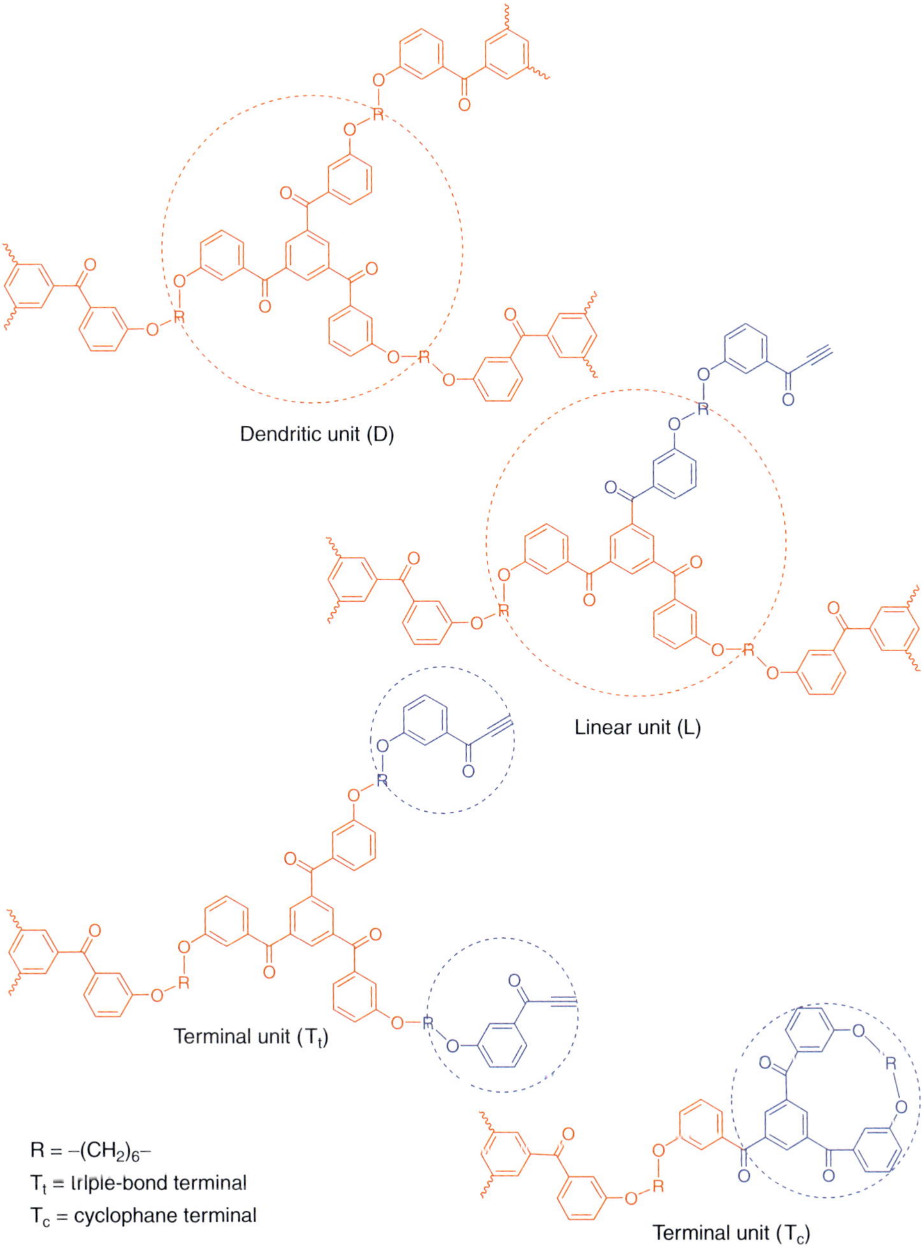

Dendritic unit (D)

Linear unit (L)

Terminal unit (T_t)

Terminal unit (T_c)

R = $-(CH_2)_6-$
T_t = triple-bond terminal
T_c = cyclophane terminal

Scheme 4.11 Structures of D, L and T units in *hb*-P**41**(6).

resonance peaks in the ^1H NMR spectra, the degrees of branching of the polymers (DBs) are estimated to be 78–100%. These DB values are much higher than those of the "conventional" hyperbranched polymers, which are commonly in the neighborhood of 50% [59].

4.2.3.3 Properties

The great advantage of these systems is that there are no metallic residues left behind after polymerization because the catalysts and solvents can be removed from the polymer products with ease. Taking advantage of this feature, we generated three-dimensional polymer nanostructures by carrying out the diyne polymerizations in the presence of an anodic aluminum oxide (AAO) template. The nanostructured polymers were freed by breakage or dissolution of the AAO template in aqueous sodium hydroxide solutions. Examples of the scanning electron microscopy (SEM) images of such a polymer, *hb*-P**41**(6), are shown in Figure 4.8. The *hb*-PArP adopts well the shapes and sizes of the pores of the template and forms micrometer-long polymer nanotubes, as can be clearly seen from the image given in Figure 4.8d.

Due to its high photosusceptibility, benzophenone has been introduced into biological and synthetic polymers to serve as a photo-cross-linking agent, and its photoreactions in various polymer matrixes such as PS, PMMA and PC have been well documented [60]. The *hb*-PAkPs contain numerous aroyl-benzene units and are thus expected to show high photo-cross-linking efficiencies. Indeed, when a thin film of *hb*-P**46** on a glass plate is exposed to a hand-held UV lamp at room temperature, it is readily cross-linked. The cross-linking reaction may proceed via a well-established radical mechanism [60]: the carbonyl groups abstract hydrogen atoms from the

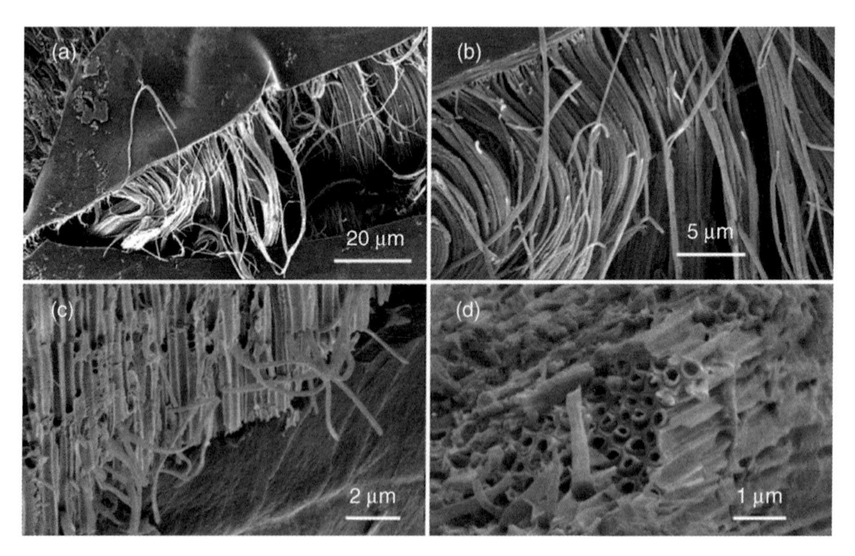

Figure 4.8 SEM micrographs of the nanotubes of *hb*-P**41**(6) prepared inside an AAO template with a pore size of ~250 nm.

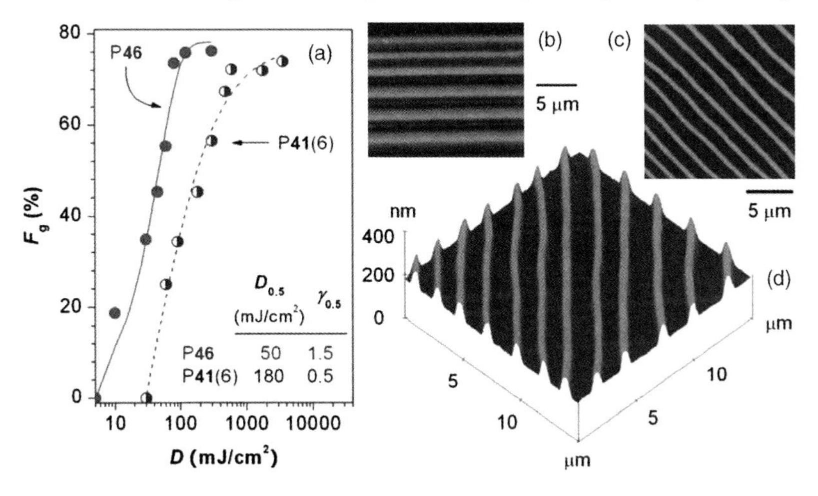

Figure 4.9 (a) Plots of gel fractions (F_g) of *hb*-P**41** (6) and *hb*-P**46** films versus exposure doses (*D*). (b–d) Atomic force microscopy images of micro- and nano-scale patterns obtained from the thin films of *hb*-P**46** exposed to 1 J cm^{-2} of UV irradiation (λ = 365 nm).

benzyl units to generate benzyl radicals, whose coupling or combination leads to cross-linking and hence gel formation [57].

Figure 4.9a shows the effect of radiation dose on the gel formation of *hb*-PAkP films after they have been exposed to a weak UV light with a power of ~1 mW cm^{-2}. Although the photoreaction conditions have not been optimized, the polymers already exhibit much higher sensitivities ($D_{0.5}$ = 43–180 mJ cm^{-2}) than those of commercial poly(amic ester)-based photoresists ($D_{0.5}$ = 650–700 mJ cm^{-2}) [61]. Well-resolved patterns with line widths of ~1.0 μm are readily formed when a film of *hb*-P**46** is exposed to a UV dose of 1 J cm^{-2}. Patterns with submicron resolution (line width down to 500 nm) are also achieved, as demonstrated by the example given in Figure 4.9d. Evidently, *hb*-P**46** is an excellent photoresist material.

Our previous studies have shown that hyperbranched organometallic polymers containing ferrocenyl units can be transformed into nanostructured magnetic ceramics in high yields [62–64]. The ferrocene-containing *hb*-PAkPs synthesized in this study are thermally stable and photonically cross-linkable. We envision that the *hb*-PAkPs may be used to create magnetic patterns. We developed a two-step process for the pattern generation. In the first step, microstructured patterns are created by developing a thin film of *hb*-P**46**X in 1,2-dichloroethane, after the polymer film has been exposed to UV irradiation through a copper negative mask for 30 min (Figure 4.10a). In the second step, the organometallic patterns are transformed into magnetic ceramics by pyrolyzing the microgrids at 1000 °C for 1 h under nitrogen. As can be seen from Figure 4.10b, the negative-tone photoresist has been successfully transformed into well-defined and fine-patterned iron-containing ceramics with excellent shape retention. The patterned ceramic has been found to exhibit unique soft ferromagnetism.

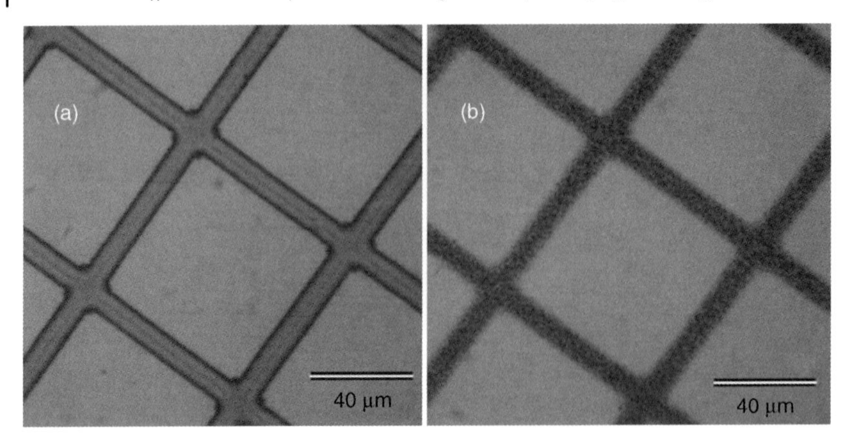

Figure 4.10 Photos of (a) the micropattern fabricated by the photolysis of ferrocene-containing *hb*-P**46**X and (b) the magnetic pattern generated by the pyrolysis of the micropattern under nitrogen at 1000 °C.

4.2.4 Hyperbranched Polytriazoles (hb-PTAs)

Acetylenes and azides can undergo 1,3-dipolar cycloaddition to form triazoles. This reaction has recently been promoted by Sharpless *et al.* as "click chemistry" [65]. Researchers have tried to prepare *hb*-PTAs by "click polymerization" of AB_2-type monomers, where A and B denote azide and acetylene groups, respectively [66]. The azidodiyne monomers carrying mutually reactive acetylene and azide groups are, however, difficult to prepare and handle and can self-polymerize into regiorandom *hb*-PTAs at room temperature. Attempts to control the polymer structure by using Cu(I) as catalyst yielded 1,4-regioregular *hb*-PTAs but the polymers were insoluble. Clearly, new systems need to be developed for the synthesis of *hb*-PTAs with macroscopic processability.

4.2.4.1 Synthesis

We rationalize that one possible way to gain process control and to prevent self-polymerization from occurring is to put the acetylene and azide functional groups into different monomers. This will make the preparation and purification of the monomers easier and meanwhile help extend their shelf-lives. A series of diazides and triynes were thus prepared and then tested as monomers for click polymerizations (Scheme 4.12). Thermally induced 1,3-dipolar polycycloadditions of the diazides with the triynes yielded regiorandom but soluble *hb*-PTAs [67]. The thermal click polymerizations are, however, slow, taking 3 days to produce appreciable amounts of polymeric products, although they are faster than the self-polymerizations of the AB_2 monomers [66] and the linear 1,3-dipolar polycycloadditions of the $A_2 + B_2$ monomers [68–70], which take more than 1 week, sometimes as long as 10 days, to produce some satisfactory amounts of polymers.

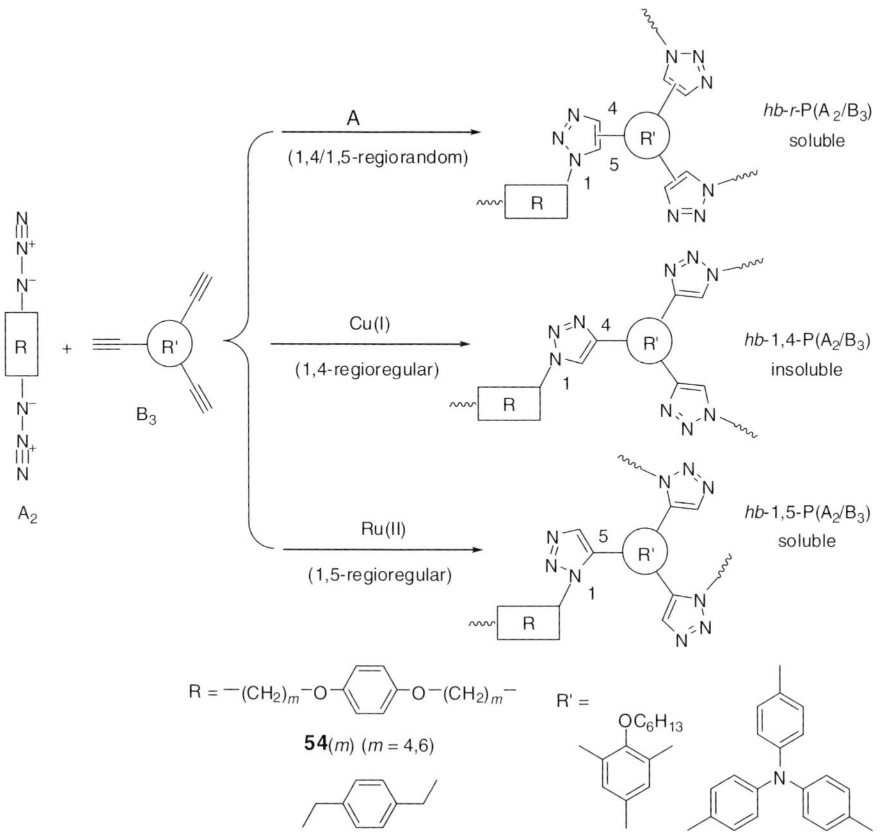

Scheme 4.12 Syntheses of *hb*-PTAs by click polymerizations.

Similar to the AB_2-type monomers discussed above, the $A_2 + B_3$ monomers can also be polymerized by the copper(I)-catalyzed click reaction into 1,4-regioregular polymers but the resultant *hb*-1,4-PTAs are again insoluble in common organic solvents. Isomerically pure yet completely soluble *hb*-PTAs with a sole 1,5-regiostructure are successfully synthesized when $Cp^*RuCl(PPh_3)_2$ is used as catalyst. The ruthenium(II)-catalyzed click polymerization proceeds very rapidly, giving an *hb*-1,5-PTA with a molecular weight of 9400 in 75% yield in as short as 30 min.

4.2.4.2 **Structures**

One possible reason for the insolubility of the *hb*-1,4-PTAs may be due to the *in situ* complexation of the copper(I) catalyst with the triazole rings formed in the click polymerization, which has cross-linked the *hb*-1,4-PTA spheres [71]. This process is unlikely to occur in the Ru(II)-catalyzed system. Another possible cause for the difference in the polymer solubility is the difference in the architectural structures of the 1,4- and 1,5-disubstituted triazole rings. As can be seen from the

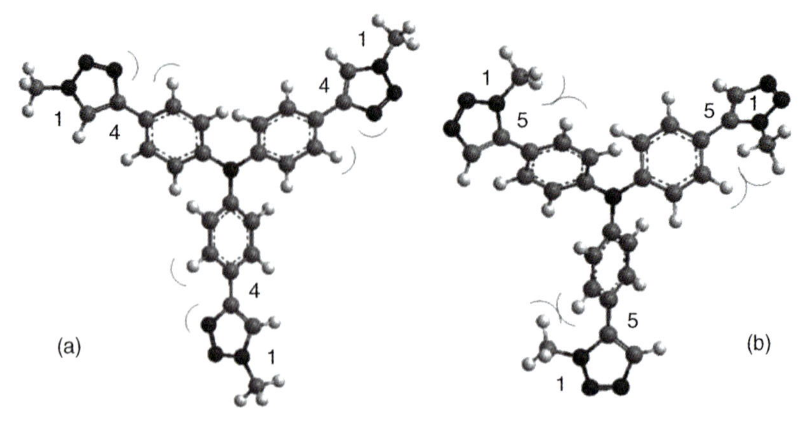

Figure 4.11 Simulated conformations of model compounds with (a) 1,4- and (b) 1,5-disubstitution patterns: (a) tris[4-(1-methyl-1,2,3-triazol-4-yl)phenyl]amine (1,4-isomer) and (b) tris[4-(1-methyl-1,2,3-triazol-5-yl)phenyl]amine (1,5-isomer).

computer-simulated conformations of the model compounds given in Figure 4.11, the 1,4-triazole ring experiences little steric interaction with the neighboring phenyl ring and can thus take a relatively planar confirmation. On the other hand, the steric repulsion between the 1,5-triazole and phenyl rings is strong, which forces the 1,5-isomer to adopt a more twisted molecular conformation. The structural planarity of the 1,4-isomer may facilitate π–π stacking of the aromatic units, thus making the polymer difficult to dissolve.

4.2.4.3 **Properties**

The *hb*-PTAs are light-emitting, film-forming, and photonically curable. Utilizing these properties, we tried to use the polymers to create fluorescent images. The negative-tone pattern generated by the photolysis of the regiorandom *hb*-*r*-PTA containing triphenylamine core, that is, *hb*-*r*-P[**54**(4)/**35**], emits a white light, whereas the pattern generated by its regioregular congener, *hb*-1,5-P[**54**(4)/**35**], gives blue fluorescence (Figure 4.12). The polymers are synthesized from the same comonomer

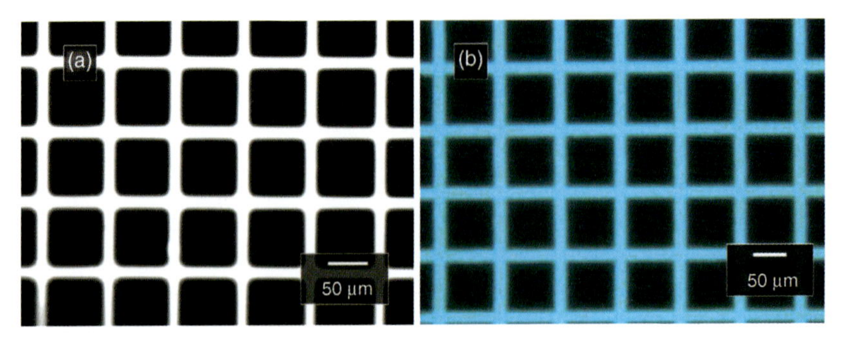

Figure 4.12 Negative patterns generated by photolyses of (a) *hb*-*r*-P[**54**(4)/**35**] and (b) *hb*-1,5-P[**54**(4)/**35**]; photos taken under a fluorescence microscope.

pairs of **54**(4) and **35** and are different only in the regioisomeric structures of their repeat branches. It is truly remarkable that this seemingly small change in the molecular structure has brought about such a large change in the light-emitting behavior of the polymers. Evidently, the structural alteration in the disubstitution has resulted in changes in the effective conjugation lengths and the packing modes of the polymers, which offers a new means for fine tuning photonic and electro-optical properties of the hyperbranched polymers in the technologically useful solid state.

4.2.5 Hyperbranched Poly(Aryleneethynylene)s (*hb*-PAEs)

Repetitive coupling of acetylenes with aryl halides is another effective way to build hyperbranched architecture. This type of polycoupling is often catalyzed by palladium complexes in the presence of amines and has been widely used for the preparation of linear polymers and perfect dendrimers [72, 73]. Employing this reaction protocol, *hb*-PAEs have been prepared from AB_2-type monomers [74, 75]. Putting mutually reactive halide and acetylene groups into a single monomer is, however, a nontrivial job due to the involved synthetic problems such as undesired self-oligomerization. Purification of the desired monomers is often troublesome and ends with low isolation yields. Separation of the halide and acetylene functional groups into different monomers allows easier synthetic access and offers a greater variety of monomer choices but meanwhile invites the risk of cross-linking reactions [76]. Control of the polycoupling conditions is thus a necessity if one intends to synthesize processable *hb*-PAEs with desired structures and properties.

4.2.5.1 Synthesis

We worked on the optimization of polycoupling conditions by varying such parameters as reaction time, monomer concentration, and addition mode of comonomers, in an effort to control the polymer structures and to obtain soluble polymeric products. Under the optimized conditions, soluble *hb*-PAEs containing luminophores such as anthracene- and fluorene-functionalized *hb*-P(**57**/**58**) are obtained from the cross-coupling polymerization of diyne **57** with triiodide **58** (Scheme 4.13) [77]. Similarly, soluble *hb*-PAEs functionalized by azo-cored push–pull nonlinear optical (NLO) chromophores were obtained from the palladium-catalyzed polycoupling of diethynylazobenzene **59** with triiodoarenes **60** and **61** [78].

4.2.5.2 Properties

Second-order NLO materials have attracted much attention due to their photonic applications. A major effort in the area is to efficiently translate a large molecular first hyperpolarizability value into high second harmonic generation (SHG) coefficient (d_{33}). The greatest obstacle has been the chromophoric aggregation in the thin films. The NLO dyes are usually highly polarized by the push–pull interactions. During the film formation, the chromophores with large dipole moments tend to pack compactly owing to the strong intermolecular electrostatic interactions, leading to diminishing or cancellation of the NLO effects in the solid state [79]. Hyperbranched polymers

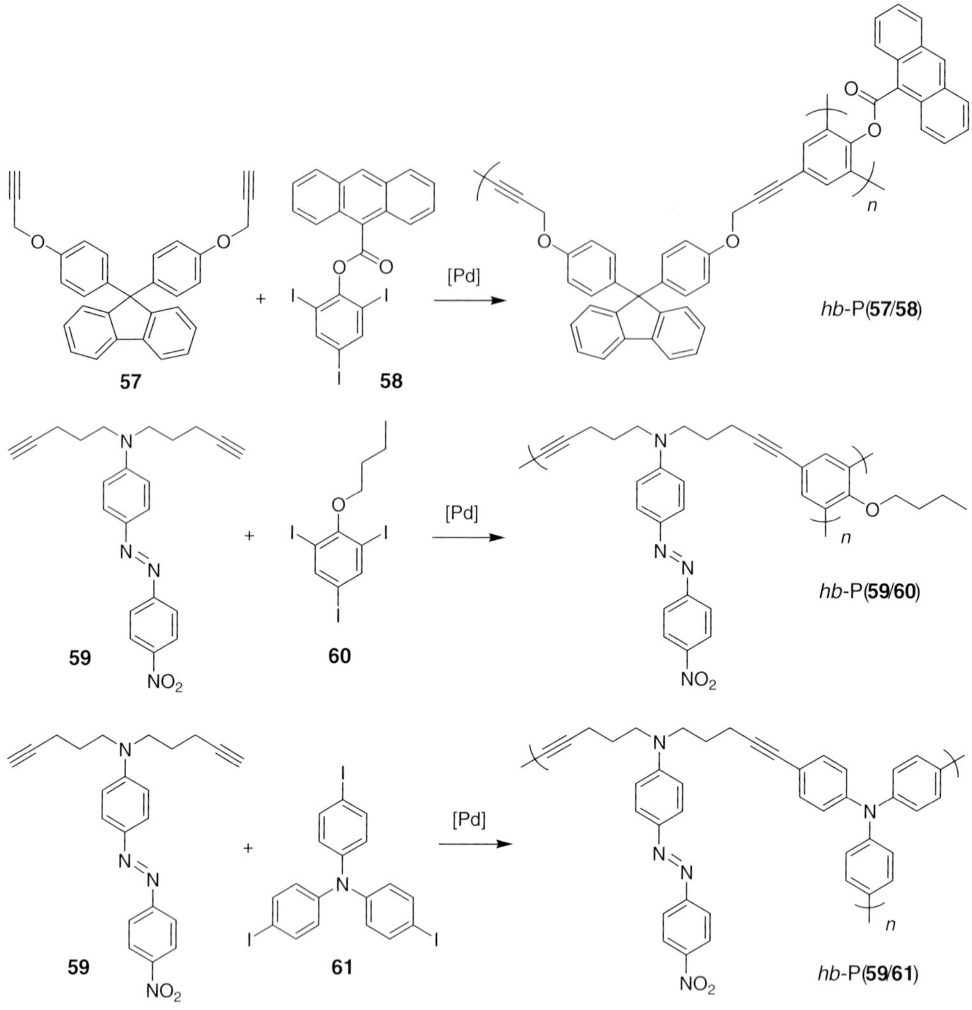

Scheme 4.13 Syntheses of functional *hb*-PAEs by cross-coupling polymerization.

should be ideal matrix materials as they offer three-dimensional spatial separation of the NLO chromophores in the spherical architecture, and their void-rich topological structure should help minimize optical loss in the NLO process.

The *hb*-PAEs containing the NLO-active azo dyes are film-forming and morphologically stable ($T_g > 180\,°C$). Their poled thin films exhibit high SHG coefficients (d_{33} up to $177\,pm\,V^{-1}$), thanks to the chromophore-separation and site-isolation effects of the hyperbranched structures of the polymers in the three-dimensional

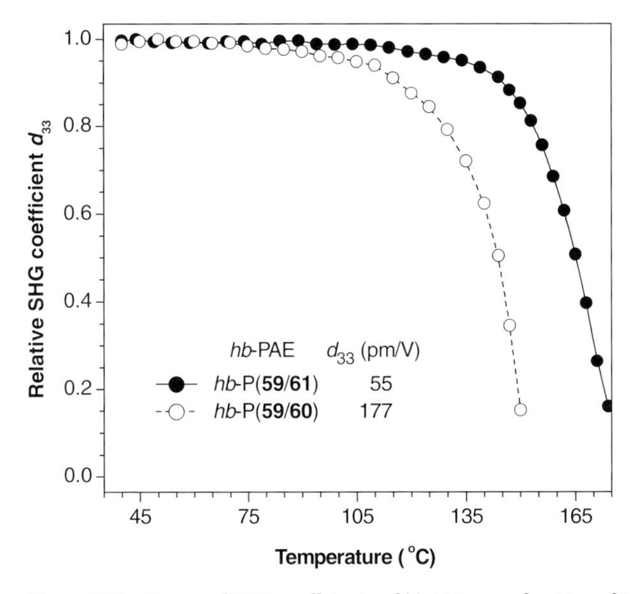

Figure 4.13 Decays of SHG coefficients of *hb*-PAEs as a function of temperature.

space (Figure 4.13) [78]. The optical nonlinearity of the poled film of *hb*-P(**59**/**61**) is thermally stable with no drop in d_{33} observable when heated to a temperature as high as 152 °C, due to the facile cross-linking of the multiple triple bonds in the *hb*-PAE at a moderate temperature of 88 °C.

4.2.6 Hyperbranched Polydiynes (*hb*-PDYs)

The *hb*-PAEs discussed above possess only isolated triple bonds. We are interested in the synthesis of *hb*-PDYs containing two triple bonds in one repeat unit, because such polymers may exhibit novel properties associated with the diyne functionality. The rich reactivity of the diyne group, for example, may endow the *hb*-PDYs with photosusceptibility, thermal curability, and metal-coordinating capability. To synthesize the *hb*-PDYs, we took an A_3 homo-coupling approach: we knitted triyne monomers by using an oxidative polycoupling methodology [80].

4.2.6.1 Synthesis

Carbon-rich *hb*-PDYs containing functional groups such as ether, amine and phosphorus oxide are readily synthesized from homo-polycoupling of their corresponding aromatic triyne monomers (Scheme 4.14) [81]. To prevent network formation by uncontrolled cross-linking, the polymerization reactions are terminated by pouring the reaction mixtures into acidified methanol before they reach the gel points. Another way to ensure solubility is to copolymerize the triynes with appropriate amounts of monoynes such as **XI**.

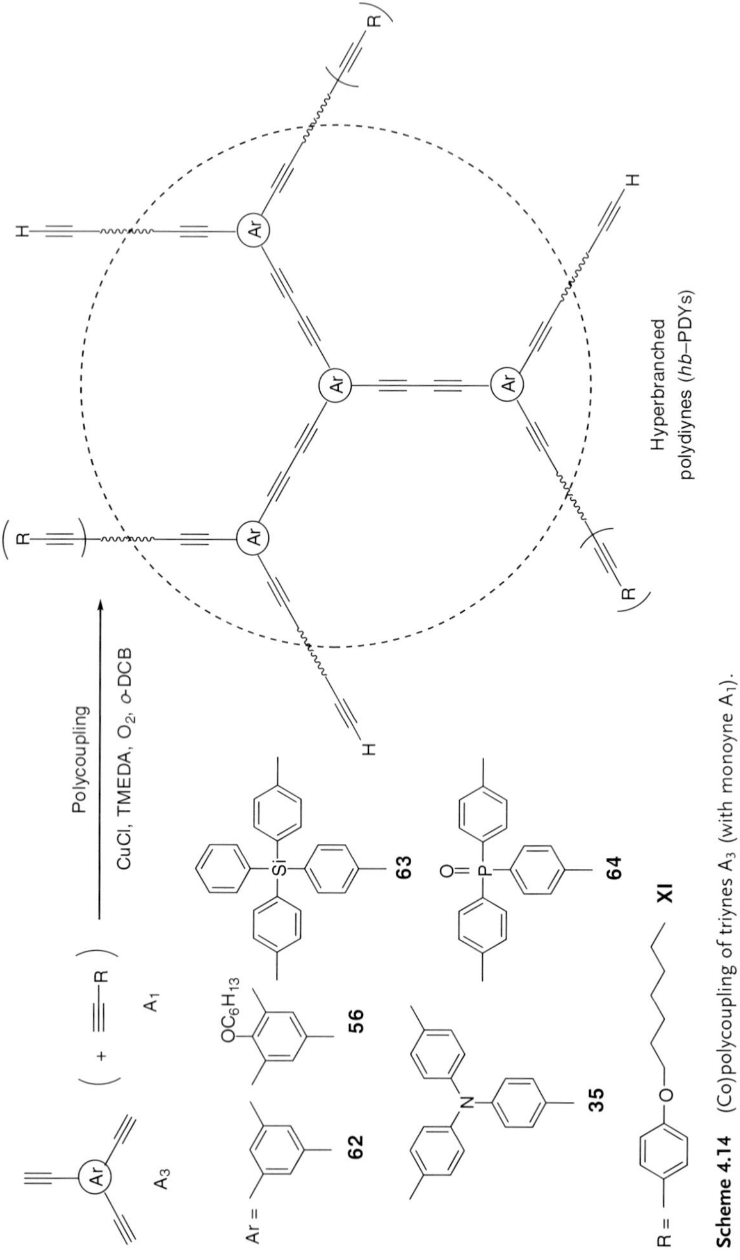

Scheme 4.14 (Co)polycoupling of triynes A_3 (with monoyne A_1).

Spectroscopic analyses reveal that both the homo- and copolymers contain terminal monoyne triple bonds, which enable the peripheries of the *hb*-PDYs to be decorated by end-capping reactions. This is demonstrated by the coupling of *hb*-P**35** with aryl iodides **XII** and **XIII** (Scheme 4.15). The polymer end-capped by phenyl groups (*hb*-P**35**-**XII**) becomes only partially insoluble after purification, possibly due to the π–π stacking-induced supramolecular aggregation during precipitation and drying of the product of the polymer reaction. Product *hb*-P**35**-**XIII** is, however, completely soluble, thanks to the long *n*-dodecyloxy group of the end-capping agent. No signal of resonance of the terminal acetylene proton is observed in the ^{1}H NMR spectrum, confirming the completion of the end-capping reaction.

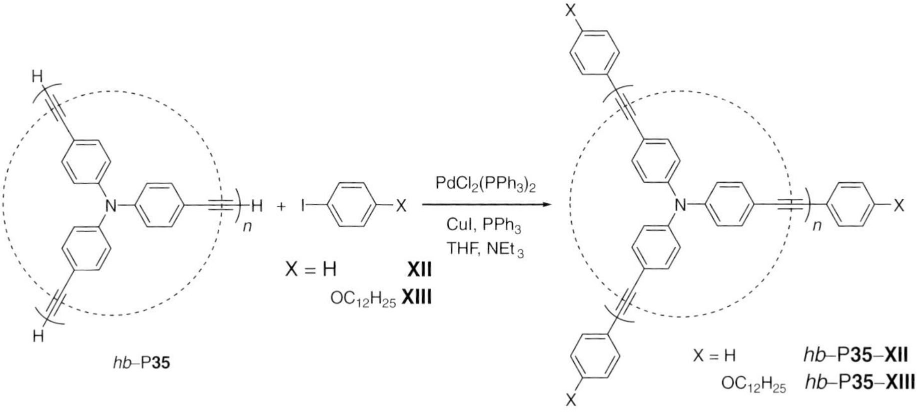

Scheme 4.15 Peripheral end-capping through terminal acetylene coupling with aryl iodide.

4.2.6.2 **Thermal Curing**

Diyne molecules readily oligomerize upon heating [16]. Many prepolymers carrying monoyne end groups have been transformed into thermoset networks [82]. The *hb*-PDYs contain diyne groups in the cores and monoyne groups on the shells and are thus expected to be thermally curable. When *hb*-P**63**-**XI** is heated in a DSC cell, it starts to release heat at ~200 °C due to commencement of thermally induced cross-linking reactions (Figure 4.14). The second heating scan of the DSC analysis gives almost a flat line parallel to the abscissa in the same temperature region, suggesting that all the acetylene triple bonds have reacted during the first heating scan. The cross-linking reaction of *hb*-P**35** starts from ~150 °C and peaks at ~204 °C. The homopolymer begins to cure at lower temperatures in comparison to its copolymer congener, because the former carries more reactive terminal acetylene triple bonds. When the terminal acetylene groups of *hb*-P**35** are fully end-capped by phenyl groups, the resultant *hb*-P**35**-**XII** contains only internal acetylene groups, which need higher temperatures to initiate and complete the thermal curing reactions. This further manifests the effect of the acetylene reactivity on the thermal curability of the *hb*-PDYs.

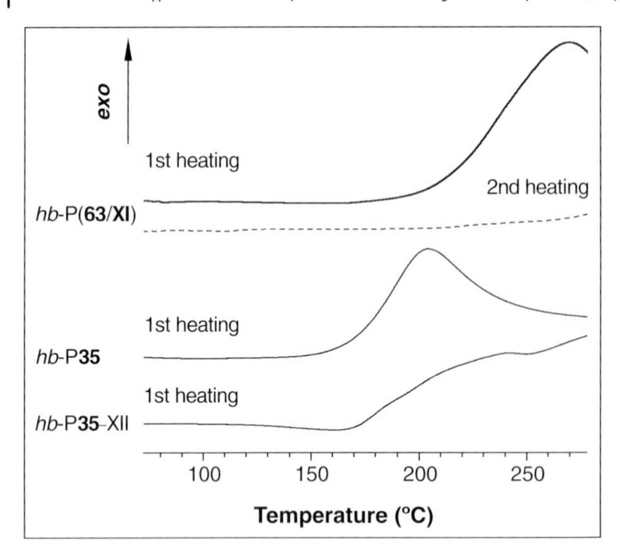

Figure 4.14 DSC thermograms of *hb*-PDYs measured at a scan rate of 10 °C min^{-1} under nitrogen.

4.2.6.3 **Micropattern Formation**

The diyne units can be induced to cross-link not only thermally but also photonically. A negative pattern is generated after development of a UV-irradiated thin film of *hb*-P**35** through a copper negative mask (Figure 4.15). The *hb*-PDY synthesized from the triphenylamine-containing monomer emits strong green light when observed under a fluorescence microscope. This is remarkable, considering that many conjugated polymers show weak emissions in the solid state or when fabricated into thin films due to the nonradiative energy transfer caused by π–π stacking of the polymer chains or defect formation [83]. Since conventional photoresists, such as SU-8, are generally nonluminescent and can thus only be used as a passive material, the bright emission of the *hb*-PDY may allow it to find applications as an active matrix for fabrication of liquid-crystal displays, light-emitting diodes, and other photonic devices.

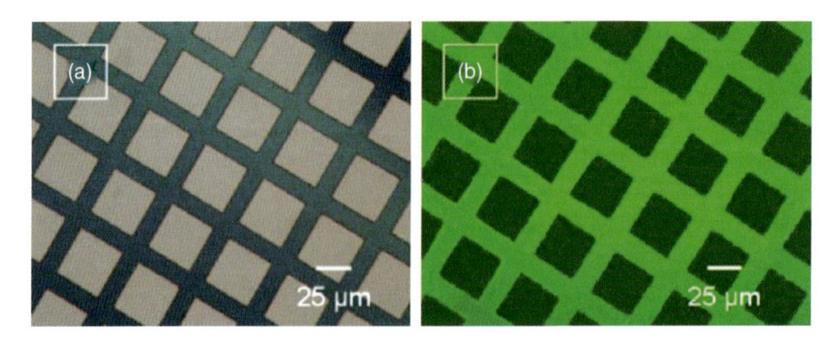

Figure 4.15 (a) Optical and (b) fluorescent micrographs of photopatterns generated by photolysis of *hb*-P**35**.

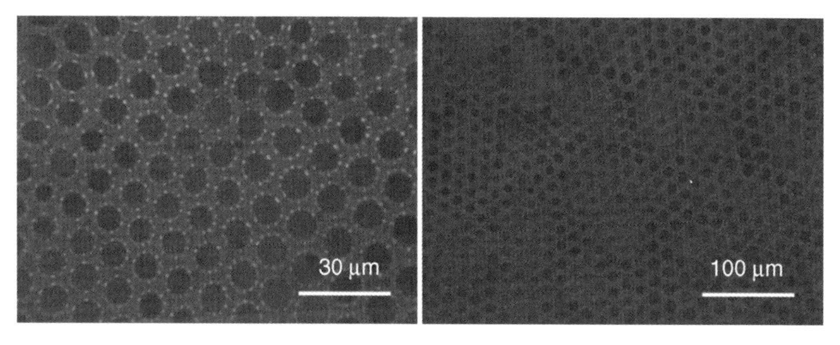

Figure 4.16 Optical micrographs of breath figures of *hb*-P**35** obtained from its CS_2 solutions by blow drying in a stream of moist air.

Template methods have been widely used to fabricate three-dimensional nano- and microstructured patterns. The breath figure process is a simple way to generate large arrays of patterned assemblies. It has been reported that star-shaped polymers and block copolymers form honeycomb morphologies through an evaporation-induced assembly process [84, 85]. Our and other research groups have recently shown that neither star-shaped nor block structure is necessarily needed for the formation of the well-defined assembling morphologies [86–89].

We tried to assemble structures of *hb*-PDYs by employing the breath figure process. Figure 4.16 shows the photographs of the patterned structures of *hb*-P**35** formed by blowing a stream of moist air over its CS_2 solutions. Hexagonally ordered hollow bubble arrays with an average void size of $\sim$10 μm are obtained over a large area. Similar to the patterns generated by the UV irradiation through a copper negative mask, the honeycomb patterns obtained from the breath figure process are light-emitting when observed under a fluorescence microscope and can thus potentially be used as the active layers in the fabrication of optical and photonic devices.

4.2.6.4 **Metal Complexation**

Acetylene molecules are versatile ligands in organometallic chemistry [16]. For example, acetylene–metal complexes can be formed through covalent interactions of one triple bond with $Co_2(CO)_8$ [90] and two triple bonds with $CpCo(CO)_2$ [91]. The *hb*-PDYs contain numerous acetylenic triple bonds and can thus be readily metallized by complexation with the cobalt carbonyls (Scheme 4.16). Upon admixing *hb*-P**35** and the cobalt carbonyls in THF at room temperature, the solution color changed from yellow to brown, accompanied by CO gas evolution. The mixtures remained homogenous toward the ends of the reactions and the reaction products were purified by pouring the THF solutions into hexane. The polymer complexes are stable in air, and the incorporation of the cobalt metal is verified by spectral analyses [92].

The as-prepared solution of complex *hb*-P**65** can be readily spin coated onto glass or silicon wafer to give a homogenous, yellowish brown-colored thin film. Photoirradiation of the film of the metallized polymer through a copper negative mask causes decomposition of the cobalt carbonyl complex. The film is readily photobleached,

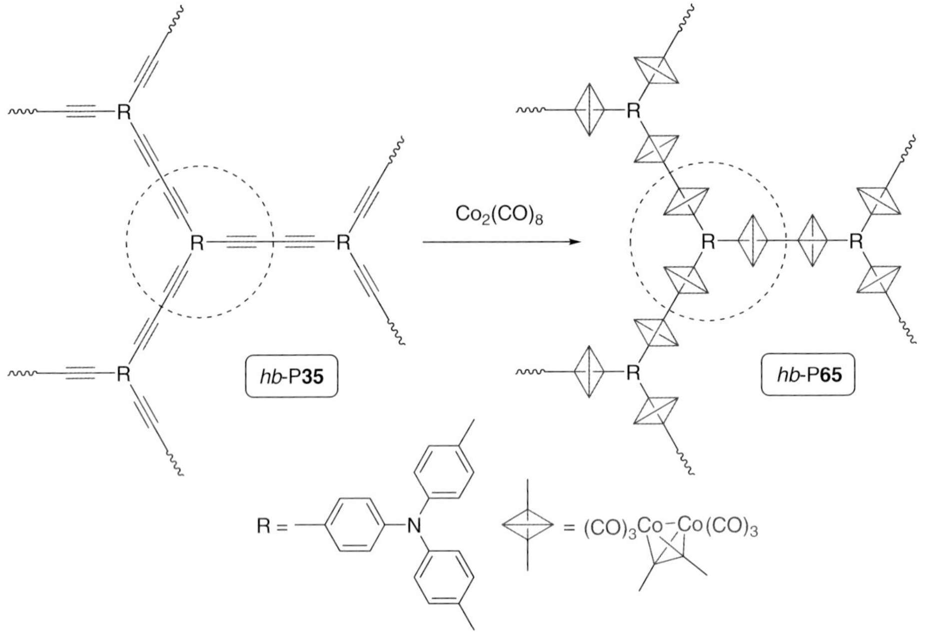

Scheme 4.16 Preparation of cobalt-containing polymer *hb*-P**65** via metallization of *hb*-P**35**.

generating a faithful copy of the two-dimensional pattern of the photomask without going through any developing processes (Figure 4.17). The magnified image given in Figure 4.17b clearly reveals the sharp edges of the photopattern [93, 94].

Inspired by the UV light-induced color change of *hb*-P**65**, we further studied its optical properties in more detail. Similar to its parent *hb*-P**35** [81], the metallized

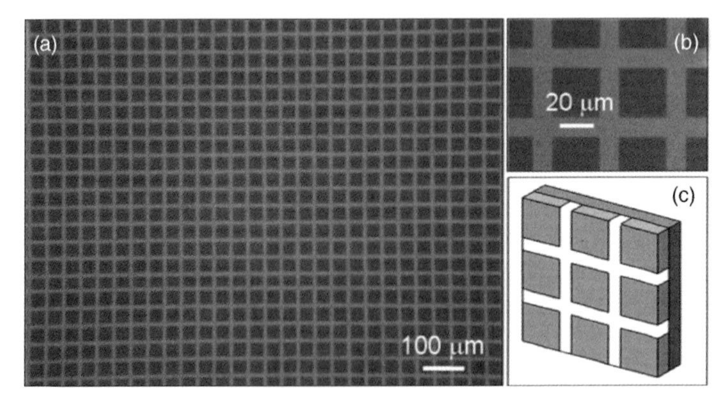

Figure 4.17 (a) Optical micrograph of a two-dimensional photonic pattern generated by UV irradiation of *hb*-P**65** through a Cu-negative mask, (b) image with a high magnification, and (c) model of the photo-pattern, with dark and white colors denoting unexposed and exposed areas, respectively.

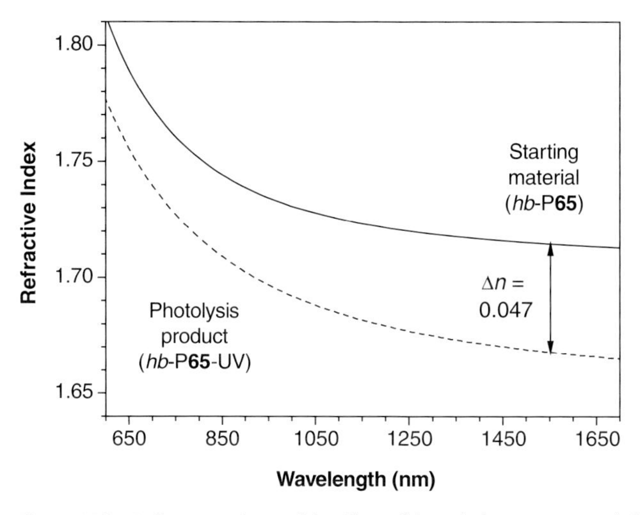

Figure 4.18 Refractive indexes of thin films of the cobalt-containing polydiyne before (*hb*-P**65**) and after photolysis (*hb*-P**65**-UV).

polymer exhibits very high refractive index (RI) values ($n = 1.813$–1.714) in the spectral region of 600–1600 nm (Figure 4.18) while maintaining a high revised Abbé number ($\upsilon_D{}' = 60$). Remarkably, the RIs drop significantly after UV irradiation ($\Delta n = 0.047$). With such a big RI change and a low optical dispersion, the polymer is promising for an array of photonic applications: it may, for example, serve as photorefractive material in holographic devices and work as a high RI optical coating.

Metallic species such as iron, nickel and cobalt can catalyze the growth of carbon nanotubes (CNTs) in a chemical vapor deposition (CVD) process. Because of the ready thermal curability of the *hb*-PDYs, spin-coated films of organometallic polymers *hb*-P**65** and *hb*-P**66** are envisioned to hamper the metallic nanoclusters from agglomerating in the CVD process and hence to provide nanoscopic catalyst seeds for the CNT growth. This proves to be the case. As can be seen from Figure 4.19, uniform bundles of multi-walled CNTs are grown by the CVD process at 700 °C with acetylene gas as the carbon source. The diameters and lengths of the CNTs are alterable by varying the surface activation and growth time [95]. Thanks to the hyperbranched carbon-rich backbone structure and the thus-achieved depression in the agglomeration of the metallic catalyst seeds, the diameters of the CNTs are small in size (~15 nm) and uniform in size distribution.

4.2.6.5 **Magnetic Ceramization**

It has become clear now that the carbon-rich *hb*-PDYs are readily curable (from ~150 °C), thermally stable (up to ~550 °C), and pyrolytically carbonizable (up to ~80 wt%). Furthermore, their triple bonds are easily metallizable by complexation with cobalt carbonyls. Since the polymer complexes contain a large number of metal atoms, we tried to utilize them as precursors for fabrication of magnetic ceramics. The pyrolyses of the polydiyne-cobalt complexes *hb*-P**65** and *hb*-P**66** at 1000 °C for 1 h

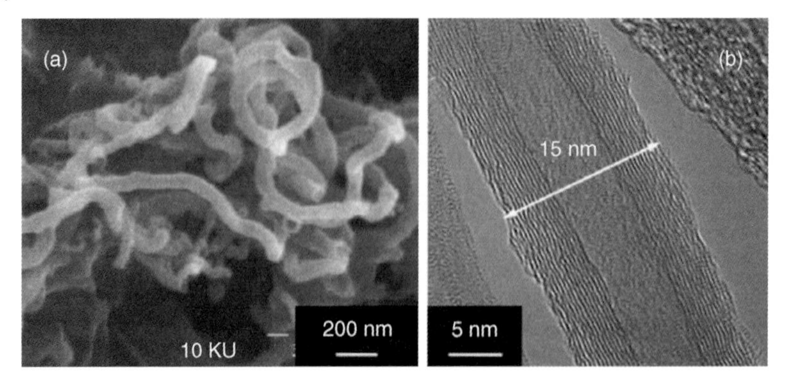

Figure 4.19 (a) SEM and (b) TEM microphotographs of the carbon nanotubes prepared by a CVD process at 700 °C on the silicon wafers spin-coated with thin films of *hb*-P**65**.

under nitrogen furnishes the ceramic products of C_xCo_y**67** and **68**, respectively, in 50–65% yields (Scheme 4.17). All the ceramics are magnetizable and can be readily attracted to a bar magnet.

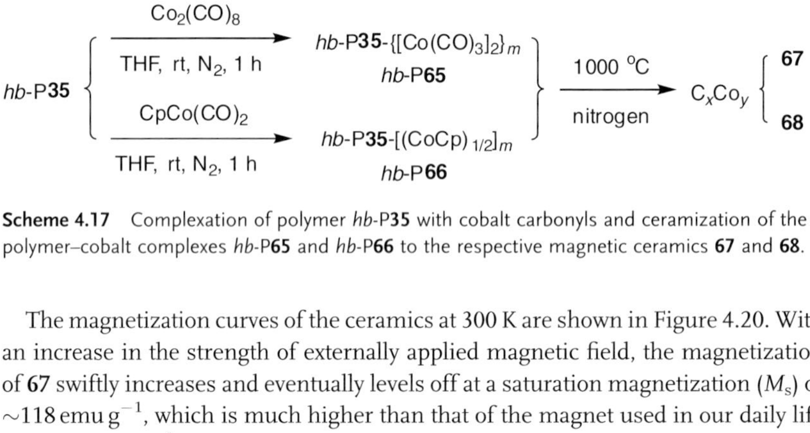

hb-P**35** —[$Co_2(CO)_8$; THF, rt, N_2, 1 h]→ *hb*-P**35**-{[Co(CO)$_3$]$_2$}$_m$ (*hb*-P**65**)

hb-P**35** —[$CpCo(CO)_2$; THF, rt, N_2, 1 h]→ *hb*-P**35**-[(CoCp)$_{1/2}$]$_m$ (*hb*-P**66**)

hb-P**65** / *hb*-P**66** —[1000 °C; nitrogen]→ C_xCo_y (**67** / **68**)

Scheme 4.17 Complexation of polymer *hb*-P**35** with cobalt carbonyls and ceramization of the polymer–cobalt complexes *hb*-P**65** and *hb*-P**66** to the respective magnetic ceramics **67** and **68**.

The magnetization curves of the ceramics at 300 K are shown in Figure 4.20. With an increase in the strength of externally applied magnetic field, the magnetization of **67** swiftly increases and eventually levels off at a saturation magnetization (M_s) of ~118 emu g^{-1}, which is much higher than that of the magnet used in our daily life ($M_s = 74$ emu g^{-1}) [96, 97]. The high M_s value of **67**, along with its powder XRD, XPS and SEM data [92], indicates that the cobalt nanocrystallites in the ceramic are well wrapped by carboneous species, which have prevented the cobalt particles from being oxidized during and after the pyrolysis and ceramization processes [97].

Clearly, *hb*-P**65** is an excellent precursor to magnetoceramics because its three-dimensional spherical cage structure has enabled the retention of the pyrolyzed species and the steady growth of the magnetic crystallites [63, 98]. The M_s value of **68** is lower (~26 emu g^{-1}), which is understandable, because the cobalt content of its precursor complex (*hb*-P**66**) is lower. The hysteresis loops of the magnetoceramics are very small. The coercivity (H_c) values of **67** and **68** are 0.058 and 0.142 kOe, respectively. The high magnetizability (M_s up to 118 emu g^{-1}) and low coercivity (H_c down to ~0.06 kOe) of **67** make it an outstanding soft ferromagnetic material [97].

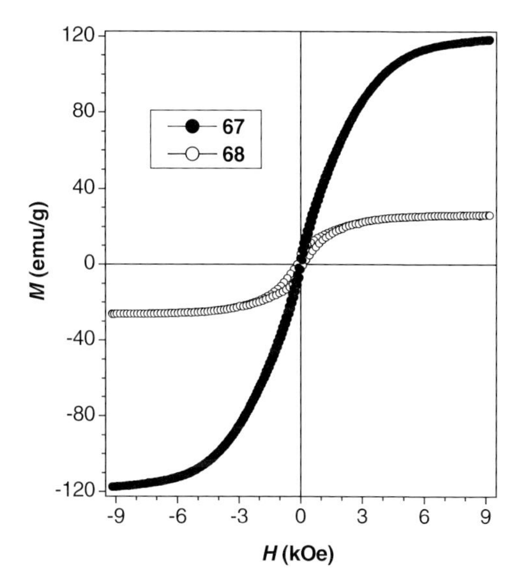

Figure 4.20 Plot of magnetization (*M*) vs. applied magnetic field (*H*) for magnetoceramics **67** and **68**.

4.3 Conclusions

In this chapter, we have briefly summarized our efforts and results on the preparation of new hyperbranched polymers, especially on the development of unimolecular A_n systems for the syntheses of a variety of functional polymers from triple-bonded acetylene monomers. Effective polymerization processes including polycyclotrimerization, polycycloaddition and polycoupling of diynes and triynes initiated by metallic and nonmetallic catalysts have been established, which have enabled the creation of hyperbranched polyphenylenes, polytriazoles and polydiynes with high molecular weights and excellent macroscopic processability in high yields. The new polymerization routes opened and the new structural insights gained in this study offer versatile synthetic tools and valuable design guidelines for further developments in this area of research.

Using the triple-bond building blocks in combination with functional groups has resulted in the creation of hyperbranched π-conjugated polymers with advanced materials properties. The carbon-rich polymers showed outstanding thermal stability. Incorporation of luminophoric units into the structures of the polymers enabled the modulation of their emission colors and efficiencies at the molecular level. The numerous aromatic rings in the *hb*-PArPs conferred strong optical limiting power on the polymers, while the poled thin films of the azo-functionalized *hb*-PAEs showed stable NLO performance with high SHG coefficients. Combination of the slim diyne linker with the polarizable triphenylamine core brought about excellent optical transparency and exceptionally high photorefractivity. The photonically susceptible benzyl, benzophenone and diyne units endowed the polymers with high photocurability and hence the potential to be used as sensitive photoresists and active matrixes in the construction of optical devices. The assemblies of hexagonal arrays of the breath figures and the micrometer-long polymer nanotubes were obtained from the dynamic and static templating processes. Complexations with cobalt carbonyls yielded spin-coatable hyperbranched organometallic polymers, whose RIs were readily tuned by UV irradiation to a very large extent. The hyperbranched polymer–cobalt complexes served as excellent precursors to soft ferromagnetic ceramics and as nanosized catalyst seeds for the fabrication of CNTs.

The simple synthesis and ready processability of the hyperbranched polymers, coupled with their unique molecular structures and useful functional properties, make them attractive and promising for an array of high-technology applications. Further studies on the syntheses, properties and applications of new acetylene-based hyperbranched polymers are under way in our laboratories.

Acknowledgments

The work described in this chapter was partially supported by the Research Grants Council of Hong Kong (602706, HKU2/05C, 603505, and 603304), the National Science Foundation of China (20634020), and the Ministry of Science and Technology of China (2002CB613401). We thank all the people involved in this project, some of whose names can be found in the references given below. B. Z. T. is grateful for support from the Cao Guangbiao Foundation of Zhejiang University.

References

1 Shirakawa, H. (2001) *Angew. Chem. Int. Ed.*, **40**, 2575.
2 MacDiarmid, A.G. (2001) *Angew. Chem. Int. Ed.*, **40**, 2581.
3 Heeger, A.J. (2001) *Angew. Chem. Int. Ed.*, **40**, 2591.
4 Masuda, T. and Higashimura, T. (1984) *Acc. Chem. Res.*, **17**, 51.
5 Masuda, T. and Higashimura, T. (1986) *Adv. Polym. Sci.*, **81**, 121.
6 Diederich, F. and Gobbi, L. (1999) *Top Curr. Chem.*, **201**, 43.
7 Watson, M.D., Fechtenkotter, A., and Mullen, K. (2001) *Chem. Rev.*, **101**, 1267.
8 Yashima, E. (2002) *Anal. Sci.*, **18**, 3.
9 Mayershofer, M.G. and Nuyken, O. (2005) *J. Polym. Sci., Part A: Polym. Chem.*, **43**, 5723.
10 Ray, C.R. and Moore, J.S. (2005) *Adv. Polym. Sci.*, **177**, 91.
11 Bunz, U.H.F. (2005) *Adv. Polym. Sci.*, **177**, 1.
12 Aoki, T., Kaneko, T., and Teraguchi, M. (2006) *Polymer*, **47**, 4867.
13 Masuda, T. (2007) *J. Polym. Sci., Part A: Polym. Chem.*, **45**, 165.
14 Akagi, K. (2007) *Bull. Chem. Soc. Jpn.*, **80**, 649.
15 Shirakawa, H., Masuda, T., and Takeda, T. (1994) in *The Chemistry of Triple-Bonded Functional Groups* (ed. S. Patai), John Wiley and Sons Ltd, Chichester. pp. 945–1016.
16 Stang, P.J. and Diederich, F. (eds) (1995) *Modern Acetylene Chemistry*, Wiley-VCH Verlag GmbH, Weinheim.
17 Patai, S. (ed.) (1978) *The Chemistry of the Carbon–Carbon Triple Bond*, John Wiley and Sons Inc., New York.
18 Lam, J.W.Y. and Tang, B.Z. (2005) *Acc. Chem. Res.*, **38**, 745.

19 Lam, J.W.Y. and Tang, B.Z. (2003) *J. Polym. Sci., Part A: Polym. Chem.*, **41**, 2607.
20 Kim, Y.H. (1998) *J. Polym. Sci., Part A: Polym. Chem.*, **36**, 1685.
21 Hult, A., Johansson, M., and Malmström, E. (1999) *Adv. Polym. Sci.*, **143**, 1.
22 Hawker, C.J. (1999) *Curr. Opin. Colloid Interface Sci.*, **4**, 117.
23 Inoue, K. (2000) *Prog. Polym. Sci.*, **25**, 453.
24 Voit, B.I. (2000) *J. Polym. Sci., Part A: Polym. Chem.*, **38**, 2505.
25 Grayson, S.M. and Frechet, J.M.J. (2001) *Chem. Rev.*, **101**, 3819.
26 Tomalia, D.T. and Frechet, J.M.J. (2002) *J. Polym. Sci., Part A: Polym. Chem.*, **40**, 2719.
27 Xu, K. and Tang, B.Z. (1999) *Chin. J. Polym. Sci.*, **17**, 397.
28 Lam, J.W.Y., Luo, J., Peng, H., Xie, Z., Xu, K., Dong, Y., Cheng, L., Qiu, C., Kwok, H.S., and Tang, B.Z. (2000) *Chin. J. Polym. Sci.*, **19**, 585.
29 Xu, K., Peng, H., Sun, Q., Dong, Y., Salhi, F., Luo, J., Chen, J., Huang, Y., Zhang, D., Xu, Z., and Tang, B.Z. (2002) *Macromolecules*, **35**, 5821.
30 Zheng, R.H., Dong, H.C., Peng, H., Lam, J.W.Y., and Tang, B.Z. (2004) *Macromolecules*, **37**, 5196.
31 Li, Z., Lam, J.W.Y., Dong, Y.Q., Dong, Y.P., Sung, H.H.Y., Williams, I.D., and Tang, B.Z. (2006) *Macromolecules*, **39**, 6458.
32 Grayson, S.M. and Frechet, J.M.J. (2001) *Macromolecules*, **34**, 6542.
33 Peng, H., Cheng, L., Luo, J.D., Xu, K.T., Sun, Q.H., Dong, Y.P., Salhi, F., Lee, P.S.P., Chen, J.W., and Tang, B.Z. (2002) *Macromolecules*, **35**, 5349.
34 Liu, J.Z., Zheng, R.H., Tang, Y.H., Haussler, M., Lam, J.W.Y., Qin, A., Ye, M.X., Hong, Y.N., Gao, P., and Tang, B.Z. (2007) *Macromolecules*, **40**, 7473.
35 Peng, H., Luo, J., Cheng, L., Lam, J.W.Y., Xu, K., Dong, Y., Zhang, D., Huang, Y., Xu, Z., and Tang, B.Z. (2002) *Opt. Mater.*, **21**, 315.
36 Xie, Z.L., Peng, H., Lam, J.W.Y., and Tang, B.Z. (2003) *Macromol. Symp.*, **195**, 179.
37 Lam, J.W.Y., Chen, J.W., Law, C.C.W., and Tang, B.Z. (2003) *Macromol. Symp.*, **196**, 289.
38 Häußler, M., Lam, J.W.Y., Zheng, R., Peng, H., Luo, J., Chen, J., Law, C.C.W., and Tang, B.Z. (2003) *C. R. Chim.*, **6**, 833.
39 Dong, H., Lam, J.W.Y., Häußler, M., Zheng, R., Peng, H., Law, C.C.W., and Tang, B.Z. (2004) *Curr. Trend. Polym. Sci.*, **9**, 15.
40 Chen, J., Peng, H., Law, C.C.W., Dong, Y.P., Lam, J.W.Y., Williams, I.D., and Tang, B.Z. (2003) *Macromolecules*, **36**, 4319.
41 Haussler, M., Qin, A., and Tang, B.Z. (2007) *Polymer*, **48**, 6181.
42 Häußler, M., Dong, H., Lam, J.W.Y., Zheng, R., Qin, A., and Tang, B.Z. (2005) *Chin. J. Polym. Sci.*, **23**, 567.
43 Häußler, M. and Tang, B.Z. (2007) *Adv. Polym. Sci.*, **209**, 1.
44 Häussler, M., Liu, J., Zheng, R., Lam, J.W.Y., Qin, A., and Tang, B.Z. (2007) *Macromolecules*, **40**, 1914.
45 Shi, J., Tong, B., Zhao, W., Shen, J., Zhi, J., Dong, Y.P., Häußler, M., Lam, J.W.Y., and Tang, B.Z. (2007) *Macromolecules*, **40**, 8195.
46 Kong, X., Lam, J.W.Y., and Tang, B.Z. (1999) *Macromolecules*, **32**, 1722.
47 Zheng, R., Dong, H., and Tang, B.Z. (2005) in *Macromolecules Containing Metal- and Metal-Like Elements*, vol. 4 (eds A. Abd-El-Azi, C. Carraher, C., Pittman, J. Sheats, and M. Zeldin), John Wiley and Sons Inc., Hoboken, p. 7.
48 Zheng, R., Häussler, M., Dong, H., Lam, J.W.Y., and Tang, B.Z. (2006) *Macromolecules*, **39**, 7973.
49 Häußler, M., Liu, J., Lam, J.W.Y., Qin, A., Zheng, R., and Tang, B.Z. (2007) *J. Polym. Sci., Part A: Polym. Chem.*, **45**, 4249.
50 Luo, J., Xie, Z., Lam, J.W.Y., Cheng, L., Chen, H., Qiu, C., Kwok, H.S., Zhan, X., Liu, Y., Zhu, D., and Tang, B.Z. (2001) *Chem. Commun.*, 1740.
51 Chen, J., Law, C.C.W., Lam, J.W.Y., Dong, Y.P., Lo, S.M.F., Williams, I.D., Zhu, D., and Tang, B.Z. (2003) *Chem. Mater.*, **15**, 1535.
52 Law, C.C.W., Chen, J., Lam, J.W.Y., Peng, H., and Tang, B.Z. (2004) *J. Inorg. Organomet. Polym.*, **14**, 39.
53 Chen, J., Xie, Z., Lam, J.W.Y., Law, C.C.W., and Tang, B.Z. (2003) *Macromolecules*, **36**, 1108.

54 Tutt, L.W. and Kost, A. (1992) *Nature*, **356**, 225.

55 Tang, B.Z., Leung, S.M., Peng, H., Yu, N.T., and Su, K.C. (1997) *Macromolecules*, **30**, 2848.

56 Balasubramanian, K., Selvaraj, S., and Venkataramani, P.S. (1980) *Synthesis*, 29.

57 Dong, H., Zheng, R., Lam, J.W.Y., Häussler, M., and Tang, B.Z. (2005) *Macromolecules*, **38**, 6382.

58 Tang, B.Z., Jim, C.K.W., Qin, A., Haeussler, M., and Lam, J.W.Y. (2007) US Patent, Appl. no. 60/933,884.

59 Jikei, M. and Kakimoto, M. (2001) *Prog. Polym. Sci.*, **26**, 1233.

60 Hasegawa, M. and Horie, K. (2001) *Prog. Polym. Sci.*, **26**, 259.

61 Kim, K.H., Jang, S., and Harris, F.W. (2001) *Macromolecules*, **34**, 8925.

62 Sun, Q., Xu, K., Peng, H., Zheng, R., Häussler, M., and Tang, B.Z. (2003) *Macromolecules*, **36**, 2309.

63 Sun, Q., Lam, J.W.Y., Xu, K., Xu, H., Cha, J.A.K., Zhang, X., Jing, X., Wang, F., and Tang, B.Z. (2000) *Chem. Mater.*, **12**, 2617.

64 Häußler, M., Sun, Q., Xu, K., Lam, J.W.Y., Dong, H., and Tang, B.Z. (2005) *J. Inorg. Organomet. Polym.*, **15**, 67.

65 Rostovtsev, V.V., Green, L.G., Fokin, V.V., and Sharpless, K.B. (2002) *Angew. Chem. Int. Ed.*, **41**, 2596.

66 Scheel, A.J., Komber, H., and Voit, B.I. (2004) *Macromol. Rapid. Commun.*, **25**, 1175.

67 Qin, A., Haeussler, M., Lam, J.W.Y., Tse, K.K.C., and Tang, B.Z. (2006) *Polym. Prepr.*, **47**, 681.

68 van Steenis, D.J.V.C., David, O.R.P., van Strijdonck, G.P.F., van Maarseveen, J.H., and Reek, J.N.H. (2005) *Chem. Commun.*, 4333.

69 Bakbak, S., Leech, P.J., Carson, B.E., Saxena, S., King, W.P., and Bunz, U.H.F. (2006) *Macromolecules*, **39**, 6793.

70 Qin, A., Jim, C.K.W., Lu, W., Lam, J.W.Y., Häuβler, M., Dong, Y., Sung, H.H.Y., Williams, I.D., Wong, G.K.L., and Tang, B.Z. (2007) *Macromolecules*, **40**, 2308.

71 Chan, T.R., Hilgraf, R., Sharpless, K.B., and Fokin, V.V. (2004) *Org. Lett.*, **6**, 2853.

72 Bunz, U.H.F. (2001) *Acc. Chem. Res.*, **34**, 998.

73 Negishi, E.I. and Anastasia, L. (2003) *Chem. Rev.*, **103**, 1979.

74 Bharathi, P. and Moore, J.S. (2000) *Macromolecules*, **33**, 3212.

75 Kim, C., Chang, Y., and Kim, J.S. (1996) *Macromolecules*, **29**, 6353.

76 Weder, C. (2005) *Chem. Commun.*, 5378.

77 Dong, Y.Q., Li, Z., Lam, J.W.Y., Dong, Y.P., Feng, X.D., and Tang, B.Z. (2005) *Chin. J. Polym. Sci.*, **23**, 665.

78 Li, Z., Qin, A., Lam, J.W.Y., Dong, Y.Q., Dong, Y.P., Ye, C., Williams, I.D., and Tang, B.Z. (2006) *Macromolecules*, **39**, 1436.

79 Burland, D.M., Miller, R.D., and Walsh, C.A. (1994) *Chem. Rev.*, **94**, 31.

80 Hay, A.S. (1998) *J. Polym. Sci., Part A: Polym. Chem.*, **36**, 505.

81 Häussler, M., Zheng, R.H., Lam, J.W.Y., Tong, H., Dong, H.C., and Tang, B.Z. (2004) *J. Phys. Chem. B*, **108**, 10645.

82 Hergenrother, P.M. (1990) in *Concise Encyclopedia of Polymer Science and Engineering* (ed. J.I. Kroschwitz), John Wiley and Sons Inc., New York, p. 5.

83 Hua, J.L., Lam, J.W.Y., Dong, H., Wu, L., Wong, K.S., and Tang, B.Z. (2006) *Polymer*, **47**, 18.

84 Widawski, G., Rawiso, M., and Francois, B. (1994) *Nature*, **369**, 387.

85 Jenekhe, S.A. and Chen, X.L. (1999) *Science*, **283**, 372.

86 Tang, B.Z. (2001) *Polym. News*, **26**, 262.

87 Salhi, F., Cheuk, K.K.L., Sun, Q., Lam, J.W.Y., Cha, J.A.K., Li, G., Li, B., Luo, J., Chen, J., and Tang, B.Z. (2001) *J. Nanosci. Nanotechnol.*, **1**, 137.

88 Srinivasarao, M., Collings, D., Philips, A., and Patel, S. (2001) *Science*, **292**, 79.

89 Peng, J., Han, Y., and Li, B. (2004) *Polymer*, **45**, 447.

90 Newkome, G.R., He, E.F., and Moorefield, C.N. (1999) *Chem. Rev.*, **99**, 1689.

91 Nishihara, H., Kurashina, M., and Murata, M. (2003) *Macromol. Symp.*, **196**, 27.

92 Häußler, M., Lam, J.W.Y., Zheng, R., Dong, H., Tong, H., and Tang, B.Z. (2005) *J. Inorg. Organomet. Polym. Mater.*, **15**, 519.

93 Häußler, M., Lam, J.W.Y., Qin, A., Tse, K.K.C., Li, M.K.S., Liu, J., Jim, C.K.W., Gao, P., and Tang, B.Z. (2007) *Chem. Commun.*, 2584.

94 Wilson, J. (2007) *Chem. Technol.*, **4**, T42.

95 Häußler, M., Tse, K.C., Lam, J.W.Y., Tong, H., Qin, A., and Tang, B.Z. (2006) *Polym. Mater. Sci. Eng.*, **95**, 213.

96 Tang, B.Z., Geng, Y., Lam, J.W.Y., Li, B., Jing, X., Wang, X., Wang, F., Pakhomov, A.B., and Zhang, X.X. (1999) *Chem. Mater.*, **11**, 1581.

97 O'Handley, R.C. (2000) *Modern Magnetic Materials: Principles and Applications*, John Wiley and Sons Inc., New York, p. 491.

98 Sun, Q., Xu, K., Peng, H., Zheng, R., Häußler, M., and Tang, B.Z. (2003) *Macromolecules*, **36**, 2309.

5 Through-Space Conjugated Polymers

Yasuhiro Morisaki and Yoshiki Chujo

5.1 Introduction

Over the past three decades, conjugated polymers have attracted considerable attention in polymer chemistry as well as in materials chemistry, due to their potential for application to opto-electronic devices such as electroluminescence (EL) devices [1], field effect transistors (FETs) [2], photovoltaic cells [3], and so on. They possess the highly favorable characteristic of versatile processability since they can form thin films by spin-coating, spray-coating, and casting; in addition, they incorporate the inherent features of organic polymers, that is, they are lightweight and flexible. In these polymeric materials, charges and energy are transferred in some layers from one side to the opposite side through a conjugated polymer backbone, as well as through aromatic rings in the polymer main chain by a hopping mechanism. To date, a large number of π-conjugated polymers, σ-conjugated polymers, and cross-conjugated polymers [4] have been synthesized. However, few studies have focused on synthesizing through-space conjugated polymers comprising layered π-electron systems in a single polymer chain. It is worth noting that layered π-electron systems are commonly found in nature. For example, layered aromatic rings can be found in DNA [5]. The π-stacked array of base pairs in double-stranded DNA plays an important role for stabilization of their higher-ordered structures along with hydrogen bonds. Light-harvesting antenna complexes in photosynthetic systems contain regular arrays of chlorophylls or bacteriochlorophylls [6]. Many crystals, as well as liquid crystals [7] such as discotic liquid crystals [8], also consist of layered π-electron systems. Thus, despite the fact that π-stacked structures are the common structure, the synthesis of a through-space conjugated polymer with a stacked array of π-electron systems has rarely been carried out in the field of synthetic polymer chemistry.

Through-space conjugated polymers are π-stacked polymers in which π-electron systems such as aromatic rings or π-electron systems are layered in the single polymer chain. This chapter focuses on the synthesis of through-space conjugated polymers in which π-electron systems such as aromatic rings are layered and π-stacked in a single polymer chain. It is divided into two main sections, which

Conjugated Polymer Synthesis. Edited by Yoshiki Chujo

ISBN: 978-3-527-32267-1

are arranged according to the classes of the layered π-electron systems in the side chain and main chain.

5.2 Through-Space Conjugated Polymers with the Layered π-Electron Systems in the Side Chain

The through-space conjugated polymers discussed in this section are synthesized by chain polymerizations of vinyl monomers possessing aromatic groups. This is a typically used synthesis strategy, and several through-space conjugated polymers can be formed with stacked π-electron systems in their side chain by cationic, anionic, and radical polymerization.

5.2.1 Polyacenaphthylene

As an example, in earlier studies, polyacenaphthylenes were prepared by chain polymerization of acenaphthene. Both cationic polymerization of acenaphthene **1** by boron trifluoride [9] and thermal polymerization [10] proceeded smoothly to obtain the corresponding polyacenaphtalenes **2** (Scheme 5.1). The polyacenaphthylene in diluted solution exhibited photoluminescence from monomer units and intramolecular excimers [11], depending on the stereoregularity [11d]. A rigid polymer main chain and partly overlapped structures between next-to-nearest neighbor naphthalene units in the polymer side chain contributed to excimer formation [11, 12].

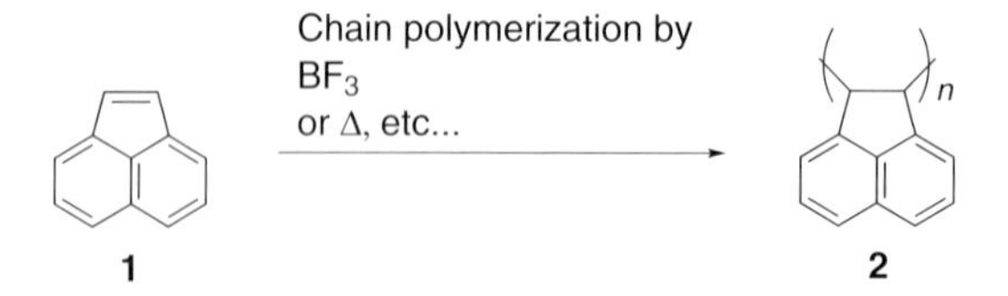

Scheme 5.1 Synthesis of polyacenaphthylene.

5.2.2 Polydibenzofulvene

Nakano and coworkers reported the polymerization of dibenzofulvene derivatives by anionic, cationic, and radical initiators to afford π-stacked polydibenzofulvenes [13]. Although 1,1-diphenylethylene is inactive during chain polymerization due to steric hindrance [14], dibenzofulvene, which has a planar fluorene ring, affords a fluorene-ring layered polymer. As shown in Scheme 5.2, the polymerization of dibenzofulvene **3** by anionic initiators such as *n*-BuLi, 9-fluorenyllithium, and *t*-BuOK proceeded smoothly at $-78\,^{\circ}C$ to yield the corresponding polymer **4**. The polymers were obtained

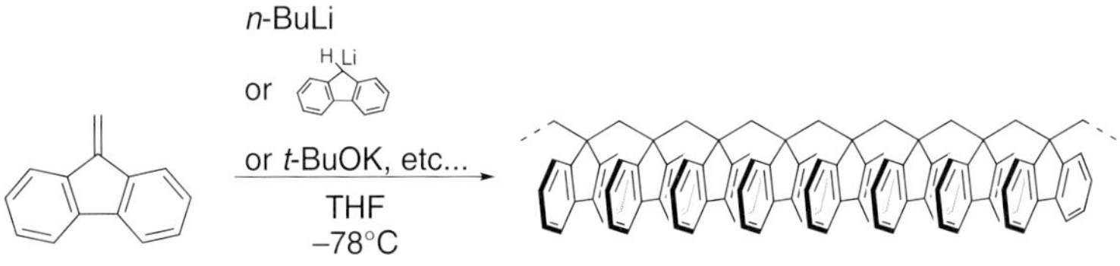

n-BuLi
or (FlLi)
or *t*-BuOK, etc...
THF
–78°C

3 → 4

THF-soluble part
n-BuLi: 13%, M_n = 770
FlLi: >99%, M_n = 810
t-BuOK: 24%, M_n = 1170

Scheme 5.2 Synthesis of polydibenzofulvene **4**.

as THF (or $CHCl_3$) soluble and insoluble parts, which had similar IR spectra. They were characterized by nuclear magnetic resonance (NMR) and matrix-assisted laser desorption/ionization-time of flight (MALDI-TOF) mass spectra. In particular, variable temperature (VT) NMR measurements carried out from −50 to 60 °C indicated that the π-stacked structure was stable in solution and did not undergo drastic conformational changes. In addition, the oligomers (dimer∼hexamer) could be separated by size exclusion chromatography (SEC) and isolated, and the X-ray crystal structure was analyzed [13b]. The single-crystal structures indicated that fluorene units of oligomers (tetramer∼hexamer) were π-stacked in the polymer chain. Rathore and coworkers also synthesized and characterized the π-stacked oligofluorenes by NMR spectroscopy and X-ray crystallography [15].

The absorption spectra of separated oligobenzofulvenes in solution exhibited hypochromic and bathochromic shifts [13b]. As the number of layered fluorene rings increased, the peak wavelength increased and the molar extinction coefficient (ε) decreased, with both quantities reaching constant values at around five fluorene rings. These results indicate that the through-space conjugation effect was extended over five successive fluorene moieties. In the fluorescence emission spectra of the oligomers, the dimer exhibited an emission peak at around 310 nm, similar to the case of monomeric fluorene; furthermore, the trimer exhibited emission peaks at 310 and 400 nm, similar to the case of a fluorene excimer. The tetramer and higher oligomers exhibited an emission peak at 400 nm. [13b]. It was reported that cation radicals were delocalized through the π-stacked fluorene rings [13g]. The hole drift mobility of the cast film of polydibenzofulvene was estimated by TOF measurement, and the estimated value was higher than that of a typical through-bond π-conjugated polymer, poly(*p*-phenylenevinylene) (PPV), and slightly lower than that of Se, which is an inorganic semiconductor [13c].

5.2.3 Polybenzofulvene

The free radical polymerization of a benzofulvene derivative, 7-methyl-1-methylene-3-phenylidene **5**, with benzoyl peroxide used as the initiator, provided the corresponding polybenzofulvene **6** in 75% yield [16], as shown in Scheme 5.3. The number-average molecular weight (M_n) and weight-average molecular weight (M_w) of polymer **6** were found to be 13 800 and 64 700, respectively. The reaction route is also depicted in Scheme 5.3, and the more reactive monobenzylic radical **5a** yielded polymer **6**. Radical reactivity, as well as less steric hindrance, was responsible for this selectivity. In a dilute solution of polymer **6**, two emission peaks were observed at 347 and 448 nm, which were assigned to monomer and excimer emissions of the benzofulvene unit, respectively.

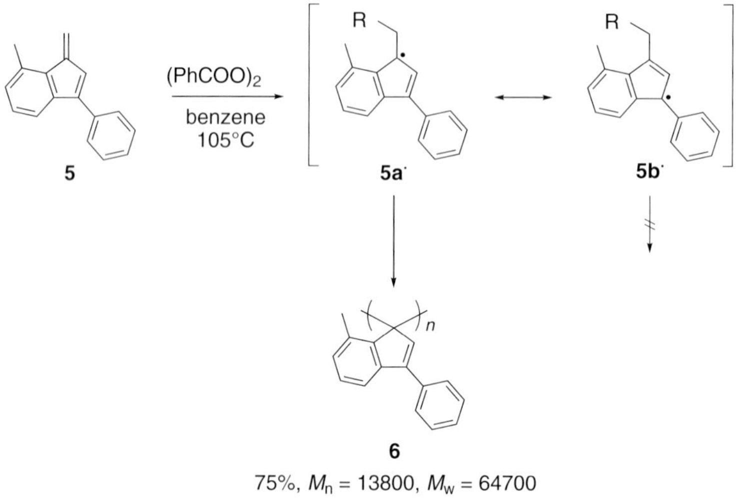

(PhCOO)$_2$, benzene, 105°C
5, 5a˙, 5b˙, 6
75%, M_n = 13800, M_w = 64700

Scheme 5.3 Synthesis of polybenzofulvene **6**.

Cappelli and coworkers also reported the polymerization of benzofulvene derivatives [17]. Benzofulvene **7**, which has a carboxylate group at the 2-position of an indene skeleton, was stable in solution. However, the polymerization of **7** started spontaneously with the removal of solvent, yielding the corresponding polybenzofulvene **8**, as shown in Scheme 5.4. The molecular weight of polymer **8** depended on the amount of monomer **7** in solution. For example, 400 mg of monomer **7** in solution provided a corresponding high-molecular-weight polymer **8** with an M_w value of 2 670 000 and an M_w/M_n value of 1.6, as determined using a system incorporating SEC and multi-angle light scattering (SEC-MALS). The structure of

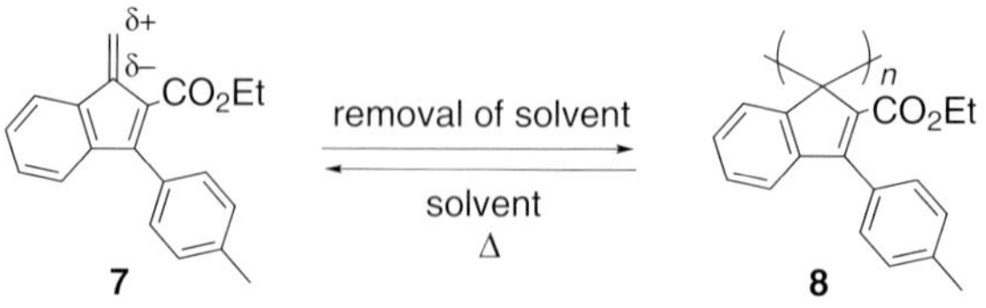

δ+, δ−, CO_2Et, removal of solvent, solvent, Δ
7, 8

Scheme 5.4 Synthesis of polybenzofulvene **8**.

polymer **8** was confirmed by ^{13}C NMR spectroscopy, indicating that, similar to the case of the aforementioned polymer **6** [17b], vinyl polymerization took place rather than diene polymerization. This vinyl polymer structure was stabilized by means of aromatic stacking interaction.

Polybenzofulvene **8** was obtained in the presence of a radical inhibitor such as 2,6-di-*tert*-butyl-*p*-cresol. The polarization of exocyclic double bonds was induced by solvent removal [18]. Thus, the authors suggested that the zwitterionic intermediate initiated the cationic polymerization leading to the polymer (Scheme 5.5). The obtained polymer was depolymerized in boiling aprotic solvents to regenerate the benzofulvene monomer, as shown in Scheme 5.4.

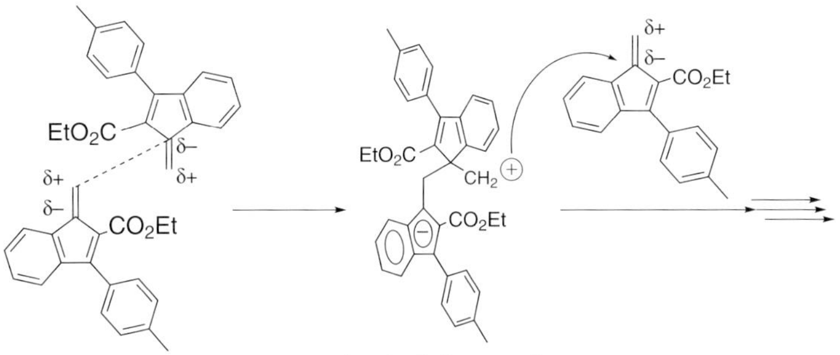

zwitterionic intermediate

Scheme 5.5 Reaction mechanism of the synthesis of polybenzofulvene.

5.2.4
Polystyrene-graft-Poly(4-Phenylquinoline)

Jenekhe and coworkers reported the synthesis of π-stacked polymers bearing the conjugated poly(4-phenylquinoline) in the side chain [19]. Incidentally, they reported that rigid oligoquinolines and polyquinolines readily form π-stacked structures in the solid state [20]. The polymer was prepared by a polymer reaction, as shown in Scheme 5.6. The treatment of poly(4-acetylstyrene) **9** ($M_n = 28\,990$ and $M_w/M_n = 1.09$) with 5-acetyl-2-aminobenzophenone in the presence of diphenyl phosphate yielded polystyrene-*graft*-poly(4-phenylquinoline) **10**. After 48 h, 2-aminobenzophenone was added to the reaction mixture as an end-capping reagent to obtain the corresponding polymer **11**. The M_n and M_w/M_n were found to be 602 100 and 1.15, respectively, and the average degree of polymerization of the poly(4-phenylquinoline) side chain was estimated to be 10 ($m = 9$, Scheme 5.6) according to the ^{1}H NMR integral ratio.

The absorption spectrum of polymer **11** exhibited vibronic peaks at 442 and 415 nm resulting from ground state electronic interactions in the π-stacked poly(4-phenylquinoline) side chain, which were not observed in the absorption spectrum of the poly(4-phenylquinoline) itself. The spectrum of **11** was red-shifted in comparison

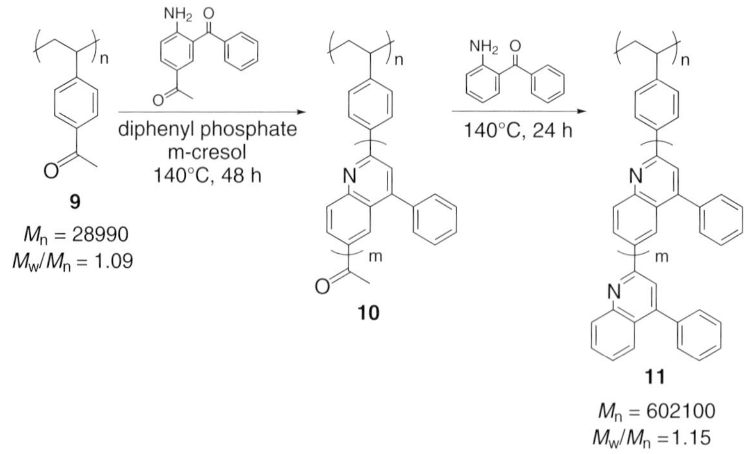

Scheme 5.6 Synthesis of polystyrene-*graft*-poly(4-phenylquinoline) **11**.

with that of poly(4-phenylquinoline). Furthermore, the absorption and emission spectra of **11** in formic acid solution were blue-shifted in comparison with those of **11** in toluene or THF solution. The strong electrostatic repulsion by protonated and pristine quinoline units resulted in interference with the π-stacking of the poly(4-phenylquinoline) side chain.

According to the atomic force microscopy (AFM) and scanning electron microscopy (SEM) images, polymer **11** formed cylindrical nanostructures with an average width and length of 17 ± 3 and 66 ± 9 nm, respectively, which were attributed to an individual single polymer chain. An electroluminescence (EL) device using polymer **11** as an emitting layer was fabricated. The device showed enhanced efficiency and brightness as compared to devices developed using poly(4-phenylquinoline); this observation can be attributed to the improved three-dimensional structural order of the thin film of π-stacked polymer.

5.2.5
[3.2]Paracyclophane-Containing Polymer

[*m.n*]Cyclophane compounds ($m \leq 3$ and $n \leq 3$) have a π-stacked structure comprising face-to-face aromatic rings [21]. Several studies have reported on the synthesis of through-space conjugated polymers containing an [*m.n*]cyclophane skeleton in their side chain [22].

Glatzhofer and coworkers reported the synthesis of a series of vinyl polymers possessing [3.2]paracyclophane units in their side chain [23]. They employed (*E,E*)-[6.2]paracyclophane-1,5-diene **12** as a monomer and constructed a [3.2]paracyclophane skeleton, as shown in Scheme 5.7 [23b]. Cationic polymerization with $BF_3 \cdot Et_2O$ or $SnCl_4$ proceeded smoothly to obtain polymer **13** in high yield with an

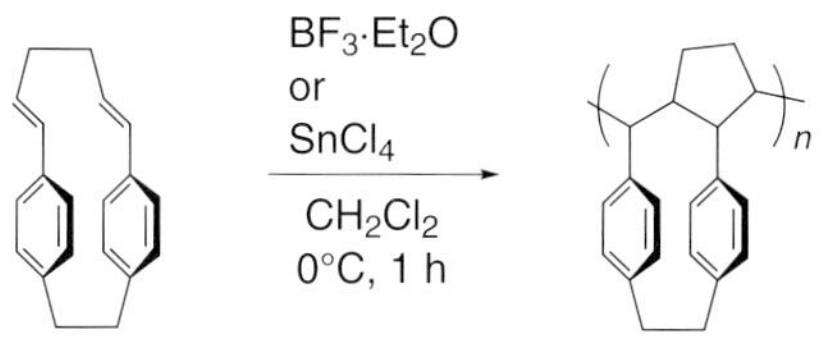

Scheme 5.7 Synthesis of [3.2]paracyclophane-containing polymer **13**.

M_w value of 12 000–67 200 (by GPC), which could be controlled by changing the feed ratio of the monomer and initiator. Radical polymerization with AIBN was also available for this monomer. However, anionic polymerization was unsuccessful. The obtained polymer **13** was thermally stable, and decomposition started above approximately 400 °C, as revealed by thermogravimetric analysis (TGA) in a nitrogen atmosphere.

Polymers **13** and **14**, which were obtained by $BF_3{\cdot}Et_2O$-catalyzed cationic cyclopolymerization [24b], were subjected to fluorescence spectroscopy [23a]. The emission λ_{max} data at various excitation wavelengths are summarized in Table 5.1, including the data of model compounds **15** [23a] and **16** [25]. Model compounds **15** and **16** exhibited emission only from the cyclophane unit at around 356 and 358 nm, respectively, while polymer **13** showed broad emission peaks at 374, 393 (strongest peak), 412, and 435 nm. These emission maxima appeared to be emission bands due to the extended excimer fluorescence from electronically interacting repeat units. Similar emission behaviors were observed for polymer **14**; however, these emission peaks were weak and less distinct, reflecting a lower degree of stereoregularity as well as a lack of complete cyclopolymerization [24].

Polymer **13** exhibited good hole mobility that was several times higher than that of poly(*N*-vinylcarbazole). The existence of layered cyclophane segments resulted in the delocalization of the generated charge carriers. The exposure of polymer **13** to I_2 vapor under reduced pressure led to an increase in the electrical conductivity from 10^{-9} S cm^{-1} to 5×10^{-4} S cm^{-1} [23a]. The cationic ($BF_3{\cdot}Et_2O$ or CF_3SO_3H) or radical (AIBN or benzoyl peroxide) cyclopolymerization of thiophene analog **17** yielded thiophenophane-containing polymer **18** (Scheme 5.8). Investigation of the hole mobility of polymer **18** indicated higher conductivity ranging from 10^{-4} to 10^{-3} S cm^{-1}, during exposure to I_2 vapor [23c].

5.2.6
Polymethylene with [2.2]Paracyclophane

Polymethylene is an interesting polymer with a substituent attached to each carbon of the polymer main chain [26]. The functional substituents in the polymethylene chain are closely packed; therefore, it was expected that the incorporation of bulky [2.2]paracyclophane into a polymethylene backbone as a pendent group would provide a

Table 5.1 Emission λ_{max} of polymers **13**[a] and **14**,[b] and compounds **15**[c] and **16**[d].

Material	Excitation λ/nm	Emission λ_{max}/nm
13	290	358
	330	374 (sh), 393, 412 (sh)
	350	393, 412, 435 (sh)
	370	412, 435 (sh)
14	313	379
	330	360 (sh), 85
	350	385 (sh), 398
	370	398 (sh), 420, 440 (sh)
15	300	356
	313	356
	330	356
	350	—
16	—	358

a) In CH_2Cl_2 (0.20 M) at room temperature [23a].
b) In CH_2Cl_2 (0.20 M) at room temperature [23a].
c) In CH_2Cl_2 (0.060 M) at room temperature [23a].
d) In ether–isopentane–ethanol (5 : 5 : 2, 1.0×10^{-3} M) at room temperature [25].

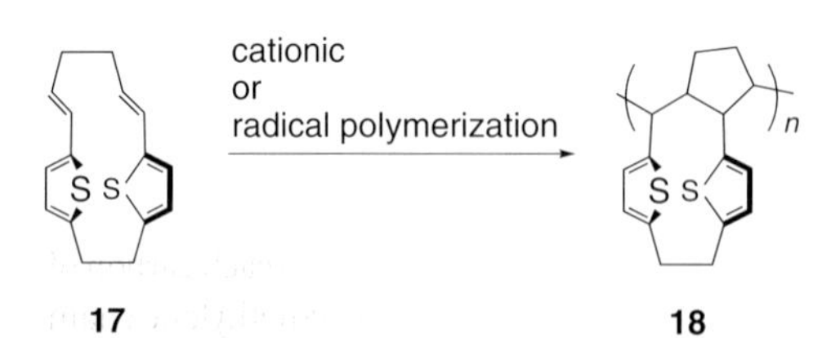

Scheme 5.8 Synthesis of [3.2](2,5)thiophenophane-containing polymer **18**.

π-stacked through-space conjugated structure. From this view point, polymethylene (*rac*)-**20** was prepared in 20% yield with an M_n value of 1100 and an M_w/M_n value of 1.6 by the $BF_3 \cdot Et_2O$-catalyzed polymerization [26d] of racemic 4-diazomethyl[2.2] paracyclophane (*rac*)-**19** [27], as shown in Scheme 5.9.

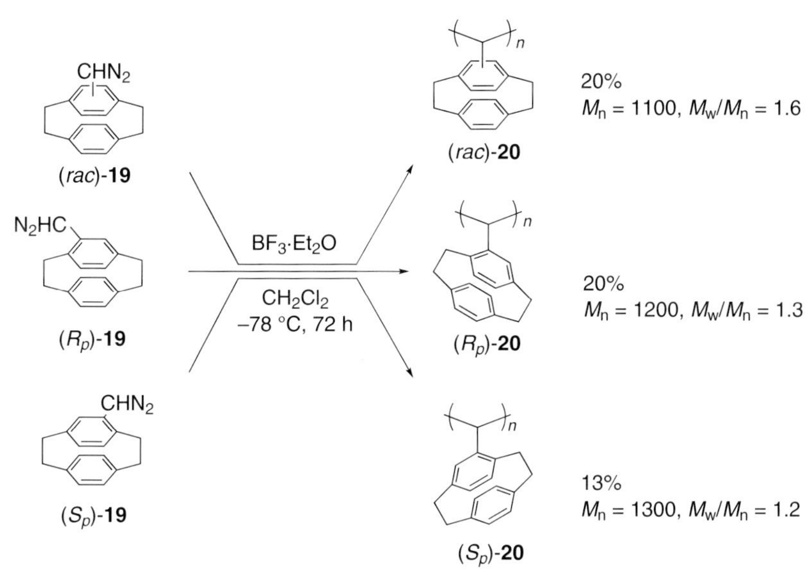

Scheme 5.9 Synthesis of [2.2]paracycophane-containing polymethylenes (*rac*)-**20**, (R_p)-**20**, and (S_p)-**20**.

Figure 5.1 shows the UV–vis absorption spectra of [2.2]paracyclophane **21**, polyethylene analogue **22** [28], and polymethylene (*rac*)-**20** in dilute $CHCl_3$ solution. The spectrum of polymethylene (*rac*)-**20** exhibited a red shift in comparison with the spectra of [2.2]paracyclophane and polyethylene, indicating electronically interacting repeat [2.2]paracyclophane units in the ground state due to their closely packed side chain.

Monosubstituted [2.2]paracyclophane is a planar chiral compound [21b]. Optically active 4-diazomethyl[2.2]paracyclophane monomers (R_p)-**19** and (S_p)-**19** were successfully prepared and polymerized to obtain optically active polymers (R_p)-**20** and (S_p)-**20**, respectively, as shown in Scheme 5.9. Figure 5.2 shows the circular dichroism (CD) spectra of polymethylenes (*rac*)-**20**, (R_p)-**20**, and (S_p)-**20**. Polymers (R_p)-**20**, and (S_p)-**20** exhibited mirror-image Cotton effects, while polymer (*rac*)-**20** did not. Based on the Cotton effect patterns [29], the optically active polymers (R_p)-**20** and (S_p)-**20** are expected to form the optically active higher-ordered structure of the polymethylene chain.

As another example of polymers with π-stacked structures in their side chain, poly (*N*-vinylcarbazole)s have been extensively studied and used for opto-electronic materials such as hole-transporting materials [30]. However, poly(*N*-vinylcarba-

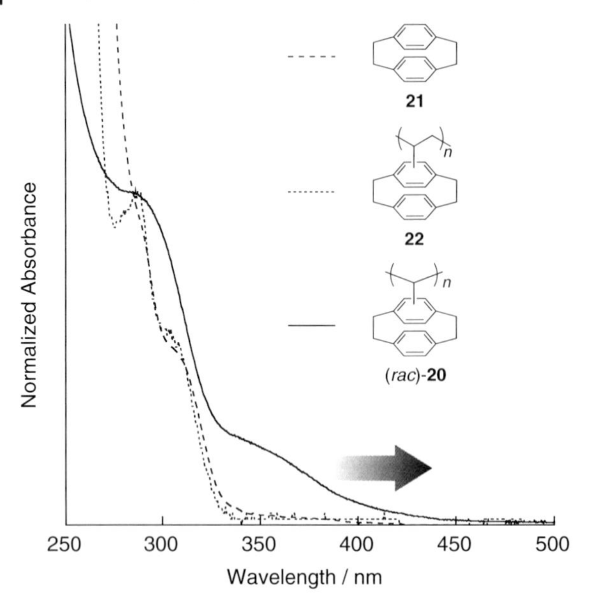

Figure 5.1 UV–vis absorption spectra of polymethylene **20**, [2.2]paracyclophane **21**, and polyethynele **22** in $CHCl_3$.

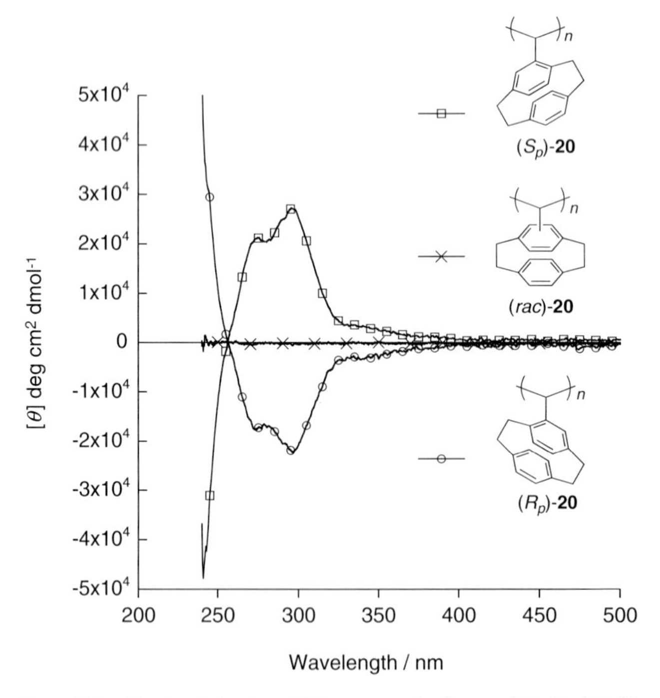

Figure 5.2 Circular dichroism (CD) spectra of polymers (R_p)-**20**, (S_p)-**20**, and (*rac*)-**20** in $CHCl_3$.

zole)s comprise partly π-stacked structures in short sequences due to their low stereoregularity.

5.3
Through-Space Conjugated Polymers with the Layered π-Electron Systems in the Main Chain

This section introduces the synthesis and properties of through-space conjugated polymers comprising layered π-electron systems. Foldamers [31] are not discussed here; rather, this section is devoted to polymers with stable and fixed π-stacked structures in their main chain. Polymers in this category are synthesized by step polymerization; in particular, palladium-catalyzed coupling reactions are powerful tools for constructing the π-conjugated framework.

5.3.1
Phenylene-Layered Polymer Based on a Norbornane Scaffold

Martínez, Barcina, and coworkers synthesized oligophenylene-layered polymers on the basis of a 7,7-disubstituted norbornane scaffold [32a]. The synthetic route is shown in Scheme 5.10. The reaction of 7,7-di(*p*-iodophenyl)norbornane **23** with 2,5-dihexylbenzene1,4-bisdiboronic acid **24** in the presence of a catalytic amount of Pd $(PPh_3)_4$ provided the corresponding through-space conjugated polymer **25** with a yield of 94%. The obtained polymer was soluble in toluene and $CHCl_3$. The average degree of polymerization and polydispersity (M_w/M_n) were calculated by GPC to be 18 and 2.4, respectively. It was reported that polymer **25** included cyclic oligomers (DP ~7 and ~10), as determined by GPC. Under diluted reaction conditions, the intensity of the GPC peak derived from the cyclic oligomers increased; however, the compounds could not be isolated.

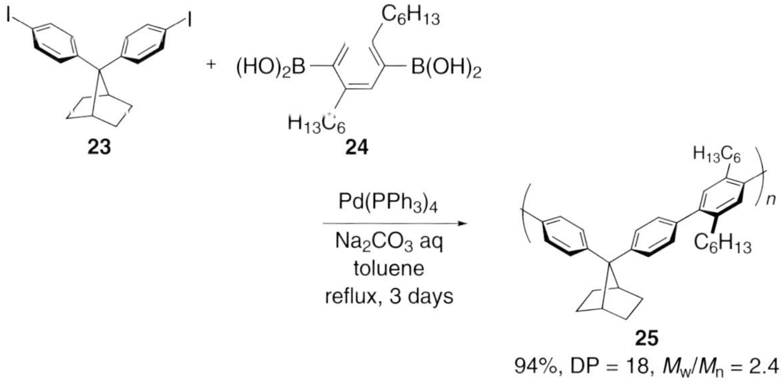

Scheme 5.10 Synthesis of oligophenylene-layered polymer **25** on the basis of a norbornane scaffold.

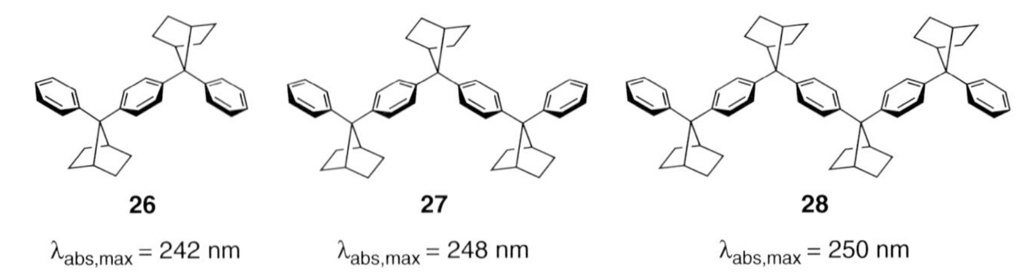

Figure 5.3 Phenylene-layered oligomers **26–28** on the basis of a norbornane scaffold.

The distance between the ipso carbons of the phenyl groups at the 7-position of the norbornane skeleton was approximately 2.46 Å, which is less than the sum of the van der Waals radii of sp^2 carbons (3.4 Å). Thus, in comparison with the UV–vis absorption spectrum of 2′,5′-dihexyl-*p*-terphenyl, that of polymer **25** revealed the occurrence of through-space conjugation. The absorption maximum of polymer **25** appeared at 268 nm, which was bathochromically shifted from that of 2′,5′-dihexyl-*p*-terphenyl. Well-defined oligomers **26–28** (Figure 5.3) comprising three, four, and five-layered phenylenes were also synthesized based on the norbornane scaffold [32b]. These oligomers showed through-space conjugation, and the effective conjugation length was estimated to be five (∼six) cofacial phenylene rings.

5.3.2
Aromatic Ring-Layered Polymers Based on a Xanthene Scaffold

5.3.2.1 [2.2]Paracyclophane-Layered Polymer

Xanthene compounds can serve as a scaffold for the construction of an aromatic ring-layered structure. The rotary motion of the two aromatic rings at the 4,5-positions of the xanthene skeleton is restricted due to steric hindrance. 2,7-Di-*tert*-butyl-4,5-diiodoxanthene **29** was polymerized with pseudo-*p*-diethynyl[2.2]paracyclophane **30** and ethynyl[2.2]paracyclophane **31** (feed ratio of **29** : **30** : **31** = 10 : 9 : 2) in the presence of a catalytic amount of $Pd(PPh_3)_4$/CuI to obtain the corresponding [2.2]paracyclophane-layered polymer **32** in 79% yield (Scheme 5.11) [33b]. The M_n value of polymer **32** was estimated to be 4100, according to the ^{1}H NMR integral ratio, indicating that an average of eight [2.2]paracyclophane units, in other words, sixteen benzenes, were layered in the polymer main chain. The polymer was soluble in common organic solvents such as THF, $CHCl_3$, and toluene. A thin film of polymer **32** could be readily obtained by spin-coating from a toluene solution.

Polymers **32a–c** were prepared by changing the feed ratio of the monomers, and the UV–vis absorption maxima of **32a–c** and model compound **33** were compared (Figure 5.4) [33b]. The absorption spectra of polymers **32a–c** were almost identical and independent of the number of layered [2.2]paracyclophane units. They exhibited absorption maxima at around 330 nm, while polymer **33** exhibited its absorption maximum at a shorter wavelength of 320 nm. It was considered that even

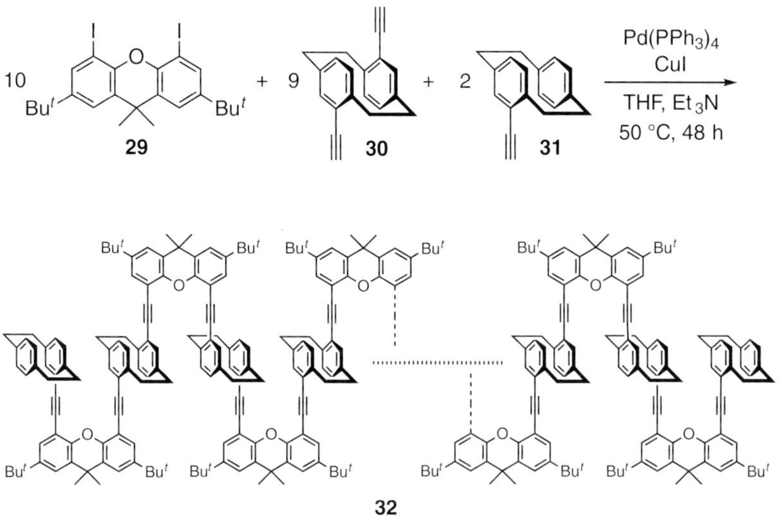
10
Bu^t
Bu^t
29
+ 9
30
+ 2
31
Pd(PPh3)4
CuI
THF, Et3N
50 °C, 48 h
32

Scheme 5.11 Synthesis of [2.2]paracyclophane-layered polymer **32** on the basis of a xanthene scaffold.

polymer **32c** with $M_n = 2100$ had sufficiently extended through-space conjugation in the ground state.

Polymers **34a–c** end-capped with anthracene groups [33b] and **35a–c** end-capped with *p*-nitrophenyl groups [33c] were synthesized (Figure 5.5). Figure 5.6a shows the UV–vis absorption spectrum and emission spectrum of polymer **32a** ($M_n = 4100$),

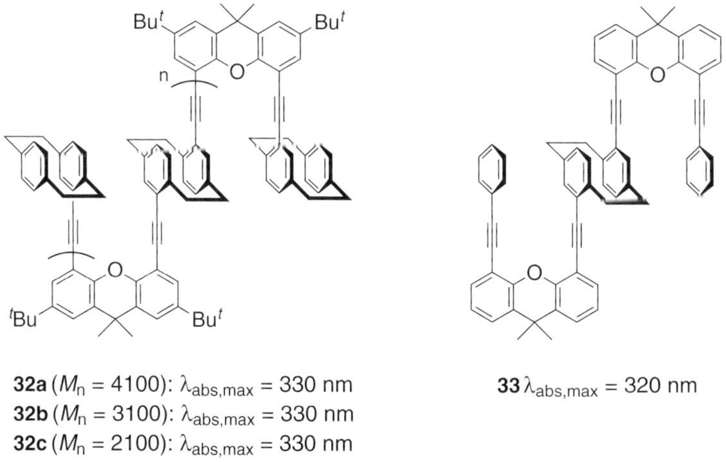
Bu^t
Bu^t
n
^tBu
Bu^t
32a (Mn = 4100): λabs,max = 330 nm
32b (Mn = 3100): λabs,max = 330 nm
32c (Mn = 2100): λabs,max = 330 nm
33 λabs,max = 320 nm

Figure 5.4 [2.2]Paracyclophane-layered polymers **32a–c** and compound **33**.

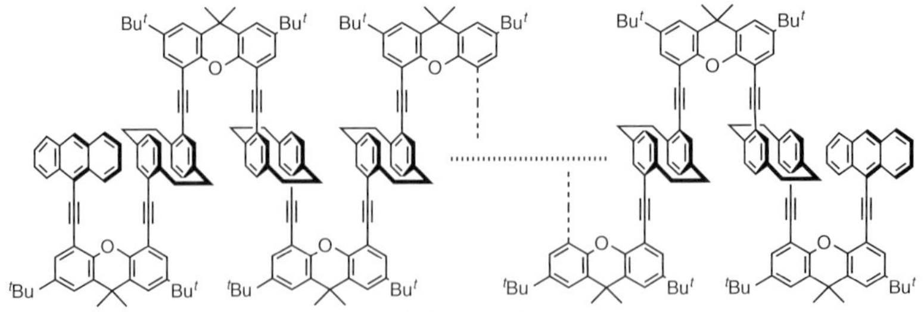

34a (M_n = 7500)
34b (M_n = 4200)
34c (M_n = 2600)

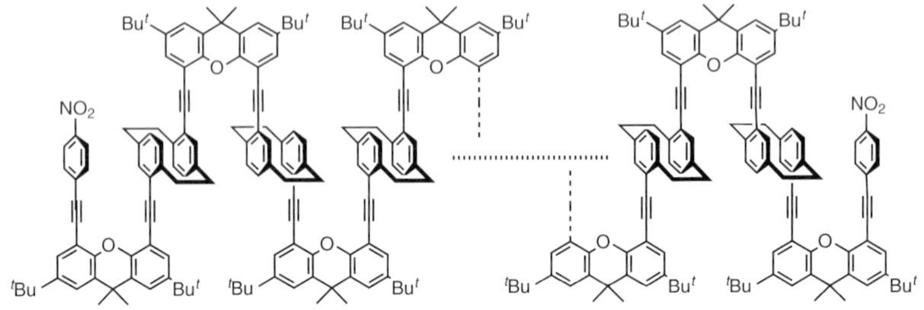

35a (M_n = 5750)
35b (M_n = 3000)
35c (M_n = 1700)

Figure 5.5 [2.2]Paracyclophane-layered polymers **34a-c** end-capped with anthracenes and **35a-c** end-capped with nitrobenzenes.

and Figure 5.6b shows those of polymer **34b** ($M_n = 4200$). Polymer **32a** exhibited the characteristic π–π^* absorption band of the layered [2.2]paracyclophane units (Figure 5.6a), while polymer **34b** exhibited a sharp absorption peak at around 270 nm and a broad absorption peak at around 400 nm, which were assigned to the absorption band of the end-capping anthracene units and the absorption band of the layered [2.2]paracyclophane units, respectively (Figure 5.6b).

As shown in Figure 5.6a, polymer **32a** showed emission at around 400 nm for an excitation wavelength of 334 nm, which was attributed to the emission from the layered [2.2]paracyclophane units. On the other hand, polymer **34b** showed emission at around 450 nm with a vibrational structure at an excitation of 334 nm. At this excitation wavelength, only the layered [2.2]paracyclophane units were effectively excited, because the end-capping anthracene units did not have an absorption band at around 334 nm. Thus, in the case of polymer **34b**, a strong emission peak was observed at 450 nm, which was attributed to the terminal anthracene units, instead of emission from the layered [2.2]paracyclophanes observed at around 400 nm. Time-resolved fluorescence studies of polymer **34b** [33b] confirmed that the emission from the cyclophane units decreased and that from the anthracene units increased simultaneously. Figure 5.6a and b show good overlap between the emission peak

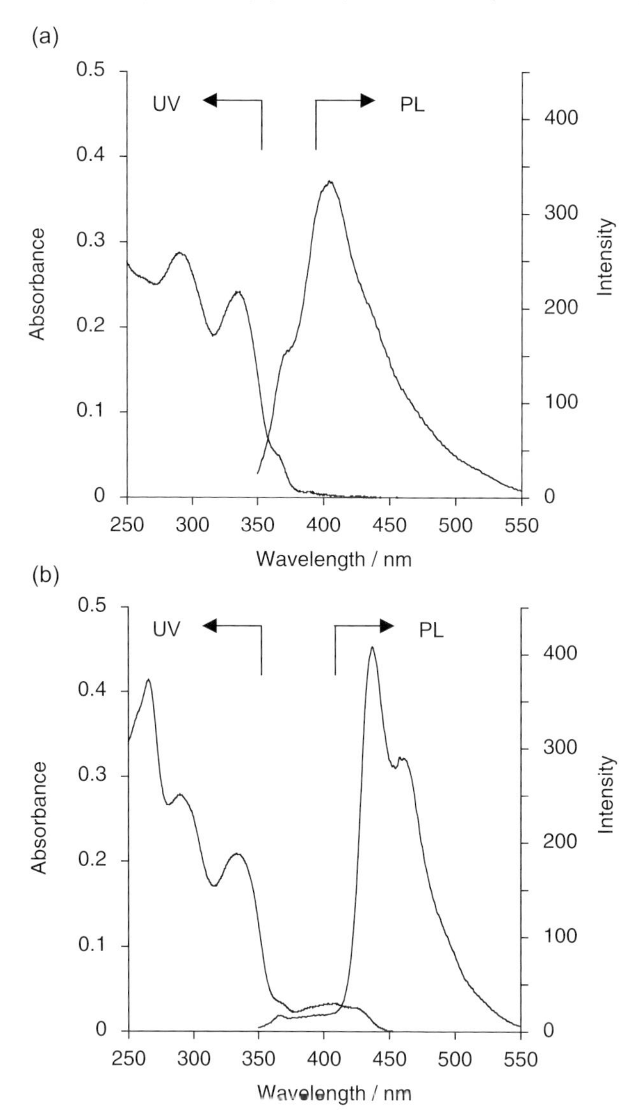

Figure 5.6 (a) UV–vis absorption and fluorescence emission spectra of polymer **32a**, and (b) UV–vis absorption and fluorescence emission spectra of polymer **34b** in $CHCl_3$.

of the layered-cyclophane units (at 400 nm in Figure 5.6a) and the absorption peak of the 9-ethynylanthracene moieties (at around 400 nm in Figure 5.6b), indicating fluorescence resonance energy transfer (FRET) [34] from the cyclophane units to the end-capping anthracenes. The efficiency of this energy transfer from the layered cyclophane units to the anthracene units was calculated to be 57%.

The fluorescence emission spectra of polymers **35a–c** and **32a** in dilute $CHCl_3$ solution are shown in Figure 5.7. The peak intensities of polymers **35a–c** were

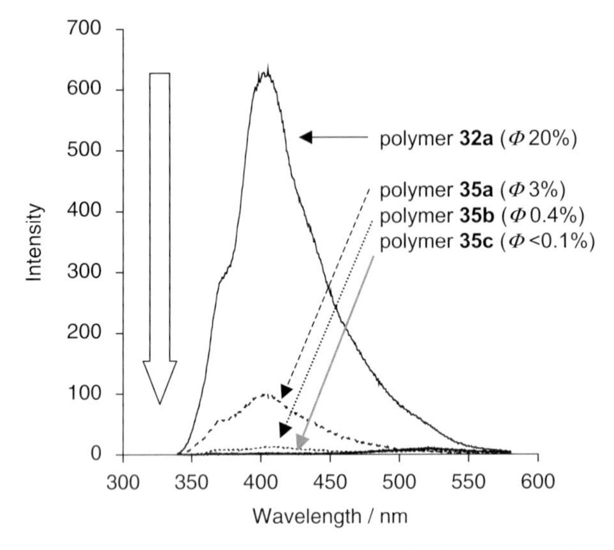

Figure 5.7 Fluorescence emission spectra of polymers **32a** and **35a–c** in $CHCl_3$.

considerably lower than the peak intensity of polymer **32a**. The end-capping *p*-nitrophenyl groups of polymers **35a–c** effectively quenched the photoluminescence from the layered [2.2]paracyclophanes via the through-space interactions. Solvent effects (cyclohexane, $CHCl_3$, THF, CH_2Cl_2, and DMF) were not observed in the emission-quenching study, which indicated that energy transfer predominated over electron transfer.

5.3.2.2 Oligophenylene-Layered Polymers

As mentioned above, this method of using xanthene as a scaffold leads to the introduction of not only [2.2]paracyclophane but also various aromatic compounds as stacked aromatic rings in the polymer backbone. Oligophenylene- [35], carbazole- [36], and oligothiophene-layered polymers were synthesized; for example, the synthetic route for biphenylene-layered polymers is shown in Scheme 5.12 [35]. 9,9-Didodecyl-4,5-diiodoxanthene **33**, 4,4′-biphenyldiboronic acid **34**, and *p*-nitrophenylboronic acid **35** were polymerized with a $Pd(OAc)_2$/S-Phos (S-Phos: 2-dicyclohexylphosphino-2′,6′-dimethoxybiphenyl) catalytic system [37] to produce the corresponding polymer **36** in 40% yield with an M_n value of 6250. The two end-capping nitrobenzene units in polymer **36** effectively quenched the photoluminescence of the layered biphenylenes by intramolecular photo-excited energy transfer from the layered biphenylenes to the end-capping nitrobenzene units.

5.3.2.3 Ferrocene-Layered Polymers Based on a Naphthalene Scaffold

The distance between the 4- and 5-positions of xanthene is approximately 4.5 Å, while that between the 1- and 8-positions of naphthalene is approximately 2.5 Å. When a naphthalene scaffold is used, a closer proximity of the layered aromatic rings can be achieved. However, the reactive efficiency of polymerization is expected to be

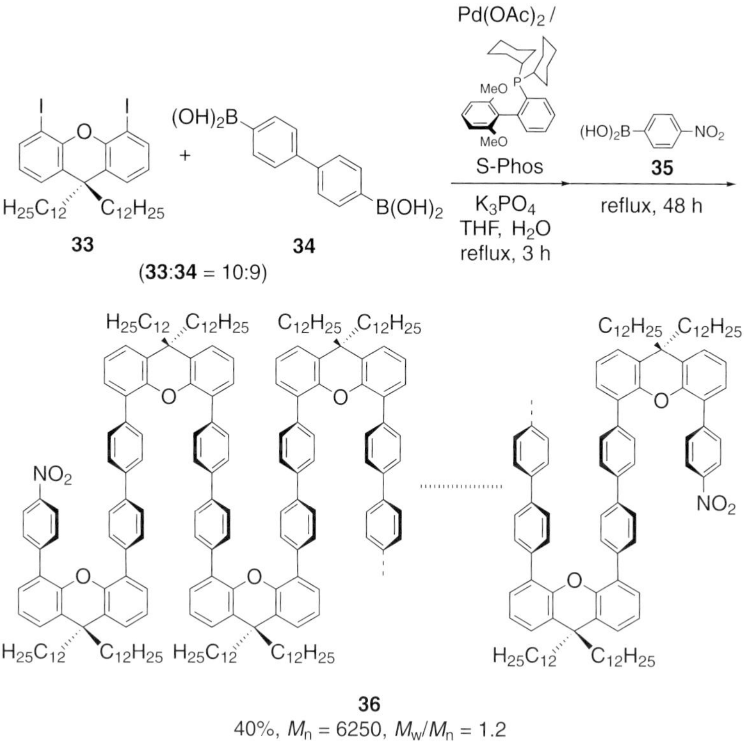

Scheme 5.12 Synthesis of biphenylene-layered polymer **36** on the basis of a xanthene scaffold.

relatively low due to steric hindrance. Rosenblum, Foxman, and coworkers reported the synthesis of π-stacked metallocene polymers based on the naphthalene scaffold [38]. The reaction of 1,8-diiodonaphthalene with 1,1′-ferrocenyldizinc chloride by the Negishi coupling reaction was also successful in obtaining a polymer with a relatively low M_n value of 3600. On the other hand, as shown in Scheme 5.13, the treatment of 1,1′-bis(8-cyclopentadienyl-1-naphthyl)-3,3′-bis(2-octyl)ferrocene **37** with $NaN(SiMe_3)_2$ and $FeCl_2$ yielded the corresponding ferrocene-layered polymer **38**, which had a higher molecular weight. The addition of $CuCl_2$ and $NiBr_2$, instead of $FeCl_2$, provided a ferrocene–cobaltocene-layered polymer and a ferrocene–nickelocene-layered polymer, respectively [33b,c].

5.3.3 Cyclophane-Containing Through-Space Conjugated Polymers

5.3.3.1 [2.2]Metacyclophane-Containing Through-Space Conjugated Polymers

In 1985, Mizogami and Yoshimura synthesized through-space conjugated polymers containing [2.2]metacyclophane in the main chain, as shown in Scheme 5.14 [39a]. Pseudo-*p*-dihydroxy[2.2]metacyclophane **39** was dimerized in the absence of a catalyst to obtain compound **40** in the form of bright reddish orange whiskers. The

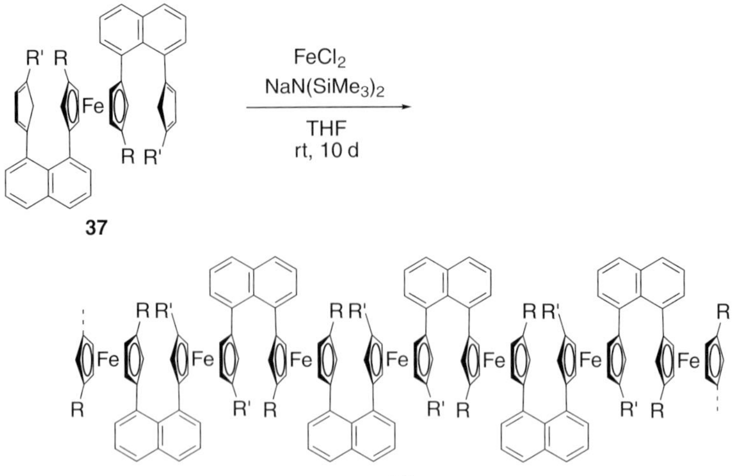

FeCl2
NaN(SiMe3)2
THF
rt, 10 d

37

38
R = 2-octyl, R' = H: 71%, M_n = 18000 (by ^{1}H NMR)
R, R' = 2-octyl: 40%, M_n = 139000 (by GPC)

Scheme 5.13 Synthesis of ferrocene-layered polymer **38** on the basis of a naphthalene scaffold.

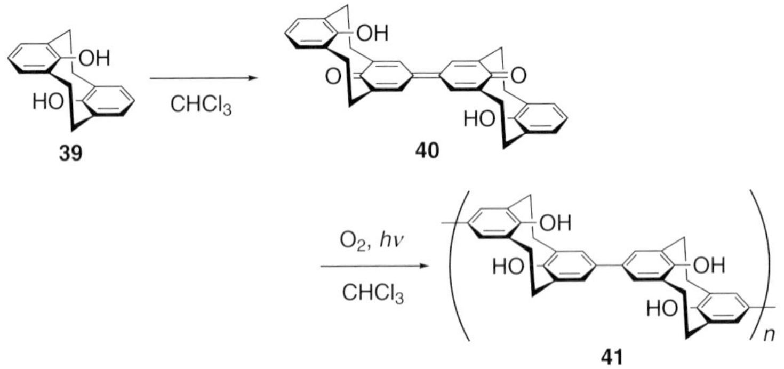

CHCl3
39
40
O2, hν
CHCl3
41

Scheme 5.14 Synthesis of [2.2]metacyclophane-containing polymer **41**.

oxidative coupling reaction of **40** in the presence of O_2 under sunlight yielded polymer **41**, which was not dissolved in any solvent. The structure of polymer **41** was characterized by IR spectroscopy and elemental analysis. Polymer **41** exhibited an increase in electrical conductivity from 10^{-9} S cm^{-1} to 10^{-1} S cm^{-1} when subjected to chemical doping with H_2SO_4 vapor.

Scheme 5.15 shows the synthetic route to diphenoquinone-layered polymer **42** [39b]. The polymerization of **39** in the presence of an excess amount of $FeCl_3$ was carried out to obtain the corresponding polymer **42** quantitatively in the form of bright orange–red powders. The structure of this polymer was determined by IR spectroscopy and elemental analysis. The electrical conductivity of a pressed pellet of

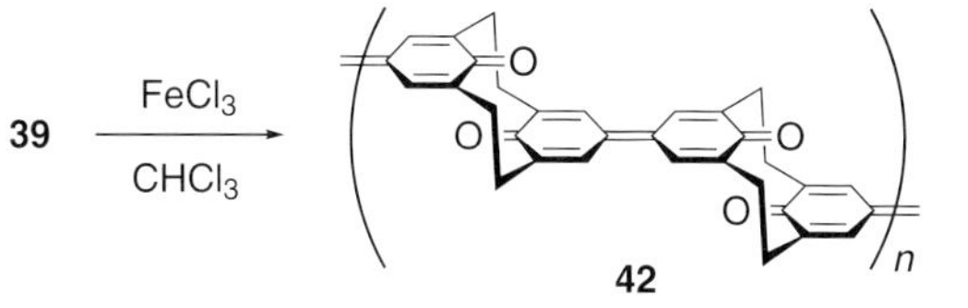

39 $\xrightarrow[\text{CHCl}_3]{\text{FeCl}_3}$ 42

Scheme 5.15 Synthesis of diphenoquinone-layered polymer **42**.

polymer **42** was measured, and the conductivities of the pristine sample and H_2SO_4-vapor-doped sample were $10^{-9}\,S\,cm^{-1}$ and $10^{-4}\,S\,cm^{-1}$, respectively.

The synthesis and properties of dithia[3.3]metacyclophane- and dithia[3.3](2,6)pyridinophane-containing conjugated polymers have also been reported [40]. These dithia[3.3]metaphanes show conformational flexibility, which is manifested in the rapid *syn–anti* ring inversion of two aromatic rings in solution [21a, 41]; therefore, they do not form stable π-stacked structures. The through-space conjugation of this class of polymers was ineffective in the ground state. Dithia[3.3](2,6)pyridinophane-containing polymer had the ability to coordinate to transition metals such as palladium, and the polymer complex acted as a recyclable catalyst for the Heck coupling reaction [40d].

5.3.3.2 [2.2]Paracyclophane-Containing Through-Space Conjugated Polymers

Palladium-catalyzed coupling reactions have been widely used for the synthesis of a wide variety of conjugated polymers. Pseudo-*p*-disubstituted[2.2]paracyclophanes can be employed as components of conjugated polymers [22, 42]. Schemes 5.16–5.18 show the synthesis of [2.2]paracyclophane-containing poly(*p*-arylene-ethynylene) **45** [43a], poly(*p*-arylenevinylene) **47** [41b], and poly(*p*-arylene) **50** [43c]. Thus, various [2.2]paracyclophane-containing through-space conjugated polymers [44] have been synthesized by using pseudo-*p*-disubstituted[2.2]paracyclophanes such as pseudo-*p*-diethynyl[2.2]paracyclophane **43** [44c], pseudo-*p*-divinyl[2.2]paracyclophane **46** [45],

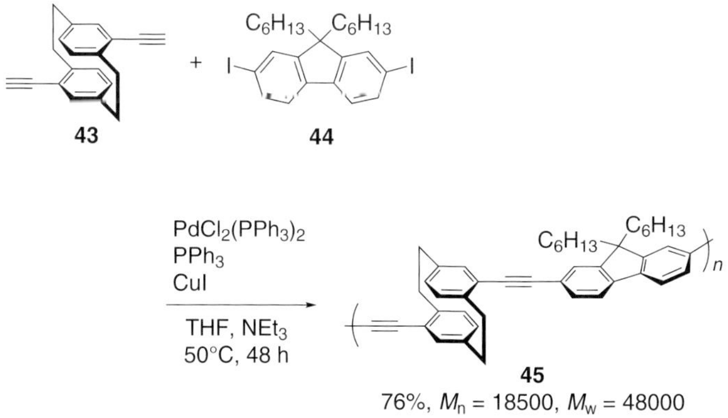

43 + 44 (C_6H_{13}, C_6H_{13})

$PdCl_2(PPh_3)_2$
PPh_3
CuI
THF, NEt_3
50°C, 48 h

45
76%, M_n = 18500, M_w = 48000

Scheme 5.16 Synthesis of [2.2]paracyclophane-containing poly(*p*-arylene-ethynylene) **45** by the Sonogashira–Hagihara reaction.

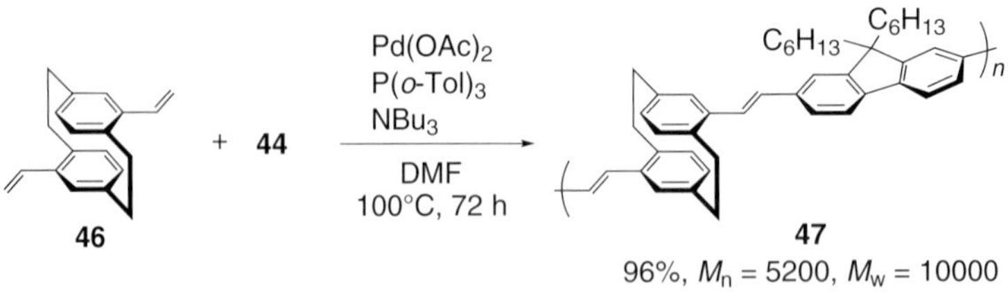

Pd(OAc)$_2$ P(o-Tol)$_3$ NBu$_3$ DMF 100°C, 72 h
46 + **44** → **47**
96%, M_n = 5200, M_w = 10000

Scheme 5.17 Synthesis of [2.2]paracyclophane-containing poly(p-arylenevinylene) **47** by the Heck reaction.

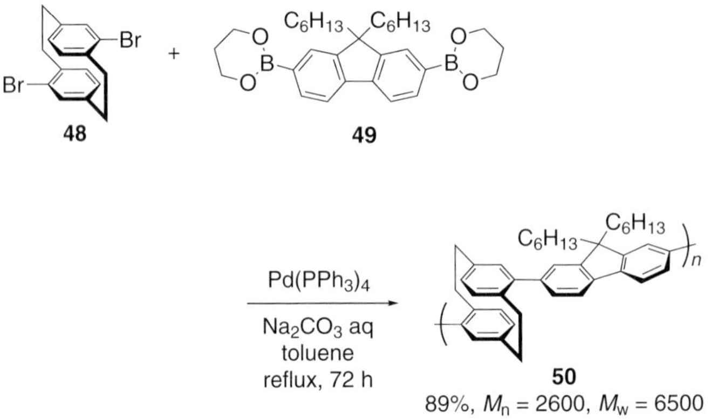

48 + **49** → Pd(PPh$_3$)$_4$, Na$_2$CO$_3$ aq, toluene, reflux, 72 h → **50**
89%, M_n = 2600, M_w = 6500

Scheme 5.18 Synthesis of [2.2]paracyclophane-containing poly(p-arylene) **50** by the Suzuki-Miyaura reaction.

and pseudo-p-dibromo[2.2]paracyclophane **48** [46], as the key monomers, and the representative polymers **51–56** are shown in Figure 5.8 [44]. These polymers were soluble in common organic solvents such as THF, $CHCl_3$, CH_2Cl_2, and toluene due to their alkyl side chains; therefore, thin films were readily prepared by casting and spin-coating methods. They also exhibited good thermal stability, and, for example, the 10% weight loss temperature of polymer **50** was found to be approximately 430 °C, as revealed by thermogravimetric analysis [43c].

[2.2]Paracyclophane leads to the construction of the π-stacked structure of π-electron systems; that is, [2.2]paracyclophane-containing through-space conjugated polymer **51** comprises the layered π-electron system of compound **57** (Figure 5.9). The physical properties depend on this π-electron system. The UV–vis absorption spectra of polymer **51** and compound **57** in dilute $CHCl_3$ solution are shown in Figure 5.10. Both absorption spectra exhibited two absorption peaks derived from the π–π^* band of a phenylene–ethynylene unit. The absorption maximum and absorption edge of polymer **51** appeared at 387 and around 440 nm, respectively, which were red-shifted in comparison with those of compound **57**. This result suggests that the [2.2]paracyclophane-containing conjugated polymer exhibited an extension of the π-conjugation length via through-space conjugation and the through-bond conjugation of sp- and sp^2-carbon frameworks. In general, the absorption maxima of

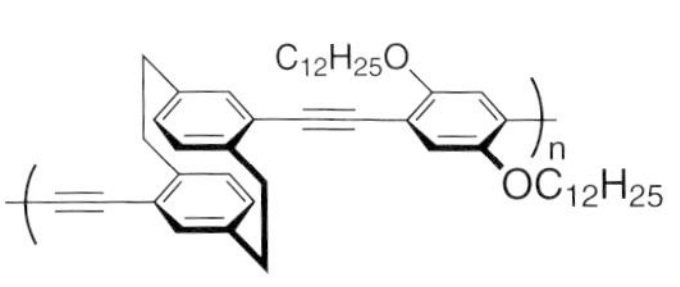

51
$M_n = 14500$, $M_w = 35500$

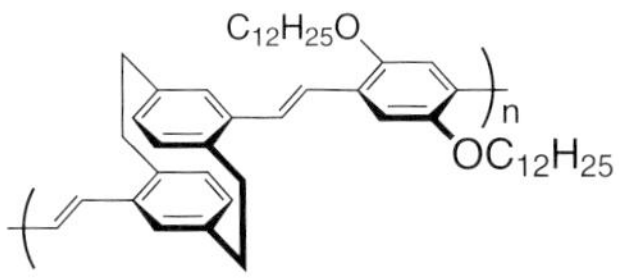

52
$M_n = 4200$, $M_w = 8300$

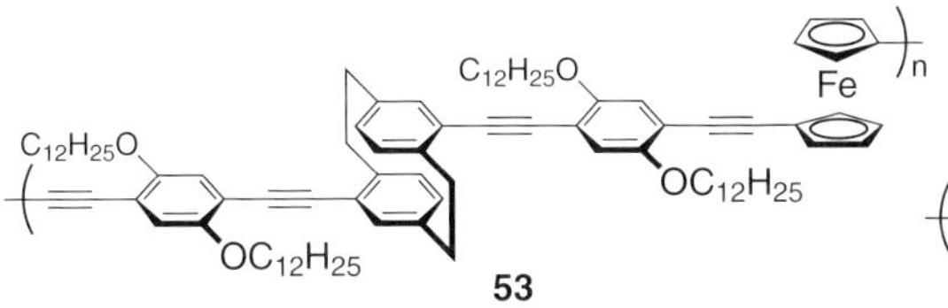

53
$M_n = 24500$, $M_w = 66300$

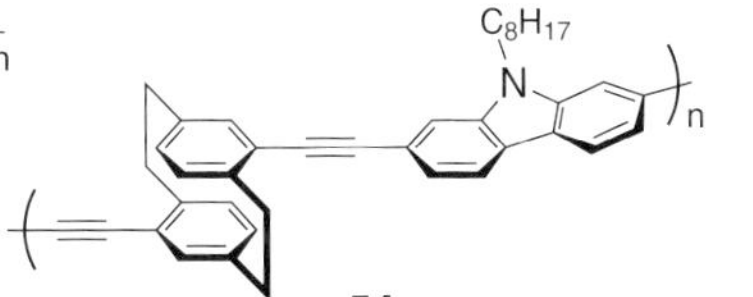

54
$M_n = 4400$, $M_w = 8500$

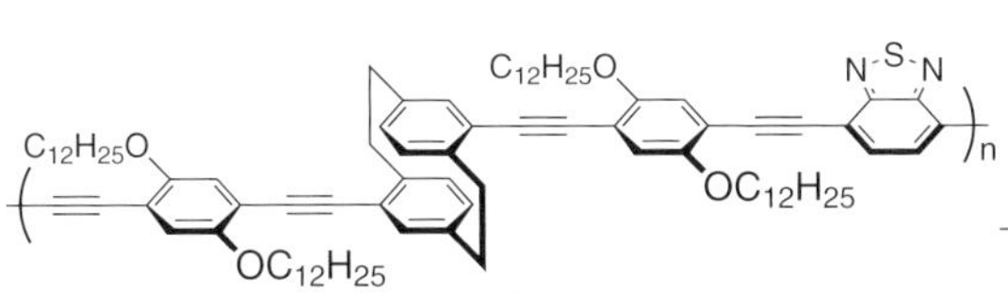

55
$M_n = 29700$, $M_w = 76800$

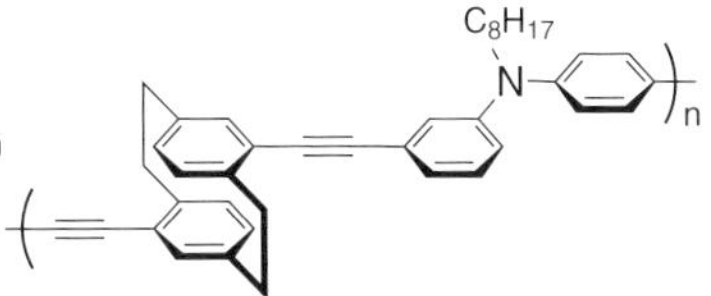

56
$M_n = 4300$, $M_w = 13300$

Figure 5.8 [2.2]Paracyclophane-containing through-space conjugated polymers **51–56**.

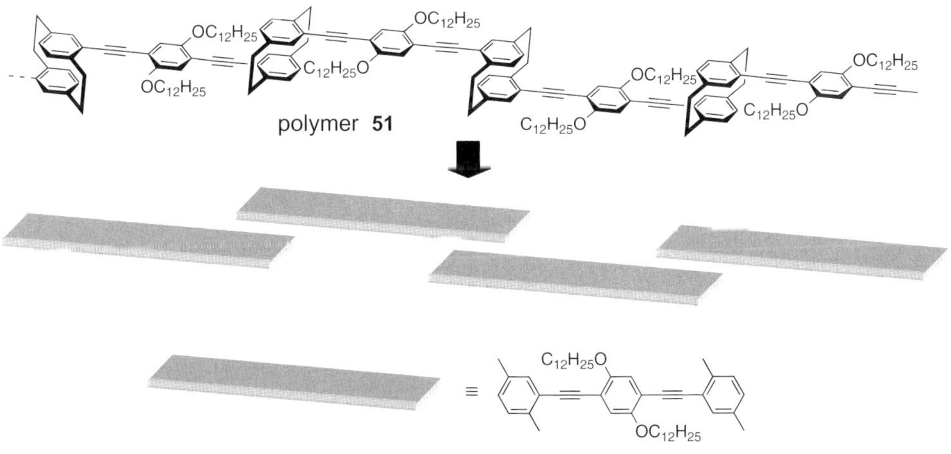

polymer **51**

57

Figure 5.9 π-Stacked structure of the π-electron system of compound **57**.

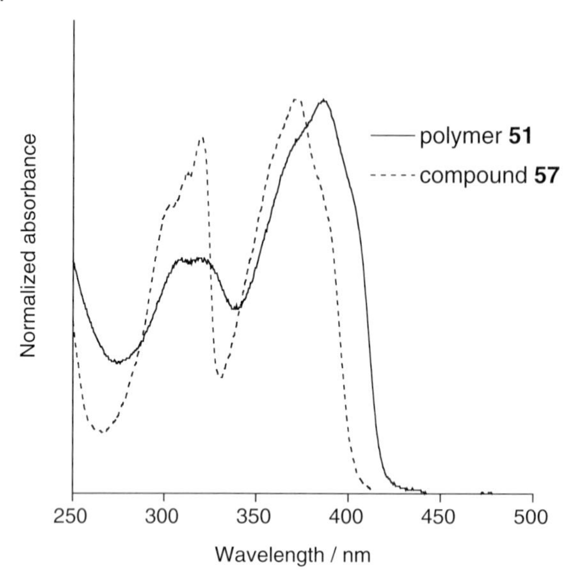

Figure 5.10 UV–vis absorption spectra of polymer **51** and compound **57** in $CHCl_3$.

typical PPEs appear at wavelengths of above 400 nm [47]. The through-bond conjugation is more effective for the extension of a conjugation length than the through-space conjugation.

Figure 5.11 shows the fluorescence emission spectra of polymer **51** and compound **57** in dilute $CHCl_3$ (1.0×10^{-7} M) at the excitation wavelength of each

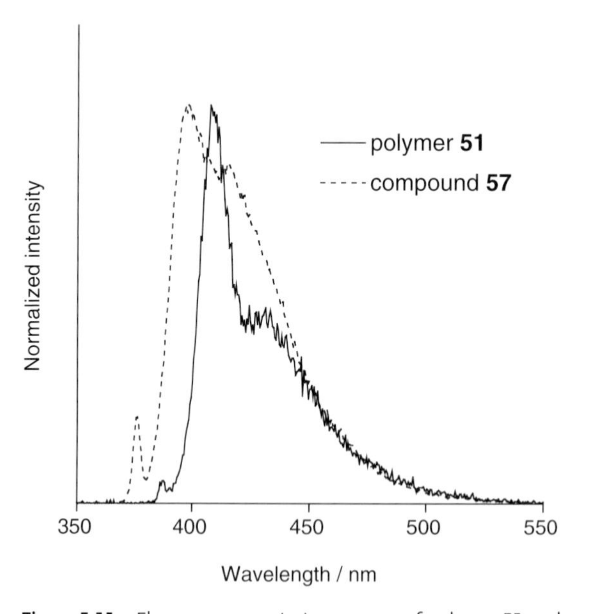

Figure 5.11 Fluorescence emission spectra of polymer **51** and compound **57** in $CHCl_3$.

absorption maximum. They emitted bright blue light, and their fluorescence quantum efficiencies were found to be 0.82 and 0.86, respectively. The spectrum of polymer **51** exhibited the appearance of a clear vibrational structure in spite of the π-stacked structure. It has been reported that cyclophane compounds have two types of emission mechanisms – emission from "the chromophore state" and emission from "the phane state" – depending on the conjugation length of one of the two π-systems facing each other and the type of overlap between them [21b, 45, 48]. The emission behavior shown in Figure 5.11 implies that the emission of polymer **51** was derived from the monomer state rather than the phane state.

Through-space conjugated polymer **60** comprising pseudo-*ortho*-linked [2.2]paracyclophane units was synthesized by a Sonogashira coupling reaction between pseudo-*ortho*-diethynyl[2.2]paracyclophane **58** [49] and diiodobenzene **59**, as shown in Scheme 5.19 [50]. Polymer **60** was obtained with a low M_n value of 3800 and an M_w value of 4900 due to the folded structure of the pseudo-*ortho*-linked [2.2]paracyclophane. The absorption spectrum of polymer **60** in dilute $CHCl_3$ also exhibited a typical π–π^* band of poly(*p*-arylene ethynylene)s, and its λ_{max} was observed at 377 nm. An emission peak in $CHCl_3$ solution (1.0×10^{-7} M) was observed at around 430 nm with a clear vibrational structure, and the emission efficiency was calculated to be 0.86. These optical profiles of pseudo-*ortho*-linked polymer **60** were similar to those of pseudo-*para*-linked polymer **51**, because both polymers consist of the same π-electron system, that is, the model compound **57**.

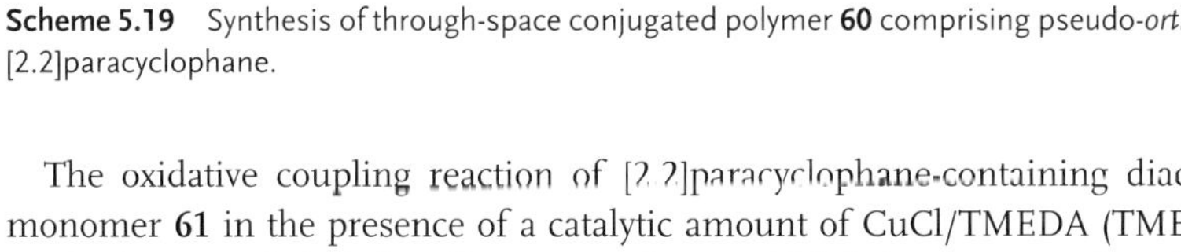

Scheme 5.19 Synthesis of through-space conjugated polymer **60** comprising pseudo-*ortho*-linked [2.2]paracyclophane.

The oxidative coupling reaction of [2.2]paracyclophane-containing diacetylene monomer **61** in the presence of a catalytic amount of CuCl/TMEDA (TMEDA: *N*,*N*,*N*′,*N*′-tetramethylethylenediamine) under O_2 yielded the corresponding through-space conjugated polymer **62** with butadiyne units, as shown in Scheme 5.20 [51]. An emission peak for polymer **62** was observed around 440 nm with a vibrational structure in dilute $CHCl_3$ solution (1.0×10^{-5} M), and its fluorescence quantum efficiency was estimated to be 0.45.

Conjugated polymer **63**, comprising disilylene and [2.2]paracyclophane units in a polymer backbone, was synthesized by the polycondensation reaction of 1,2-dichloro1,1,2,2-tetrahexyldisilane with bifunctional lithium acetylide from **43** and butyllithium (Scheme 5.21) [52]. The obtained polymer exhibited the extension of the conjugation length via Si–Si through-bond conjugation and the through-space conjugation.

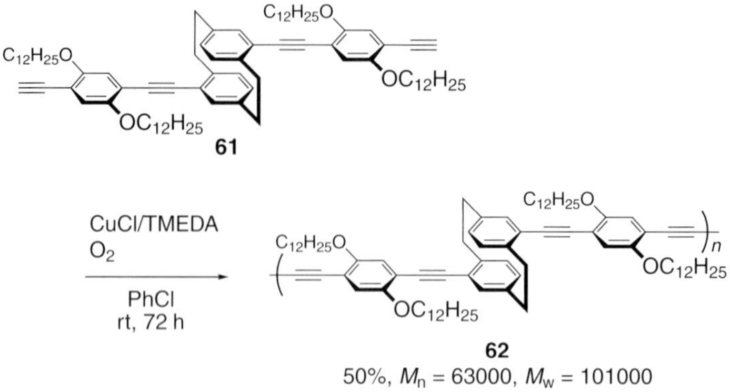

61

CuCl/TMEDA
O_2
PhCl
rt, 72 h

62
50%, M_n = 63000, M_w = 101000

Scheme 5.20 Synthesis of through-space conjugated polymer **62** by the oxidative coupling reaction.

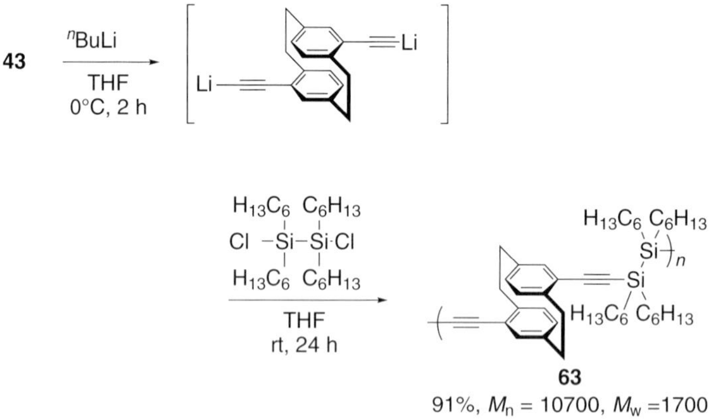

43

^{n}BuLi
THF
0°C, 2 h

$Cl-Si(C_6H_{13})_2-Si(C_6H_{13})_2-Cl$
THF
rt, 24 h

63
91%, M_n = 10700, M_w =1700

Scheme 5.21 Synthesis of through-space σ-π-conjugated polymer **63**.

Chujo and Naka reported the cycloaddition polymerization of aldothioketenes to obtain dithiafulvene-containing conjugated polymers [53]. This reaction could be applied to [2.2]paracyclophane monomer **43**, and the corresponding through-space conjugated polymer **64** consisting of dithiafulvene and [2.2]paracyclophane units in the main chain was synthesized in moderate yield (57%) with an M_n value of 7200, as shown in Scheme 5.22 [54]. This polymer had an extended conjugation length via the through-space interaction of the [2.2]paracyclophane unit. A thin film of polymer **64** exhibited electrical conductivities of 2×10^{-5} and 2×10^{-3} S cm^{-1} when doped with TCNQ and I_2, respectively, which were higher than those of the through-bond dithiafulvene analogue [53b].

As shown in Scheme 5.23, cyano-substituted poly(*p*-arylenevinylene) **67** containing [2.2]paracyclophane units in the main chain was prepared with a yield of 96% and an M_n value of 8000 by the Knoevenagel reaction of pseudo-*p*-bis(cyanomethyl)[2.2]

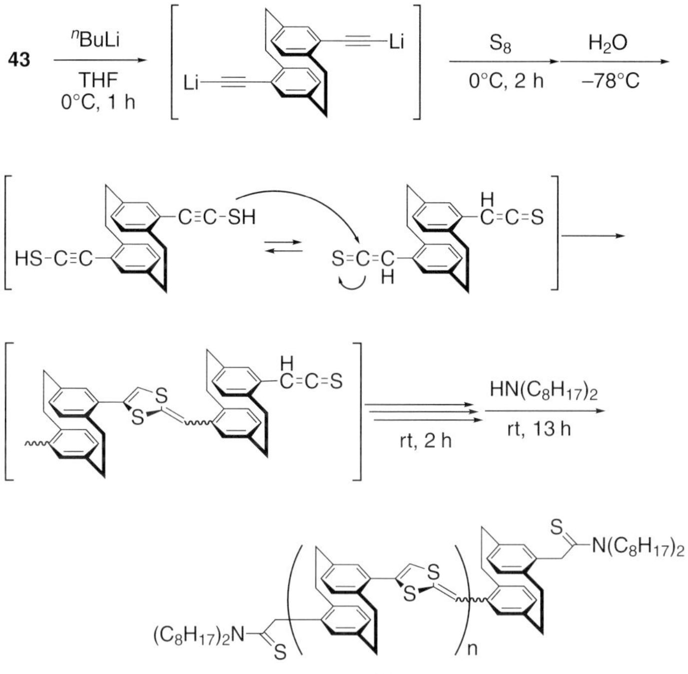

64
57%, $M_n = 7200$, $M_w = 9300$

Scheme 5.22 Synthesis of through-space conjugated polymer **64** by cycloaddition polymerization.

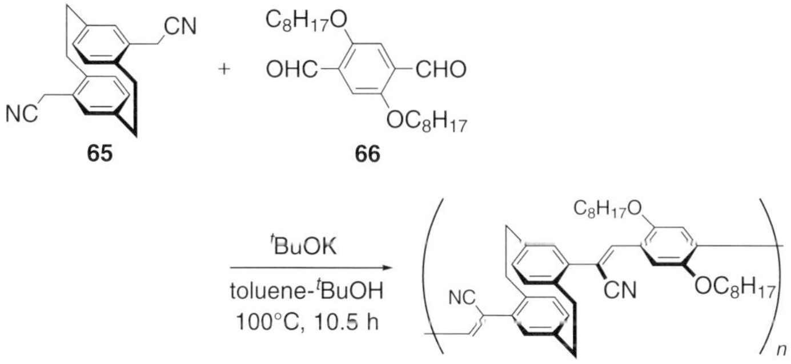

67
96%, $M_n = 8000$, $M_w = 18000$

Scheme 5.23 Synthesis of through-space conjugated polymer **67** by the Knoevenagel reaction.

paracyclophane **65** with 1,4-diformyl-2,5-dioctyloxybenzene **66** [55]. Polymer **67** exhibited an absorption maximum at 436 nm in dilute $CHCl_3$, which was attributed to the π–π^* transition bands of the conjugated poly(*p*-arylenevinylene) backbone. The absorption maximum ($\lambda_{max} = 436$ nm) of **67** exhibited clear bathochromic shifts

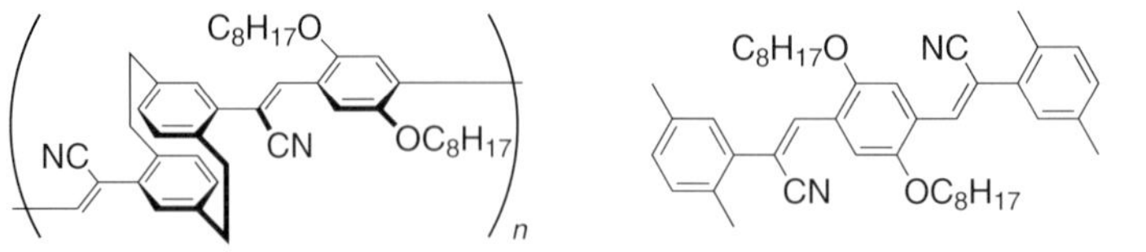

67 $\lambda_{abs,max}$ = 436 nm
$\lambda_{em,max}$ = 510 nm (Φ_F = 0.32)

68 $\lambda_{abs,max}$ = 413 nm
$\lambda_{em,max}$ = 491 nm (Φ_F = 0.30)

Figure 5.12 [2.2]Paracyclophane-containing through-space conjugated polymer **67** and model compound **68**.

relative to that of model compound **68** (λ_{max} = 413 nm, Figure 5.12), indicating the extension of the conjugation length as a result of through-space conjugation. The fluorescence emission spectra of polymer **67** and compound **68** were almost identical, and the quantum efficiencies of **67** and **68** were 32% and 30%, respectively. Excimer emission in dilute solution was not observed in the case of polymer **67**; this is because the emission of [2.2]paracyclophane-containing through-space conjugated polymers occurs from the monomer state rather than the phane state (excimer state) as discussed above.

The electrochemical polymerization of oligothienyl-substituted [2.2]paracyclophanes has been carried out independently by two research groups [56, 57]. As shown in Scheme 5.24, [2.2]paracyclophane-containing through-space conjugated polymer **69** was obtained by the electrochemical polymerization of monomer **68** during cycling past 1.0 V (vs. Ag/Ag^+) or by applying a potential at 0.9~1.2 V (vs. Ag/Ag^+) [56a]. A film of the corresponding polymer **69** was formed on the electrode surface. Thus, polymers **70–73** were prepared by electrochemical polymerization (Figure 5.13), and polymer **73** was also obtained by an oxidative coupling reaction in the presence of $FeCl_3$ [57b]. Polymer **73** was soluble in organic solvents such as hexane and $CHCl_3$ because of the long alkyl substituent, and the M_w value of the $CHCl_3$ soluble part was found to be 50 000. The cyclic voltammograms of the polymers showed a reversible redox wave with an $E_{1/2}$ value of approximately 1.0 V.

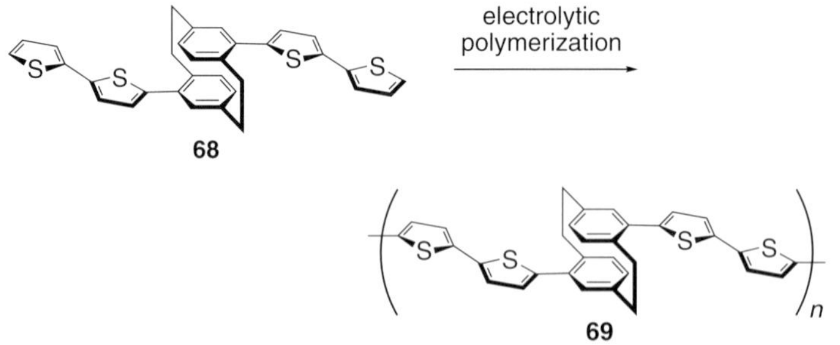

electrolytic polymerization

68

69

Scheme 5.24 Synthesis of through-space conjugated polymer **69** by electrochemical polymerization.

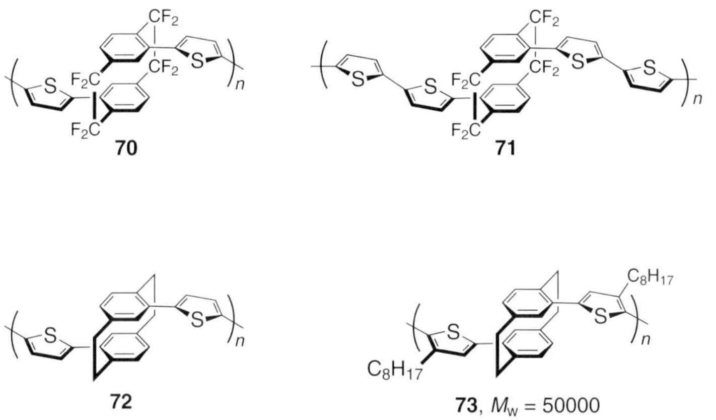

Figure 5.13 Oligothiophene-containing through-space conjugated polymers **70–73**.

5.4 Conclusion

The synthesis of through-space conjugated polymers consisting of a π-stacked structure has been introduced in this chapter. The number of through-space conjugated polymers discussed is limited. However, various interesting polymers composed of π-stacked arrays have recently been prepared by polycondensation reactions and chain polymerizations of vinyl monomers. From a synthetic point of view, a method of synthesizing aromatic ring-layered polymers based on a xanthene scaffold is available for constructing the π-stacked structure of a variety of aromatic ring and π-electron systems. The application of through-space conjugated polymers in optoelectronic devices such as field emission transistors, photovoltaic devices, and electroluminescence devices has not yet been extensively studied. Through-space conjugated polymers can potentially be used in these devices, as well as in a single molecule device, because the presence of the π-stacked structure and the orientation of the aromatic rings will continue to play an important role in the occurrence of effective charge or energy transfers in organic devices. The next key target in the field of through-space conjugated polymers is the development of a method for controlling the higher-ordered structures of this class of polymers. This development is anticipated in the near future.

Acknowledgments

We gratefully acknowledge financial support from the Ministry of Education, Culture, Sports, Science and Technology, Japan; Grant-in Aid for Young Scientists (B) (No. 16750096) and Grant-in Aid for Young Scientists (A) (No. 21685012). We also appreciate Research Fellowships from the Japan Society for the Promotion of Science for Young Scientists.

References

1 (a) Kraft, A., Grimsdale, A.C., and Holmes, A.B. (1998) *Angew. Chem. Int. Ed.*, **37**, 402–428; (b) Friend, R.H., Gymer, R.W., Holmes, A.B., Burroughes, J.H., Marks, R.N., Taliani, C., Bradley, D.D.C., Dos Santos, D.A., Bredas, J.L., Lögdlund, M., and Salaneck, W.R. (1999) *Nature*, **397**, 121–128; (c) Mitschke, U. and Bauerle, P. (2000) *J. Mater. Chem.*, **10**, 1471–1507; (d) Bernius, M.T., Inbasekaran, M., O'Brien, J., and Wu, W. (2000) *Adv. Mater.*, **121**, 1737–1750; (e) Akcelrud, L. (2003) *Prog. Polym. Sci.*, **28**, 875–962; (f) Müellen, K. and Scherf, U. (eds) (2006) *Organic Light Emitting Devices: Synthesis, Properties and Application*, Wiley-VCH Verlag GmbH, PL Weinheim, Germany.

2 (a) Sirringhaus, H. (2005) *Adv. Mater.*, **17**, 2411–2425; (b) Groza, J.R. and Locklin, J.J. (eds) (2007) *Organic Field-Effect Transistors*, CRC Press Taylor & Francis Group, New York;(c) Murphy, A.R. and Fréchet, J.M.J. (2007) *Chem. Rev.*, **107**, 1066–1096; (d) Takimiya, K., Kunugi, Y., and Otsubo, T. (2007) *Chem. Lett.*, **36**, 578–583; (e) Anthony, J.E. (2008) *Angew. Chem. Int. Ed.*, **47**, 452–483.

3 (a) O'Regan, B. and Graetzel, M. (1991) *Nature*, **353**, 737–740; (b) Brabec, C., Dyakonov, V., and Scherf, U. (eds) (2008) *Organic Photovoltaics: Materials, Device Physics, and Manufacturing Technologies*, Wiley-VCH Verlag GmbH, PL Weinheim, Germany.

4 (a) Salaneck, W.R., Clark, D.T., and Samuelsen, E.J. (eds) (1991) *Science and Applications of Conducting Polymers*, Adam Hilger, Bristol; (b) Nalwa, H.S. (1997) *Handbook of Organic Conductive Molecules*, John Wiley and Sons Ltd, Chichester; (c) Skotheim, T.A., Elsenbaumer, R.L., and Reynolds, J.R. (eds) (2006) *Handbook of Conducting Polymers*, 3rd edn, Marcel Dekker, PL New York.

5 (a) Watson, J.D. and Crick, F.H.C. (1953) *Nature*, **171**, 737–738; (b) Maiya, B.G. and Ramasama, T. (2001) *Curr. Sci.*, **80**, 1523–1530; (c) Lewis, F.D., Letsinger, R.L., and Wasielewski, M.R. (2001) *Acc. Chem. Res.*, **34**, 159–170; (d) Schuster, G.B. (2000) *Acc. Chem. Res.*, **33**, 253–260; (e) Delaney, S. and Barton, J.K. (2003) *J. Org. Chem.*, **68**, 6475–6483; (f) Barton, J.K., Carell, T., Dohno, C., Fiebig, T., Geacintov, N.E., Ito, T., Kawai, K., Lewis, F.D., Majima, T., and O'Neil, M.P. (2005) *Charge Transfer in DNA: From Mechanism to Application* (ed. A. Wagenknecht), Wiley-VCH Verlag GmbH, Weinheim, Germany, pp. 1–223.

6 (a) McDermott, G., Prince, S.M., Freer, A.A., Hawthornthwaite-Lowless, A.M., Papiz, M.Z., Cogdell, R.J., and Isaacs, N.W. (1995) *Nature*, **374**, 517–521; (b) Karrasch, S., Bullough, P.A., and Ghosh, R. (1995) *EMBO J.*, **14**, 631–638; (c) Koepke, J., Hu, X., Muenke, C., Schulten, K., and Michel, H. (1996) *Structure*, **4**, 581–597; (d) McLuskey, K., Prince, S.M., Cogdell, R.J., and Isaacs, N.W. (2001) *Biochemistry*, **40**, 8783–8789; (e) Roszak, A.W., Howard, T.D., Southall, J., Gardiner, A.T., Law, C.J., Isaacs, N.W., and Cogdell, R.J. (2003) *Science*, **302**, 1969–1972.

7 (a) Demus, D., Goodby, J.W., Gray, G.W., Spiess, H.W., and Vill, V. (eds) (1998) *Handbook of Liquid Crystals: Low Molecular Weight Liquid Crystals II*, Wiley-VCH Verlag GmbH, Weinheim, Germany; (b) Kato, T. (2002) *Science*, **295**, 2414–2418.

8 (a) Herwig, P., Kayser, C.W., Müllen, K., and Spiess, H.W. (1996) *Adv. Mater.*, **8**, 510–513; (b)van de Craats, A.M., Warman, J.M., Fechtenkötter, A., Brand, J.D., Harbison, M.A., and Müllen, K. (1999) *Adv. Mater.*, **11**, 1469–1472; (c) Shklyarevskiy, I.O., Jonkheijm, O., Stutzmann, N., Wasserberg, D., Wondergem, H.J., Christianen, P.C.M., Schenning, A.P.H.J., de Leeuw, D.M., Tomoviú, Z., Wu, J., and Müllen, K. (2005) *J. Am. Chem. Soc.*, **127**, 16233–16237; (d) Xiao, S., Myers, M., Miao, Q., Sanaur, S., Pang, K., Steigerwald, M.L., and Nuckolls, C. (2005) *Angew. Chem. Int. Ed.*, **44**, 7390–7394.

9 (a) Story, V.M. and Canty, G. (1964) *J. Res. Mater. Bur. Stand.*, **68**, 165–171; (b) Flowers, R.G. and Miller, H.F. (1947) *J. Am. Chem. Soc.*, **69**, 1388–1389;

(c) Imoto, M. and Takemoto, K. (1955) *J. Polym. Sci.*, **15**, 271–276.

10 (a) Dziewonski, K. and Leyko, Z. (1914) *Ber. Dtsch. Chem. Ges.*, **47**, 1679–1690; (b) Dziewonski, K. and Stolyhwo, T. (1924) *Ber. Dtsch. Chem. Ges.*, **57**, 1531–1540; (c) Jones, J.I. (1951) *J. Appl. Chem.*, **1**, 568–576; (d) Kaufman, M. and Williams, A.F. (1951) *J. Appl. Chem.*, **1**, 489–503; (e) Mohorcic, G. (1957) *Bull. Sci. (Zagreb)*, **3**, 105–106; (f) Manecke, G. and Danhäuser, J. (1962) *Makromol. Chem.*, **56**, 208–223; (g) Barrales-Rienda, J.M., and Pepper, D.C. (1967) *Polymer*, **8**, 337–349.

11 (a) Schneider, F. and Springer, J. (1971) *Makromol. Chem.*, **146**, 181–193; (b) David, C., Lempereur, M., and Geuskens, G. (1972) *Eur. Polym. J.*, **8**, 417–427; (c) David, C., Piens, M., and Geuskens, G. (1972) *Eur. Polym. J.*, **8**, 1019–1031; (d) Wang, Y.-C. and Morawetz, H. (1975) *Makromol. Chem. Suppl.*, **1**, 283–295; (e) Phillips, D., Roberts, A.J., and Soutar, I. (1980) *J. Polym. Sci. Polym. Lett. Ed.*, **18**, 123–129; (f) Smith, T.A. and Ghiggino, K.P. (2006) *Polym. Int.*, **55**, 772–779.

12 Mendicuti, F., Kulkarni, R., Patel, B., and Mattice, W.L. (1990) *Macromolecules*, **23**, 2560–2566.

13 (a) Nakano, T., Takewaki, K., Yade, T., and Okamoto, Y. (2001) *J. Am. Chem. Soc.*, **123**, 9182–9183; (b) Nakano, T. and Yade, T. (2001) *J. Am. Chem. Soc.*, **125**, 15474–15484; (c) Nakano, T., Yade, T., Yokoyama, M., and Nagayama, N. (2004) *Chem. Lett.*, **33**, 296–297; (d) Nakano, T., Yade, T., Fukuda, Y., Yamaguchi, T., and Okumura, S. (2005) *Macromolecules*, **38**, 8140–8148; (e) Yade, T. and Nakano, T. (2006) *J. Polym. Sci. Part A: Polym. Chem.*, **44**, 561–572; (f) Nakano, T. and Yade, T. (2008) *Chem. Lett.*, **37**, 258–259.

14 (a) Evans, A.G. and George, D.B. (1961) *J. Chem. Soc.*, 4653–4659; (b) Evans, A.G. and George, D.B. (1962) *J. Chem. Soc.*, 141–146; (c) Yuki, H., Hotta, J., Okamoto, Y., and Murahashi, S. (1967) *Bull. Chem. Soc. Jpn.*, **40**, 2659–2663; (d) Richards, D.H. and Scilly, N.F. (1969) *J. Polym. Soc., Polym. Lett.*, **7**, 99–101.

15 Rathore, R., Abdelwashed, S.H., and Guzei, I.A. (2003) *J. Am. Chem. Soc.*, **125**, 8712–8713.

16 Londergan, T.M., Teng, C.J., and Weber, W.P. (1999) *Macromolecules*, **32**, 1111–1114.

17 (a) Cappelli, A., Peicot Mohr, G., Anzini, M., Vomero, S., Donati, A., Casolaro, M., Mendichi, R., Giorgi, G., and Makovec, F. (2003) *J. Org. Chem.*, **68**, 9473–9476; (b) Cappelli, A., Anzini, M., Vomero, S., Donati, A., Zetta, L., Mendichi, R., Casolaro, M., Lupetti, P., Salvatici, P., and Giorgi, G. (2005) *J. Polym. Sci., Part A Polym. Chem.*, **43**, 3289–3304; (c) Cappelli, A., Galeazzi, S., Giuliani, G., Anzini, M., Donati, A., Zetta, L., Mendichi, R., Aggravi, M., Giorgi, G., Paccagnini, E., and Vomero, S. (2007) *Macromolecules*, **40**, 3005–3014; (d) Cappelli, A., Simone, G., Giuliani, G., Anzini, M., Aggravi, M., Donati, A., Zetta, L., Boccia, A.C., Mendichi, R., Giorgi, G., Paccagnini, E., and Vomero, S. (2008) *Macromolecules*, **41**, 2324–2334.

18 Bergmann, E.D. (1968) *Chem. Rev.*, **68**, 41–84.

19 Jenekhe, S.A., Alam, M.M., Zhu, Y., Jiang, S., and Shevade, A.V. (2007) *Adv. Mater.*, **19**, 536–542.

20 (a) Shetty, A.S., Liu, E.B., Lachicotte, R.J., and Jenekhe, S.A. (1999) *Chem. Mater.*, **1**, 2292–2295; (b) Zhu, Y., Alam, M.M., and Jenekhe, S.A. (2003) *Macromolecules*, **36**, 8958–8968; (c) Tonzola, C.T., Alam, M.M., Bean, B.A., and Jenekhe, S.A. (2004) *Macromolecules*, **37**, 3554–3563.

21 (a) Vögtle, F. (1993) *Cyclophane Chemistry: Synthesis, Structures and Reactions*, John Wiley & Sons Ltd, Chichester; (b) Gleiter, R. and Hopf, H. (eds) (2004) *Modern Cyclophane Chemistry*, Wiley-VCH Verlag GmbH, Weinheim, Germany,(c) Hopf, H. (2008) *Angew. Chem. Int. Ed.*, **47**, 9808–9812.

22 Morisaki, Y. and Chujo, Y. (2008) *Prog. Polym. Sci.*, **33**, 346–364.

23 (a) Glatzhofer, D.T. and Longone, D.T. (1986) *J. Polym. Sci., Part A Polym. Chem.*, **24**, 947–954; (b) Longone, D.T. and Glatzhofer, D.T. (1986) *J. Polym. Sci., Part A Polym. Chem.*, **24**, 1725–1733; (c) Ulański, J., Kubacki, J., Glowacki, I.,

Kryszewski, M., and Glatzhofer, D.T. (1992) *J. Appl. Polym. Sci.*, **44**, 2103–2106; (d) Guerrero, D.J. and Glatzhofer, D.T. (1994) *J. Polym. Sci., Part A Polym. Chem.*, **32**, 457–464.

24 (a) Nishimura, J. and Yamashita, S. (1982) in *Cyclopolymerization and Polymers with Chain-Ring Structures* (eds G.B. Butler and J.E. Kresta), American Chemical Society, Washington, D.C., pp. 177–195;(b) Furukawa, J. and Nishimura, J. (1976) *J. Polym. Sci., Polym. Lett. Ed.*, **14**, 85–90; (c) Furukawa, J. and Nishimura, J. (1976) *J. Polym. Sci. Polym. Symp.*, **56**, 437–446; (d) Nishimura, J. and Yamashita, S. (1979) *Polym. J.*, **11**, 619–627; (e) Nishimura, J., Mimura, M., Nakazawa, N., and Yamashita, S. (1980) *J. Polym. Sci., Polym. Lett. Ed.*, **18**, 2071–2084; (f) Nishimura, J., Furukawa, M., Yamashita, S., Inazu, T., and Yoshino, T. (1981) *J. Polym. Sci., Polym. Lett. Ed.*, **19**, 3257–3268.

25 Otsubo, T., Kitasawa, M., and Misumi, S. (1979) *Bull. Chem. Soc. Jpn.*, **52**, 1515–1520.

26 (a) Heerwein, H. (1948) *Angew. Chem.*, **60**, 78;(b) Kantor, S.W. and Osthoff, R.C. (1953) *J. Am. Chem. Soc.*, **75**, 931–932; (c) Feltzin, J., Restaino, A.J., and Mesrobian, R.B. (1955) *J. Am. Chem. Soc.*, **77**, 206–210; (d) Bawn, C.E.H., Ledwith, A., and Matthies, P. (1958) *J. Polym. Sci.*, **33**, 21–26; (e) Cowel, G.W. and Ledwith, A. (1970) *Quart. Rev.*, **24**, 119;(f) Imoto, M. and Nakaya, T. (1972) *J. Macromol. Sci., Rev. Macromol. Chem.*, **C7**, 1–48; (g) Mucha, M. and Wunderlich, B. (1974) *J. Polym. Sci., Polym. Phys. Ed.*, **12**, 1993; (h) Ihara, E., Haida, N., Iio, M., and Inoue, K. (2003) *Macromolecules*, **36**, 36–41; (i) Ihara, E., Fujioka, M., Haida, N., Itoh, T., and Inoue, K. (2005) *Macromolecules*, **38**, 2101–2108; (j) Ihara, E., Nakada, A., Itoh, T., and Inoue, K. (2006) *Macromolecules*, **39**, 6440–6444; (k) Ihara, E., Kida, M., Fujioka, M., Itoh, T., and Inoue, K. (2007) *J. Polym. Sci., Part A: Polym. Chem.*, **45**, 1536–1545; (l) Liu, L., Song, Y., and Li, H. (2002) *Polym. Int.*, **51**, 1047–1049; (m) Hetterscheid, D.G., Hendriksen, C., Dzik, W.I., Smits, J.M.M., Eck, E.R.H.v., Rowan, A.E., Busico, V., Vacatello, M., Castelli, V.V.A., Segre, A., Jellema, E., Bloemberg, T.G., and Bruin, B.d. (2006) *J. Am. Chem. Soc.*, **128**, 9746–9752; (n) Noels, A.F. (2007) *Angew. Chem. Int. Ed.*, **46**, 1208–1210;
(o) Bai, J., Burke, L.D., and Shea, K.J. (2007) *J. Am. Chem. Soc.*, **129**, 4981–4991.

27 Wada, N., Morisaki, Y., and Chujo, Y. (2009) *Macromolecules*, **42**, 1439–1442.

28 Iwatsuki, S., Itoh, T., Kubo, M., and Okuno, H. (1994) *Polym. Bull.*, **32**, 27–34.

29 (a) Weigang, O.E. Jr. and Nugent, M.J. (1969) *J. Am. Chem. Soc.*, **91**, 4556–4558; (b) Rosini, C., Ruzziconi, R., Superchi, S., Fringuelli, F., and Piermatti, O. (1998) *Tetrahedron: Asymmetry*, **9**, 55–62.

30 (a) Pearson, J.M. and Stolka, M. (1981) *Poly(N-vinylcarbazole)*, Gordon and Breach, New York, NY;(b) Grazulevicius, J.V., Strohriegl, P., Pielichowski, J., and Pielichowski, K. (2003) *Prog. Polym. Sci.*, **28**, 1297–1353.

31 (a) Hill, D.J., Mio, M.J., Prince, R.B., Hughes, T.S., and Moore, J.S. (2001) *Chem. Rev.*, **101**, 3893–4011; (b) Cubberley, M.S. and Iverson, B.L. (2001) *Curr. Opin. Chem. Biol.*, **5**, 650–653.

32 (a) García Martínez, A., Osío Barcina, J., de Fresno Cerezo, A., Schlüter, A.-D., and Frahn, J. (1999) *Adv. Mater.*, **11**, 27–31; (b) Caraballo Martínez, N., del Rosario Colorado Heras, M., Mba Blázquez, M., Osío Barcina, J., García Martínez, A., and del Rosario Torres Salvador, M. (2007) *Org. Lett.*, **9**, 2943–2946.

33 (a) Morisaki, Y. and Chujo, Y. (2005) *Tetrahedron Lett.*, **46**, 2533–2537; (b) Morisaki, Y., Murakarni, T., and Chujo, Y. (2008) *Macromolecules*, **41**, 5960–5963; (c) Morisaki, Y., Murakarni, T., and Chujo, Y. (2009) *J. Inorg. Organomet. Polym. Mater.*, **19**, 104–112; (d) Morisaki, Y., Murakarni, T., Sawamura, T., and Chujo, Y. (2009) *Macromolecules.*, **42**, 3656–3660.

34 Förster, T. (1946) *Naturwissenschaften*, **33**, 166–175.

35 Morisaki, Y., Imoto, H., Miyake, J., and Chujo, Y. (2009) *Macromol. Rapid Commun.*, **30**, 1094–1100.

36 Morisaki, Y., Fernandes, J.A., Wada, N., and Chujo, Y. (2009) *J. Polym. Sci., Part A: Polym. Chem.*, **47**, 4279–4288.

37 (a) Miyaura, N., and Suzuki, A. (1979) *J. Chem. Soc., Chem. Commun.*, 866–867;

(b) Miyaura, N. and Suzuki, A. (1995) *Chem. Rev.*, **95**, 2457–2483; (c) Walker, S.D., Barder, T.E., Martinelli, J.R., and Buchwald, S.L. (2004) *Angew. Chem., Int. Ed.*, **43**, 1871–1876.

38 (a) Arnold, R., Matchett, S.A., and Rosenblum, M. (1988) *Organometallics*, **7**, 2261–2266; (b) Nugent, H.M., Rosenblum, M., and Klemarczky, P. (1993) *J. Am. Chem. Soc.*, **115**, 3848–3849; (c) Rosenblum, M., Nugent, H.M., Jang, K.-S., Labes, M.M., Cahalane, W., Klemarczyk, P., and Reiff, W.M. (1995) *Macromolecules*, **28**, 6330–6342; (d) Hudson, R.D.A., Foxman, B.M., and Rosenblum, M. (1999) *Organometallics*, **18**, 4098–4106.

39 (a) Mizogami, S. and Yoshimura, S. (1985) *J. Chem. Soc. Chem. Commun.*, 1736–1738; (b) Mizogami, S. and Yoshimura, S. (1985) *J. Chem. Soc. Chem. Commun.*, 427–428.

40 (a) Morisaki, Y., Ishida, T., and Chujo, Y. (2009) *C. R. Chim.*, **12**, 332–340; (b) Morisaki, Y., Ishida, T., and Chujo, Y. (2003) *Polym. J.*, **35**, 501–506; (c) Morisaki, Y., Ishida, T., and Chujo, Y. (2006) *Polym. Bull.*, **57**, 623–630; (d) Morisaki, Y., Ishida, T., and Chujo, Y. (2006) *Org. Lett.*, **8**, 1029–1032.

41 (a) Semmelhack, M.F., Harrison, J.J., Young, D.C., Gutiérrez, A., Rafii, S., and Clardy, J. (1985) *J. Am. Chem. Soc.*, **107**, 7508–7514; (b) Mitchell, R.H. (2002) *J. Am. Chem. Soc.*, **124**, 2352–2357.

42 (a) Morisaki, Y. and Chujo, Y. (2006) *Angew. Chem. Int. Ed.*, **45**, 6430–6437; (b) Morisaki, Y., and Chujo, Y. (2009) *Bull. Chem. Soc. Jpn.*, **82**, 1070–1082.

43 (a) Morisaki, Y. and Chujo, Y. (2002) *Chem. Lett.*, 194–195; (b) Morisaki, Y. and Chujo, Y. (2004) *Macromolecules*, **37**, 4099–4103; (c) Morisaki, Y. and Chujo, Y. (2005) *Bull. Chem. Soc. Jpn.*, **78**, 288–293.

44 (a) Morisaki, Y. and Chujo, Y. (2002) *Macromolecules*, **35**, 587–589; (b) Morisaki, Y., Ishida, T., and Chujo, Y. (2002) *Macromolecules*, **35**, 7872–7877; (c) Morisaki, Y. and Chujo, Y. (2003) *Macromolecules*, **36**, 9319–9324; (d) Morisaki, Y., Wada, N., and Chujo, Y. (2005) *Polymer*, **46**, 5884–5889; (e) Morisaki, Y., Ishida, T., Tanaka, H., and Chujo, Y. (2004) *J. Polym. Sci. Part A. Polym. Chem.*, **42**, 5891–5899; (f) Morisaki, Y., Wada, N., and Chujo, Y. (2005) *Polym. Bull.*, **53**, 73–80.

45 Bazan, G.C., Oldham, W.J. Jr., Lachicotte, R.J., Tretiak, S., Chernyak, V., and Mukamel, S. (1998) *J. Am. Chem. Soc.*, **120**, 9188–9204.

46 Reich, H.J. and Cram, D.J. (1969) *J. Am. Chem. Soc.*, **91**, 3527–3533.

47 (a) Bunz, U.H.F. (2000) *Chem. Rev.*, **100**, 1605–1644; (b) Weder, C. (ed.) (2005) *Poly (arylene ethynylene)s: From Synthesis to Application, Adv. Polym. Sci.*, vol. 177, Springer, PL Berlin.

48 (a) Oldham, W.J., Miao, Y.J., Laghicotte, R.J., and Bazan, G.C. (1998) *J. Am. Chem. Soc.*, **120**, 419–420; (b) Wang, S., Bazan, G.C., Tretiak, S., and Mukamel, S. (2000) *J. Am. Chem. Soc.*, **122**, 1289–1297.

49 (a) Pye, P.J., Rossen, K., Reamer, R.A., Tsou, N.N., Volante, R.P., and Reider, P.J. (1997) *J. Am. Chem. Soc.*, **119**, 6207–6208; (b) Braddock, D.C., MacGilp, I.D., and Perry, B.G. (2002) *J. Org. Chem.*, **67**, 8679–8681; (c) Bondarenko, L., Dix, I., Hinrichs, H., and Hopf, H. (2004) *Synthesis*, **16**, 2751–2759.

50 Morisaki, Y., Wada, N., Arita, M., and Chujo, Y. (2009) *Polym. Bull.*, **62**, 305–314.

51 Morisaki, Y. and Chujo, Y. (2002) *Polym. Bull.*, **49**, 209–215.

52 Morisaki, Y., Fujimura, F., and Chujo, Y. (2003) *Organometallics*, **22**, 3553–3557.

53 (a) Naka, K., Uemura, T., and Chujo, Y. (1998) *Macromolecules*, **31**, 7570–7571; (b) Naka, K., Uemura, T., and Chujo, Y. (1999) *Macromolecules*, **32**, 4641–4646; (c) Naka, K., Uemura, T., and Chujo, Y. (2000) *Polym. J.*, **32**, 435–439.

54 Morisaki, Y., Lin, L., and Chujo, Y. (2009) *Polym. Bull.*, **62**, 737–747.

55 (a) Morisaki, Y., Lin, L., and Chujo, Y. (2009) *Chem. Lett.*, **38**, 734–735; (b) Morisaki, Y., Lin, L., and Chujo, Y. (2009) *J. Polym. Sci. Part A: Polym. Chem.*, **47**, 5979–5988.

56 (a) Guyard, L. and Audebert, P. (2001) *Electrochem. Commun.*, **3**, 164–167; (b) Guyard, L., Audebert, P., Dolbier, W.R. Jr., and Duan, J.-X. (2002) *J. Electroanal. Chem.*, **537**, 189–193.

57 (a) Salhi, F., Lee, B., Metz, C., Bottomley, L.A., and Collard, D.M. (2002) *Org. Lett.*, **4**, 3195–3198; (b) Salhi, F. and Collard, D.M. (2003) *Adv. Mater.*, **15**, 81–85.

6
Fully Conjugated Nano-Sized Macrocycles: Syntheses and Versatile Properties

Masayoshi Takase and Masahiko Iyoda

6.1
Introduction

One of the fascinating subjects in materials, chemical, and physical sciences is the creation of fully π-conjugated macromolecules with well-defined shapes, since such materials have potential applications in organic electronics [1]. There are three main structural categories of these macromolecules: linear π-conjugated oligo-/polymers [2], two-dimensional (2D) polycyclic aromatic hydrocarbons (PAHs) [3], and π-conjugated macrocycles [3]. With recent developments in synthetic and analytical techniques, not only their intrinsic characters, such as electronic and photophysical properties in the bulk state, but also their supramolecular nanostructures are of major interest from the fundamental and materials science viewpoints. In particular, for applications in field-effect transistors (FET), photovoltaic devices, organic light emitting diodes (OLEDs), charge/ion/gas storage devices, as well as in catalysis, one-dimensional (1D) and 2D supramolecular nanostructures are especially appealing [4]. In order to control these nanostructures, it would be desirable to utilize largely π-extended giant molecules, because it is easy to construct and predict the nanostructures formed following a bottom-up strategy.

Among them, cyclic structures possessing shape-persistent, non-collapsible, and fully π-conjugated backbones are expected to build up columnar 1D nanotubes and 2D porous networks [5]. Fundamentally, macrocycles are regarded as an infinite π-conjugation model with an inner cavity, and thus these conjugated π-systems have attracted considerable attention from the viewpoints of their unusual optical and magnetic behavior based on their effective circular conjugations [2, 3]. Moreover, since macrocycles have interior and exterior sites, site-specific substitution at either or both sites can make the structure attractive for materials science [5]. Based on these perspectives, relatively small shape-persistent π-conjugated cyclic compounds are known as host molecules for fullerenes [6] and ionic species [7]. However, not very many nanoscale giant macrocycles have been reported so far, although they can be recognized as one of the fundamentally most important classes of intermediate

Conjugated Polymer Synthesis. Edited by Yoshiki Chujo

ISBN: 978-3-527-32267-1

structures between linear π-conjugated oligo-/polymers and infinite π-conjugated cyclic structures.

Generally, repeated, and thus tedious, synthetic procedures are necessary in order to build up cyclic structures, which often hampers experimental exploration of such materials. In this regard, molecular design and synthesis are key issues. A variety of different aromatic building blocks have so far been chosen, such as benzene, thiophene, pyrrole, carbazole, fluorene, pyridine, and others, leading to the establishment of structure–property correlations of the bulk materials, single molecules, and supramolecular structures at different nanoscales. In contrast, the conjugation among these building blocks via carbon–carbon triple/double bonds and carbon–nitrogen double bonds is provided by various methods such as transition metal-catalyzed cross coupling, alkyne/alkene metathesis, and dynamic condensation of carbonyl and amine groups with or without template assistance. It should be pointed out that the selection of synthetic methods is the most important issue in order to create expanded π-systems.

As there are many reviews on π-conjugated macrocycles [3], we here focus on the recent developments in planar or nearly planar giant π-conjugated macrocycles with particular regard to the synthetic methods and their versatile properties. Therefore, the synthesis and properties of large non-conjugated cyclic structures are not included in this chapter. Schlüter's cyclic oligophenylene [8], the cyclic porphyrin arrays of Sanders [9], Lindsey [10], Gossauer [11] and Osuka [12], Sessler's cyclo[n]pyrroles [13], and, recently, Höger's molecular spoked wheels [14] are all giant macrocycles, and masterpieces of this work, but are not dealt with here. However, they are included in other reviews [3a,b,l,m]. Furthermore, expanded porphyrins and porphyrinoids are a structurally interesting class of novel planar/non-planar aromatic macrocycles, some even recognized as Möbius ring systems, recently developed by Osuka [15] and Latos-Grazienski [16] *et al.* These topics are summarized in another review [17] and are not included in this chapter.

6.2
Synthesis of π-Conjugated Macrocycles

6.2.1
One-Pot Synthesis of Phenylacetylene Macrocycles

Various kinds of transition metal-catalyzed cross coupling reactions have been utilized to make a π-conjugated sp or sp^2 linkage, such as the Sonogashira reaction, Glaser, Hay, and Eglinton couplings, the Heck reaction, Stille coupling, McMurry coupling, and metathesis [18]. Furthermore, dynamic imine formation, Wittig/Horner, and Knoevenagel reactions can also be employed to form carbon–nitrogen or carbon–carbon double bonds [19]. Among various kinds of synthetic methods, "one-pot" reactions for the synthesis of macrocycles using repeated monomer units are an effective tool for constructing cyclic structures, because one can easily produce macrocyclic structures without troublesome multi-step synthetic

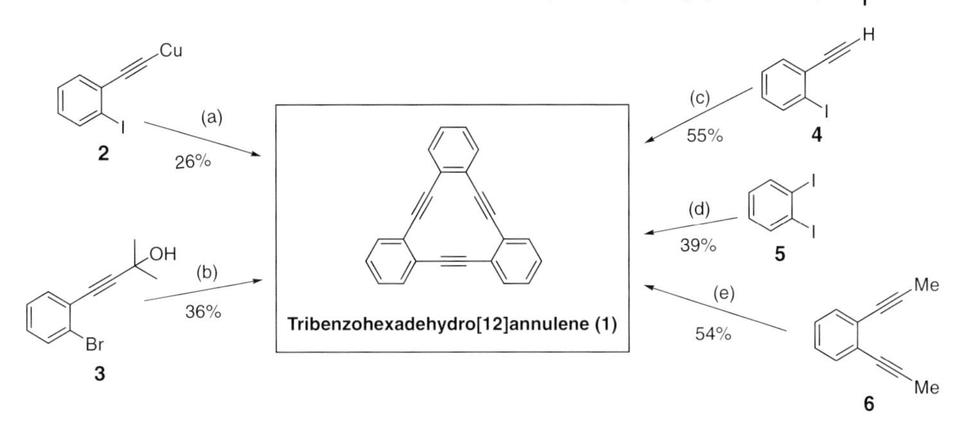

Scheme 6.1 "One-pot" synthesis of tribenzohexadehydro[12]annulene. Reagents and conditions: (a) pyridine, reflux [20a]; (b) [Pd $(PPh_3)_4$], CuI, PhH, NaOH, BTAC, 85 °C [21]; (c) CuI, PPh_3, K_2CO_3, DMF, 160 °C [22]; (d) HC≡CH (excess), [$Pd(PPh_3)_2Cl_2$], CuI, piperidine, DMF, 60 °C [23]; (e) [$(tBuO)_3W$≡CtBu], toluene, 80 °C [24].

procedures. The representative examples of "one-pot" cyclooligomerization of tribenzohexadehydro[12]annulene **1** are shown in Scheme 6.1. In 1966, Eglinton *et al.* [20a,b] reported the "one-pot" synthesis of **1** via the Castro–Stephens reaction of (*o*-iodophenyl)ethynyl copper **2** [20c]. Huynh *et al.* used *in situ* deprotection of the terminal alkyne **3** and the subsequent Sonogashira reaction to give **1** in an improved yield [21]. Later, our group further improved the synthesis using 1-ethynyl-2-iodobenzene **4** and catalytic amounts of CuI and PPh_3, together with 3 equiv of K_2CO_3 in DMF [22] and also carried out palladium-catalyzed cyclotrimerization of *o*-diiodobenzene **5** in the presence of acetylene gas [23]. In 2003, Vollhardt *et al.* used alkyne metathesis of 1,2-di(prop-1-ynyl)benzene **6** with tungsten reagent $(tBuO)_3W$≡C*t*Bu to form **1** in 54% yield [24], which method was debuted by Bunz *et al.* for the synthesis of π-conjugated macrocycles, as described below.

Other macrocycles **7**–**14** which were synthesized by one-pot cyclooligomerization of a repeated monomer unit are shown in Figure 6.1 [25–33]. As can easily be recognized, the formation of diynes has, in most cases, been employed for the cyclizations, indicating the utility of the Glaser-type diyne formation. To achieve high yields or even simply get the cyclic structure, however, it is difficult to prepare preorganized rigid structures or to design building blocks to shape the direction of cyclization. In a few case the structures are formed in high yields, 52% for **11** and 60–65% for **13**, although for both cases multi-step syntheses are involved in preparing the precursors.

Interesting approaches were demonstrated for the syntheses of **16**, **18**, and **20**, as shown in Scheme 6.2. Syntheses of *meta*-phenyleneacetylene macrocycles **16** and **18** were carried out from various precursors such as a linear oligomer (60–80%) [34], a half-cycle (45%) [32, 35], or (3-iodophenyl)ethynyl copper (4.6%) [36]. Based on the pioneering work of polymer synthesis by Weiss and Müllen *et al.* using alkyne metathesis [37], Bunz *et al.* developed the synthesis of *meta*-phenyleneacetylene

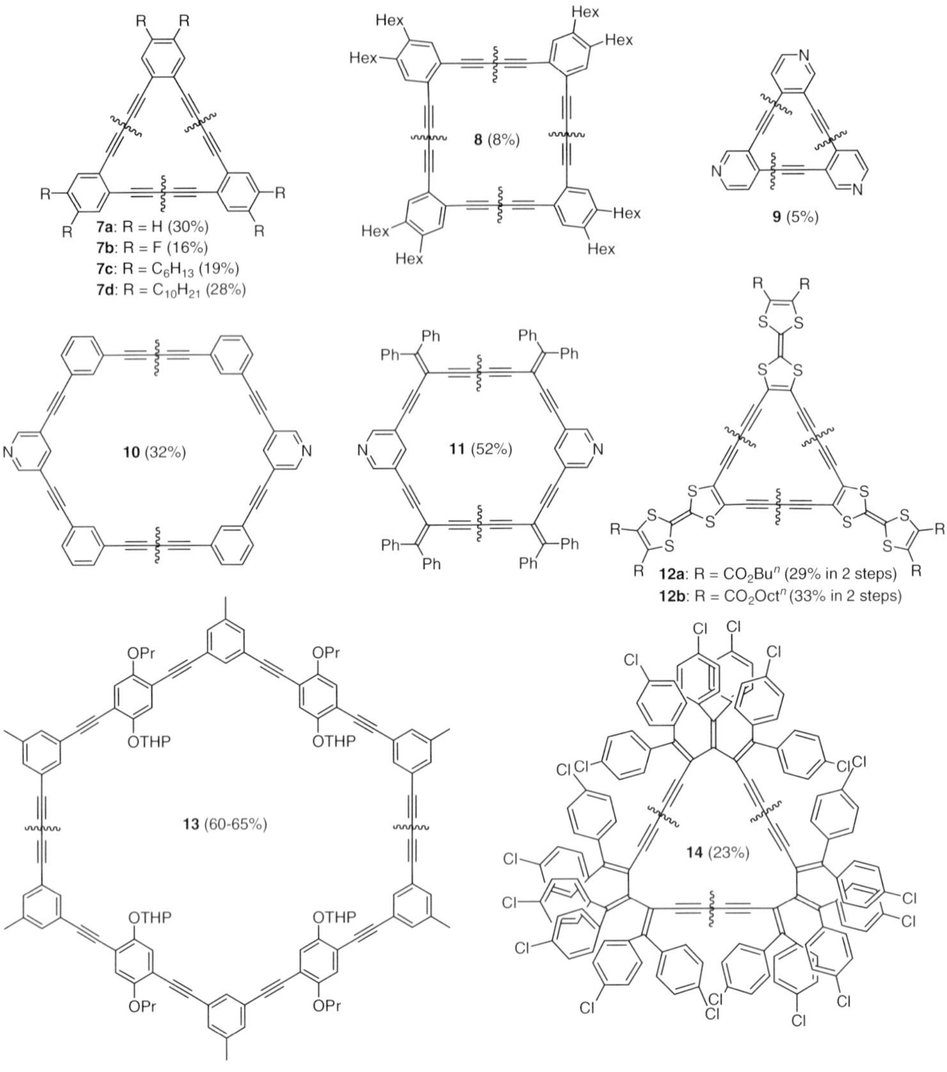

7a: R = H (30%)
7b: R = F (16%)
7c: R = C_6H_{13} (19%)
7d: R = $C_{10}H_{21}$ (28%)

8 (8%)

9 (5%)

10 (32%)

11 (52%)

12a: R = CO_2Bu^n (29% in 2 steps)
12b: R = CO_2Oct^n (33% in 2 steps)

13 (60-65%)

14 (23%)

Figure 6.1 Other examples of "one-pot" synthesis for conjugated cycles **7a**,**7b** [25], **7c** [26], **7d** [27], **8** [26], **9** [28], **10** [29], **11** [30], **12** [31], **13** [32], **14** [33]. The yields from their precursors are indicated in parentheses.

macrocycles **16** with *in situ* catalysts [38]. Although the yields (0.5–6%), and quantities (2.3–30 mg) of the target compounds are quite low, this was an important first step for the synthesis of highly symmetrical cyclic structures. By using highly active Mo(VI) alkylidyne catalysts prepared by a reductive recycle strategy [39a,b], the yields of **18** and **20** were markedly improved by Moore *et al.*, with thermodynamically stable macrocycles being preferentially formed [39c,d] (Scheme 6.2). The macrocycle **20b,** composed of 3,6-diethynylated carbazole, self-assembled to form nanofibril through

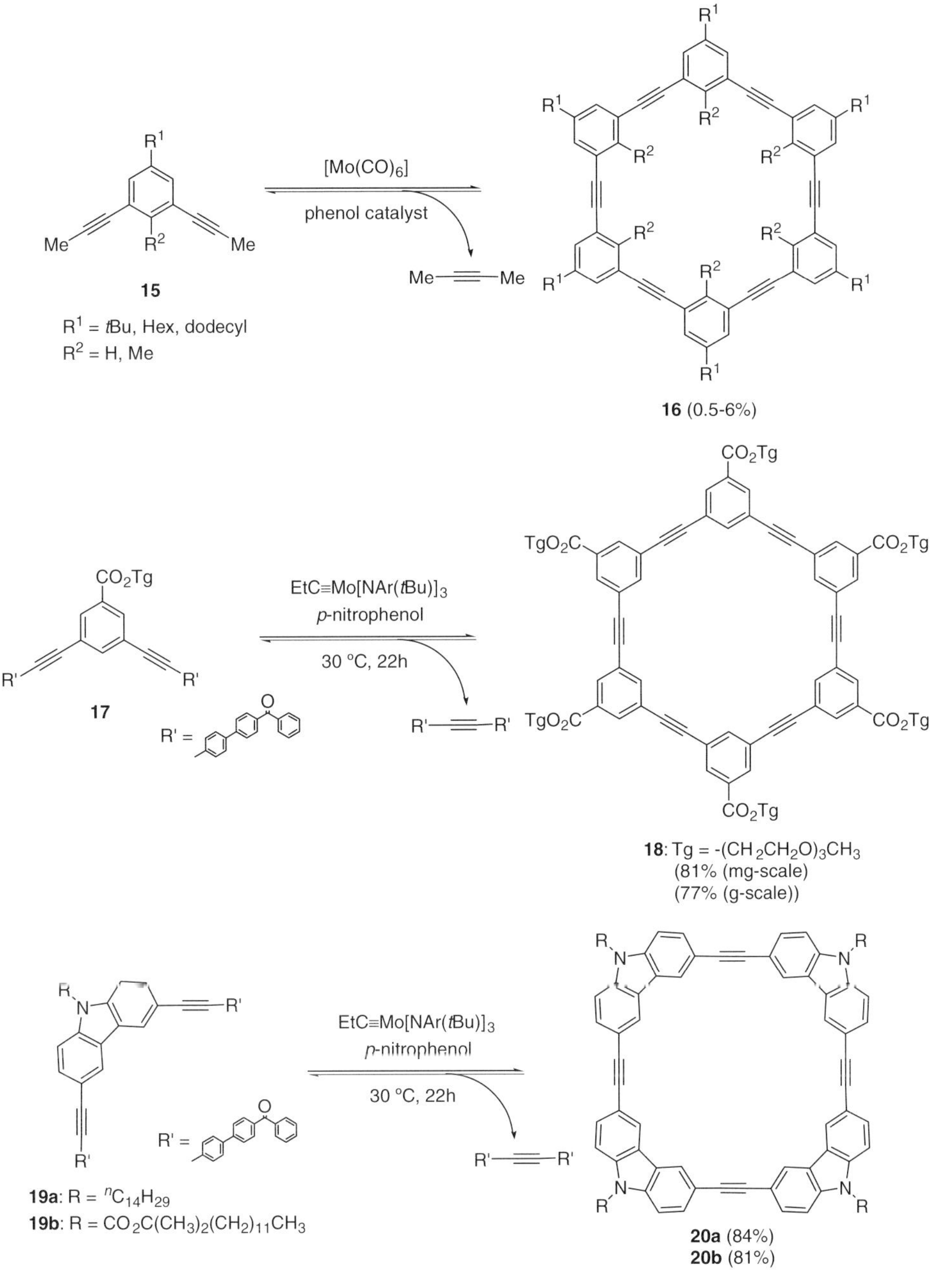

R^1 = tBu, Hex, dodecyl
R^2 = H, Me

15 → 16 (0.5-6%): $[Mo(CO)_6]$, phenol catalyst; − Me—≡—Me

17 → 18: EtC≡Mo[NAr(tBu)]$_3$, p-nitrophenol, 30 °C, 22h; − R'—≡—R'

18: Tg = $-(CH_2CH_2O)_3CH_3$
(81% (mg-scale)
(77% (g-scale))

19 → 20: EtC≡Mo[NAr(tBu)]$_3$, p-nitrophenol, 30 °C, 22h; − R'—≡—R'

19a: R = $^nC_{14}H_{29}$
19b: R = $CO_2C(CH_3)_2(CH_2)_{11}CH_3$

20a (84%)
20b (81%)

Scheme 6.2 "One-pot" synthesis of conjugated cycles **16**, **18**, and **20** via alkyne metathesis.

a sol–gel process owing to the strong columnar stacking, and the obtained nanofibril film was able to detect explosives such as 2,4-dinitrotoluene (DNT) and 2,4,6-trinitrotoluene (TNT) by monitoring a fluorescence intensity [40].

Imine-containing macrocycles **23** were also synthesized by condensation of dialdehyde **21a** and diamine **22a** in refluxing methanol [41a] or by imine metathesis of oligomers **21b** and **22b**, which selectively produced cyclic dimer **23** [41b] (Scheme 6.3). Although tribenzododecadehydro[18]annulenes **7** and *meta*-phenyleneacetylene macrocycles **16** were first synthesized using a stepwise route which includes the preparation of linear precursors followed by cyclization, one-pot synthesis usually surpasses the stepwise route [42, 43].

6.2.2
Template Synthesis of Macrocycles

Along with one-pot synthesis, template-directed synthesis of macrocycles has also been developed to improve the yields and selectivity of cyclization processes. Well known examples are seen in the synthesis of cyclic porphyrin arrays that employ ligand coordination to the central metal of the porphyrin [9, 10b,c, 11c, 31]. Sanders *et al.* reported that the yields of cyclooligomerization of **24** to give porphyrin dimer **25** and trimer **26** increased in the presence of a suitable template from 20–25% to 72% and 30–35% to 52%, respectively (Scheme 6.4) [9].

Höger *et al.* applied this strategy for the syntheses of *m*-phenyleneacetylene **31**, as depicted in Scheme 6.5 [35]. Thus, monomer **32** was connected to 1,3,5-benzenetricarboxyric acid by Mitsunobu reaction, and the resultant precursor **29** was subjected to Eglinton–Glaser coupling under high dilution conditions, resulting in the formation of the cyclic trimer **30** in high yield (94%). In marked contrast, Eglinton–Glaser coupling of monomer **33** under similar conditions gave a mixture of cyclization products in 98% total yield which contained the cyclic trimer in 20–25% yield. Because of the similar physical properties of the cyclic tri-, tetra-, and pentamers, they could not be fully separated by recrystallization or column chromatography. Hydrolysis of the template-bound macrocycle **30** proceeded smoothly under alkaline conditions to afford **31** in quantitative yield [44, 45].

Since template-directed synthesis involves a low concentration of reactive compounds and a high local concentration of functional groups, it can provide a selective and efficient reaction for the preparation of macrocycles. It should be emphasized, however, that the design of the central template and peripheral wings is the key step to accessing the desired macrocycles.

6.2.3
Synthesis of Conjugated Thiophene, Porphyrin, and Heteroarylene Macrocycles

As described in the previous section, macrocycles containing benzenes, porphyrins, or related heteroarylenes have been extensively synthesized to create new functional materials for research on structure–property relationships, 1D/2D nanostructures, and single molecule electronics. Oligothiophenes are well-studied π-conjugated

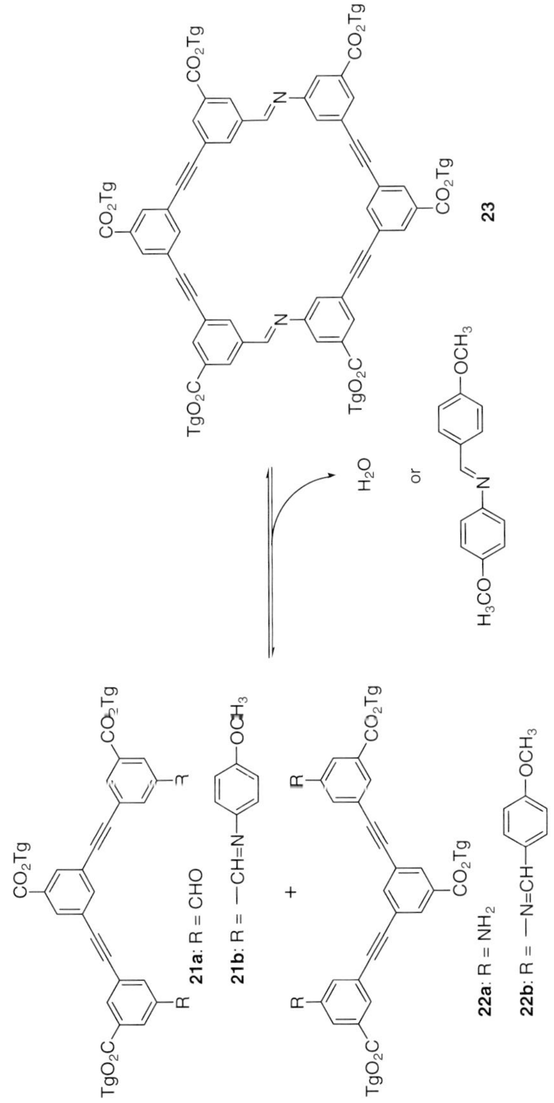

21a: R = CHO

21b: R = —CH=N—C_6H_4—OCH_3

+

22a: R = NH_2

22b: R = —N=CH—C_6H_4—OCH_3

H_2O

or

H_3CO—C_6H_4—N=CH—C_6H_4—OCH_3

23

Scheme 6.3 Synthesis of imine-containing macrocycle **23**.

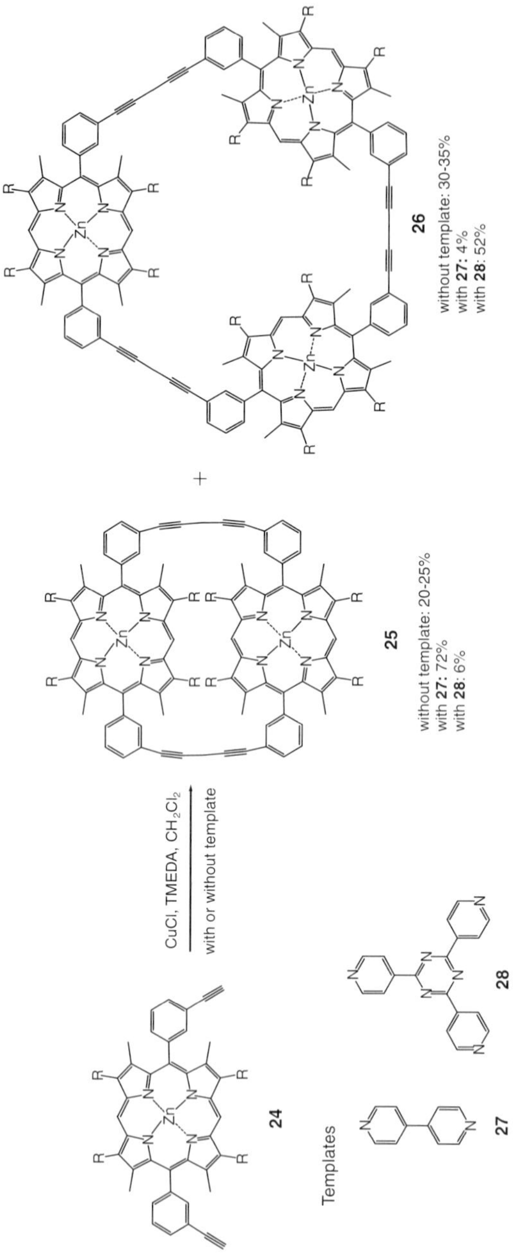

24

CuCl, TMEDA, CH_2Cl_2

with or without template

25

without template: 20-25%
with **27**: 72%
with **28**: 6%

+

26

without template: 30-35%
with **27**: 4%
with **28**: 52%

Templates

27

28

Scheme 6.4 Template synthesis of **25** and **26**.

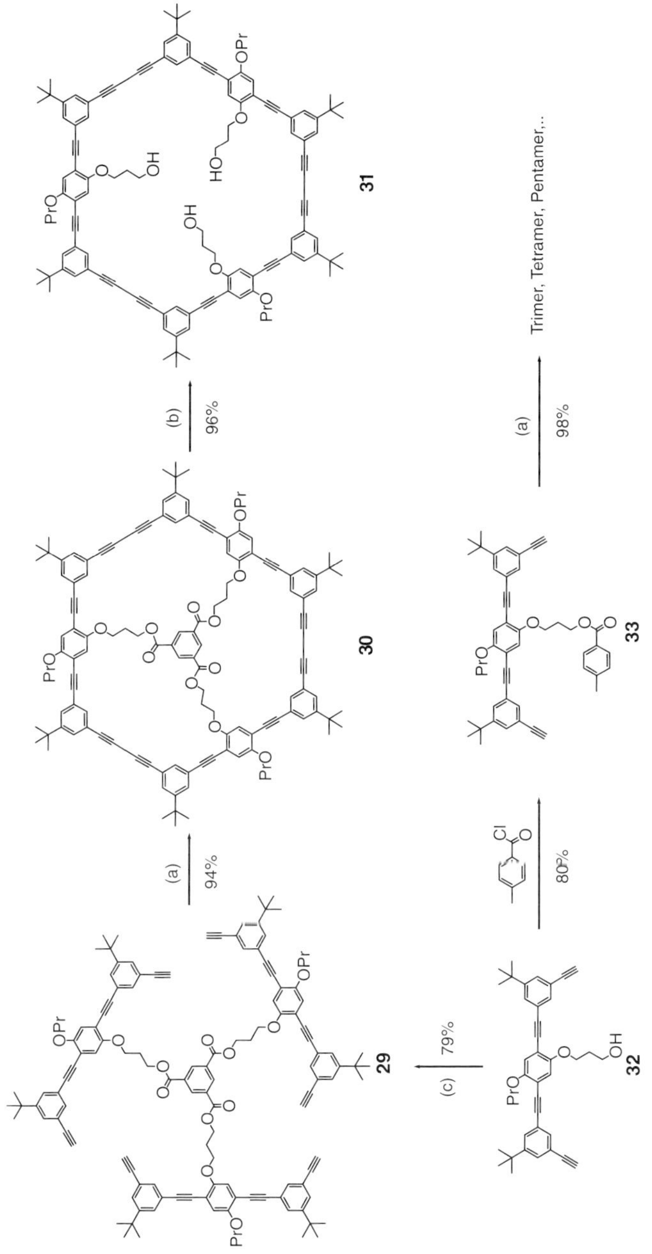

Scheme 6.5 Template synthesis of macrocycle 31. Reagents and conditions: (a) CuCl/$CuCl_2$, pyridine, rt; (b) NaOH, LiOH, MeOH, H_2O, THF, 80 °C; (c) 1,3,5-benzenetricarboxylic acid, DEAD, PPh_3, THF, rt.

oligomers because of their excellent characteristics in organic electronics in addition to the versatility in chemical modification of the thiophene ring [46, 47a]. Macrocycles possessing 2,5-thienylenes and π-spacers are intrinsically expected to offer effective electrical conjugations compared with previously reported *m*-phenyleneacetylenes [34, 47a,b]. Moreover, though no host–guest complexes derived from the macrocycles of oligothiophenes have been fully characterized, the sulfur atoms located in the inner cavity of the macrocycle can incorporate small ions or molecules. Based on these advantages, several kinds of cyclic oligothiophenes have been recently synthesized by Bäuerle [5b, 48, 49], Mayor [50], and our [51, 52] groups.

In 2000, Bäuerle *et al.* reported the first synthesis of fully conjugated macrocyclic oligothiophenes **36** ($n = 1$–3) from thiophenediynes **34** (Scheme 6.6) [49]. Oxidative coupling of **34** produced cyclic oligomers **35** in 2–12% yields, and these were converted into macrocyclic oligothiophenes **36** in 7–27% yields. Parallel to the synthesis of **36**, Mayor and Didschies reported one of the most exciting giant macrocycles **38**, using multi-step synthesis starting from **37** [50]. This molecule **38** has a diameter of about 12 nm but exhibits cyclic conjugation via 2,5-diethynylthiophene units.

Recently, Bäuerle *et al.* developed a new methodology for the synthesis of macrocyclic oligothiophenes by way of platinum intermediates [49]. As shown in Scheme 6.7, thiophenediynes were first treated with a platinum complex to produce the metalacyclic intermediates **39** and **40**. Reductive elimination of platinum, followed by treatment with Na_2S, resulted in the formation of macrocyclic oligothiophenes **36** and **41** in moderate yields.

π-Expanded cyclic oligothiophenes composed of thiophene, acetylene, and ethylene units have been employed by our group to construct molecular wires [51], because long-chain molecules can be easily synthesized from expanded oligothiophenes. Therefore, when both terminal carbons of linear oligothiophenes are connected, giant macrocycles with a round shape can be formally constructed. Our synthetic approach to π-extended oligothiophenes consists of the use of McMurry coupling as a key reaction to construct giant π-conjugated macrocycles in a single-step procedure [51, 52]. As shown in Scheme 6.8, McMurry coupling of the dialdehyde **42** under moderate dilution conditions produced dimer **43**, trimer **44**, tetramer **45**, pentamer **46**, and hexamer **47** in 32%, 9.4%, 6.2%, 3.9%, and 2.3% yields, respectively [51]. All macrocyclic oligothiophenes are stable redox-active compounds.

Similarly to the synthesis of **43–47**, McMurry coupling of the 6-thiophene dialdehyde **48** proceeded smoothly to form dimer **49**, trimer **50**, tetramer **51**, and pentamer **52** in 39%, 8.3%, 2.5%, and 1.2% yields, respectively (Scheme 6.9) [52]. Although the yields of higher oligomers showed a steep decrease, tetramer **51** and pentamer **52** were fully characterized and used for measurements of functional properties.

Porphyrin-containing giant macrocycles are of current interest not only as model compounds of light-harvesting antenna complexes but also as chromophores possessing extraordinary electro-optical and nonlinear optical properties [9–12, 31,m]. However, most of them lack a complete π-conjugation pathway around the whole

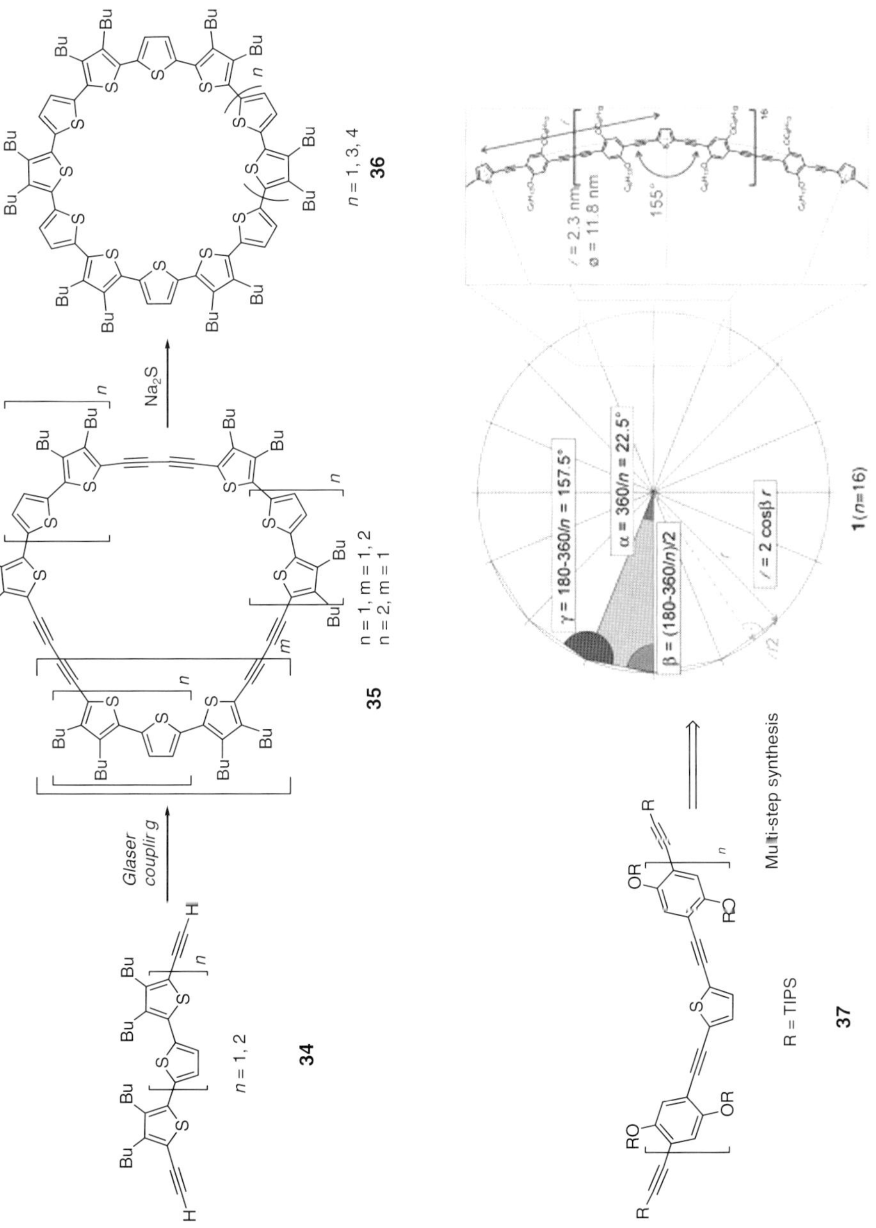

Scheme 6.6 Syntheses of fully conjugated macrocycles **35**, **36**, and **38**.

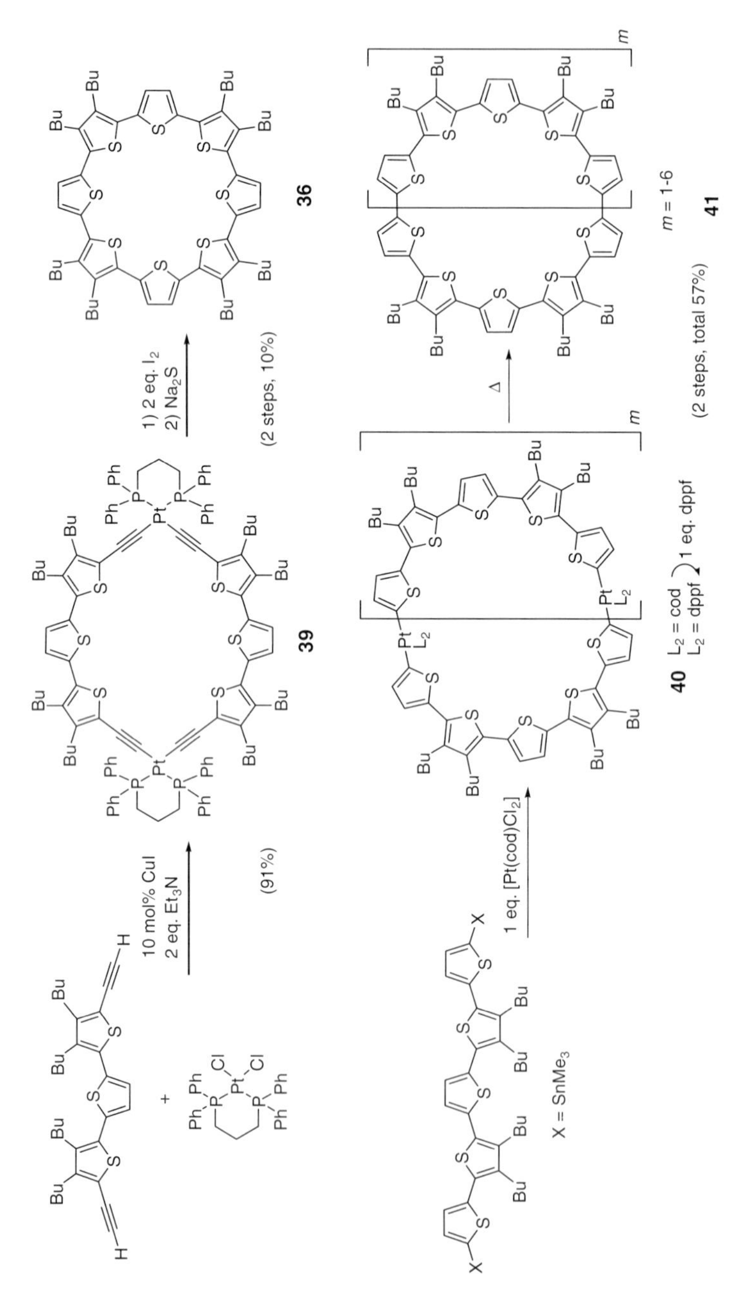

Scheme 6.7 Synthesis of macrocyclic oligothiophenes **36** and **41** via platinum complexes **39** and **40**.

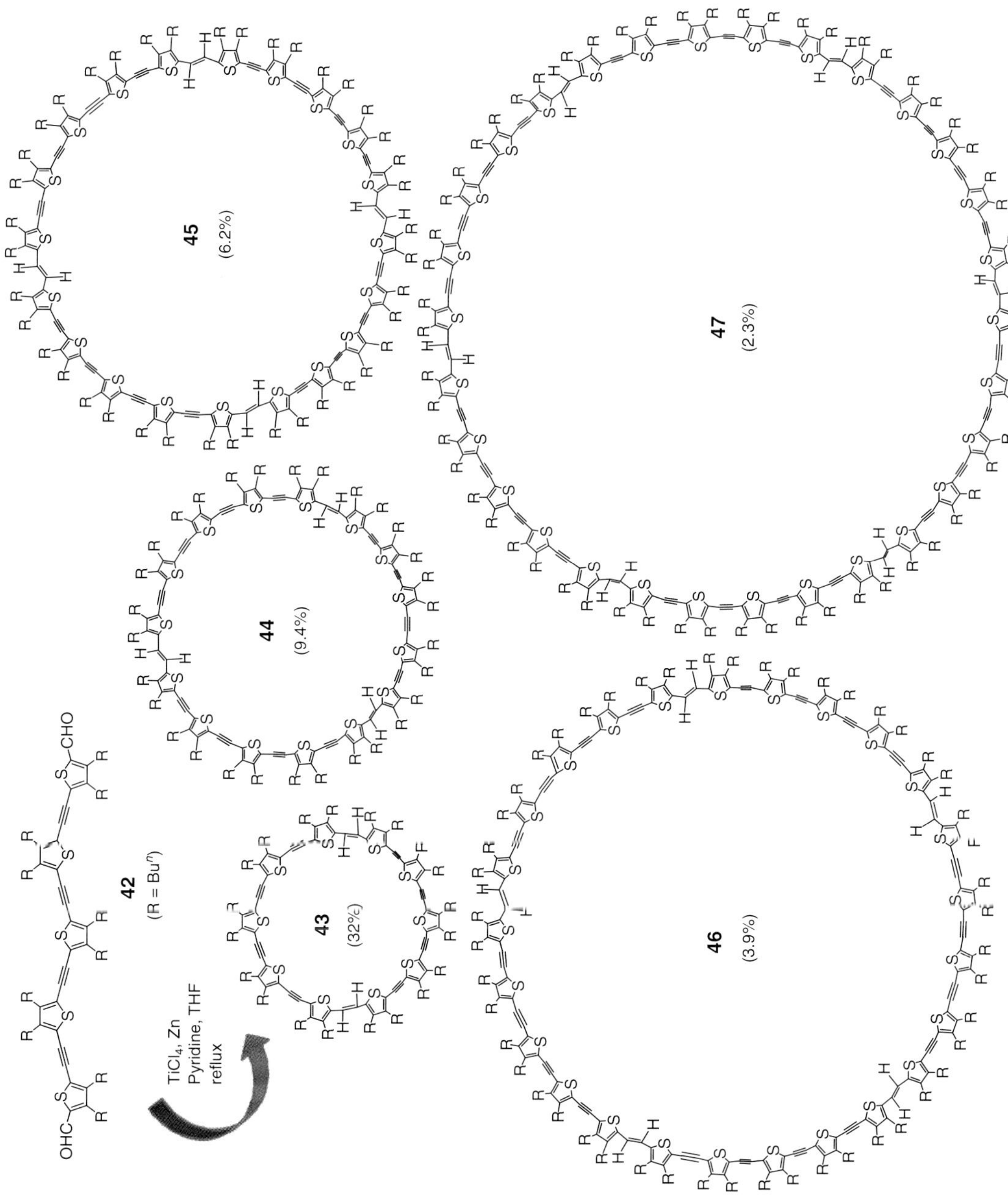

Scheme 6.8 McMurry coupling reaction of **42**.

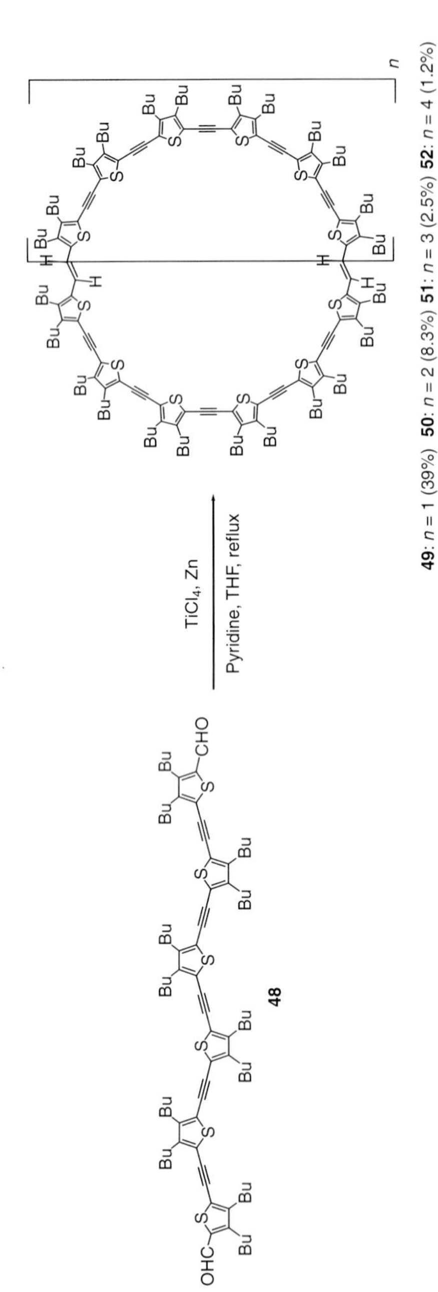

Scheme 6.9 Synthesis of giant macrocyclic oligothiophenes **49–52**.

macrocycle owing to their aryl-based *meso*-linkages. In this regard, two kinds of giant π-conjugated porphyrin macrocycles have been reported so far.

Sugiura *et al.* reported the synthesis of square-shaped porphyrin tetramer **53** and dodecamer **54** (Figure 6.2) by connecting two *meso*-positions with diacetylene linkages. One-pot Glaser–Hey coupling of 5,10-diaryl-15,20-diethynyl nickel porphyrin at 2.6 mM concentration gave the cyclized tetramer **53** in 22% yield [53], while a similar procedure was also applied to the synthesis of the larger dodecamer **54** where orthogonally-arranged porphyrin trimer gave the target in 9% yield [54]. This large square **54** was visualized and its cyclic structure was confirmed by ultrahigh-vacuum scanning tunneling microscopy (UHV-STM) on a Cu(111) surface.

Recently, Anderson *et al.* reported the template synthesis of π-conjugated porphyrin nanorings **58** and **62a,b** using pyridine–zinc porphyrin complexation (Scheme 6.10). Oxidative cyclization of the linear zinc porphyrin octamer **55** was achieved in the presence of the complementary octadentate ligand **56** to give the cyclic octamer **58** in 14% isolated yield (55% conversion by gel permeation chromatography (GPC)) [55a]. On the other hand, coupling of the linear octamer **55** without template **56** gave polymeric products which eluted faster than **55** on analytical GPC. Since the equilibrium constants K_f for the complexes **55·56** and **55·58** (**57**) are too large to be evaluated, competition experiments were employed by using pyridine, resulting in K_f values of $1.4 \times 10^{28}\,M^{-1}$ and $1.3 \times 10^{37}\,M^{-1}$ for **55·56** and **57**, respectively. In contrast to the predicted rigidity of ethynyl porphyrin oligomers, it was suspected that the octamer **58** was not highly strained, and this view was also supported by the similarity of the absorption spectra of **55** and **58**.

By using symmetric hexapyridyl template **60** and porphyrin dimer **59a,b**, the synthesis of even more strained porphyrin hexamers **62a,b** was achieved [55b]. The K_f value for the complex **61** ($7 \times 10^{38}\,M^{-1}$) is higher than that for **57**, and it also has effective molarity (340 ± 60 M), which indicates complementarity of templates to the porphyrin rings estimated from K_f and the binding constant of one arm of the template for one site on the octa- or hexamers, is much higher than that for **57** (5.4 M). The effective molarity for forming **61** is an extremely high value for a noncovalent self-assembly process [56], and it is consistent with the fact that pyridine is not able to displace the template. More effective π-conjugation for the ring structure could be observed than for its linear analogue, probably due to the rigid geometry of the ring and the lack of the end-effects of the linear analogue.

Compared to a number of planar π-conjugated macrocycles, belt-shaped chromophores with an effective π-conjugation are still limited due to lack of synthetic methodology [3d, 6, 57]. However, unusual structures with a curved π-surface possessing radially orientated p orbitals are not only esthetically appealing, but can be envisioned as potential precursors for carbon nanotubes and related tube structures.

Müllen *et al.* reported the synthesis of carbazole and fluorene dodecamers **66a,b** by direct Ar–Ar coupling with the aid of porphyrin templates (Scheme 6.11). Tris(2,7-carbazole)s **63** were connected by Mitsunobu esterification to the tetraphenylporphyrin template **64** to afford the precursor **65** in 73% yield, and then the cyclization of **65** was performed by nickel-mediated Yamamoto coupling under high dilution conditions (8×10^{-5} M) with a microwave reactor to give the cyclized product in 9%

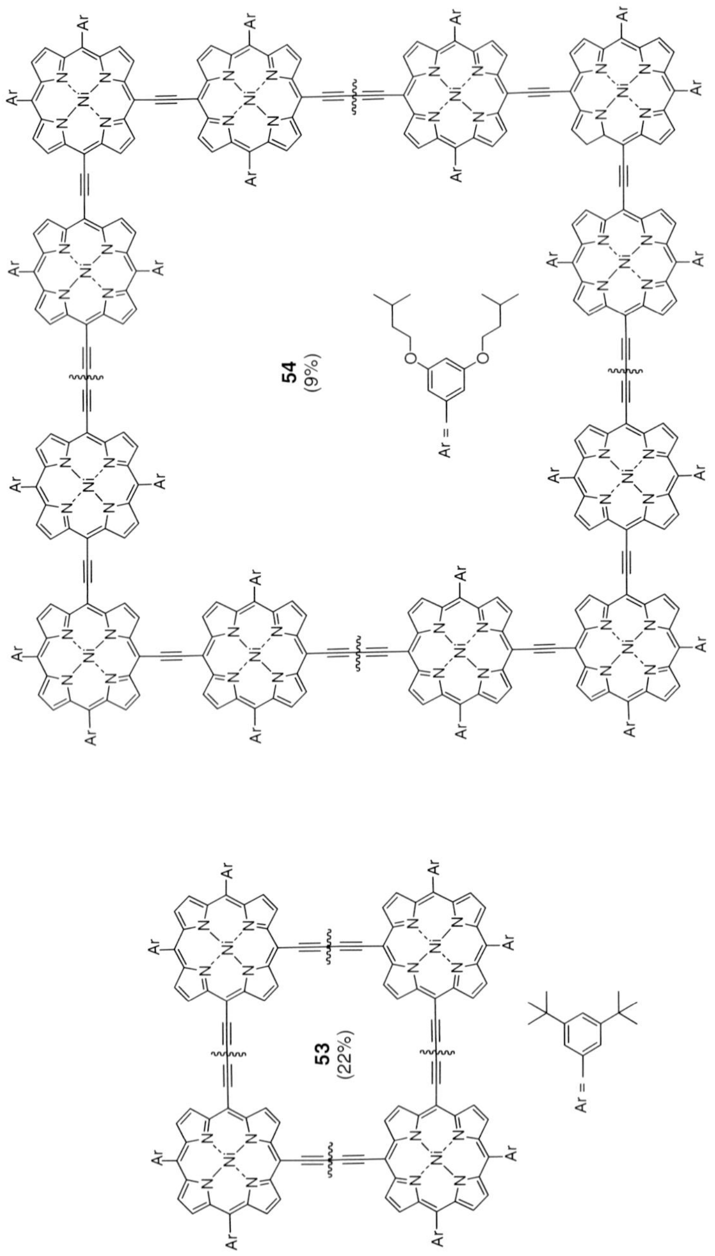

Figure 6.2 Porphyrin squares **53** and **54**. The yields from their precursors are indicated in parentheses.

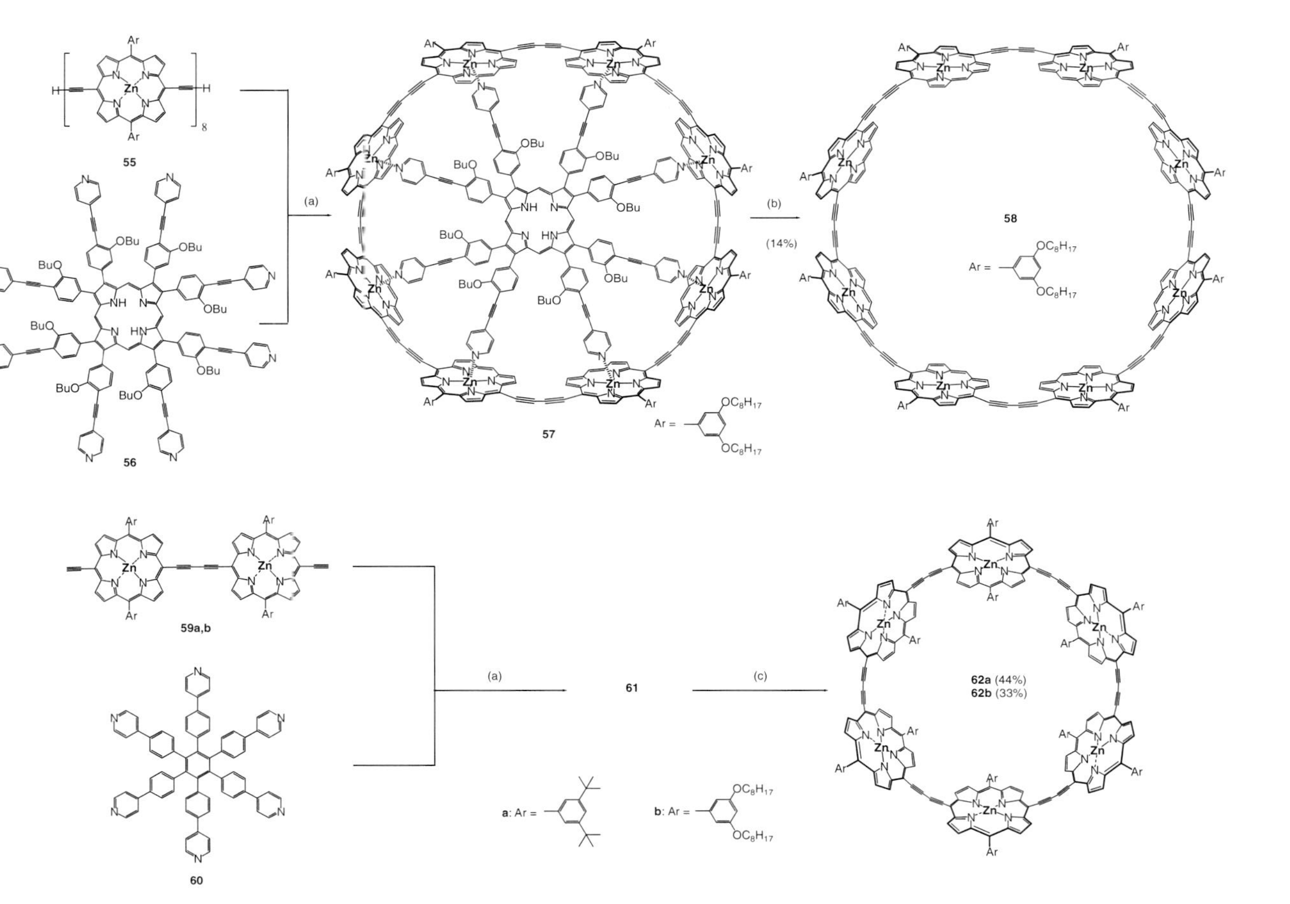

Scheme 6.10 Synthesis of porphyrin[n] nanorings **58** and **62a,b**. Reagents and conditions: (a) $[Pd(PPh_3)_2Cl_2]$, CuI, iPr_2NH, I_2, air, 60 °C; (b) pyridine; (c) DABCO.

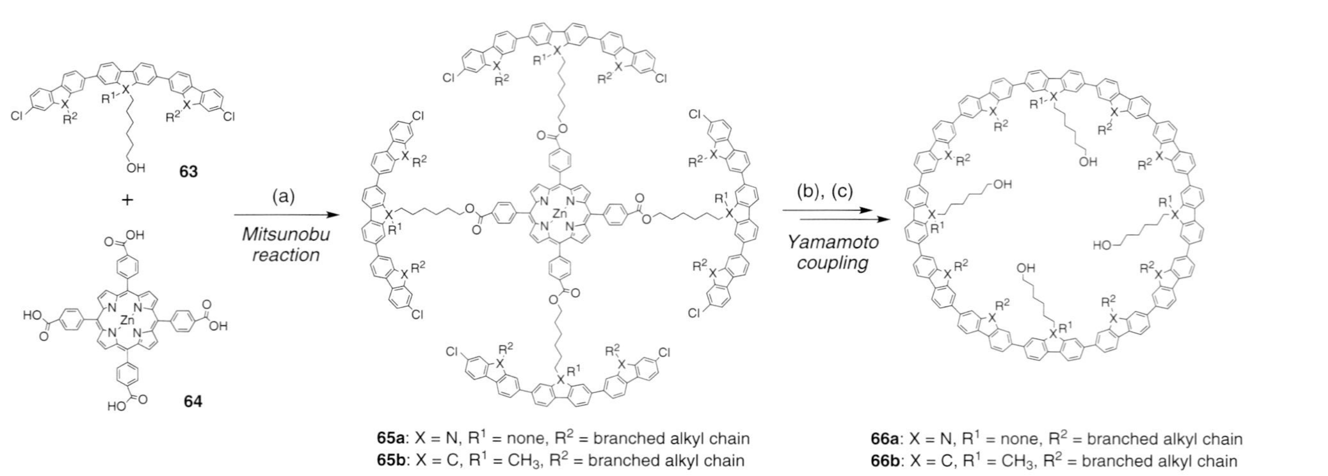

Scheme 6.11 Synthesis of carbazole and fluorine dodecamers **66a,b**. Reagents: (a) PPh_3, DEAD, THF; (b) $[Ni(COD)_2]$, bpy, COD, toluene, THF or DMF; (c) KOH, H_2O, MeOH, THF.

yield. The porphyrin template was readily removed by successive hydrolysis with KOH to generate the fully conjugated macrocyclic carbazole dodecamer **66a** [58a]. Fluorene dodecamer **66b** was also synthesized by a similar synthetic strategy [58b].

MacLachlan *et al.* reported a number of Schiff-base (imine)-containing macrocycles **72–75** with *o*-phenylenediamine **67** and 3,6-diformylcatechol **68** or bis(salicylaldehyde)s **69–71** (Scheme 6.12) [59a,b]. These macrocycles are capable of both binding transition metal atoms and assembling into tubes, possibly leading to catalytic nanomaterials or ion-conducting channels. In fact, the small cycle **72** obtained from a one-pot condensation reaction in high yield (70%) induced ionic assemblies in the presence of small cations like sodium and cesium ions. Similarly, larger π-conjugated macrocyles **73–75** were also synthesized in high yields (40–68%), and their interesting photophysical properties were reported.

6.3 Isolation and Self-Association in Solution and in the Solid State

In order to isolate π-conjugated macrocycles, gel permeation chromatography (GPC) or high pressure liquid chromatography (HPLC) are widely employed as well as conventional column chromatography. The hydrodynamic volumes of cyclic structures and corresponding linear oligomers with similar molecular weights are considerably different. Cyclic structures attaining a larger volume are eluted faster than their linear analogues, and normally these cyclic structures and low molecular weight linear oligomers give sharp peaks in GPC compared to higher molecular weight oligo/polymers. Therefore, a recycling GPC system plays an important role in the purification of monodisperse π-conjugated macrocycles and oligomers. In fact, GPC has been thought to be an appropriate technique to obtain highly purified linear/circular π-conjugated oligomers [53–55] and its explicit role has been emphasized in some reports [60].

Many fundamental aspects of the electronic and structural properties of π--conjugated macrocycles are not well established. Usually mass spectrometry (MS), ^{1}H and ^{13}C NMR, UV–vis, mp, IR, and GPC techniques play important roles for their structural proofs, and some of the compounds are characterized by elemental analysis, single crystal X-ray structures and STM images [3k]. Amongst these, in the case of the above mentioned one-pot synthesis, MS information is practically the most important. Furthermore, since macrocycles have structures with higher symmetry than the starting precursors and unwanted linear oligomeric side products, NMR gives information on the symmetry and the absence of end groups like terminal acetylenes [53, 54] and *meso*-positions of porphyrins [61]. Based on the recent development of NMR techniques, self-diffusion coefficients (D) have also been adopted by using pulsed-field gradient (PFG) NMR spectroscopy in order to deepen the conformational investigations in solution, where D values reflect hydrodynamic volumes of macrocycles and their aggregates [62].

In solution, self-assembly of π-conjugated macrocycles and other π-systems is conventionally investigated by chemical shift changes in NMR spectroscopy owing to

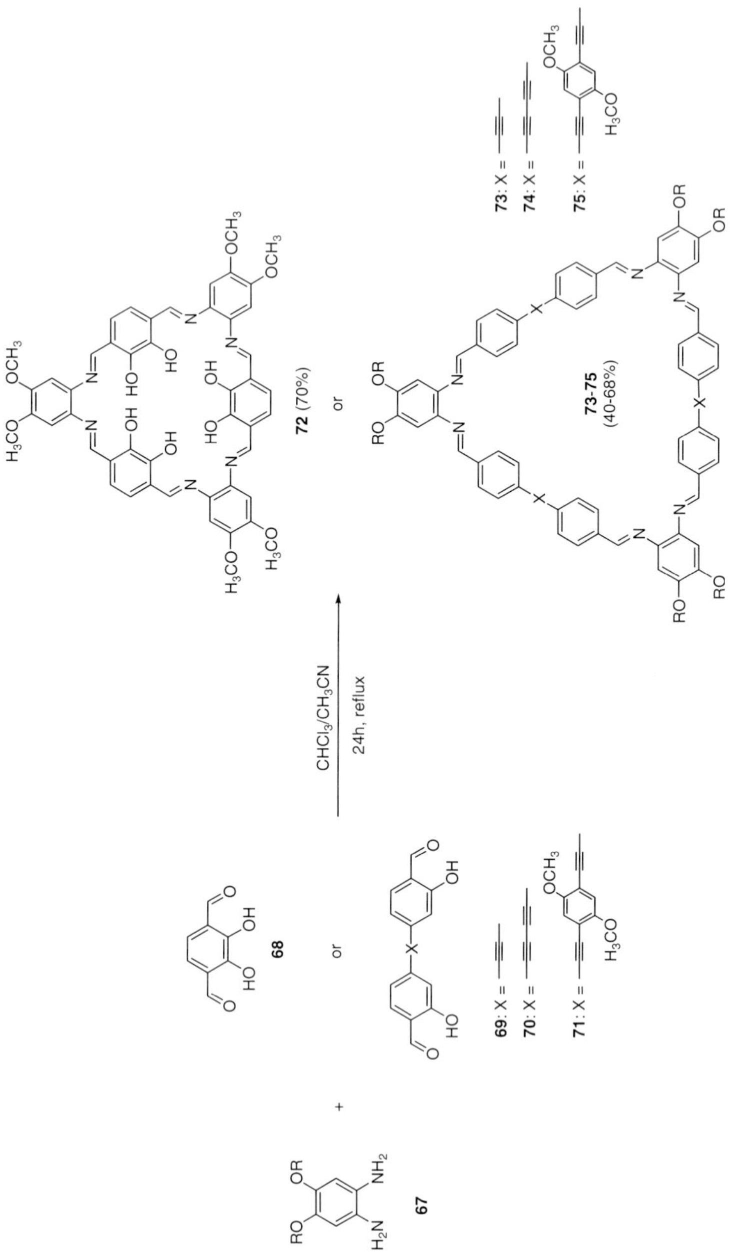

Scheme 6.12 Synthesis of Schiff-base macrocycles **72–75**.

stacking interactions of the aromatic rings. With this technique, the stoichiometry of aggregates is evaluated, and thus association constants and thermodynamic parameters like enthalpy and entropy for self-assembly can be determined, based on the theoretical models [3i, 63]. While NMR spectroscopy needs relatively small amounts of the sample and is readily available, vapor pressure osmometry (VPO) is also employed to check the reliability of the results obtained from NMR experiments. This is because NMR simply reflects short-range effects of the aromatic ring stacking, and, moreover, in the case of higher aggregation, it gives rise to significant line broadening. [3i, 63, 64].

Scanning tunneling microscopic (STM) investigations can be used not only to visualize the molecules and confirm their proposed sizes and shapes, but also to characterize 2D self-assembly on a surface like highly oriented pyrolytic graphite (HOPG) [4a,c,f,5a,b,c] and, furthermore, in applications for single-molecule electronics [65].

6.4 Versatile Properties of Giant π-Conjugated Macrocycles

While a variety of synthetic methods for nano-sized π-conjugated macrocycles have recently been developed, their versatile properties remain to be completely investigated. Therefore, only a limited number of properties of giant macrocycles such as self-aggregation, remarkable two-photon absorption properties, and the formation of host–guest complexes on the surface have been revealed, and will be discussed in this section.

Recently, functional nanostructures have been actively investigated by a number of groups in different fields of materials science. Although polymers have been frequently employed to construct nanostructures such as nanofibers, nanostructures of π-conjugated macrocycles have scarcely been reported to date. Interestingly, nano-sized macrocycles show size-dependent formation of nanostructures. The giant macrocycles **43–47** composed of the same building blocks are nearly round, and have molecular diameters of 3–8 nm with inner cavities of diameter 1.7–6 nm. As shown in Figure 6.3, **43**–**47** self-aggregate in the solid state to form either single crystals, fibrous materials, or nanoparticles. Thus, **43** affords single crystals from a chloroform–heptane solution, whereas **44** and **45** form well-defined fibrous structures of thickness100–200 nm under similar conditions. In contrast, **46** and **47** having conformational flexibility form chained lumps of size 300–800 nm. Because **44–47** possess both solvophilic and solvophobic moieties, the amphiphilic properties result in microscopic separation, which leads to the formation of nanofibers from **44** and **45** and nanoparticles from **46** and **47**. Since **43–47** are oligothiophenes, they exhibit multi-step reversible redox behavior with fairly low first oxidation potentials, reflecting their cyclic conjugation. As a result, doping **43–47** with iodine vapor forms semiconductors ($\sigma_{rt} = 1.96–2.63 \times 10^{-3}\ S\ cm^{-1}$).

Absorption and emission spectra of giant macrocycles **43–47** and **49–52** exhibit a unique feature (Figure 6.4 and Table 6.1). Although a series of linear oligo(thienylene-

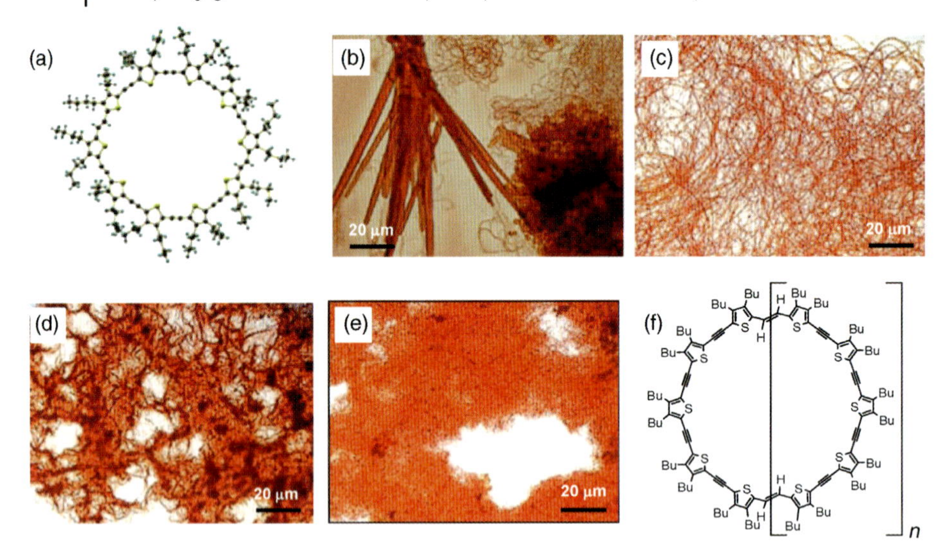

Figure 6.3 X-ray structure of **43** (a) and microscopic images of **44** (b), **45** (c), **46** (d), and **47** (e) together with the general formula of **43–47**.

ethynylene)s up to 16-mer exhibited near saturation for the absorption maximum at the octamer stage [66, 67], the macrocyclic oligothiophenes **43–47**, **49** and **50** except for **51** and **52** exhibit a red-shift of the longest absorption maxima with increasing ring size, reflecting almost full conjugation through the rings. In the case of emission spectra, linear oligo(thienylene-ethynylene)s were reported to exhibit two major emission bands based on the vibronic 0–0 and 0–1 transitions [67]. As shown in Table 6.1, the fluorescence spectra of **43–47** and **49–52** show two major emissions at almost the same wavelengths (559–562 and 600–606 nm) with a large Stokes shift of 72–157 nm. The separation of the two emissions in the fluorescence spectra corresponds to the vibrational energy gap.

Since macrocycles containing acetylene bridges have been reported to show large two-photon absorption cross sections (δ_{max}) [48a, 68], the two-photon properties of giant macrocycles **49–52** were investigated [52a]. The two-photon absorption cross sections (δ_{max}) are **49**: 15 100 GM, **50**: 66 700 GM, **51**: 82 600 GM, and **52**: 107 800 GM. Therefore, the increase in the ring size and π-character from **49** (72π) to **50** (108π), from **49** to **51** (144π), and from **49** to **52** (180π) result in 4.4-, 5.5-, and 7.1-fold amplifications of the maximum two-photon absorption cross section, respectively. These large enhancements of two-photon absorption cross section are due to intramolecular interactions in these giant macrocycles. Note that the increasing π-conjugation leads to an increase in the two-photon absorption cross section with magnitudes as high as 100 000 GM in the visible spectral region.

Giant macrocycles fabricated at the solid/liquid interface provide self-assembled porous networks, and the guest molecules can be deposited either into the cavity or at the rim of conjugated backbones. This will depend on the nature of the electronic properties of the host networks and the guest components. The host–guest

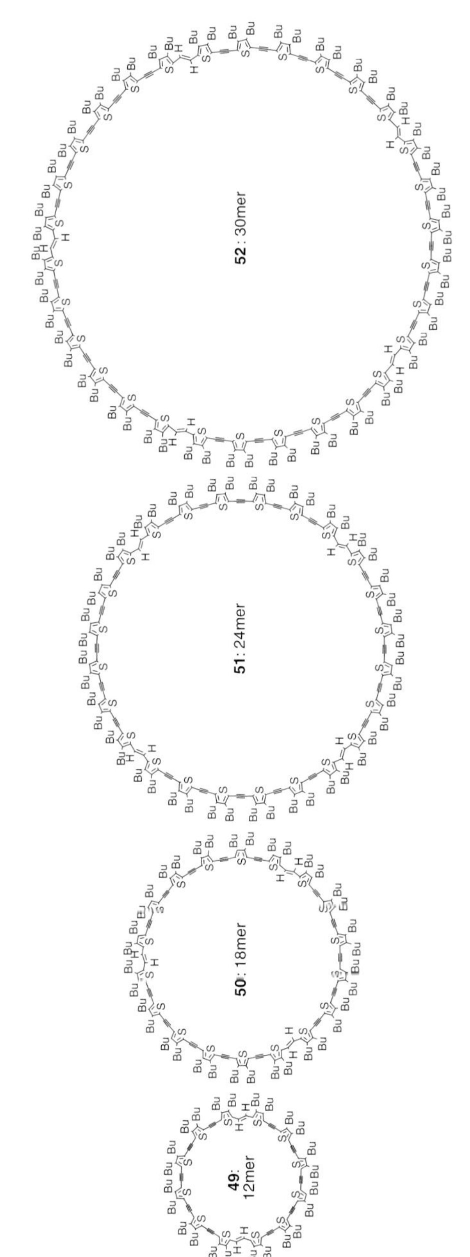

Figure 6.4 Giant macrocycles **49–52**.

Table 6.1 Absorption coefficients and fluorescence quantum yields of **43–47** and **49–52** in CH_2Cl_2[a].

Compd.	Absorption		Fluorescence	
	λ_{max} [nm]	ε [$M^{-1}cm^{-1}$]	λ_{max} [nm]	Φ_F[a]
43 (60π)	447	249 000	560, 604	0.069
49 (72π)	452	295 000	562, 606	0.11
44 (90π)	469	311 000	562, 605	0.084
50 (108π)	473	418 000	560, 603	0.085
45 (120π)	479	405 000	560, 602	0.11
51 (144π)	478	528 000	559, 600	0.10
46 (150π)	485	551 000	560, 603	0.089
52 (180π)	480	618 000	559, 600	0.095
47 (180π)	488	639 000	560, 600	0.086

a) Fluorescence quantum yields (Φ_F) were determined by comparison with quinine sulfate in 0.5 M H_2SO_4 ($\Phi_F = 0.51$).

complexation will also rely on the size complementarity between the cavity and guest components. Quite recently, Müllen *et al.* have reported the formation of the host–guest complex consisting of a hexa-*peri*-hexabenzocoronene (HBC) molecule sitting in the middle of the free cavity of a giant macrocycle **76** [69]. As shown in Scheme 6.13, this host–guest complex has been prepared by physisorption of a monolayer of the macrocycles followed by gas-phase deposition of graphene molecules by pulsed laser deposition. In this case, the size complementarity between the donor HBC and the 2,7-carbazole macrocycle **76** would be ascribed to the 1 : 1 complex formation.

6.5 Conclusion

One of the long-range goals of synthesizing π-conjugated macrocycles would be to create fine-tuned 1D, 2D, or 3D host–guest systems resulting in new properties and applications in nanomaterials. In particular, giant macrocycles having nanometer-sized structures would be of interest for structure–property relationships, 1D/2D nanostructures, and single molecule electronics owing to the efficient guest incorporation and electronic properties depending on their sizes and supramolecular structures. However, the functional properties of giant macrocycles have only recently been elucidated and employed in materials science, and hence these macrocycles remain as yet exotic and scarce in materials, chemical, and physical sciences. The work referred to in this chapter would be expected to participate in a wide range of materials science. At the more fundamental level, however, research on giant macrocycles also has synthetic importance for the creation of novel macrocyclic molecules via conventional or unconventional methods. Further development of the chemical synthesis of giant macrocycles, accompanied with physical single molecule

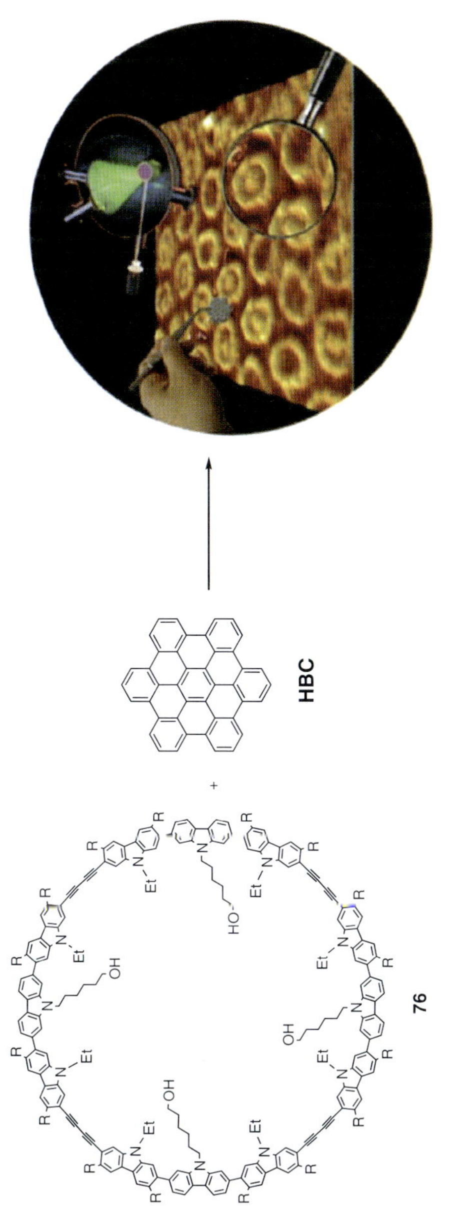

Scheme 6.13 STM image of monolayer of **76** after deposition with HBC molecules.

manipulation would realize new applications of these molecules to organic nanoelectronics in the near future.

References

1 (a) Haley, M.H. and Tykwinski, R.R.(eds) (2006) *Carbon-Rich Compounds, From Molecules to Materials*, Wiley-VCH Verlag GmbH, Weinheim; (b) Wu, J., Pisula, W., and Müllen, K. (2007) *Chem. Rev.*, **107**, 718–747; (c) Grimsdale, A.C. and Müllen, K. (2005) *Angew. Chem. Int. Ed.*, **44**, 5592–5629.

2 (a) Müllen, K. and Wegner, G. (eds) (1998) *Electronic Materials: The Oligomer Approach*, Wiley-VCH Verlag GmbH, Weinheim; (b) Skotheim, T.A., Elsenbaumer, R.L., and Reynolds, J.R. (eds) (1998) *Handbook of Conducting Polymers*, 2nd edn, Marcel Dekker, New York; (c) Grimsdale, A.C. and Müllen, K. (2006) *Adv. Polym. Sci.*, **199**, 1–82; (d) Babudri, F., Farinola, G.M., and Nao, F. (2004) *J. Mater. Chem.*, **14**, 11–34.

3 For recent reviews of π-conjugated macrocycles, see: (a) Diederich, F., Stang, P.J., and Tykwinski, R.R. (eds) (2008) *Modern Supramolecular Chemistry: Strategies for Macrocycle Synthesis*, Wiley-VCH Verlag GmbH, Weinheim; (b) Höger, S. (2005) Chapter 10, in *Acetylene Chemistry* (eds F. Diederich, P.J. Stang, and R.R. Tykwinski), Wiley-VCH Verlag GmbH, Weinheim; (c) Zhang, W. and Moore, J.S. (2006) *Angew. Chem. Int. Ed.*, **45**, 4416–4439; (d) Tahara, K. and Tobe, Y. (2006) *Chem. Rev.*, **106**, 5274–5290; (e) Spitler, E.L., Johnson, C.A. II, and Haley, M.M. (2006) *Chem. Rev.*, **106**, 5344–5386; (f) MacLachlan, M.J. (2006) *Pure Appl. Chem.*, **78**, 873–888; (g) Höger, S. (2005) *Angew. Chem. Int. Ed.*, **44**, 3806–3808; (h) Höger, S. (2004) *Chem. Eur. J.*, **10**, 1320–1329; (i) Young, D.Z. and Moore, J.S. (2003) *Chem. Commun.*, 807–818; (j) Yamaguchi, Y. and Yoshida, Z. (2003) *Chem. Eur. J.*, **9**, 5430–5440; (k) Grave, C. and Schlüter, A.D. (2002) *Eur. J. Org. Chem.*, 3075–3098; (l) Nakamura, Y., Aratani, N., and Osuka, A. (2007) *Chem. Soc. Rev.*, **36**, 831–845; (m) Sugiura, K. (2003) *Top. Curr. Chem.*, **228**, 65–85.

4 (a) Kudernac, T., Lei, S., Elemans, J.A.A.W., and De Feyter, S. (2009) *Chem. Soc. Rev.*, **38**, 402–421; (b) Zang, L., Che, Y., and Moore, J. (2008) *Acc. Chem. Res.*, **41**, 1596–1608; (c) Hoeben, F.J.M., Jonkheijm, P., Meiyer, E.W., and Schenning, P.H.J. (2005) *Chem. Rev.*, **105**, 1491–1546; (d) Yamamoto, Y., Fukushima, T., Suna, Y., Ishii, N., Saeki, A., Seki, S., Tagawa, S., Taniguchi, M., Kawai, T., and Aida, T. (2006) *Science*, **314**, 1761–1764; (e) Barth, J.V., Costantini, G., and Kern, K. (2005) *Nature*, **437**, 671–679; (f) Grave, C., Lentz, D., Schäfer, A., Samorì, P., Rabe, J.P., Franke, P., and Schülter, A.D. (2003) *J. Am. Chem. Soc.*, **125**, 6907–6918.

5 (a) Tahara, K., Lei, S., Mamdouh, W., Yamaguchi, Y., Ichikawa, T., Uji-I, H., Sonoda, M., Hirose, K., De Schryver, F.C., De Feyter, S., and Tobe, Y. (2008) *J. Am. Chem. Soc.*, **130**, 6666–6667; (b) Mena-Osteritz, E., and Bäuerle, P. (2006) *Adv. Mater.*, **18**, 447–451; (c) Pan, G.-B., Chen, X.-H., Höger, S., and Freyland, W. (2006) *J. Am. Chem. Soc.*, **128**, 4218–4219; (d) Shetty, A.S., Fischer, P.R., Stork, K.F., Bohn, P.W., and Moore, J.S. (1996) *J. Am. Chem. Soc.*, **118**, 9409–9414.

6 (a) Kawase, T. and Kurata, H. (2006) *Chem. Rev.*, **106**, 5250–5273; (b) Kawase, T. and Oda, M. (2006) *Pure Appl. Chem.*, **77**, 831–839.

7 (a) Tobe, Y., Utsumi, N., Nagano, A., and Naemura, K. (1998) *Angew. Chem. Int. Ed.*, **37**, 1285–1287; (b) Hosokawa, Y., Kawase, T., and Oda, M. (2001) *Chem. Commun.*, 1948–1949.

8 Hensel, V. and Schlüter, D. (1999) *Chem. Eur. J.*, **5**, 421–429.

9 (a) Anderson, H.L. and Sanders, J.K.M. (1989) *J. Chem. Soc., Chem. Commun.*, 1714–1715; (b) Anderson, H.L. and Sanders, J.K.M. (1990) *Angew. Chem., Int. Ed. Engl.*, **29**, 1400–1403; (c) Anderson, S., Anderson, H.L., and Sanders, J.K.M. (1992) *Angew. Chem., Int. Ed. Engl.*, **31**, 907–910; (d) Mackay, L.G., Wylie, R.S., and Sanders, J.K.M. (1994) *J. Am. Chem. Soc.*, **116**, 3141–3142.

10 (a) Wagner, R.W., Seth, J., Yang, S.I., Kim, D., Bocian, D.F., Holten, D., and Lindsey, J.S. (1998) *J. Org. Chem.*, **63**, 5042–5049; (b) Li, J., Ambroise, A., Yang, S.I., Diers, J.R., Seth, J., Wack, C.R., Bocian, D.F., Holten, D., and Lindsey, J.S. (1999) *J. Am. Chem. Soc.*, **121**, 8927–8940; (c) Yu, L. and Lindsey, J.S. (2001) *J. Org. Chem.*, **66**, 7402–7419; (d) Holten, D., Bocian, D.F., and Lindsey, J.S. (2002) *Acc. Chem. Res.*, **35**, 57–69.

11 (a) Mongin, O., Schuwey, A., Vallot, M.-A., and Gossauer, A. (1999) *Tatrahedron Lett.*, **40**, 8347–8350; (b) Rucareanu, S., Mongin, O., Schuwey, A., Hoyler, N., and Gossauer, A. (2001) *J. Org. Chem.*, **66**, 4973–4988; (c) Rucareanu, S., Schuwey, A., and Gossauer, A. (2006) *J. Am. Chem. Soc.*, **128**, 3396–3413.

12 (a) Nakamura, Y., Hwang, I.-W., Aratani, N., Ahn, T.K., Ko, D.M., Takagi, A., Kawai, T., Matsumoto, T., Kim, D., and Osuka, A. (2005) *J. Am. Chem. Soc.*, **127**, 236–246; (b) Pang, X., Aratani, N., Takagi, A., Matsumoto, T., Kawai, T., Hwang, I.-W., Ahn, T.K., Kim, D., and Osuka, A. (2004) *J. Am. Chem. Soc.*, **126**, 4468–4469; (c) Hori, T., Aratani, N., Takagi, A., Matsumoto, T., Kawai, T., Yoon, M.-C., Yoon, Z.S., Cho, S., Kim, D., and Osuka, A. (2006) *Chem. Eur. J.*, **12**, 1319–1327.

13 (a) Eller, L.R., Stepien, M., Fowler, C.J., Lee, J.T., and Sessler, J.L. (2007) *J. Am. Chem. Soc.*, **129**, 11020–11021; (b) Stepien, M., Donnio, B., and Sessler, J.L. (2007) *Angew. Chem. Int. Ed.*, **46**, 1431–1435; (c) Köhler, T., Ou, Z., Lee, J.T., Seidel, D., Lynch, V., Kadish, K.M., and Sessler, J.L. (2005) *Angew. Chem. Int. Ed.*, **44**, 83–87; (d) Seidel, D., Lynch, V., and Sessler, J.L. (2002) *Angew. Chem. Int, Ed.*, **41**, 1422–1425.

14 (a) Lei, S., Ver Heyen, A., De Feyter, S., Surin, M., Lazzaroni, R., Rosenfeldt, S., Ballauff, M., Lindner, P., Mössinger, D., and Höger, S. (2009) *Chem. Eur. J.*, **15**, 2518–2535; (b) Mössinger, D., Hornung, J., Lei, S., De Feyter, S., and Höger, S. (2007) *Angew. Chem. Int. Ed.*, **46**, 6802–6806.

15 (a) Tanaka, Y., Saito, S., Mori, S., Aratani, N., Shinokubo, H., Shibata, N., Higuchi, Y., Yoon, Z.S., Kim, K.S., Noh, S.B., Park, J.K., Kim, D., and Osuka, A. (2008) *Angew. Chem. Int. Ed.*, **47**, 681–684; (b) Saito, S., Shin, J.-Y., Lim, J.M., Kim, K.S., Kim, D., and Osuka, A. (2008) *Angew. Chem. Int. Ed.*, **47**, 9657–9660; (c) Park, J.K., Yoon, Z.S., Yoon, M.-C., Kim, K.S., Mori, S., Shin, J.-Y., Osuka, A., and Kim, D. (2008) *J. Am. Chem. Soc.*, **130**, 1824–1825; (d) Sankar, J., Mori, S., Saito, S., Rath, H., Suzuki, M., Inokuma, Y., Shinokubo, H., Kim, K.S., Yoon, Z.S., Shin, J.-Y., Lim, J.M., Matsuzaki, Y., Matsushita, O., Muranaka, A., Kobayashi, N., Kim, D., and Osuka, A. (2008) *J. Am. Chem. Soc.*, **130**, 13568–13579.

16 Stępień, M., Latos-Grażyński, L., Sprutta, N., Chwalisz, P., and Szterenberg, L. (2007) *Angew. Chem. Int. Ed.*, **46**, 7869–7873.

17 Yoon, Z.S., Osuka, A., and Kim, D. (2009) *Nature Chem.*, **1**, 113–122.

18 (a) For Ar-Ar bond formations, see: Negishi, E. (1982) *Acc. Chem. Res.*, **15**, 340–348; (b) Erdik, D. (1984) *Tetrahedron*, **40**, 641–657; (c) Iyoda, M., Otsuka, H., Sato, K., Nisato, N., and Oda, M. (1990) *Bull. Chem. Soc. Jpn.*, **63**, 80–87; (d) Miyaura, N. and Suzuki, A. (1995) *Chem. Rev.*, **95**, 2457–2483; (e) Beller, M. and Bolm, C. (eds) (1998) *Transition Metals for Organic Synthesis*, Wiley-VCH Verlag GmbH, Weinheim; (f) Hong, R., Hoen, R., Zhang, J., and Lin, G. (2001) *Synlett*, 1527–1530; (g) Krause, N. (ed.) (2002) *Modern Organocopper Chemistry*, Wiley-VCH Verlag GmbH, Weinheim; (h) Meyers, A.I., Nelson, T.D., Moorlag, H., Rawson, D.J., and Meier, A. (2004)

Tetrahedron, **60**, 4459–4473; (i) Zapf, A. and Beller, M. (2005) *Chem. Commun.*, 431–440; (j) Fürstner, A. and Martin, R. (2005) *Chem. Lett.*, **34**, 624–629; (k) Woodward, S. (2005) *Angew. Chem. Int. Ed.*, **44**, 5560–5562.

19 Maeda, T., Otsuka, H., and Takahara, A. (2009) *Prog. Polym. Sci.*, **34**, 581–604.

20 (a) Campbell, I.D., Eglinton, G., Henderson, W., and Raphael, R.A. (1966) *J. Chem. Soc., Chem. Commun.*, 87–89; (b) Solooki, D., Ferrara, J.D., Malaba, D., Bradshaw, J.D., Tessier, C.A., and Youngs, W.J. (1997) *Inorg. Synth.*, **31**, 122–128; (c) Stephens, R.D. and Castro, C.E. (1963) *J. Org. Chem.*, **28**, 3313–3315.

21 Huynh, C. and Linstrumella, G. (1988) *Tetrahedron*, **44**, 6337–6344.

22 Iyoda, M., Sirinintasak, S., Nishiyama, Y., Vorasingha, A., Sultana, F., Nakao, K., Kuwatani, Y., Mtsuyama, H., Yoshida, M., and Miyake, Y. (2004) *Synthesis*, 1527–1531.

23 Iyoda, M., Vorasingha, A., Kuwatani, Y., and Yoshida, M. (1998) *Tatrahedron Lett.*, **39**, 4701–4704.

24 Miljanić, O.S., Vollhardt, K.P.C., and Whitener, S.D. (2003) *Synlett*, 29–34.

25 Nishinaga, T., Nodera, N., Miyata, Y., and Komatsu, K. (2002) *J. Org. Chem.*, **67**, 6091–6096.

26 Zhou, Q., Carroll, P.J., and Swager, T.M. (1994) *J. Org. Chem.*, **59**, 1294–1301.

27 Zimmermann, B., Baranović, G., Štafanić, Z., and Rožman, M. (2006) *J. Mol. Struct.*, **794**, 115–124.

28 Baxter, P.N.W. and Dali-Youcef, R. (2005) *J. Org. Chem.*, **70**, 4935–4953.

29 Sun, S.-S., and Lee, A.J. (2001) *Organometallics*, **20**, 2353–232358.

30 (a) Campbell, K., McDonald, R., Branda, N.R., and Tykwinski, R.R. (2001) *Org. Lett.*, **3**, 1045–1048; (b) Campbell, K., McDonald, R., and Tykwinski, R.R. (2002) *J. Org. Chem.*, **67**, 1133–1140.

31 Enozawa, H., Hasegawa, M., Takamatsu, D., Fukui, K., and Iyoda, M. (2006) *Org. Lett.*, **8**, 1917–1920.

32 Höger, S. and Enkelmann, V. (1995) *Angew. Chem. Int. Ed.*, **34**, 2713–2716.

33 Iyoda, M., Kuwatani, Y., Yamagata, S., Nakamura, N., Todaka, M., and Yamamoto, G. (2004) *Org. Lett.*, **6**, 4667–4670.

34 Zhang, J.S., Pesak, D.J., Ludewick, J.L., and Moore, J.S. (1994) *J. Am. Chem. Soc.*, **116**, 4227–4239.

35 Höger, S. (2007) in *Functional Organic Materials* (eds T.J.J. Müller and U.H.F. Bunz), Wiley-VCH Verlag GmbH, Weinheim, pp. 225–260.

36 Staab, H. and Neunhöffer, K. (1974) *Synthesis*, 424–425.

37 Weiss, K., Michael, A., Auth, E.-M., Bunz, U.H.F., Mangel, T., and Müllen, K. (1997) *Angew. Chem. Int. Ed.*, **36**, 506–509.

38 Ge, P.-H., Fu, W., Herrmann, W.A., Herdtweck, E., Campana, C., Adams, R.D., and Bunz, U.H.F. (2000) *Angew. Chem. Int. Ed.*, **39**, 3607–3610.

39 (a) Zhang, W., Kraft, S., and Moore, J.S. (2003) *Chem. Commun.*, 832–833; (b) Zhang, W., Kraft, S., and Moore, J.S. (2004) *J. Am. Chem. Soc.*, **126**, 329–335; (c) Zhang, W. and Moore, J.S. (2004) *J. Am. Chem. Soc.*, **126**, 12796–12797; (d) Zhang, W. and Moore, J.S. (2005) *J. Am. Chem. Soc.*, **127**, 11863–11870.

40 (a) Balakrishnan, K., Datar, A., Zhang, W., Yang, X., Naddo, T., Huang, J., Zuo, J., Yen, M., Moore, J.S., and Zang, L. (2006) *J. Am. Chem. Soc.*, **128**, 6576–6577; (b) Naddo, T., Che, Y., Zhang, W., Balakrishnan, K., Yang, X., Yen, M., Zhao, J., Moore, J.S., and Zang, L. (2007) *J. Am. Chem. Soc.*, **129**, 6978–6979.

41 (a) Zhao, D. and Moore, J.S. (2002) *J. Org. Chem.*, **67**, 3548–3554; (b) Zhao, D. and Moore, J.S. (2003) *Macromolecules*, **36**, 2712–2720.

42 Moore, J.S. (1997) *Acc. Chem. Res.*, **114**, 402–413.

43 Spitler, E.L., Johnson, C.A., and Haley, M.M. (2006) *Chem. Rev.*, **106**, 5344–5386.

44 Höger, S., Meckenstock, A.-D., and Pellen, H. (1997) *J. Org. Chem.*, **62**, 4556–4557.

45 (a) Höger, S., Mackenstock, A.-D., and Müller, S. (1998) *Chem. Eur. J.*, **4**, 2423–2434; (b) Ziegler, A., Mamdouth, W., Ver Heyen, A., Surin, M., Uji-I, H., Abdel-Mottaleb, M.M.S., De Schryver, F.C., De Feyter, S., Lazzaroni, R., and Höger, S. (2005) *Chem. Mater.*, **17**, 5670–5683.

46 Roncali, J. (1992) *Chem. Rev.*, **92**, 711–738.

47 (a) Mishra, A., Ma, C.-Q., and Bäuerle, P. (2009) *Chem. Rev.*, **109**, 1141–1276; (b) Bendnarz, M., Reineker, P., Mena-Osteritz, E., and Bäuerle, P. (2004) *J. Lumin.*, **110**, 225–231.

48 (a) Bhasker, A., Ramakrishna, G., Hagedorn, K., Varnavski, O., Mena-Osteritz, E., Bäuerle, P., and Goodson, T. III (2007) *J. Phy. Chem. B*, **111**, 946–954; (b) Ammann, M., Rang, A., Schalley, C.A., and Bäuerle, P. (2006) *E. J. Org. Chem.*, 1940–1948; (c) Fuhrmann, G., Krömer, J., and Bäuerle, P. (2001) *Synth. Met.*, **119**, 125–126; (d) Krömer, J., Rios-Carreras, I., Fuhrmann, G., Musch, C., Wunderlin, M., Debaerdemaeker, T., Mena-Osteritz, E., and Bäuerle, P. (2000) *Angew. Chem. Int. Ed.*, **39**, 3481–3486.

49 (a) Zhang, F., Götz, G., Winkler, H.D.F., Schalley, C.A., and Bäuerle, P. (2009) *Angew. Chem. Int. Ed.*, **48**, 6632–6635; (b) Fuhrmann, G., Debaerdemaeker, T., and Bäuerle, P. (2003) *Chem. Commun.*, 948–949.

50 Mayor, M. and Didschies, C. (2003) *Angew. Chem. Int. Ed.*, **42**, 3176–3179.

51 (a) Nakao, K., Nishimura, M., Tamachi, T., Kuwatani, Y., Miyasaka, H., Nishinaga, T., and Iyoda, M. (2006) *J. Am. Chem. Soc.*, **128**, 16740–16747; (b) Iyoda, M. (2007) *Heteroatom Chem.*, **18**, 460–466.

52 (a) Williams-Harry, M., Bhaskar, A., Ramakrishna, G., Goodson, T. III, Imamura, M., Mawatari, A., Nakao, K., Enozawa, H., Nishinaga, T., and Iyoda, M. (2008) *J. Am. Chem. Soc.*, **130**, 3252–3253; (b) Iyoda, M. (2009) *Compt. Rend. Chim.*, **12**, 395–402.

53 Sugiura, K.-i., Fujimoto, Y., and Sakata, Y. (2000) *Chem. Commun.*, 1105–1106.

54 Kato, A., Sugiura, K.-i., Miyasaka, H., Tanaka, H., Kawai, T., Sugimoto, M., and Yamashita, M. (2004) *Chem. Lett.*, **33**, 578–579.

55 (a) Hoffmann, M., Wilson, C.J., Odell, B., and Anderson, H.L. (2007) *Angew. Chem. Int. Ed.*, **46**, 3122–3125; (b) Hoffmann, M., Kärnbratt, J., Chang, M.-H., Herz, L.M., Albinsson, B., and Anderson, H.L. (2008) *Angew. Chem. Int. Ed.*, **47**, 4993–4996.

56 (a) Krishnamurthy, V.M., Semetey, V., Brancher, P.J., Shen, N., and Whitesides, G.M. (2007) *J. Am. Chem. Soc.*, **129**, 1312–1320; (b) Ballester, P., Oliva, A.I., Costa, A., Deya, P.M., Frontera, A., Gomila, R.M., and Hunter, C.A. (2006) *J. Am. Chem. Soc.*, **128**, 5560–5569; (c) Ercolani, G. (2006) *Struct. Bonding (Berlin)*, **121**, 167–215.

57 (a) Takaba, H., Omachi, H., Yamamoto, Y., Bouffard, J., and Itami, K. (2009) *Angew. Chem. Int. Ed.*, **48**, 6112–6116; (b) Jasti, R., Bhattacharjee, J., Neaton, J.B., and Bertozzi, C.R. (2008) *J. Am. Chem. Soc.*, **130**, 17646–17647; (c) Nakamura, E., Tahara, K., and Matsuo, Y. (2003) *J. Am. Chem. Soc.*, **130**, 2834–2835; (d) Ohkita, M., Ando, K., and Tsuji, T. (2001) *Chem. Commun.*, 2570–2571.

58 (a) Jung, S.-H., Pisula, W., Rouhanipour, A., Räder, H.J., Jacob, J., and Müllen, K. (2006) *Angew. Chem. Int. Ed.*, **45**, 4685–4690; (b) Simon, S., Schmalts, B., Rouhanipour, A., Räder, H.J., and Müllen, K. (2009) *Adv. Mater.*, **21**, 83–85.

59 (a) Gallant, A. and MacLachlan, M.J. (2003) *Angew. Chem. Int. Ed.*, **42**, 5307–5310; (b) Ma, C., Lo, A., Abdolmaleki, A., and MacLachlan, M.J. (2004) *Org. Lett.*, **6**, 3841–3844.

60 (a) Wolffs, M., Korevaar, P.A., Jonkheijm, P., Henze, O., Feast, W.J., Schenning, A.P.H.J., and Meijer, E.W. (2008) *Chem. Commun.*, 4613–4615; (b) Tomović, Ž., van Dongen, J., George, S.J., Xu, H., Pisula, W., Leclère, P., Smulders, M.M.J., De Feyter, S., Meijer, E.W., and Schenning, A.P.H.J. (2007) *J. Am. Chem. Soc.*, **129**, 16190–16196.

61 Aratani, N., Osuka, A., Kim, Y.H., Jeong, D.H., and Kim, D. (2000) *Angew. Chem. Int. Ed.*, **39**, 1458–1462.

62 Hori, T., Peng, X., Aratani, N., Takagi, A., Matsumoto, T., Kawai, T., Yoon, Z.S., Yoon, M.-C., Yang, J., Kim, D., and Osuka, A. (2008) *Chem. Eur. J.*, **14**, 582–595.

63 Martin, R.B. (1996) *Chem. Rev.*, **96**, 3043–3064.

64 Tobe, Y., Utsumi, N., Kawabata, K., Nagano, A., Adachi, K., Araki, S., Sonoda, M., Hirose, K., and Naemura, K. (2002) *J. Am. Chem. Soc.*, **124**, 5350–5364.

65 (a) Müllen, K. and Rabe, J.P. (2008) *Acc. Chem. Res.*, **41**, 511–520;

(b) Jäckel, F., Watson, M.D., Müllen, K., and Rabe, J.P. (2004) *Phys. Rev. Lett.*, **92**, 188303–188303-4.

66 Tour, J.M. (2000) *Acc. Chem. Res.*, **33**, 791–804.

67 Li, J., Liao, L., and Pang, Y. (2002) *Tetrahedron Lett.*, **43**, 391.

68 Bhaskar, A., Guda, R., Haley, M.M., and Goodson, T. III (2006) *J. Am. Chem. Soc.*, **128**, 13972–13973.

69 Schmaltz, B., Rouhanipour, A., Räder, H.J., Pisula, W., and Müllen, K. (2009) *Angew. Chem. Int. Ed.*, **48**, 720–724.

7
Organoboron Conjugated Polymers

Atsushi Nagai and Yoshiki Chujo

7.1
Introduction

Since electrically conductive polyacetylenes were discovered [1], the design and synthesis of π-conjugated polymers [2] have attracted an enormous amount of research aimed at the development of functional materials exhibiting conductivity, electrochemical activity, light-emitting properties, third order nonlinear optical properties and so on. Many synthetic methods such as oxidation polymerization, organometallic polycondensation (e.g., Suzuki–Miyaura or Sonogashira–Hagihara coupling reactions, and so on), and electrochemical polymerization have been developed. Recently, a variety of unique π-conjugated systems incorporating exotic elements into π-conjugated polymer backbones have been prepared. For example, polysiloles [3], polygermoles [4], polyphospholes [5], poly(vinylene-arsine)s [6] have been reported.

The incorporation of boron, which is an exotic element with high Lewis acidity resulting from the vacant p-orbital of tricoordinate boron [7], into π-conjugated systems leads to extension of the π-conjugation via the vacant p-orbital of the boron atom. This first concept has been studied in detail theoretically by Good *et al.* [8]. They have demonstrated that the π-orbitals of the vinyl group in low molecular-weight vinylborane derivatives overlap conjugatively with the vacant p-orbital of the boron atom, based on the fact that the ^{11}B NMR chemical shifts and ultraviolet (UV) absorption data were identical to the calculated results from Hückel molecular orbital (MO) theory. Therefore, the fully aromatic organoboron polymers were expected to act as π-conjugated polymers which possess unusual electronic states with positive holes built in. Further, Salzner, Lagowski *et al.* [9] have predicted that polyboroles will show fairly small bandgaps because of their quinoidal structures in the ground state. Yamabe, Tanaka *et al.* also theoretically predicted that poly(bora-acetylene)s would show metallic conductivity without doping [10].

Due to such intriguing properties of the organoboron π-conjugated system, recently, approaches have been developed for the preparation of organoboron π-conjugated polymers. This chapter provides a survey of procedures currently

Conjugated Polymer Synthesis. Edited by Yoshiki Chujo

ISBN: 978-3-527-32267-1

available for the preparation of organoboron π-conjugated polymers, especially main-chain type conjugated polymers. Designs of the conjugated polymers are classified into two approaches: (i) tricoordinate boron-containing conjugated polymers by the polymerization or reaction with non-conjugated boron compounds and (ii) tetracoordinate boron-containing conjugated polymers by incorporation of organoboron dyes into π-conjugated polymers. Finally, unique π-conjugated polymers based on carboranes as icosahedral boron clusters are introduced as one of the hot topics in this field.

7.2 Tricoordinate Boron π-Conjugated Polymers

7.2.1 Hydroboration Polymerization of Diyne Monomers

Synthesis of π-conjugated organoboron polymers poly(*p*-phenylenevinylene-arylborane), which are structurally related to poly (*p*-phenylene-vinylene) (PPV), was first achieved [11] in our group via hydroboration polymerization [12] of bifunctional alkynes with sterically hindered arylboranes $ArBH_2$ (Ar = Mes, Trip) [13], which are known to be strong electron acceptors, as shown in Scheme 7.1. The bulky aryl groups on boron prevent nucleophilic attack and thus impart good environmental stability. The polymers are highly fluorescent with especially large Stokes shifts for the heteroaromatic donor–acceptor polymers composed of pyridine and thiophene units in the backbone. Interestingly, the polymer derived from diethynylpyridine exhibited white emission. This can be likely attributed to different regiochemistry in the hydroboration of the alkynylpyridine moiety, thus leading to a more complex polymer

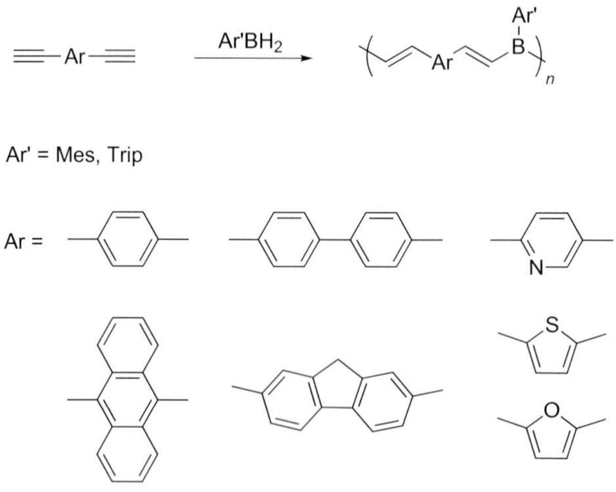

Ar'BH2

Ar'

B

Ar

n

Ar' = Mes, Trip

Ar =

N

S

O

Scheme 7.1

structure. Further, the third-order nonlinear optical properties of the polymers were evaluated by the degenerate four wave mixing method. Unusually large third-order nonlinear susceptibilities were observed ($6.87 \times 10^{-6} \sim 3.56 \times 10^{-7}$ esu). The observed large χ^3 values compared with those for low molecular weight borylated NLO chlomophores [14] reported by Marder *et al.* should be due in part to the highly extended π-conjugation. Moreover, electrical conductivity measurements on the polymer prepared from diethynylfluorene and tripylborane showed an increase in conductivity from less than 10^{-10} to 10^{-6} S cm^{-1} within 100 min upon doping with triethylamine [15]. Stationary-state photocurrent measurements with an ITO-polymer-Au cell confirmed the n-type conductivity. Transient photocurrent measurements showed that the drift mobility of an electron was approximately five times higher than that of a hole in this system.

A π-conjugated organoboron polymer that contains a ruthenium–phosphine complex in the main chain alternating with an electron-deficient boron center has been reported (Scheme 7.2). The UV–vis absorption spectrum of the polymer showed two absorption maxima at 356 and 514 nm, assignable to π–π^* and dπ–pπ transitions, respectively. Indeed, incorporation of a ruthenium alkynyl complex into the organoboron polymer main chain led to an unusually strong bathochromic shift of 141 nm relative to a repeating molecular ruthenium complex [16]. Possibly, the high electron affinity of the organoboron unit and the push–pull effect between the electron-rich ruthenium complex and the electron-deficient boron units are responsible for this unusual red-shift. Further, π-conjugated boron polymers containing platinum or palladium complexes in the backbone were also prepared by hydroboration polymerization of tetrayne/metal complexes and tripylborane [17]. While the related platinum and palladium polymers did not show similar red-shift MLCT (metal-to-ligand charge transfer interaction) bands to the ruthenium-phosphine polymer, the polymers with higher molecular weight showed high solubility toward common organic solvents and intense luminescence at 490 nm.

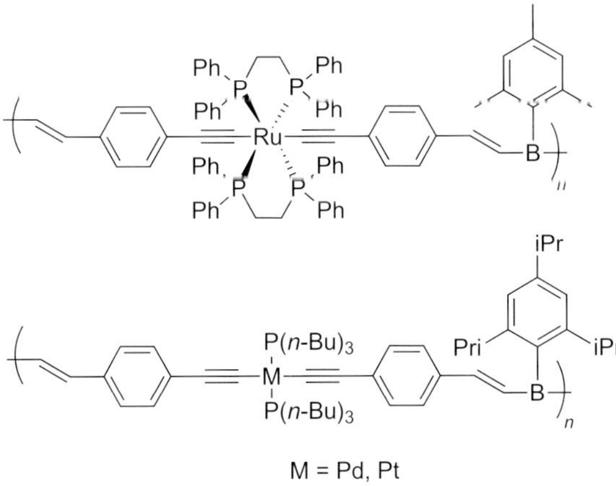

Ph Ph P P Ph Ph Ru Ph Ph P P Ph Ph B n

iPr P(n-Bu)3 M P(n-Bu)3 Pri iPr B n

M = Pd, Pt

Scheme 7.2

The π-conjugated diorganoboron polymers poly(*p*-phenylenevinylene-diboraanthracene), which are structurally related to poly(*p*-phenylene-anthracene), were recently reported by Holthausen, Jäkle, Wagner *et al.* [18]. These unique polymers are formed by hydroboration polymerization of bifunctional alkynes with ladder-like polymeric diboraanthracene, as shown in Scheme 7.3. The X-ray crystal structure analysis of diboraanthracene reveals a ladder polymeric structure that occurs through three-center two-electron B–H–B bonds. The polymer derived from 1,4-diethynyl-2,5-bis(hexyloxy)benzene shows gel formation in organic solvents and the gel reversibly transforms back into clear liquid. A solution of the polymer displays absorption maxima at 410, 349, and 290 nm, and the solution and thin film of the polymer emit intense green light at 518 nm, excited at the longest-wavelength absorption. Further, a distinct red-shift of emission was observed with increasing solvent polarity.

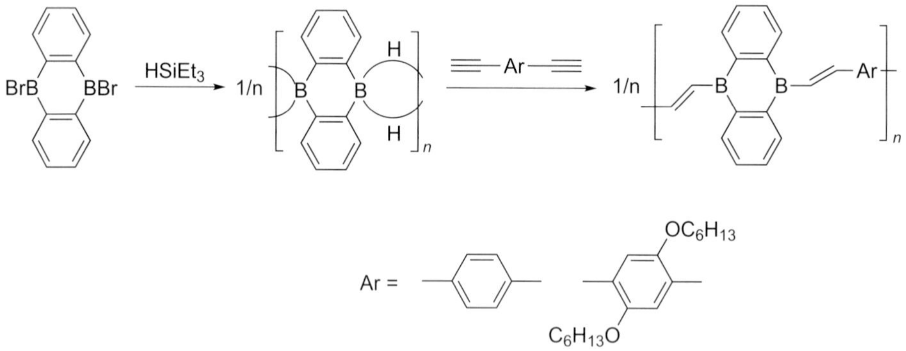

BrB BBr $\xrightarrow{HSiEt_3}$ 1/n [B(H)B(H)]$_n$ $\xrightarrow{\equiv-Ar-\equiv}$ 1/n [–CH=CH–B B–CH=CH–Ar–]$_n$

Ar = C$_6$H$_4$; C$_6$H$_2$(OC$_6$H$_{13}$)$_2$ (OC$_6$H$_{13}$, C$_6$H$_{13}$O)

Scheme 7.3

7.2.2
Tin–Boron Exchange Polymerization of Bis(Trimethylsilyl) Monomer

Synthesis of the π-conjugated organoboron polymers embedded in a polythiophene backbone has been reported by Jäkle *et al.* [19]. The polymers are formed under mild conditions through tin–boron exchange reactions (Scheme 7.4). Intriguingly, the photophysical properties of the polymers are responsible for the nature of the aryl substituent on boron. Thus, the polymers with several phenyl groups on boron are color-tunable from blue to dark-red luminescence. Especially, the attachment of aminobithiophene on boron leads to orange–red emission, which is attributed to charge transfer in this donor–π-acceptor type structure. Further, the polymers were determined to be predominantly linear in nature, the end-groups and the repeating units were analyzed by ^{1}H NMR and MALDI-TOF-TOF measurements, respectively.

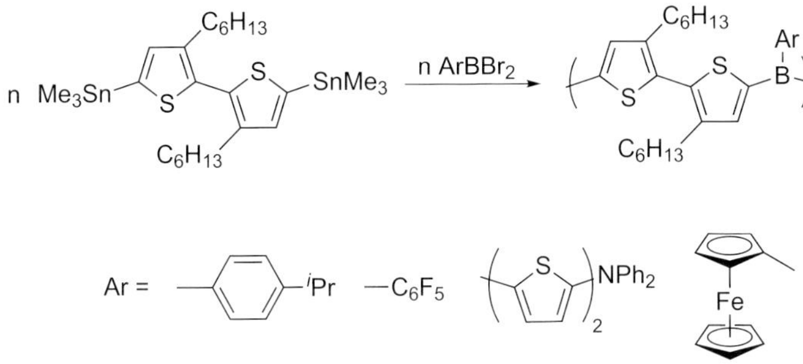

Scheme 7.4

7.2.3
Haloboration–Phenylboration Polymerization of Diyne Monomers

Haloboration–phenylboration polymerization of aromatic diyne monomers with bromodiphenylborane was carried out to afford unique π-conjugated organoboron polymers [20], as shown in Scheme 7.5. The polymers were fusible and soluble in common organic solvents including THF, benzene, toluene, $CHCl_3$, and CH_2Cl_2, and so on. The molecular weights of the polymers were estimated, by gel permeation chromatographic analysis, to be several thousands. Efficient bathochromic shifts of the absorption maxima of the polymers relative to the respective monomers were observed, indicating that π-conjugation is occurring through the vacant p-orbital of the boron.

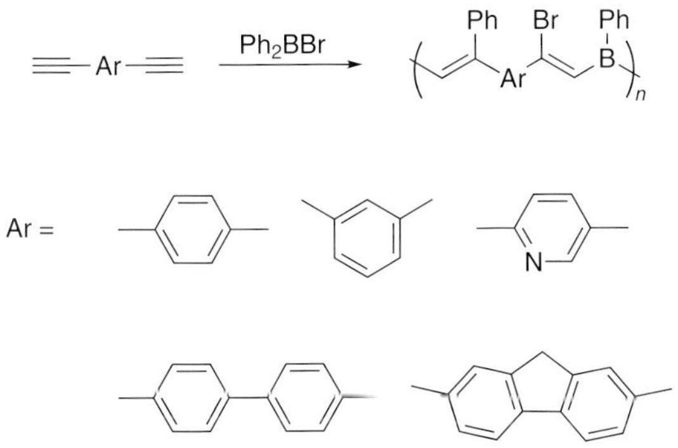

Scheme 7.5

7.2.4
Polycondensation of Aryldimethoxyborane Using Grignard and Organolithium Reagents

Poly(*p*-phenylene-borane)s [21], which are structurally related to poly(*p*-phenylene), were prepared by polycondensation of aryldimethoxyboranes and bifunctional

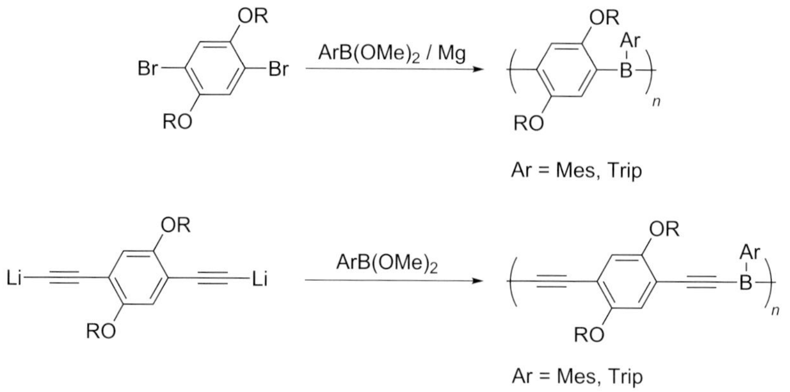

OR
Br Br $ArB(OMe)_2$ / Mg
RO
OR Ar B n
RO
Ar = Mes, Trip

OR
Li Li $ArB(OMe)_2$
RO
OR Ar B n
RO
Ar = Mes, Trip

Scheme 7.6

Grignard reagents that were generated *in situ* (Scheme 7.6). The polymers show absorption maxima in the range 359–367 nm in $CHCl_3$ solution and emit blue–green light upon excitation at 350 nm. This intriguing synthetic technique provides a novel methodology for the preparation of organoboron conjugated polymers, which is a useful alternative to hydroboration polymerization. The polymers are expected to be novel n-type conjugated polymers with fairly high air and moisture stability, and higher thermal stability is also expected due to the absence of a retrohydroboration (β-elimination) process during their thermal degradation.

Polycondensation of lithiated diacetylenes with aryldimethoxyboranes was used to afford unique conjugated polymers poly(ethynylene-*p*-phenylene-ethynylene-borane)s [22] (Scheme 7.6). The molecular weight of the polymer was found to be 2700. The polymers show absorption maxima at 397 nm and emit visible blue light, excited at the absorption maxima.

7.2.5 Stepwise Reaction from Boraanthracene

A series of multifunctional anthrylboranes, such as a dendrimer [23] with up to six boron centers and divergently extended π-conjugation via the boron atom have been reported by Yamaguchi, Tamao *et al.* (Scheme 7.7). The lowest unoccupied molecular orbital (LUMO) over three identical π-systems is extended through the boron atom. X-ray crystallography of tri-9-anthrylborane shows that the boron atom and three carbons adjacent to the boron atom were in completely planar geometry. The absorption maxima of the dendrimers show bathochromic shifts with increasing number of anthracene units, indicating extension of π-conjugation in three directions. In cyclic voltammograms of the dendrimers, irreversible reduction peaks were observed. With an increasing number of dianthrylboryl or dimesitylboryl groups, the reduction potentials shifted to the positive direction. The reduction potentials were comparable to or lower than those for π-electron deficient π-conjugated polymers such as poly(2,2′-bithiazole-5,5′-diyl) [24] and poly(quinoxaline-2,6-diyl) [25].

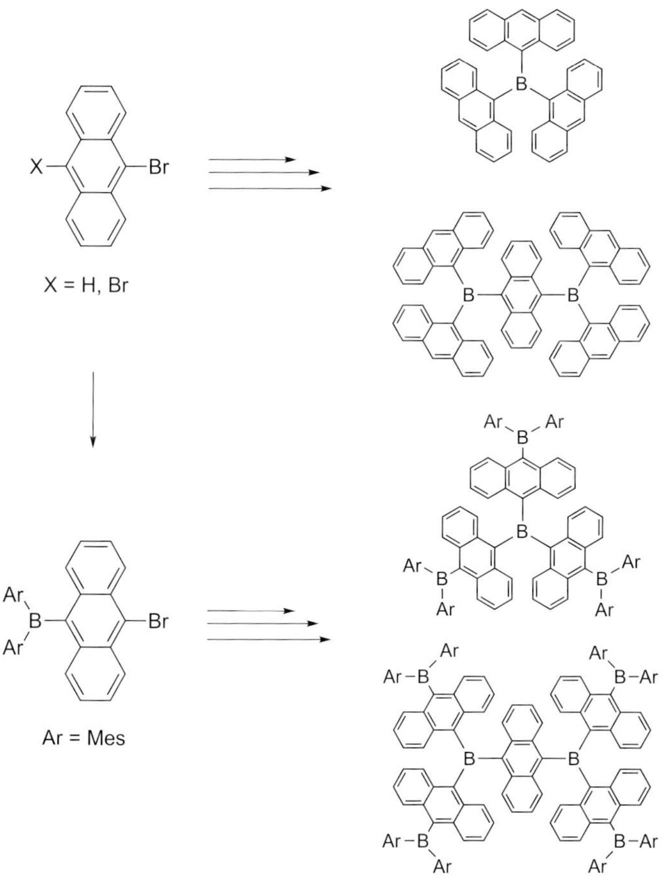

X = H, Br

Ar = Mes

Scheme 7.7

7.3 Tetracoordinate Boron π-Conjugated Polymers

7.3.1 π-Conjugated Poly(Cyclodiborazane)s

Cyclodiborazanes consisting of four-membered boron nitrogen rings are easily prepared by the dimerization of iminoborane formed by monohydration of cyano groups. Therefore, the hydroboration or allylboration of dicyano compounds allows efficient preparation of air stable poly(cyclodiborazane)s [26]. Generally, although fully aromatic poly(cycloborazane)s do not show extension of the π-conjugation length, significant extension of π-conjugation was observed in donor–acceptor type poly(cycloborazane)s having thienylene or 2,5-dialkoxy-1,4-phenylene units (Scheme 7.8) [27]. Both the absorption edges and the maxima of the polymers are dramatically red-shifted relative to model compounds or the respective monomers,

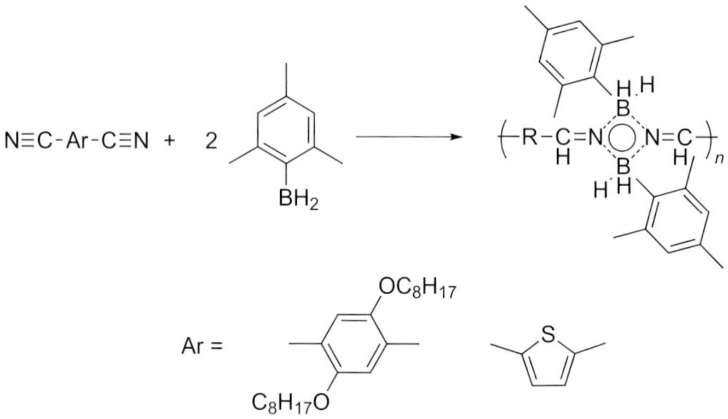

Scheme 7.8

demonstrating that π-conjugation is certainly extended beyond the repeating units. The polymers emit intense blue–green emission when irradiated with UV light or near-UV light, and the fluorescence emission wavelength is dependent on the solvent, indicating the existence of some intramolecular charge transfer (ICT) interaction due to the electron-accepting cyclodiborazane units or interunit conjugation along the main chain. Continuously alternating polymers with oligothiophene and cyclodiborazane sequences have also been reported. The emission wavelength and quantum yield were finely tunable by varying the oligothiophene chain length [28]. Moreover, σ–π conjugated poly(cyclodiborazane)s have also been prepared by hydroboration polymerization of dicyano monomer including a disilanylene unit [29].

Chiral poly(cycloborazane)s were successfully prepared by hydroboration polymerization of aromatic dicyano monomers with different enantio-camphenylboranes as chiral boranes [30], as depicted in Scheme 7.9. Only a polymer derived from 2,5-dialkoxy-1,4-dicyanobenzene exhibits intense blue–green light emission from charge-transfer interaction by incorporation of a donor–acceptor system into the polymer backbone. The emission behavior of all polymers was examined by density-

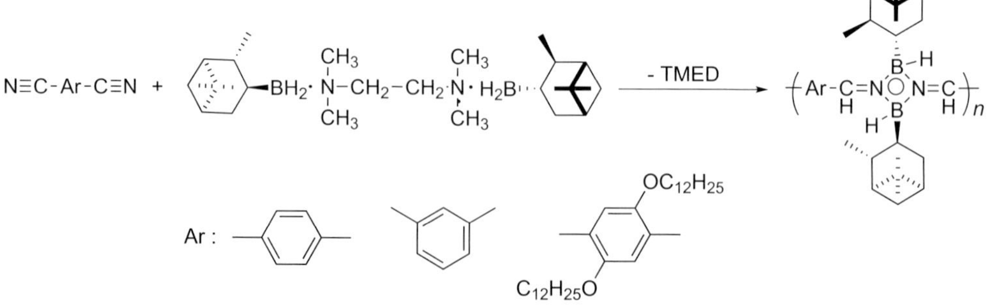

Scheme 7.9

functional theory (DFT) calculations of the designed model compounds, corresponding to polymer repeating units. Circular dichroism (CD) spectra of the polymers were observed at around 300–350 nm, attributed to cyanobenzene moieties, and the polymers further possessed high values of the specific rotations, possibly suggesting that the chiral poly(cycloborazane)s have some higher order structures, such as a helical conformation, due to the asymmetric borane.

7.3.2 Poly(Pyrazabole)s

Pyrazabole is a highly stable boron heterocycle in which a variety of functional groups can be readily introduced. Several applications of pyrazaboles have been reported, including their use as possible building blocks for discotic liquid crystals [31] or as good bridges for ansa-ferrocenes to form active container molecules for supramolecular applications [32].

Poly(pyrazabole)s were prepared by organometallic polycondensation of 2,6-diisopyrazabole with aromatic diyne monomers using Sonogashira coupling [33] (Scheme 7.10). The absorption maxima of the polymers were not particularly red-shifted in comparison with those for the aromatic monomers, indicating that the π-conjugation was poorly expanded via the repeating units. This is due to the small electronic interaction between the π-orbitals of pyrazole. However, their optical transparency would be advantageous for various applications as optical materials. The polymers are highly stable against air and moisture, and showed purple to green fluorescence emission with good quantum yields when irradiated with UV light. The observed fluorescence emission peaks were composed of multi-emission peaks, probably due to various conformations of the pyrazabole units such as flat, boat, chair, and so on. The presence of pyrazabole in the π-conjugated polymer backbone was crucial for the observed emission behavior. Further, incorporation of various electron-withdrawing groups into the main chain leads to low electron density on the boron atom, and the fluorescence emission wavelength being shifted to shorter wavelength [34]. Therefore, tuning of emission wavelength was possible by introducing designed aromatic groups having various electron densities.

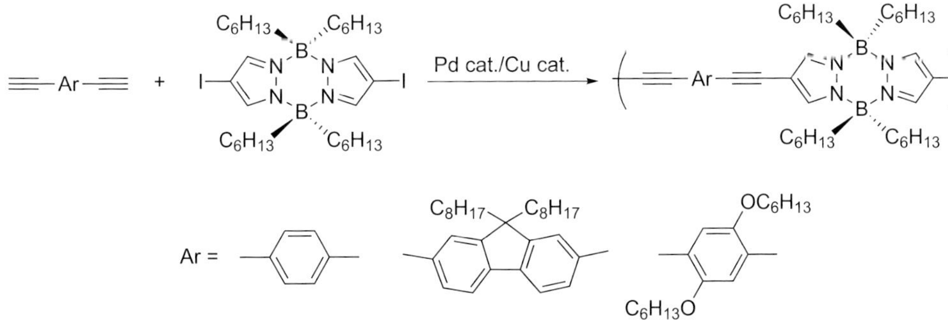

C_6H_{13} C_6H_{13} B N N Ar + I N N I B C_6H_{13} C_6H_{13} Pd cat./Cu cat. Ar n

Ar = C_8H_{17} C_8H_{17} OC_6H_{13} $C_6H_{13}O$

Scheme 7.10

Transition-metal-containing poly(pyrazabole)s were also prepared by copper-catalyzed dehydro-halogenation [35]. MLCT absorption was observed only in the case of platinum-containing polymers and not for palladium and nickel-containing polymers.

7.3.3 π-Conjugated Organoboron Quinolate Polymers

Organoboron quinolates are valuable light-emitting materials, potentially useful in organic light-emitting devices (OLEDs). Organoboron quinolates with high luminescent efficiency and high thermal stability were first reported by Wang *et al.* [36]. Di- and triboron-type quinolates [37] were also designed, and improved thermal stability and unique photophysical properties originating from their anisotropic shape was reported. Jäkle *et al.* also reported polystyrene-supported organoboron quinolates. Side chain-type organoboron quinolates were synthesized via polymer reaction of 8-hydroxyquinoline and thienyl-substituted poly(borylstyrene) [38]. Emission color-tuning from blue to red was achieved by treatment with various 8-hydroxyquinoline derivatives [39].

The first synthesis of boron quinolate polymers having a π-conjugated backbone was recently reported by our group [40] (Scheme 7.11). Boron quinolate monomer attached to diiodophenyl groups on the boron was polymerized by Sonogashira coupling with aromatic diyne monomers to produce the conjugated polymers incorporating organoboron quinolate into the main chain. No bathochromic shifts of the absorption and fluorescence maxima of the polymers were observed, as compared with the diiodo-type boron quinolate monomer, because *p*-arylene-ethynylene moieties were directly attached to boron atoms. However, the polymers emit intense blue–green light with good quantum yields over a wide range of excitation wavelengths. The emission color of the polymers was freely tuned by substituting methyl groups into the quinolate ligands [41]. Further, incorporation of the more stable organoboron aminoquinolate, with a similar structure to boron quinolate, into

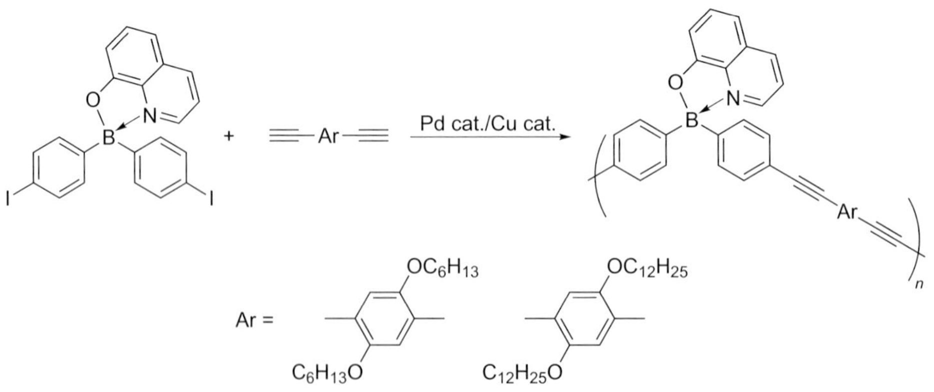

Scheme 7.11

π-conjugated polymers led to higher molar absorption coefficients while retaining good quantum yields [42]. More recently, novel organoboron quinoline-8-thiolate and quinoline-8-selenolate polymers have also been prepared by the Sonogashira coupling of diiodo-type boron quinoline-8-thiolate and selenolate monomers (Scheme 7.12) [43]. Accordingly, increasing the atomic number of the Group 16 atom adjacent to the boron atom caused a shift in emission to longer wavelength and a decrease in the quantum yields for both the model compounds and the polymers. However, the polymers showed high refractive indices ($n_d > 1.66$).

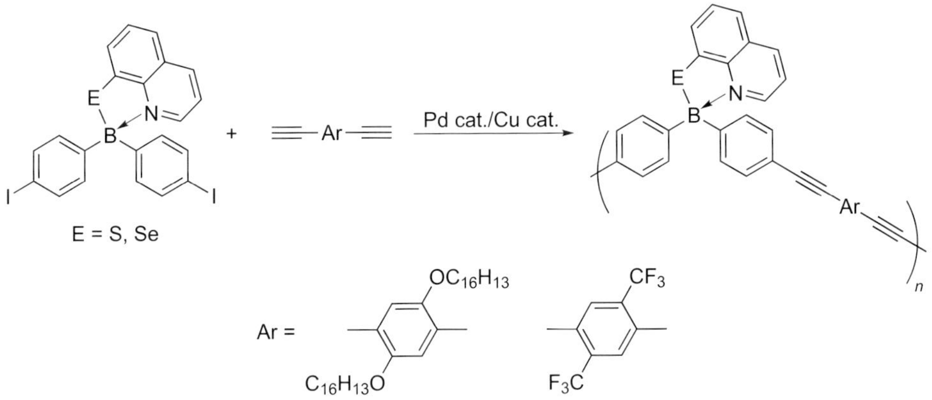

E = S, Se

Pd cat./Cu cat.

Ar = ($OC_{16}H_{13}$ / $C_{16}H_{13}O$ substituted phenylene; CF_3 / F_3C substituted phenylene)

Scheme 7.12

The organoboron quinolate polymer linked to a quinolate ligand was also prepared via the polymer reaction between the polymer precursor and triphenylborane [44], as shown in Scheme 7.13. The synthesis begins with Sonogashira coupling to generate the polymeric precursor. Deprotection followed by treatment with triphenylborane affords the organoboron quinolate polymer in high yield and moderate molecular weight. A dramatic bathochromic shift of the absorption and fluorescence emission maxima was observed, while the quantum yield of the polymer was not significantly high in comparison with that of polymer precursor. The molar absorption coefficient of the polymer at the excitation wavelength was remarkably improved by incorporation of the quinolate structure into the π-conjugated polymer backbone,

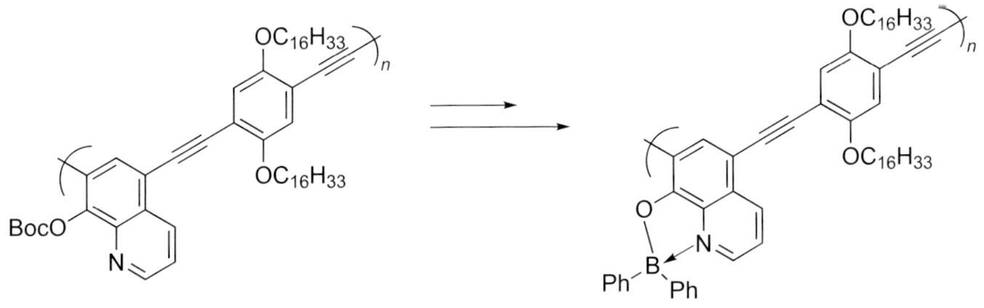

Scheme 7.13

suggesting that the conjugated main-chain behaves as a light harvesting antenna for organoboron quinolate units.

Organoboron quinolate coordination polymers were also reported by Jäkle *et al.* [45]. The coordination polymers with both the boron and quinolate ligands embedded in the polymer backbone were prepared under exceptionally mild conditions via a polycondensation reaction involving boron-induced ether cleavage (Scheme 7.14). The fluorescence properties are tuned by varying the conjugated bridge connecting the quinolate groups. Thus, the degree of conjugation of the linker critically influences the nature of the frontier orbitals, as demonstrated by DFT calculations for model compounds. The polymer with the biphenyl linker displayed strong yellow–green emission, whereas the thiophene–phenylene–thiophene linker showed an unusual concentration-dependent dual emission as a result of excimer formation.

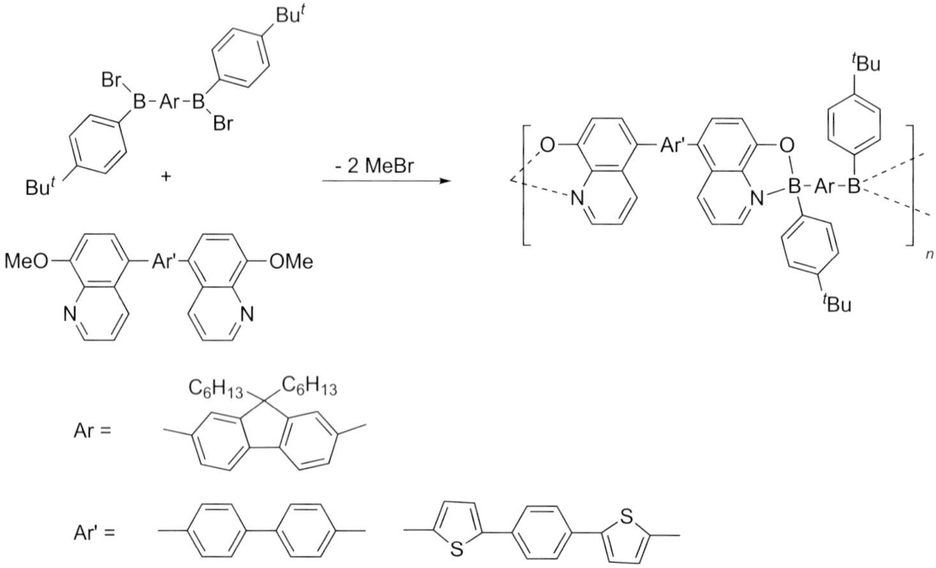

Scheme 7.14

7.3.4
π-Conjugated Organoboron Diketonate Polymer

Organoboron diketonate is another boron light-emitting material exhibiting large molar absorption coefficients and high quantum yields [46]. Conjugation of boron diketonate arises due to inhibition of non-radiative dissipation through O–H stretching modes from tautomerization between ketone and enol structures by forming a stable six-membered boron-chelating ring. The boron complex shows no loss of the unique characteristics of the diketonate fluorophore, such as high fluorescence quantum yield, and high photostability, in contrast with a heavy-metal complex. Fraser *et al.* have recently studied the preparation of boron diketonate-end-functionalized polylactides [47], poly(ε-caprolactone) [48], and their block copolymers [49] exhibiting interesting phosphorescence by ring-opening polymerization of lactides or ε-caprolactone using boron diketonate with a hydroxyl group as an initiator.

The first main chain-type organoboron 1,3-diketonate polymers were prepared by the highly efficient chelating reaction of a 1,3-diketone-based *p*-phenylene-ethynylene derivative as a polymeric ligand with the appropriate organoboron reagents (Scheme 7.15) [50]. The polymers showed significantly red-shifted absorption and fluorescence emission maxima relative to a polymeric ligand, and the emission maxima in films were dramatically red-shifted to longer wavelength as compared with those in solution. The polymers with diphenyl and pentafluorodiphenyl groups attached to the boron emit orange and red light, respectively, which is excited at the respective absorption maxima attributable to the *p*-phenylene-ethynylene moiety, as a result of extended π-conjugation along the polymeric ligand.

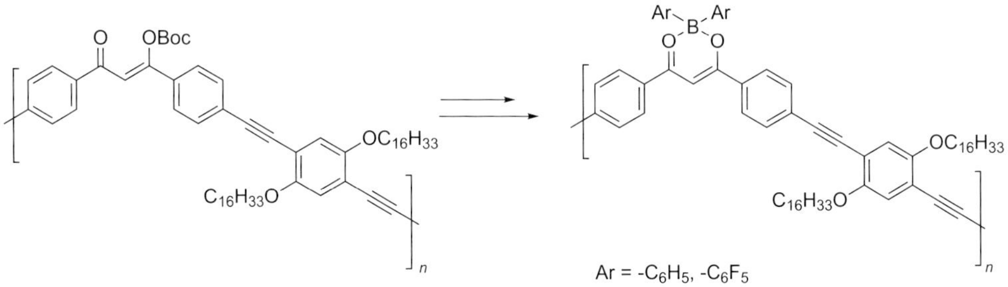

Ar = $-C_6H_5$, $-C_6F_5$

Scheme 7.15

7.3.5
π-Conjugated BODIPY-Based Polymers

4,4-Difluoro-4-bora-3a,4a-diaza-s-indacene (BODIPY)-based polymers are an important class of highly luminescent materials potentially useful in biochemical labeling, photonic molecular systems, laser dyes, organogelators, and light-emitting devices [51]. This is a result of their valuable characteristics, such as relatively higher photostability, and photoproperties (such as narrower absorption bands, sharper emission, and higher quantum yields, etc.) relative to fluorescein and other organoboron dyes, for example, boron quinolate and boron diketonate. García-Moreno, Amat-Guerri *et al.* reported that novel BODIPY-based polymers were prepared by free-radical copolymerization of methyl methacrylate with BODIPY monomers having the 8-position substituted by long alkyl chains [52] or phenyl groups [53], tethered with a methacryloyloxy group, and exhibited efficient and high photostability towards a laser in liquid solution as well as in solid polymeric matrices.

Incorporation of the BODIPY unit into π-conjugated polymer backbones was first reported by our group [54]. The organoboron π-conjugated polymers containing BODIPY units were prepared by Sonogashira coupling of BODIPY-based monomer having bis-iodophenyl and decyl groups with diyne monomers, (Scheme 7.16). The characterization by means of scanning electron microscopy (SEM) and transmission electron microscopy (TEM) revealed the strong tendency of the obtained polymers to self-assemble into particles in solution and as-cast films on the μm – nm scale. Especially, the polymer derived from 1,4-diethynylbenzene showed the presence of nm-sized particles and μm-sized fiber-like structures formed by aggregation of each particle. Further, in $CHCl_3$, gelation by three-dimensional aggregation of each fiber was observed at room temperature after 24 h. The electronic properties of the poly(*p*-arylene-ethynylene)s displayed higher energy transfer efficiency than π-conjugated linkers to BODIPY moieties, However, no red-shift of absorption and emission maxima from extended π-conjugation was observed, because *p*-arylene-ethynylene moieties were directly attached to boron atoms.

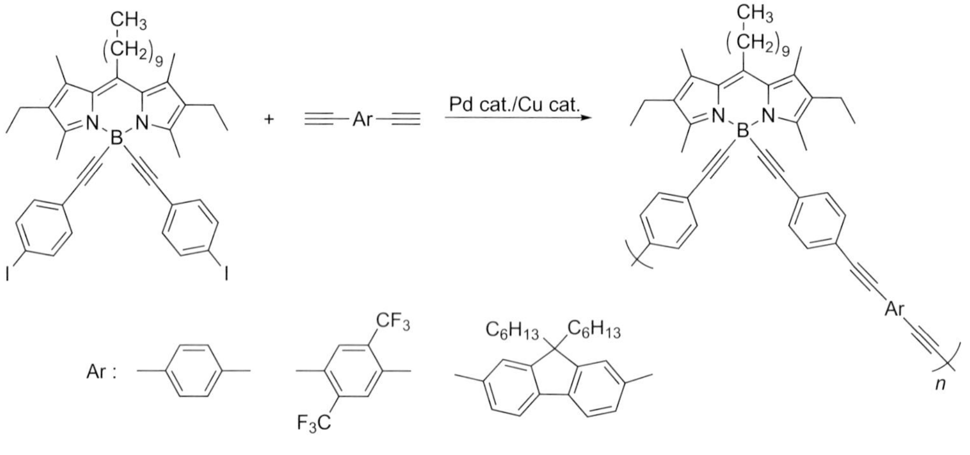

Scheme 7.16

Recently, Liu *et al.* reported the synthesis of BODIPY polymeric dyes attached directly to BODIPY cores, with triple bond [55] or fluorene [56] connections between the BODIPY cores and the aromatic units (Scheme 7.17). Absorption and fluorescence emission maxima of the polymers are dramatically shifted to the long wavelength region. The polymers having a *p*-arylene-ethynylene skeleton emit red or deep-red light (excited at the absorption maxima) and the fluorescence lifetimes are from 2.8 to 3.8 ns. Further, the BODIPY-based polymers displayed sensitive fluorescent responses to fluoride and cyanide anion through their multivalent interactions but are not responsive to chloride, bromide, and iodide anions.

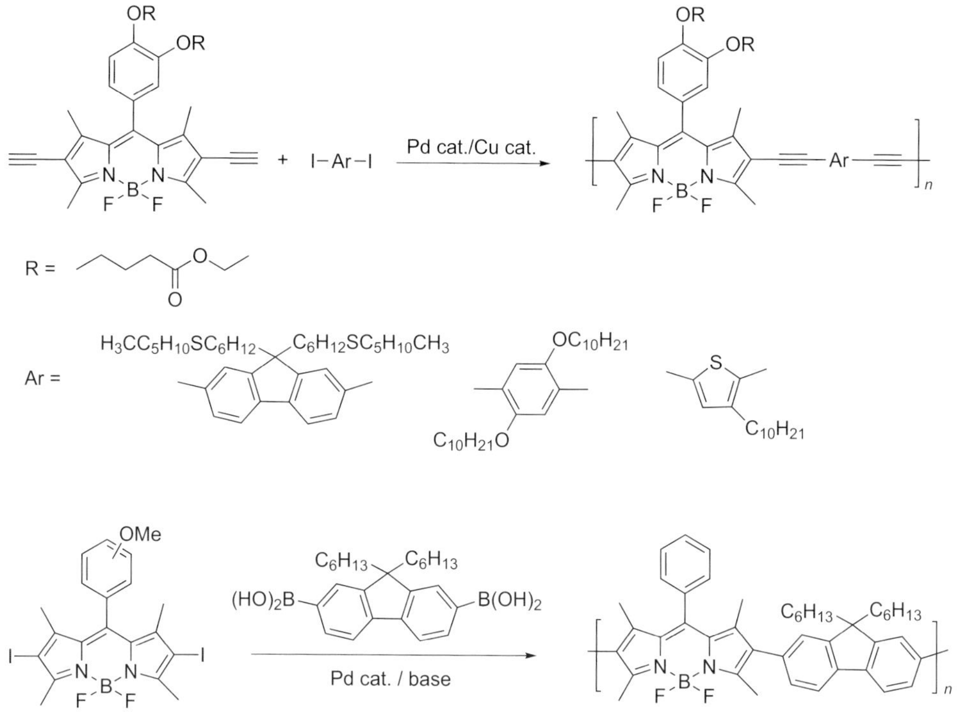

Scheme 7.17

7.4
π-Conjugated Carborane-Based Polymers

Carborane, composed of ten boron atoms and two carbon atoms, is an icosahedral boron cluster potentially useful in the field of medicinal chemistry and material science as a result of its rich boron content and high thermal or chemical stability, derived from three-center two-electron bonds and consequent three-dimensional delocalization of skeleton electrons [57]. The special reactivity and electron-withdrawing nature of carboranes can be regard as three-dimensional aromaticity [58]. Tour *et al.* described *p*-carboranes connected with *p*-phenylene-ethynylene as carborane-wheeled nanocars and showed the occurrence of a bathochromic absorption shift, indicating extension of the π-conjugation length with the incorporation of *p*-carborane [59].

Recently, a series of *o*-carborane or *m*-carborane-based π-conjugated polymers have also been prepared through Sonogashira coupling of diiodo-*o*-carborane or diiodo-*m*-carborane monomers with aromatic dyne monomers [60] (Scheme 7.18). Bathochromic shifts of the absorption maxima of the *o*-carborane or *m*-carborane

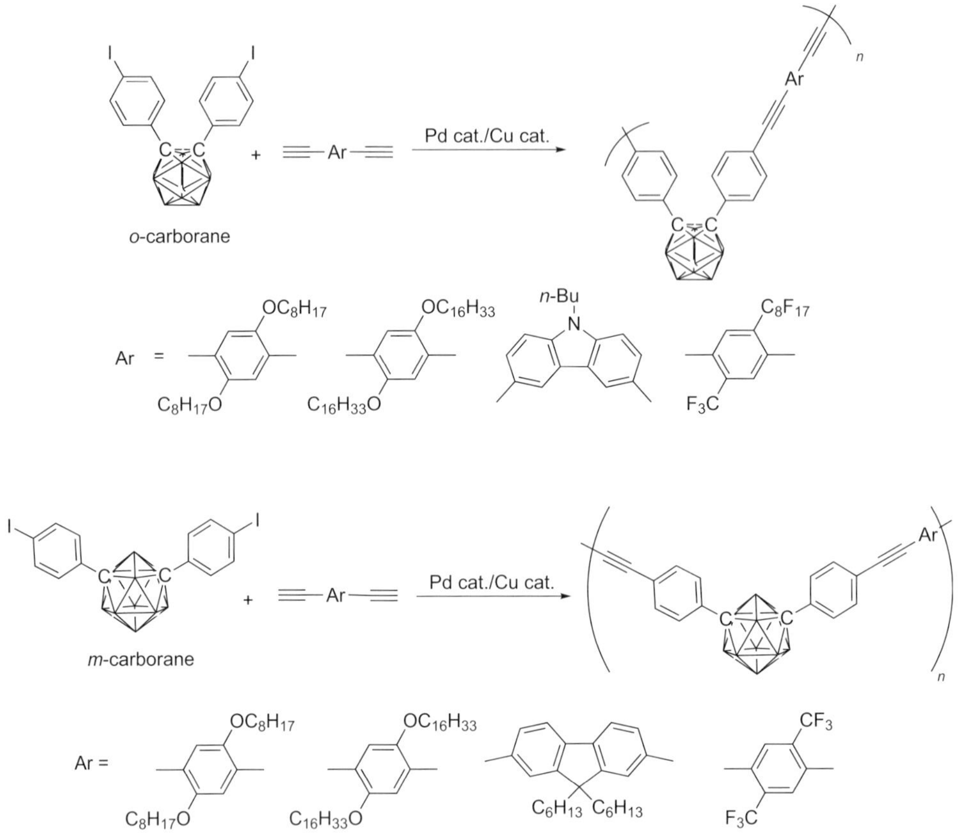

Scheme 7.18

polymers were observed in dilute solution due to extension of the π-conjugation length of the *p*-phenylene-ethynylene segment via *o*-carborane or m-carborane moieties in the polymer backbones. The fluorescence emission of the *o*-carborane polymers was observed in mixed solvent of tetrahydrofuran/water (1/99 v/v), while the emission of the polymers was not observed in organic solvents, indicating aggregation-induced emission (AIE) in water dispersion. The variable C–C bond in the *o*-carborane cluster efficiently quenches the fluorescence from *p*-phenylene-ethynylene segments in solution. In contrast, *m*-carborane polymers exhibited intense blue emission in solution due to the absence of variable C–C bond dissipation of the excited states in solution.

7.5
Conclusions

A varied array of new synthetic routes for the incorporation of boron atom into π-conjugated polymers backbone has recently become available, since the widely extended organoboron π-conjugated system was first reported in 1998. The enormous variety of potential applications of organoboron π-conjugated polymers is only beginning to be fully explored. For example, in the field of light-emitting devices, organoboron π-conjugated polymers are not only intriguing as emitters, but also as electron transporting materials with high electron affinity. In the near future, industrial applications of organoboron π-conjugated materials will be found via the design and the unique electronic properties of the polymer skeletons.

References

1 (a) Ito, T., Shirakawa, H., and Ikeda, S. (1974) *J. Polym. Sci., Polym. Chem. Ed.*, **12**, 11; (b) Shirakawa, H., Lois, E.J., MacDiarmid, A.G., Chiang, C.K., and Heeger, A.J. (1997) *J. Chem. Soc., Chem. Commun.*, 578.

2 Skotheimm, T.A. (ed.) (1986) *Handbook of Conductive Polymers, vol.* **I** and **II**, Mecrel Dekker, New York.

3 (a) Shinar, J., Ijadimaghsoodi, S., Ni, Q.X., Pang, Y., and Barton, T.J. (1989) *Synth. Met.*, **28**, C593; (b) Hong, S.Y. and Marynick, D.S. (1995) *Macromolecules*, **28**, 4991; (c) Tamao, K., Ohno, S., and Yamaguchi, S. (1996) *Chem. Commun.*, 1973; (d) Sohn, H.L., Huddleston, R.R., Powell, D.R., West, R., Oka, K., and Xu, Y.H. (1999) *J. Am. Chem. Soc.*, **121**, 2935; (e) Yamaguchi, S., Jin, R.Z., Itami, Y., Goto, T., and Tamao, K. (1999) *J. Am. Chem. Soc.*, **121**, 10420; (f) Kim, B.H. and Woo, H.G. (2002) *Organometallics*, **21**, 2796; (g) Boydston, A.J., Yin, Y., and Pagenkopf, B.L. (2004) *J. Am. Chem. Soc.*, **126**, 3724.

4 (a) Lucht, B.L., Buretea, M.A., and Tilley, T.D. (2000) *Organometallics*, **19**, 3469; (b) Sohn, H., Sailor, M.J., Magde, D., and Trogler, W.C. (2003) *J. Am. Chem. Soc.*, **125**, 3821.

5 (a) Hay, C., Hissler, M., Fischmeister, C., Rault-Berthelot, J., Toupet, L., Nyulaszi, L., and Reau, R. (2001) *Chem. Eur. J.*, **7**, 4222; (b) Hay, C., Fave, C., Hissler, M., Rault-Berthelot, J., and Reau, R. (2003) *Org. Lett.*, **5**, 3467; (c) Hissler, M., Pyer, P.W., and Reau, R. (2003) *Coord. Chem. Rev.*, **244**, 1; (d) Na, H.S., Morisaki, Y., Aiki, Y., and Chujo, Y. (2007) *Polym. Bull.*, **58**, 645; (e) Morisaki, Y., Na, H.S., Aiki, Y., and Chujo, Y. (2007) *Polym. Bull.*, **58**, 777.

6 (a) Naka, K., Umeyama, T., and Chujo, Y. (2002) *J. Am. Chem. Soc.*, **124**, 6600; (b) Umeyama, T., Naka, K., Nakahashi, A., and Chujo, Y. (2004) *Macromolecules*, **37**, 1271; (c) Umeyama, T., Naka, K., and Chujo, Y. (2004) *Macromolecules*, **37**, 5952.

7 Jäkle, F. (2005) Boron: organoboranes, in *Encyclopedia of Inorganic Chemistry*, 2nd edn (ed. R.B. King), John Wiley and Sons Ltd, Chichester.

8 (a) Good, C.D. and Ritter, D.M. (1962) *J. Am. Chem. Soc.*, **84**, 1162; (b) Good, C.D. and Ritter, D.M. (1962) *J. Chem. Eng, Data*, **1**, 416.

9 Salzner, U., Lagowski, J.B., Pickup, P.G., and Poirier, R.A. (1998) *Synth. Met.*, **96**, 177.

10 Tanaka, K., Ueda, K., Koike, T., Ando, M., and Yamabe, T. (1985) *Phys. Rev. B.*, **32**, 4279.

11 Matsumi, M., Naka, K., and Chujo, Y. (1998) *J. Am. Chem. Soc.*, **120**, 5112.

12 (a) Chujo, Y., Tomita, I., Hachiguchi, Y., Tanigawa, H., Ihara, E., and Saegusa, T. (1991) *Macromolecules*, **24**, 345; (b) Chujo, Y., Tomita, I., Murata, N., Mauemann, H., and Saegusa, T. (1992) *Macromolecules*, **25**, 27; (c) Chujo, Y., Takizawa, N., and Sakurai, T. (1994) *J. Chem. Soc., Chem. Commun.*, 227.

13 Smith, K., Pelter, A., and Jin, Z. (1994) *Angew. Chem. Int. Ed. Engl.*, **33**, 851.

14 Yuan, Z., Taylor, N.-J., Ramachandram, R., and Marder, T.B. (1996) *Appl. Organomet. Chem.*, **10**, 305.

15 (a) Kobayashi, H., Sato, N., Ichikawa, Y., Miyata, M., Chujo, Y., and Masuyama, T. (2003) *Synth. Met.*, **135/136**, 393; (b) Sato, N., Ogawa, H., Matsumoto, F., Chujo, Y., and Matsuyama, T. (2005) *Synth. Met.*, **154**, 113.
16 Matsumi, N., Chujo, Y., Lavastre, O., and Dixneuf, P.H. (2001) *Organometallics*, **20**, 2425.
17 Matsumoto, F., Matsumi, N., and Chujo, Y. (2001) *Polym. Bull.*, **46**, 257.
18 Lorbach, A., Bolte, M., Li, Haiyan, Lerner, H.-W., Holthausen, M.C., Jäkle, F., and Wanger, M. (2009) *Angew. Chem. Int. Ed*, **48**, 1.
19 Sundararaman, A., Victor, M., Varughese, R., and Jäkle, F. (2005) *J. Am. Chem. Soc.*, **127**, 13748.
20 Miyata, M., Matsumi, N., and Chujo, Y. (1999) *Polym. Bull.*, **23**, 687.
21 Matusmi, N., Naka, K., and Chujo, Y. (1998) *J. Am. Chem. Soc.*, **120**, 10776.
22 Matsumi, N., Umeyama, T., and Chujo, Y. (2000) *Polym. Bull.*, **44**, 431.
23 Yamaguchi, S., Akiyama, S., and Tamao, K. (2000) *J. Am. Chem. Soc.*, **122**, 6335.
24 Curtis, M.D., Cheng, H., and Nanos, J.I. (1998) *Macromolecules*, **31**, 205.
25 Yamamoto, T., Sugiyama, K., Kushida, T., Inoue, T., and Kanbara, T. (1996) *J. Am. Chem. Soc.*, **118**, 3930.
26 (a) Chujo, Y., Tomita, I., Murata, N., Mauermann, H., and Saegusa, T. (1992) *Macromolecules*, **25**, 27; (b) Chujo, Y., Tomita, I., Asano, T., Mauermann, H., and Saegusa, T. (1994) *Polym. J.*, **26**, 85; (c) Matusmi, N., Naka, K., and Chujo, Y. (1998) *Polym. J.*, **30**, 833.
27 Matsumi, N., Umeyama, T., and Chujo, Y. (2000) *Macromolecules*, **33**, 3956.
28 Miyata, M., Matsumi, N., and Chujo, Y. (2001) *Maromolecules*, **34**, 7331.
29 Fang, M.C., Watanabe, A., and Matsuda, M. (1996) *Polymer*, **37**, 163.
30 Nagai, A., Miyake, J., Kokado, K., Nagata, Y., and Chujo, Y. (2009) *Macromolecules*, **42**, 1560.
31 Trofimenko, S. (1967) *J. Am. Chem. Soc.*, **89**, 3802.
32 (a) Barberá, J., Giménez, R., and Serrano, J.L. (1994) *Adv. Mater.*, **6**, 470; (b) Barberá, J., Giménez, R., and Serrano, J.L. (2000) *Chem. Mater.*, **12**, 481; (c) Jäkle, F., Priermeier, T., and Wanger, M. (1996) *Organometallics*, **15**, 3165; (d) Herdweek, E., Jäkle, F., Opromolla, G., Spigler, M., Waner, M., and Zanello, P. (1996) *Organometallics*, **15**, 5524.
33 Matsumoto, F. and Chujo, Y. (2003) *Macromolecules*, **36**, 5516.
34 Matsumoto, F., Nagata, Y., and Chujo, Y. (2005) *Polym. Bull.*, **53**, 155.
35 Matsumoto, F. and Chujo, Y. (2006) *Pure Appl. Chem.*, **78**, 1407.
36 (a) Wu, Q., Esteghamatian, M., Hu, N.-X., Popovic, Z., Enright, G., Tao, Y., D'Iorio, M., and Wang, S. (2000) *Chem. Mater.*, **12**, 79; (b) Liu, S.-F., Seward, C., Aziz, H., Hu, N.-X., Popovic, Z., and Wang, S. (2000) *Organometallics*, **19**, 5709; (c) Cui, Y., Liu, Q.-D., Bai, D.-R., Jia, W.-L., Tao, Y., and Wang, S. (2005) *Inorg. Chem.*, **44**, 601.
37 Cui, Y. and Wang, S. (2006) *J. Org. Chem.*, **71**, 6485.
38 Qin, Y., Pagba, C., Piotrowaik, P., and Jäkle, F. (2002) *J. Am. Chem. Soc.*, **124**, 12672.
39 Qin, Y., Kiburu, I., Shah, S., and Jäkle, F. (2006) *Macromolecules*, **39**, 9041.
40 Nagata, Y. and Chujo, Y. (2007) *Macromolecules*, **40**, 6.
41 Nagata, Y. and Chujo, Y. (2008) *Macromolecules*, **41**, 2809.
42 Nagata, Y. and Chujo, Y. (2008) *Macromolecules*, **41**, 3488.
43 Tokoro, Y., Nagai, A., Kokado, K., and Chujo, Y. (2009) *Macromolecules*, **42**, 2988.
44 Nagata, Y., Otaka, H., and Chujo, Y. (2008) *Macromolecules*, **41**, 737.
45 Li, H. and Jakle, F. (2009) *Macromolecules*, **42**, 3448.
46 (a) Maeda, H., Mihashi, Y., and Haketa, Y. (2008) *Org. Lett.*, **10**, 3179; (b) Ono, K., Yoshikawa, K., Tsuji, Y., Yamaguchi, H., Uozumi, R., Tomura, M., Taga, K., and Saito, K. (2007) *Tetrahedron*, **63**, 9354; (c) Cogné-Laage, E., Allemand, J.-F., Ruel, O., Baudin, J.-B., Croquette, V., Blanchard-Desce, M., and Jullien, L. (2004) *Chem. Eur. J.*, **10**, 1445; (d) Halik, M. and Hartmann, H. (1999) *Chem. Eur. J.*, **5**, 2511.
47 (a) Zhang, G., Chen, J., Pyane, S.J., Kooi, S.E., Demas, J.N., and Fraser, C.L. (2007) *J. Am. Chem. Soc.*, **129**, 8942;

(b) Zhang, G., Kooi, S.E., Demas, J.N., and Fraser, C.L. (2008) *Adv. Mater.*, **20**, 2099; (c) Pfister, A., Zhang, G., Zareno, J., Horwitz, A.F., and Fraser, C.L. (2008) *ACS Nano*, **2**, 1252.

48 Zhang, G., Clair, T.L.S., and Fraser, C.L. (2009) *Macromolecules*, **42**, 3092.

49 Zhang, G., Fiore, G.L., Clair, T.L.S., and Fraser, C.L. (2009) *Macromolecules*, **42**, 3162.

50 Nagai, A., Kokado, K., Nagata, Y., and Chujo, Y. (2008) *Macromolecules*, **41**, 8295.

51 (a) Valeur, B. (2002) *Molecular Fluorescence: Principles and Applications*, Wiley-VCH Verlag GmbH, Weinheim, Germany; (b) Lakowicz, J.R. (1994) *Probe Design and Chemical Sensing* (ed. J.R. Lakowicz), Topics in Fluorescence Spectroscopy, vol. **4**, Plenum, New York; (c) Ulrich, G., Ziessel, R., and Harriman, A. (2008) *Angew. Chem., Int. Ed.*, **47**, 1184.

52 Amart-Guerri, F., Liras, M., Carrascoso, M.L., and Sastre, R. (2003) *Photochem. Photobiol.*, **77**, 577.

53 Gracía-Moreno, I., Costela, A., Campo, L., Sastre, R., Amart-Guerri, F., Liras, M., Arbeloa, F.L., Prieto, J.B., and Arbeloa, I.L. (2004) *J. Phy. Chem. A*, **108**, 3315.

54 Nagai, A., Miyake, J., Kokado, K., Nagata, Y., and Chujo, Y. (2008) *J. Am. Chem. Soc.*, **130**, 15276.

55 Douru, V.R., Vegesna, G.K., Velayudham, S., Green, S., and Liu, H. (2009) *Chem. Mater.*, **21**, 2130.

56 Meng, G., Velayudham, S., Smith, A., Luck, R., and Liu, H. (2009) *Macromolecules*, **42**, 1995.

57 (a) Williams, R.E. (1992) *Chem. Rev.*, **92**, 177; (b) Leites, L.A. (1992) *Chem. Rev.*, **92**, 279; (c) Grimes, R.N. (2000) *Coord. Chem. Rev.*, **200–202**, 773.

58 (a) Endo, Y., Sawabe, T., and Taoda, Y. (2000) *J. Am. Chem. Soc.*, **122**, 180; (b) Nakamura, H., Kamakura, K., and Onagi, S. (2006) *Org. Lett.*, **8**, 2095; (c) Li, Y., Carroll, P.J., and Sneddon, L.G. (2008) *Inorg. Chem.*, **47**, 9193.

59 (a) Morin, J.-F., Sasaki, T., Shirai, Y., Guerrero, J.M., and Tour, J.M. (2007) *J. Org. Chem.*, **72**, 9481; (b) Morin, J.-F., Shirai, Y., and Tour, J.M. (2006) *Org. Lett.*, **8**, 1713.

60 (a) Kokado, K. and Chujo, Y. (2009) *Macromolecules*, **42**, 1418; (b) Kokado, K., Tokoro, Y., and Chujo, Y. (2009) *Macromolecules*, **42**, 2925.

8
Recent Developments in π-Conjugated Macromolecules with Phosphorus Atoms in the Main Chain

Paul W. Siu and Derek P. Gates

8.1
Introduction

In this chapter, selected recent developments in the field of π-conjugated polymers featuring phosphorus atoms in the main chain will be outlined. Although not comprehensive, this review will provide examples of the current state-of-the-art in π-conjugated phosphorus macromolecules. Particular emphasis will be placed on the synthetic routes to and structures of these macromolecules. For additional information, the reader is also referred to several excellent reviews on phosphorus- and related p-block element-containing π-conjugated polymers [1, 2].

8.2
Poly(Phosphole) and Related Polymers

Phosphole is an interesting 4π electron antiaromatic system which may be incorporated into a π-conjugated "butadiene"-like polymer with phosphorus imparting an inductive effect on the electronic structure. Due to their novel electronic properties, molecular phosphole-based materials have been studied extensively and have been the subject of several reviews [3–8]. In contrast, the development of methods to incorporate phosphole moieties into π-conjugated macromolecules is less well-established. In this section, recent developments in the synthesis of π-conjugated polymers containing phosphole moieties will be outlined.

The first steps toward linking phospholes into long chain poly(phosphole)s were proposed by Mathey and coworkers in 1990 [9]. The same group reported that employing the sequential lithiation and copper(II) chloride coupling chains of phospholes facilitated the isolation of tetramer **1** [10]. Crystallographic characterization of this oligomer revealed that the C–C bridging bonds were not shortened, leading the authors to conclude that π-conjugation between the directly linked phosphole rings is unlikely. Subsequent work on phosphole-containing

Conjugated Polymer Synthesis. Edited by Yoshiki Chujo

ISBN: 978-3-527-32267-1

macromolecules has focused on systems where the phosphole moieties are separated by a π-conjugated spacer group.

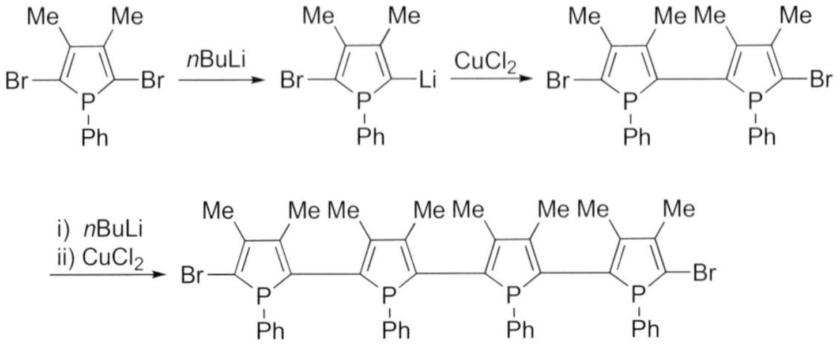

The first phosphole-containing long chain macromolecules were reported by Tilley and coworkers in 1997 [11]. These researchers employed a zirconocene diyne-coupling strategy to afford cyclic zirconacyclopentadiene polymers. The skeletal ring-atom replacement of the zirconocene moieties by phosphorus afforded the first phosphole-containing polymers where the phosphole units were spaced by π-conjugated arylene moieties. Remarkably, these fascinating poly(phosphole-*p*-biphenylene)s **2** exhibit blue photoluminescence.

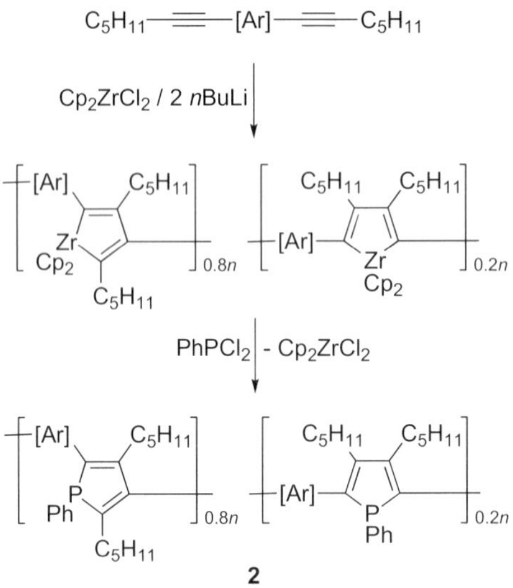

A few years after this initial report, the electropolymerization of dithiophene phospholes was reported by Réau and coworkers [12]. The monomers for these studies, feature phosphole cores with terminal thienyl moieties and are prepared following the aforementioned zirconocene coupling of alkynes. The resultant poly[(2,5-dithieny)phosphole] **3** is readily functionalized at phosphorus to afford derivatives that exhibit tunable low optical band gaps [13]. Importantly, related short chain phosphole oligomers have been employed as the emissive layer in light emitting diodes (LEDs) [14].

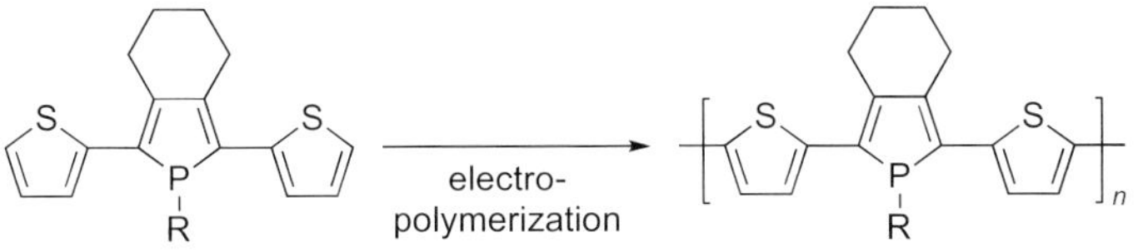

electro-
polymerization

3

The same group has prepared an AuCl-protected monomer by the facile complexation of the phosphole monomer with (tetrahydrothiophene)AuCl. Electropolymerization of this monomer afforded hybrid poly(phosphole)-poly(thiophene) **4** [15]. The uncomplexed phosphole polymer **5** can be accessed from insoluble **4** by immersion of a thin film of **4** into a solution of PPh_3. Polymer **5** was shown to be an effective sensor for elemental chalcogens. Since tricoordinate phospholes exhibit rather low decomposition temperatures, their fabrication into organic LEDs, which involves thermal evaporation of the low-molecular weight oligomers, is hindered [16]. The protection of the phosphorus(III) center through gold(I) complexation or through oxidation with elemental sulfur imparts significantly improved thermal stability, thus facilitating the fabrication of single- and multilayer organic LEDs with white emissive properties [16, 17].

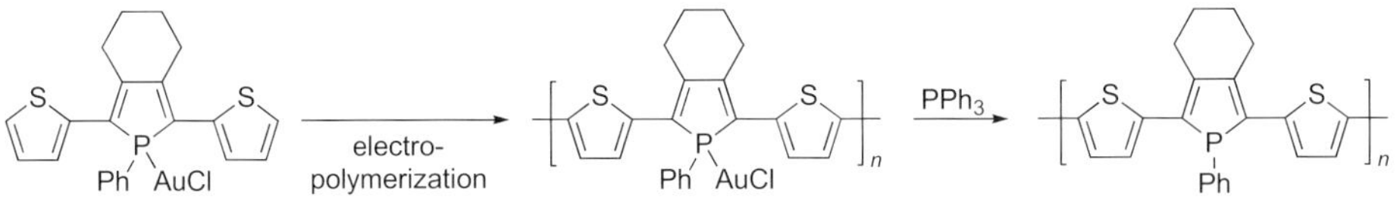

electro-
polymerization

PPh_3

4 **5**

The electropolymerization of dithienophosphole sulfide with 3,4-ethylenedioxythiophene (EDOT) resulted in the electrochromic copolymer **6**, which showed increased thermal stability when compared to pure PEDOT [18]. Moreover, this novel polymer showed a narrower optical band gap relative to the corresponding phosphole homopolymer. The copolymer **6** displays blue-to-black electrochromic behavior.

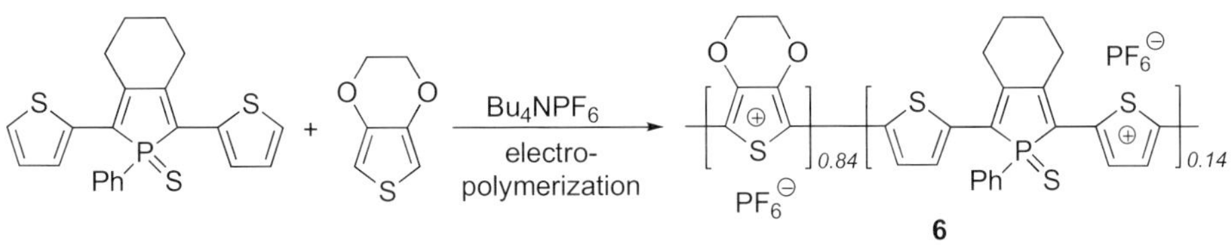

Bu_4NPF_6
electro-
polymerization

6

Alternatively, EDOT can be introduced into the polymer chain by the direct electropolymerization of a phosphole monomer containing terminal EDOT substituents to afford EDOT–phosphole copolymer **7** [19].

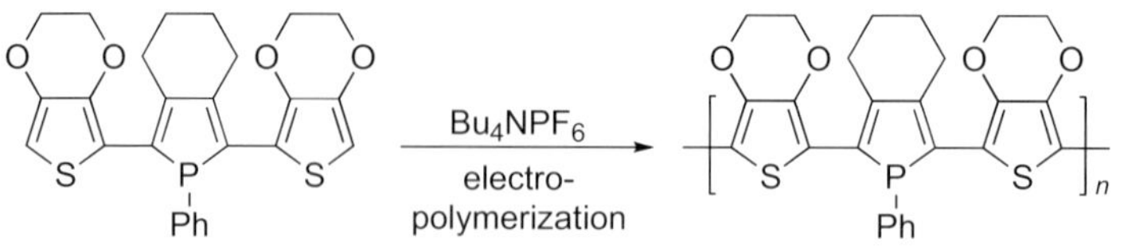

Bu_4NPF_6
electro-
polymerization

7

Dithienophospholes and related derivatives are attracting considerable attention as tunable semiconducting materials in organic field-effect transistor (OFET) devices [20–22]. For example, several dithienophosphole polymers **8** have been prepared by the tetramethylpiperidine-1-oxyl (TEMPO) initiated radical copolymerization of phosphole-functionalized styrene and styrene [23]. The resultant high molecular weight copolymer **8** exhibited a strong blue photoluminescence in solution while the corresponding phosphine oxide was red-shifted with respect to **8**.

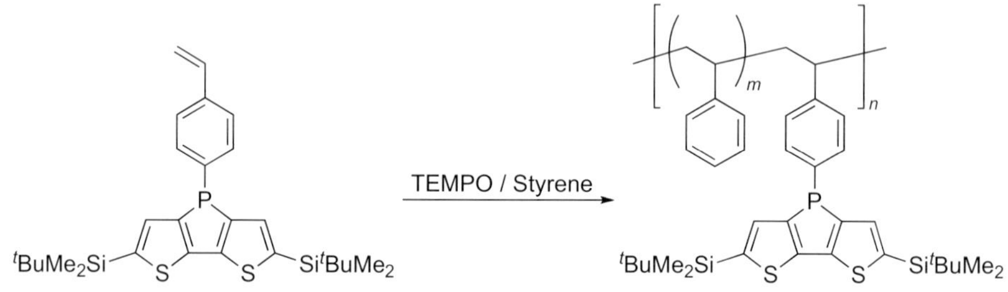

TEMPO / Styrene

8

A metal-catalyzed cross-coupling strategy has been developed to prepare macromolecules with dithienophosphole moieties in the backbone. For example, the Pd-catalyzed Stille cross-coupling of tributyltin-substituted dithienophosphole and diiodobenzene afforded fascinating π-conjugated poly(dithienophosphole)s **9** [24]. These novel polymers showed significant red shifts in their absorbance maxima when compared to the corresponding phosphole monomers. Significantly, these polymers were highly luminescent, emitting yellow–green light.

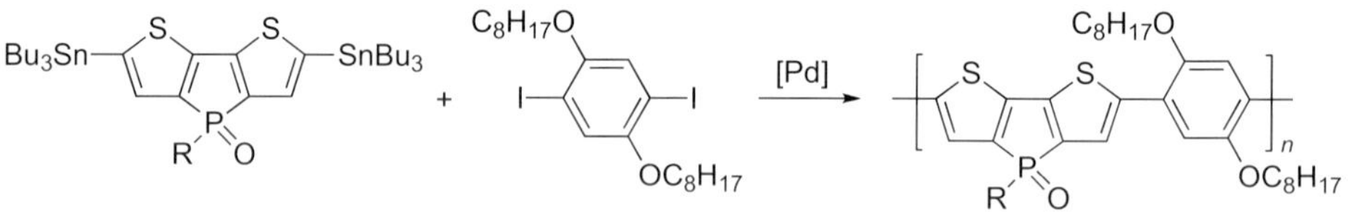

[Pd]

9

Another successful strategy that has been employed to afford poly(dithienophosphole)s involves the Suzuki–Miyaura coupling of dibromo phospholium with bis (boronic acid) [25]. The absorbance maximum of thin films of polyelectrolyte **10** are red shifted in comparison to that of **10** in solution, which suggested intermolecular π-stacking in the solid state.

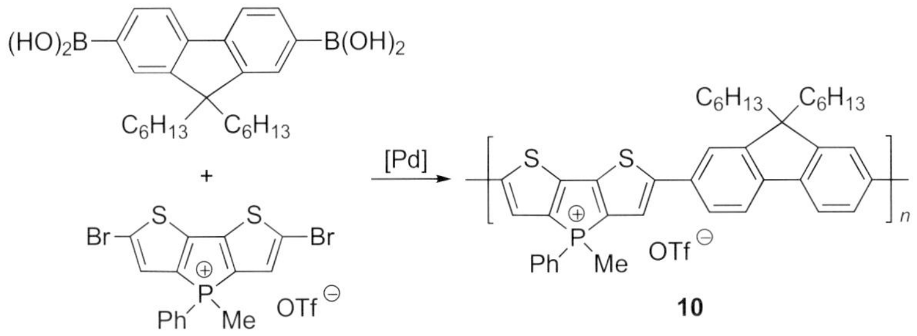

$(HO)_2B$ $B(OH)_2$ C_6H_{13} C_6H_{13} + Br S S Br P⊕ Ph Me OTf⊖ [Pd] → C_6H_{13} C_6H_{13} S S P⊕ Ph Me OTf⊖ n **10**

Dithienophosphole-fluorene copolymers (**11**, **12** and **13**) and fluorene-phosphafluorene copolymer (**14**) have also been prepared following a similar Suzuki–Miyaura coupling strategy [26–28]. The presence of fluorene units in the main chain facilitates more extended π-conjugation within the phosphole framework. As a consequence, the dithienophosphole–fluorene copolymer **11** showed a significant red shift in its absorbance maximum when compared to the dithienophosphole monomer [26]. The incorporation of electron-donating EDOT units into the dithienophosphole–fluorene copolymer results in a blue shift in the absorbance maximum for **12**, whereas incorporation of electron-accepting benzothiadiazole (BTD) units results in a red shift for **13** [27]. These macromolecules are attractive for applications in polymer LEDs and solar cells. Interestingly, when fluorene–phosphafluorene copolymers **14** were incorporated into a polymer LED device, the unoxidized polymer showed blue emission while the oxidized derivative exhibited white emission [28].

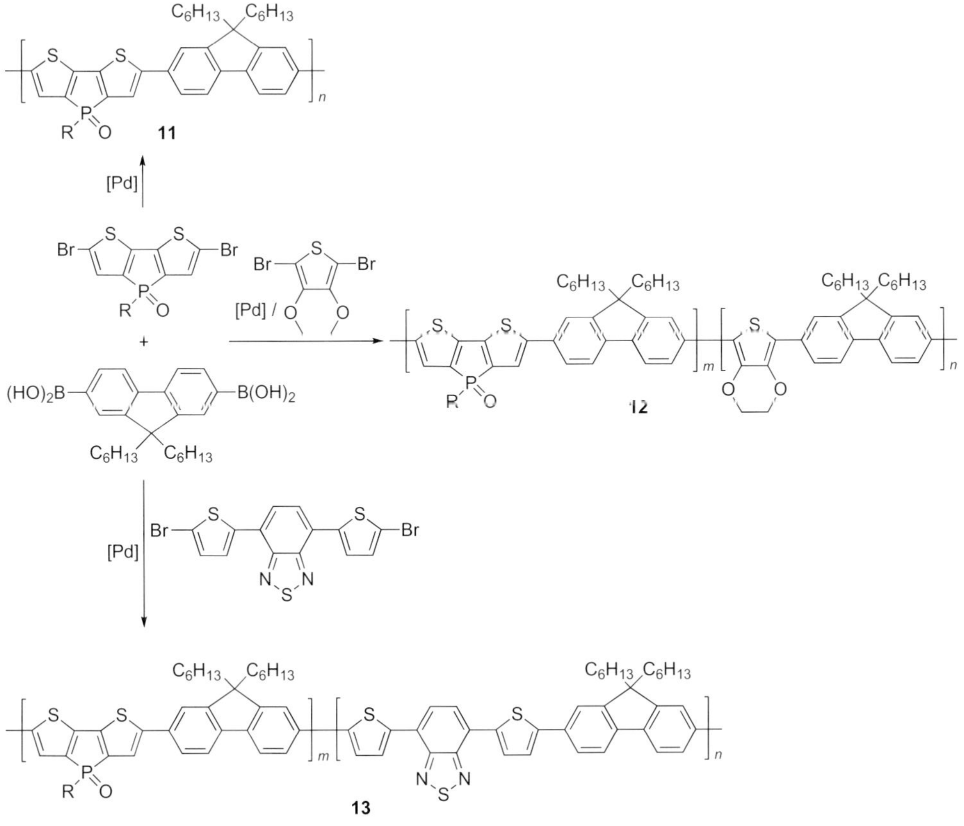

C_6H_{13} C_6H_{13} S S P R O n **11**

[Pd]

Br S S Br P R O + $(HO)_2B$ $B(OH)_2$ C_6H_{13} C_6H_{13}

Br S Br O O [Pd] / → C_6H_{13} C_6H_{13} S S P R O m S O O C_6H_{13} C_6H_{13} n **12**

[Pd] Br S S Br N S N

C_6H_{13} C_6H_{13} S S P R O m S S N S N C_6H_{13} C_6H_{13} n **13**

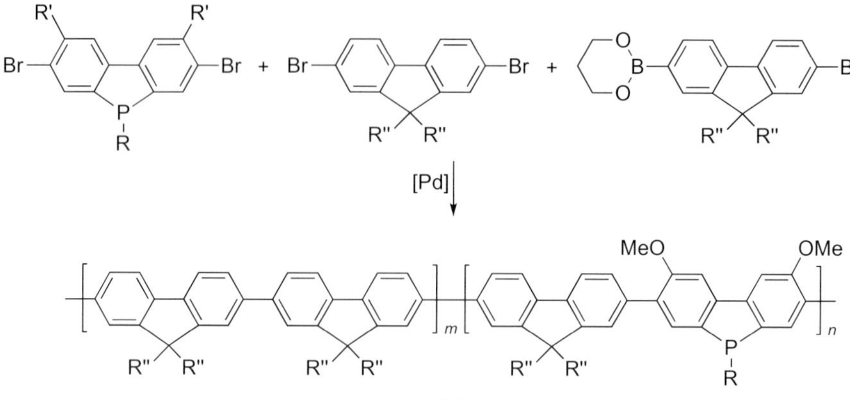

Baumgartner and coworkers reported the dehydrogenative homocoupling of dithienophospholes bearing Me_2Si–H moieties in the presence of a platinum catalyst to afford polymer **15** [29]. A bathochromic shift was noted in the excitation spectrum for **15**, when compared to the corresponding monomer. Interestingly, the Si–H functionality can be further exploited through the Pt-catalyzed hydrosilylation of bisalkynes to form dithienophosphole silyl vinylene polymers **16**.

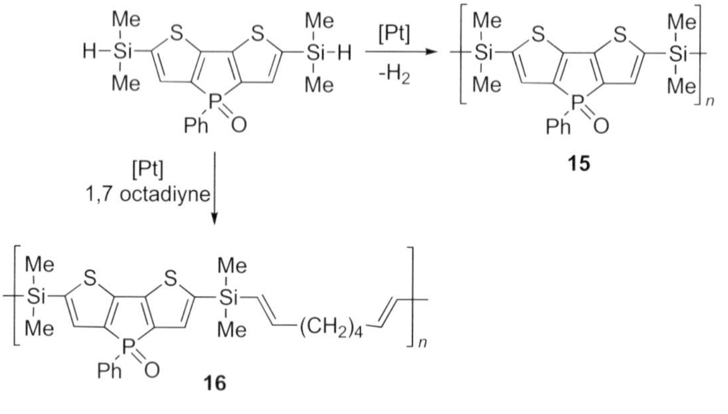

New π-conjugated phosphole polymers of type **17** containing a variety of aryl spacer groups have been prepared using the palladium/copper catalyzed cross-coupling [30–33]. These novel polymers exhibit photoluminescence and their UV–vis absorption maxima are bathochromically shifted relative to the phosphole monomers.

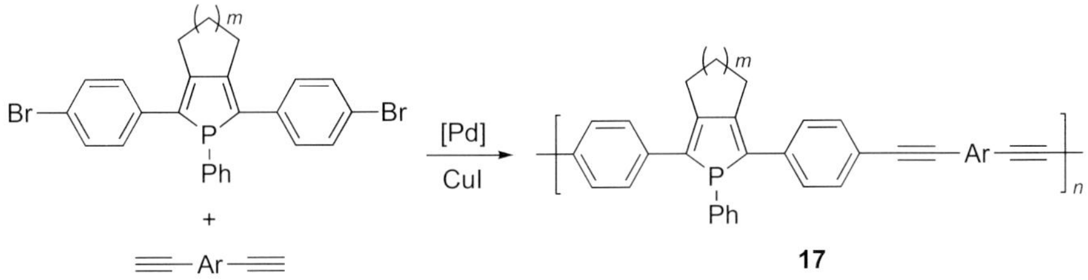

8.3 Poly(*p*-Phenylenephosphine) and Related Polymers

Poly(*p*-phenylene-*P*-alkylphosphines) **18** and **19** have been prepared using palladium- or nickel-mediated coupling routes [34, 35]. UV–vis spectroscopic studies on the polymers showed red shifts in their absorbance maxima, compared to their related molecular model compounds. Cyclic voltammetry revealed that the oxidation potentials of polymers **18** and **19** are lower than their molecular model compounds. Both studies suggested the presence of electronic delocalization through the phosphorus atoms in these materials.

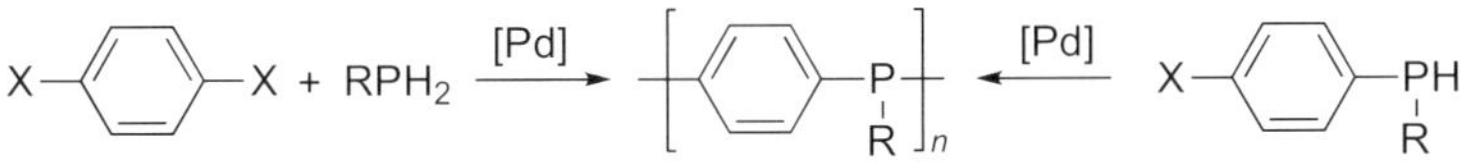

Lucht and coworkers have also reported hybrid polyaniline-poly-*p*-phenylene phosphine polymers **20** that are prepared following an analogous palladium-catalyzed coupling strategy [36]. Once again, electrochemical and UV–vis–NIR spectroscopic studies on the polymers and the related molecular model compounds supported the presence of electronic delocalization through the phosphorus centers. This is further evidenced by studies on the oxides of **20**. The oxides of **20** showed higher oxidation potentials and blue-shifted absorbance maxima in their UV–vis spectra when compared to unoxidized **20**, which suggests that the presence of a phosphine oxide moiety in the main chain breaks electronic delocalization in the backbone.

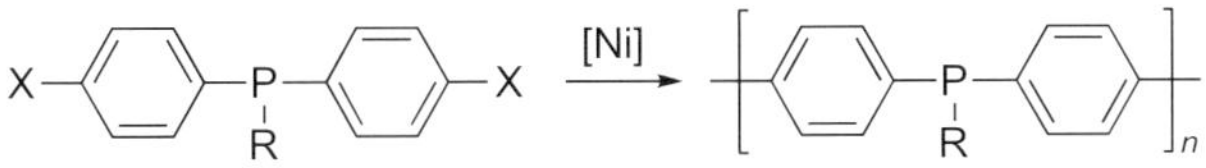

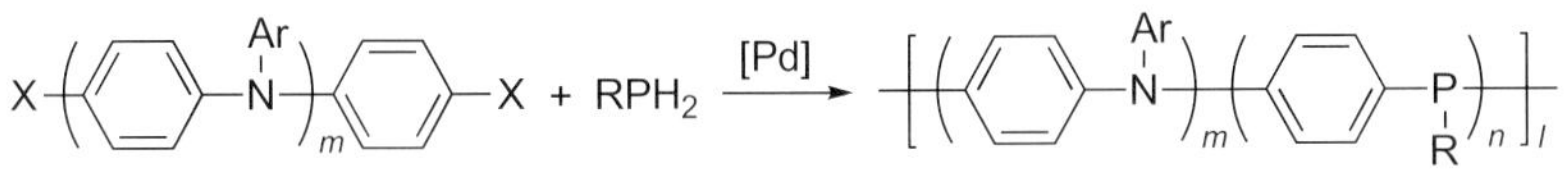

8.4
Poly(Vinylenephosphine)s and Related Polymers

Manners and coworkers have employed the anionic ring-opening polymerization of strained 1-phenyl-2,3-dimethylphosphirene to afford poly(vinylenephosphine) **21** [37]. These polymers are readily oxidized with elemental sulfur to the corresponding air-stable phosphine sulfide. GPC analysis of the oxidized polymers revealed modest molecular weights for these interesting polymers. Unfortunately, unoxidized **21** could not be analyzed by GPC due to interaction of the phosphine moieties with the columns.

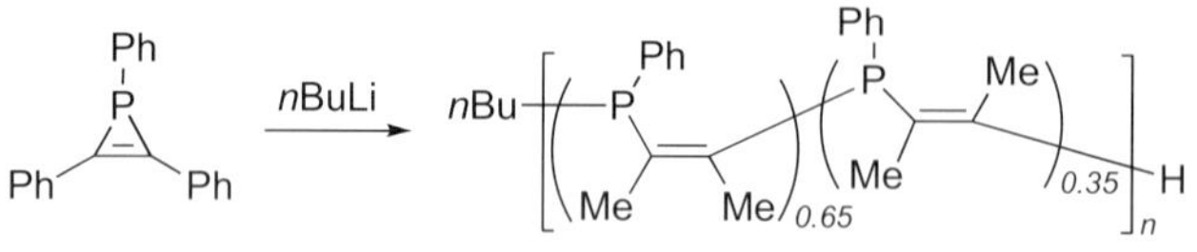

21

An alternate route to poly(vinylene phosphine) reported by Naka and coworkers involves the radical initiated copolymerization of cyclic polyphosphine with phenylacetylene [38, 39]. This ring-collapsed radical alternating polymerization (RCRAC), used extensively with cyclic arsines and stibines, affords π-conjugated phosphine polymer **22**. Analysis of **22** by NMR spectroscopy led the researchers to conclude that trans-isomers were predominant in the backbone of **22**. Significantly, polymer **22** is photoluminescent with emission in the blue–green region of the visible spectrum.

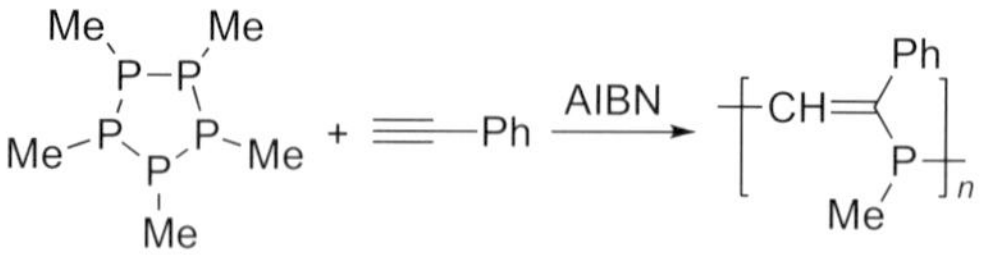

22

Stephan and coworkers have very recently reported the catalytic hydrophosphination of alkyne-functionalized secondary phosphines to afford hybrid arylene vinylene phosphine π-conjugated macromolecules **23** [40, 41]. Based on the data from NMR, GPC and MALDI-TOF MS, these materials were concluded to be primarily cyclic in nature with at least eight repeat units.

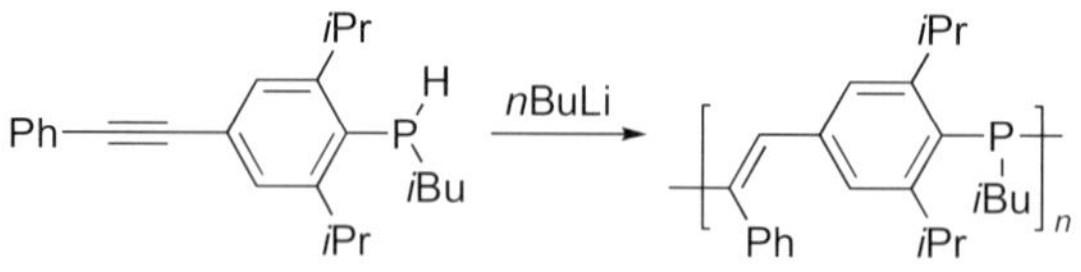

23

Although not a phosphine polymer, also relevant to this section is the hydrophosphorylation of diynes with bifunctional P–H phosphine oxides that affords a series of poly(arylenevinylene phosphine oxide)s [42]. The regioselectivity and stereoselectivity of addition was highly dependent on the catalyst used. Interestingly, [RhBr

$(PPh_3)_3]$ gave Markovnikov addition to afford *trans*-alkenylene polymers **24**, while geminal anti-Markovnikov polymers **25** were obtained when $[Ni(PPhMe_2)_4]$ was used to catalyze the hydrophosphorylation reaction.

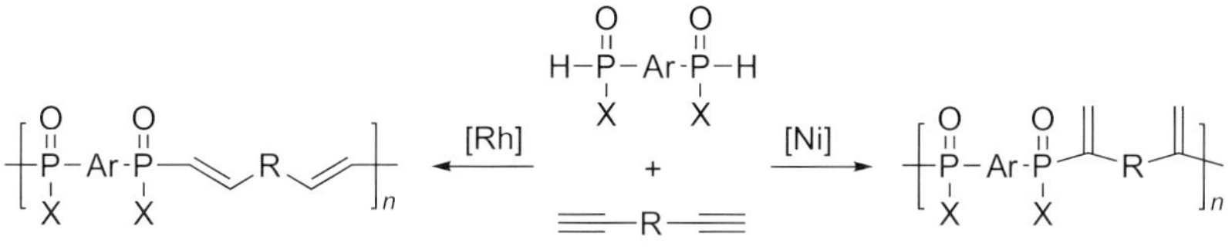

24 **25**

8.5
Poly(*p*-Phenylenephosphaalkene)s and Related Polymers

The earliest work on incorporating π-conjugated spacers between phosphaalkene moieties involved the preparation of bis(phosphaalkene)s spaced by phenylene moieties. Several bis(phosphaalkene) isomers (e.g., **26**) were prepared by this route and their use as ligands for palladium(II) was investigated [43–45]. No polymers were reported; however, these studies provide evidence for π-conjugation between P=C moieties through an arylene group.

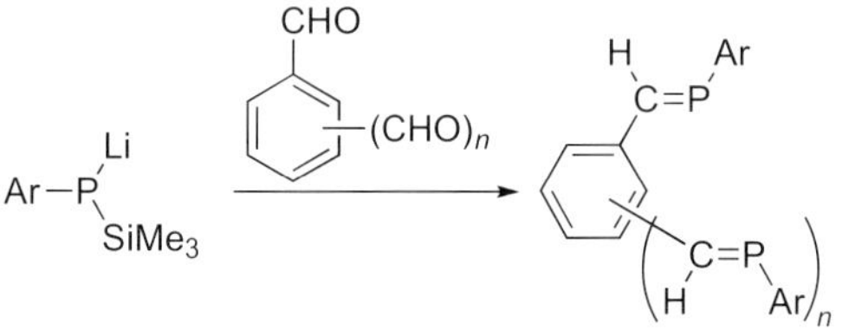

26

Gates and coworkers have reported the first examples of π-conjugated P=C-containing analogues of poly(*p*-phenylenevinylene) using a condensation polymerization strategy between a bis(trimethylsilyl)phosphine and an acid chloride [46]. Yellow poly(*p*-phenylenephosphaalkene) **27** of moderate molecular weight was the first polymer with low-coordinate phosphorus atoms in the main chain. A series of model compounds and polymers were prepared and it was determined that modifying steric constraints provided a degree of stereochemical control [47]. In particular, employing bulky *P*-aryl substituents preferentially affords the *Z*-phosphaalkene isomer with *trans* phenylene moieties. The *Z*-isomer of **27** displays a dramatic red shift in its UV–vis spectrum as compared to the corresponding *Z* model compound.

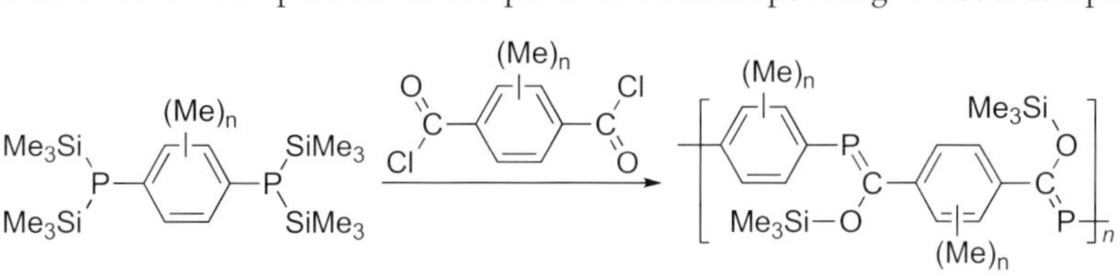

27

Protasiewicz and coworkers have developed a phospha-Witting strategy to synthesize new bis(phosphaalkene)s using sterically hindered substituents [48, 49]. A phospha-Wittig reagent was prepared *in situ* by the reduction of a bis(dichlorophosphine) with zinc metal in the presence of excess trimethylphosphine. Reaction of the phospha-Wittig reagent with various aldehydes resulted in bis(phosphaalkene) **28**.

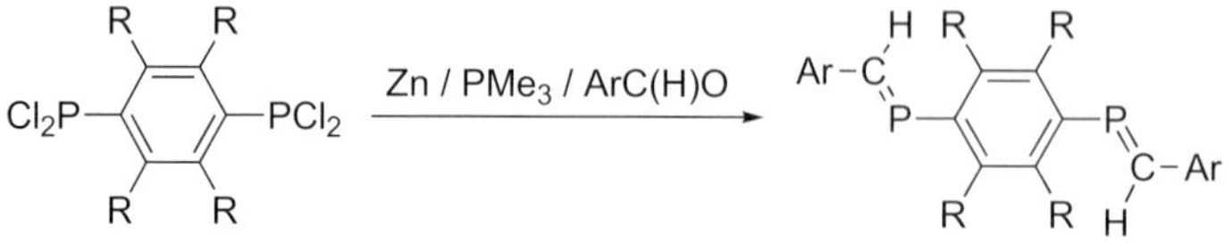

Zn / PMe$_3$ / ArC(H)O

28

Following the phospha-Wittig route, the same group have discovered an alternate synthesis to poly(*p*-phenylenephosphaalkene) [50]. A bi-functional phospha-Wittig reagent was prepared *in situ* and reacted with various dialdehydes to afford *E*-poly(*p*-phenylenephosphaalkene)s **29** with C–H substituents. The absorbance band maximum for the soluble polymer **29** was identical to the small molecule models, suggesting that the presence of the bulky 2,3,5,6-tetraaryl-substituted phenylene spacer might partially disrupt the π-conjugation. Remarkably, this polymer is fluorescent and an emission maximum was observed in the green region.

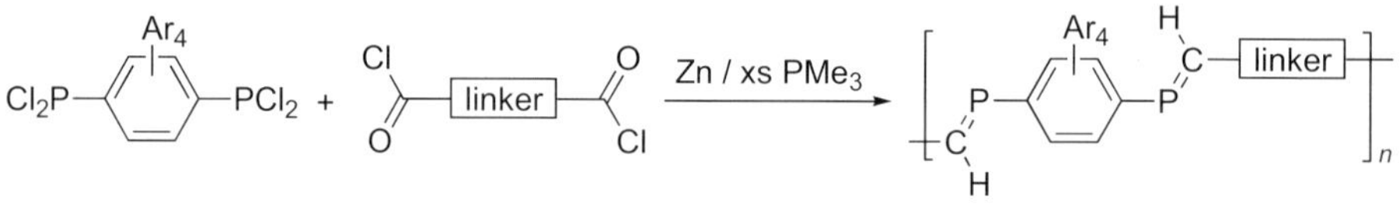

Zn / xs PMe$_3$

29

Novel π-conjugated macromolecule **30**, which contains two phosphaalkene units, was reported by Yoshifuji and coworkers [51]. Polymer **30** was shown to be an effective chelating ligand for transition metals and a palladium(II) complex was prepared by treating **30** with $PdCl_2(MeCN)_2$. The UV–vis spectra of both **30** and palladium(II)-complexed **30** showed bathochromic shifts, relative to the molecular model compound of **30**.

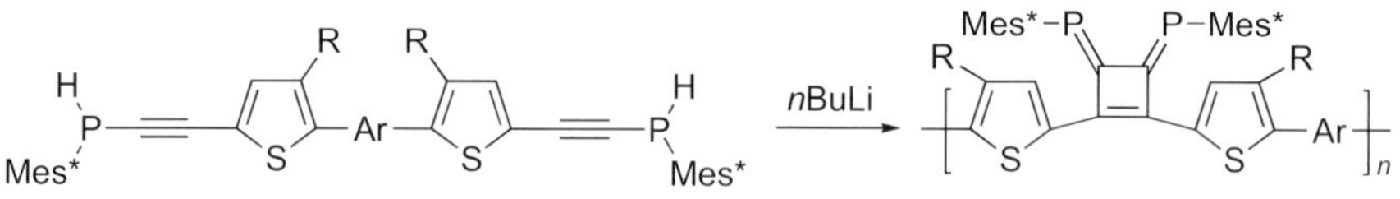

*n*BuLi

30

Very recently, a series of fascinating π-conjugated *C*-acetylenic phosphaalkene oligomers (**31**) have been synthesized [52]. The acetylenic terminal trimethylsilyl groups have been exploited for copper-catalyzed Eglinton reactions resulting in the first oligoacetylene containing a heavy main-group element that is part of the π-conjugated system [52]. Although not polymeric, bis(phosphaalkene) **32** exibits a significant red shift in its UV–vis absorption spectrum compared with the all-carbon analogue. Polymers containing these functionalities are expected to possess fascinating photophysical properties.

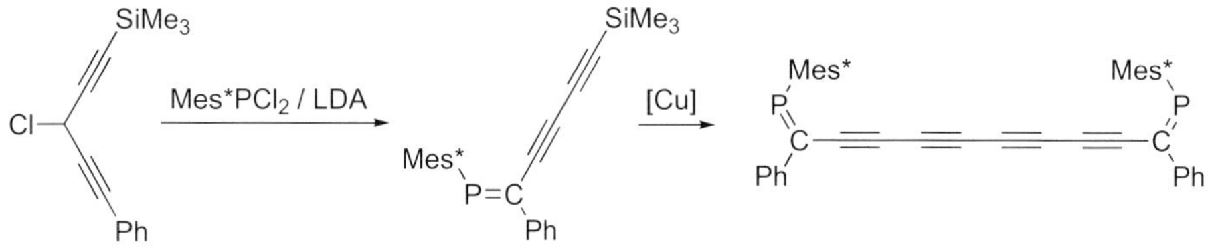

8.6 Poly(*p*-Phenylenediphosphene)s and Related Polymers

There has been a longstanding interest in the development and study of isolable species featuring P=P bonds, diphosphenes [53]. The evaluation of the π-conjugation properties between diphosphene moieties spaced by an aromatic spacer was facilitated with the preparation of the first structurally characterized bis(diphosphene) **33** by Protasiewicz and coworkers [48]. The P=P bond was stabilized by the presence of the sterically demanding 2,6-dimesitylphenyl ligand on phosphorus. In a related study, Pietschnig and coworkers utilized a bulkier ligand, 2,4,6-tris(bis(trimethylsilyl) methyl)phenyl, to afford the first ferrocenylene-bridged bis(diphosphenes) **34** [54]. Evidence from UV–vis spectroscopy and *ab initio* calculations suggests the presence of electronic interaction between the P=P and the ferrocene unit.

Particularly noteworthy is the preparation of diphosphene analogues of poly(*p*-phenylenevinylene) (PPV) [55]. Diphosphene-PPV **35** can be prepared by irradiating a solution of diphospha-Wittig reagent; however, the preferred method involves heating a diphospha-Wittig reagent neat to 250 °C for about 2 min. The UV–vis absorption and emission spectra of **35** are consistent with a π-conjugated system. The absorption spectrum of polymer **35** exhibited a typical n–π* transition for diphosphenes and the π–π* transition lies in between those observed for unsubstituted *E*-poly(phenylenevinylene) and the more related poly(phenylenevinylene-alt-2,5-dihexyloxyphenylenevinylene).

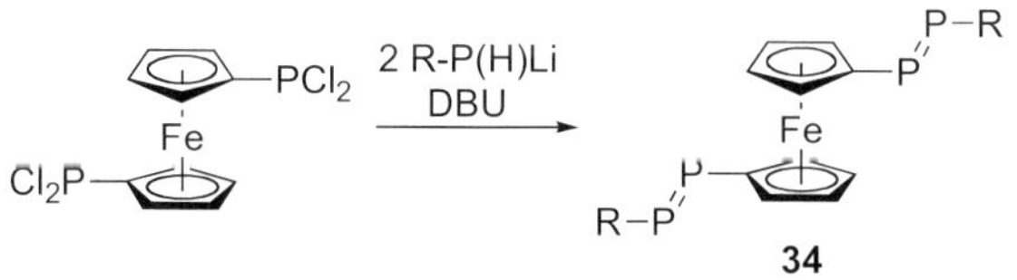

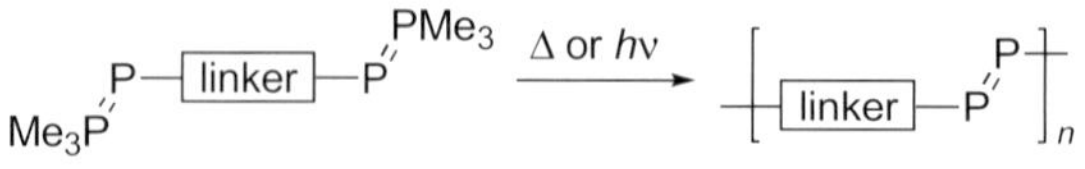

8.7 Summary

This chapter summarizes some of the recent highlights in the development of π-conjugated polymers containing phosphorus centers on the polymer backbone. This relatively new field, now early in its second decade, holds great promise for the development of materials with exciting photophysical and electronic properties and potential specialty applications.

References

1 Baumgartner, T. and Reau, R. (2006) *Chem. Rev.*, **106**, 4681–4727.
2 (a) See, for example: Hissler, M., Dyer, P.W., and Reau, R. (2003) *Coord. Chem. Rev.*, **244**, 1–44; (b) Hissler, M., Dyer, P.W., and Reau, R. (2005) *Top. Curr. Chem.*, **250**, 127–163; (c) Hissler, M., Lescop, C., and Reau, R. (2005) *C. R. Chim.*, **8**, 1186–1193; (d) Hissler, M., Lescop, C., and Reau, R. (2005) *J. Organomet. Chem.*, **690**, 2482–2487; (e) Gates, D.P. (2005) *Top. Curr. Chem.*, **250**, 107–126; (f) Hobbs, M.G., and Baumgartner, T. (2007) *Eur. J. Inorg. Chem.*, 3611–3628;(g) Hissler, M., Lescop, C., and Reau, R. (2008) *C. R. Chim.*, **11**, 628–640;(h) Crassous, J. and Reau, R. (2008) *Dalton Trans.*, 6865–6876.
3 Crassous, J. and Reau, R. (2008) *Dalton Trans.*, 6865–6876.
4 Hissler, M., Lescop, C., and Reau, R. (2008) *C. R. Chim.*, **11**, 628–640.
5 Hobbs, M.G. and Baumgartner, T. (2007) *Eur. J. Inorg. Chem.*, 3611–3628.
6 Hissler, M., Dyer, P.W., and Reau, R. (2005) *Top. Curr. Chem.*, **250**, 127–163.
7 Hissler, M., Lescop, C., and Reau, R. (2005) *C. R. Chim.*, **8**, 1186–1193.
8 Hissler, M., Lescop, C., and Reau, R. (2005) *J. Organomet. Chem.*, **690**, 2482–2487.
9 Bevierre, M.O., Mercier, F., Ricard, L., and Mathey, F. (1990) *Angew. Chem., Int. Ed.*, **29**, 655–657.
10 Deschamps, E., Ricard, L., and Mathey, F. (1994) *Angew. Chem., Int. Ed.*, **33**, 1158–1161.
11 Mao, S.S.H. and Tilley, T.D. (1997) *Macromolecules*, **30**, 5566–5569.
12 Hay, C., Fischmeister, C., Hissler, M., Toupet, L., and Reau, R. (2000) *Angew. Chem., Int. Ed.*, **39**, 1812–1815.
13 Hay, C., Hissler, M., Fischmeister, C., Rault-Berthelot, J., Toupet, L., Nyulaszi, L., and Reau, R. (2001) *Chem.-Eur. J.*, **7**, 4222–4236.
14 Fave, C., Cho, T.Y., Hissler, M., Chen, C.W., Luh, T.Y., Wu, C.C., and Reau, R. (2003) *J. Am. Chem. Soc.*, **125**, 9254–9255.
15 Sebastian, M., Hissler, M., Fave, C., Rault-Berthelot, J., Odin, C., and Reau, R. (2006) *Angew. Chem., Int. Ed.*, **45**, 6152–6155.
16 Su, H.C., Fadhel, O., Yang, C.J., Cho, T.Y., Fave, C., Hissler, M., Wu, C.C., and Reau, R. (2006) *J. Am. Chem. Soc.*, **128**, 983–995.
17 Fadhel, O., Gras, M., Lemaitre, N., Deborde, V., Hissler, M., Geffroy, B., and Reau, R. (2009) *Adv. Mater.*, **21**, 1261–1265.
18 Carrasco, P.M., Pozo-Gonzalo, C., Grande, H., Pomposo, J.A., Cortazar, M., Deborde, V., Hissler, M., and Reau, R. (2008) *Polym. Bull.*, **61**, 713–724.
19 de Talance, V.L., Hissler, M., Zhang, L.Z., Karpati, T., Nyulaszi, L., Caras-Quintero,

D., Bauerle, P., and Reau, R. (2008) *Chem. Commun.*, 2200–2202.
20 Dienes, Y., Eggenstein, M., Karpati, T., Sutherland, T.C., Nyulaszi, L., and Baumgartner, T. (2008) *Chem.-Eur. J.*, **14**, 9878–9889.
21 Ren, Y., Dienes, Y., Hettel, S., Parvez, M., Hoge, B., and Baumgartner, T. (2009) *Organometallics*, **28**, 734–740.
22 Dienes, Y., Englert, U., and Baumgartner, T. (2009) *Z. Anorg. Allg. Chem.*, **635**, 238–244.
23 Baumgartner, T., Neumann, T., and Wirges, B. (2004) *Angew. Chem., Int. Ed.*, **43**, 6197.
24 Baumgartner, T., Bergmans, W., Karpati, T., Neumann, T., Nieger, M., and Nyulaszi, L. (2005) *Chem.-Eur. J.*, **11**, 4687–4699.
25 Durben, S., Dienes, Y., and Baumgartner, T. (2006) *Org. Lett.*, **8**, 5893–5896.
26 Dienes, Y., Durben, S., Karpati, T., Neumann, T., Englert, U., Nyulaszi, L., and Baumgartner, T. (2007) *Chem.-Eur. J.*, **13**, 7487–7500.
27 Durben, S., Nickel, D., Kruger, R.A., Sutherland, T.C., and Baumgartner, T. (2008) *J. Polym. Sci., Part A: Polym. Chem.*, **46**, 8179–8190.
28 Chen, R.F., Zhu, R., Fan, Q.L., and Huang, W. (2008) *Org. Lett.*, **10**, 2913–2916.
29 Baumgartner, T. and Wilk, W. (2006) *Org. Lett.*, **8**, 503–506.
30 Morisaki, Y., Aiki, Y., and Chujo, Y. (2003) *Macromolecules*, **36**, 2594–2597.
31 Morisaki, Y., Na, H.S., Aiki, Y., and Chujo, Y. (2007) *Polym. Bull.*, **58**, 777–784.
32 Na, H.S., Morisaki, Y., Aiki, Y., and Chujo, Y. (2007) *Polym. Bull.*, **58**, 645–652.
33 Na, H.S., Morisaki, Y., Aiki, Y., and Chujo, Y. (2007) *J. Polym. Sci., Part A: Polym. Chem.*, **45**, 2867–2875.
34 Lucht, B.L. and St Onge, N.O. (2000) *Chem. Commun.*, 2097–2098.
35 Jin, Z. and Lucht, B.L. (2002) *J. Organomet. Chem.*, **653**, 167–176.
36 Jin, Z. and Lucht, B.L. (2005) *J. Am. Chem. Soc.*, **127**, 5586–5595.
37 Vanderark, L.A., Clark, T.J., Rivard, E., Manners, I., Slootweg, J.C., and Lammertsma, K. (2006) *Chem. Commun.*, 3332–3333.
38 Naka, K., Umeyama, T., Nakahashi, A., and Chujo, Y. (2007) *Macromolecules*, **40**, 4854–4858.
39 Naka, K. (2008) *Polym. J.*, **40**, 1031–1041.
40 Greenberg, S., Gibson, G.L., and Stephan, D.W. (2009) *Chem. Commun.*, 304–306.
41 Greenberg, S. and Stephan, D.W. (2009) *Inorg. Chem.*, **48**, 8623–8631.
42 Han, L.-B., Huang, Z., Matsuyama, S., Ono, Y., and Zhao, C.-Q. (2005) *J. Polym. Sci., Part A: Polym. Chem.*, **43**, 5328.
43 Jouaiti, A., Geoffroy, M., Terron, G., and Bernardinelli, G. (1992) *J. Chem. Soc., Chem. Commun.*, 155–156.
44 Jouaiti, A., Geoffroy, M., Terron, G., and Bernardinelli, G. (1995) *J. Am. Chem. Soc.*, **117**, 2251–2258.
45 Kawanami, H., Toyota, K., and Yoshifuji, M. (1996) *Chem. Lett.*, 533–534.
46 Wright, V.A. and Gates, D.P. (2002) *Angew. Chem., Int. Ed.*, **41**, 2389–2392.
47 Wright, V.A., Patrick, B.O., Schneider, C., and Gates, D.P. (2006) *J. Am. Chem. Soc.*, **128**, 8836–8844.
48 Shah, S., Concolino, T., Rheingold, A.L., and Protasiewicz, J.D. (2000) *Inorg. Chem.*, **39**, 3860–3867.
49 Dutan, C., Shah, S., Smith, R.C., Choua, S., Berclaz, T., Geoffroy, M., and Protasiewicz, J.D. (2003) *Inorg. Chem.*, **42**, 6241–6251.
50 Smith, R.C., Chen, X.F., and Protasiewicz, J.D. (2003) *Inorg. Chem.*, **42**, 5468–5470.
51 Kawasaki, S., Ujita, J., Toyota, K., and Yoshifuji, M. (2005) *Chem. Lett.*, **34**, 724–725.
52 Schafer, B., Oberg, E., Kritikos, M., and Ott, S. (2008) *Angew. Chem., Int. Ed.*, **47**, 8228–8231.
53 Yoshifuji, M., Shima, I., Inamoto, N., Hirotsu, K., and Higuchi, T. (1981) *J. Am. Chem. Soc.*, **103**, 4587–4589.
54 Moser, C., Nieger, M., and Pietschnig, R. (2006) *Organometallics*, **25**, 2667–2672.
55 Smith, R.C. and Protasiewicz, J.D. (2004) *J. Am. Chem. Soc.*, **126**, 2268–2269.

9
Organo-Arsenic, Phosphorus, and Antimony Conjugated Polymers

Kensuke Naka and Yoshiki Chujo

9.1
Introduction

The incorporation of metallic or metalloid elements such as S, Se, PR, SiR_2, GeR_2, and BR into an unsaturated polymer main chain appears to be a promising way to produce materials with useful properties [1–6]. Heteroaromatic polymers, such as polythiophenes and polysiloles, have also been intensively investigated and have found wide applications. Organoboron π-conjugated polymers highly extended through the empty p-orbital of the boron atom emerged as a new class of π-conjugated materials [6]. Among various types of such polymers, the simplest one is the heteroatom-containing polyvinylenes, of which properties might be attractive. However, no example of these polymers, except poly(vinylenesulfide), had been reported [7–9] before our reports, because of synthetic difficulties.

The introduction of inorganic elements with coordination ability into the polymer main chains results in polymer ligands which can keep transition metals along the backbones. Phosphorus, arsenic, antimony and bismuth are Group 15 elements, which have lone pairs on their atoms. The name "pnictogen" is sometimes used for elements of this group except for nitrogen, like the name "chalcogen" for Group 16 elements. The electronegativity of pnictogens decreases and the metallic character increases down the series phosphorus to bismuth. The structures of neutral pnictogen compounds in their +III oxidation states are pyramidal, and the fourth tetrahedral position is occupied by lone pairs of electrons. The bond angles are smaller than the ideal tetrahedral angle, 109.5°. In the case of the heavier elements, antimony or bismuth, the angles are close to 90°, which suggests substitution via p orbitals with little s character and the increased s character of the lone pair rather than sp^3 hybridized orbitals. The +III oxidation states of pnictogens normally act as Lewis bases or donors due to the presence of the lone pairs. Alkyl derivatives are stronger donors than aryl derivatives. Arsenic and bismuth exhibit lower basicities than phosphorus and antimony. Therefore, arsenic compounds have poorer coordination abilities than the phosphorus and stibine analogues. The +III pnictogens can also exhibit Lewis acidities due to the available empty d orbitals.

Conjugated Polymer Synthesis. Edited by Yoshiki Chujo

ISBN: 978-3-527-32267-1

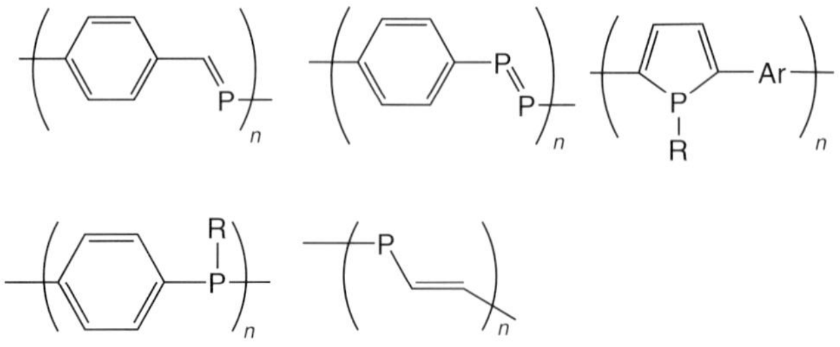

Figure 9.1 Recent examples of phosphorus-containing conjugated polymers.

9.2
Survey of Group 15 Element-Containing Polymers

Recently, several phosphorus-containing conjugated polymers with attractive properties have been prepared (Figure 9.1). The first π-conjugated P=C-containing and P=P-containing polymers have been synthesized [5, 10–13]. Poly(*p*-phenylene-phosphine)s were synthesized by palladium-catalyzed condensation and show some degree of π-conjugation along the polymer chain [14, 15]. Phosphole-containing polymers have also been prepared [16, 17]. The anionic ring opening polymerization of phosphirene provided a poly(vinylene-phosphine) [18].

Organoarsenic chemistry has a long history that dates back to the synthesis and discovery in 1760 of the first organometallic compound, $Me_2AsAsMe_2$ [19]. The discovery of the medicinal action of organoarsenicals on syphilis in 1910 led to a rapid expansion of the work on arsenic derivatives. Since an alternative cure was developed in 1940s, less attention has been paid to chemotherapy roles and more to the structures, stereochemistry and donor properties of organoarsenic compounds. In the abundant accumulation of organoarsenic chemistry, the incorporation of arsenic into polymer backbones has been limited [20–22]. No arsenic-containing conjugated polymers have been reported except by us.

Although several polymers containing antimony in the main chain have been reported [23–28], most of the polymers were synthesized by polycondensation. Moreover, there was no organoantimony polymer except for poly(*p*-phenylene triphenylantimony) [28] whose degree of polymerization was two. The antimony elements in those polymers were, however, all pentavalent, because trialkyl-/triarylantimony dichloride/dinitrate were used as comonomers in the previous polymerization.

9.3
Carbon–Main Group Element Bond Formation Via Bismetallation

Carbon–element bond formation should be a key point of synthetic methodology for incorporating a wide variety of metallic or metalloid elements into polymer main chains. Classical methodologies usually include electrophilic or nucleophilic substitution, which have provided wide varieties of heteroatom-containing π-conjugated polymers. Addition of metal halides to carbon–carbon unsaturated bonds, that is, hydrometallation, is one of the most useful atom-economical methods for carbon–heteroatom bond formation [29]. Synthesis of π-conjugated organoboron polymers was achieved by hydroboration polymerization [6, 30].

Radical bismetallation to carbon–carbon triple bonds, based on hemolytic cleavage of heteroatom–heteroatom single bonds, is one of the most useful and highly atom-economical methods for carbon–heteroatom bond formation. In the radical bismetallation, of which fewer examples are known than the metal-catalyzed version, the anti-form is mainly obtained [31–37]. In this case, irradiation or heating is often used for the initiation. Ogawa *et al.* reported photoinduced bisselenation and bistelluration of alkynes with organic diselenides and ditellurides [37]. Radical reactions of tetraphenyldiarsine or tetraphenyldiphosphine and phenylacetylene (initiated by AIBN or irradiation) provide the corresponding 1,2-disubstituted ethylenes [35]. Recently, stereospecific diphosphination of alkynes has been reported for a general route to backbone-functionalized 1,2-diphosphinoethenes, which have recently attracted increasing attention in the field of nanoarchitectures [38, 39]. However, application of bismetallation to polymer syntheses is very limited. With this background, we have developed the radical polymerization of homocyclic compounds of Group 15 elements with acetylenic compounds to produce poly(vinylene-arsine, -phosphine, and -stibine)s (Scheme 9.1).

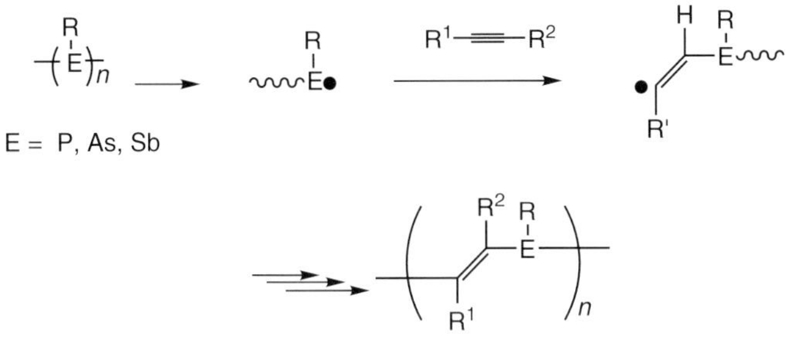

E = P, As, Sb

Scheme 9.1 Radical polymerization of homocyclic compounds of Group 15 elements with acetylenic compounds.

9.4 Homocyclic Compounds of Group 15 Elements

First, we discuss the formation and reactions of the cyclooligoarsines and their phosphorus and antimony analogues. Although a first organoarsenic homocycle, *cyclo*-$(PhAs)_6$ [40, 41], was synthesized by Michaelis and Schulte in 1881 [42], and the chemotherapeutic effects of "Salvarsan" and their derivatives were discovered [43], only a few reactions of the organoarsenic homocycles with organic compounds [44–46] have been developed compared to the extensive studies on their coordination chemistry as ligands in transition metal complexes [47]. No radical reactions of the cyclooligoarsines had been reported before our previous study. Revaluation of the organoarsenic homocycles would open a unique chemistry.

The bond energies of element–carbon bonds and element–hydrogen bonds decrease in the sequence P, As, Sb and Bi. The single bond energies of element–element bonds also decrease from phosphorus to bismuth: P–P 201 kJ mol^{-1}, As–As 146 kJ mol^{-1}, Sb–Sb 121 kJ mol^{-1}, and Bi–Bi 105 kJ mol^{-1} [48]. This sequence accounts for the number of phosphorus–phosphorus or arsenic–arsenic bonded compounds that are known, including $(RP)_n$ and $(RAs)_n$ rings, in comparison to few well-defined compounds with bismuth–bismuth bonds. The weakness of the heavier element–element bond also leads to the transformation of cyclic oligomers $(RE)_n$ [49]. The ring–ring transformation occurs more easily in the order phosphorus to bismuth. Although ring equilibria of cyclooligophosphines and cyclooligoarsines proceed at elevated temperature, the transformations between cyclooligostibines or cyclooligobithmuthines occur at or below room temperature. The compounds containing antimony or bismuth should be treated at lower temperature in an inert atmosphere, or they may need substitution by bulky groups.

The reduction of sodium methylarsonate and phenylarsonic acids with hypophosphorus acid yielded the five-membered and six-membered arsenic ring compounds, that is, pentamethylpentacycloarsine (*cyclo*-$(MeAs)_5$) and hexaphenylhexacycloarsine (*cyclo*-$(PhAs)_5$), resepectively [50, 51], but no other rings or chains of arsenic were formed by this reduction, suggesting that these ring structures are stable compared to other forms containing arsenic–arsenic bonds. The five-membered cyclooligoarsine, *cyclo*-$(MeAs)_5$, was obtained as a yellow liquid with high viscosity (Scheme 9.2).

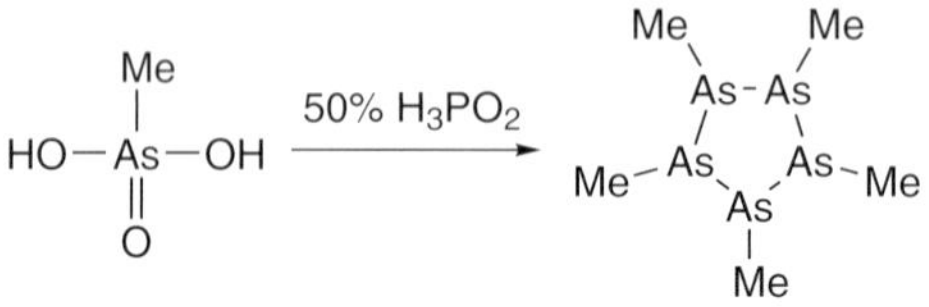

Scheme 9.2 Synthesis of pentamethylpentacycloarsine (*cyclo*-$(MeAs)_5$).

After *cyclo*-$(MeAs)_5$ was left under a nitrogen atmosphere at 25 °C for several days, a purple–black precipitate appeared. The formation of the precipitate was accelerated by heating in the presence of 2,2′-azobisisobutyronitrile (AIBN) or by irradiation with an incandescent lamp. The resulting precipitate was insoluble in any solvent and fumed in air. According to the literature [52, 53], it should be a linear poly(methylarsine) with a ladder structure. The arsenic–arsenic bond of the ring compound was cleaved spontaneously, and then the open-chain oligoarsines stacked with each other to form the ladder structure (Scheme 9.3). The electronic and polymeric structures of the ladder polyarsine were theoretically investigated, based on first-principles electronic structure calculations [54]. The isolated polyarsine ladder chain prefers to be planar. The laddering causes the band overlap of the valence band and

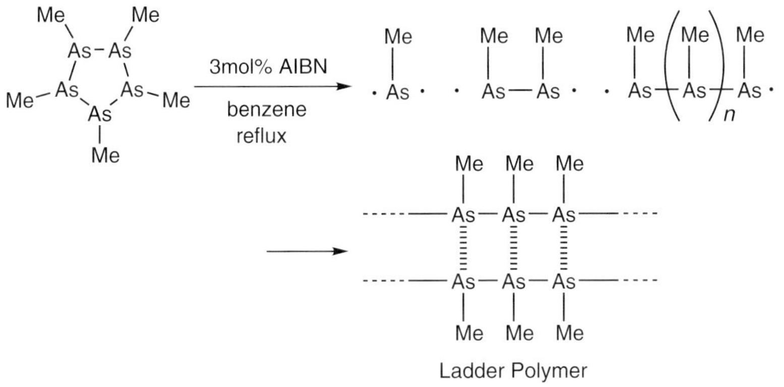

Scheme 9.3 Formation of a ladder polyarsine from *cyclo*-$(MeAs)_5$.

conduction band, and produces a metallic electronic structure with the help of the excess electrons of the arsenic atom.

The phenyl-substituted cyclooligoarsine, *cyclo*-$(PhAs)_6$, was obtained as a pale yellow crystal (Scheme 9.4) [55]. The crystal showed poor solubility in benzene. When *cyclo*-$(PhAs)_6$ was treated with 3 mol% of AIBN in refluxing benzene, the heterogeneous reaction mixture became clear within 30 min. The stable ring structure was collapsed and open-chain arsenic oligomers might be formed in the solution. By cooling the clear solution to room temperature, *cyclo*-$(PhAs)_6$ was regenerated as a light yellow powder. The catalytic amount of AIBN was enough to homogenize the mixture, in other words, to collapse all of the stable six-membered ring structures, suggesting that the produced arsenic radical also contributed to the destruction of *cyclo*-$(PhAs)_6$. In contrast to *cyclo*-$(MeAs)_5$, *cyclo*-$(PhAs)_6$ was stable at room temperature in an air atmosphere in the solid state, and showed no reaction even in refluxing benzene for several hours.

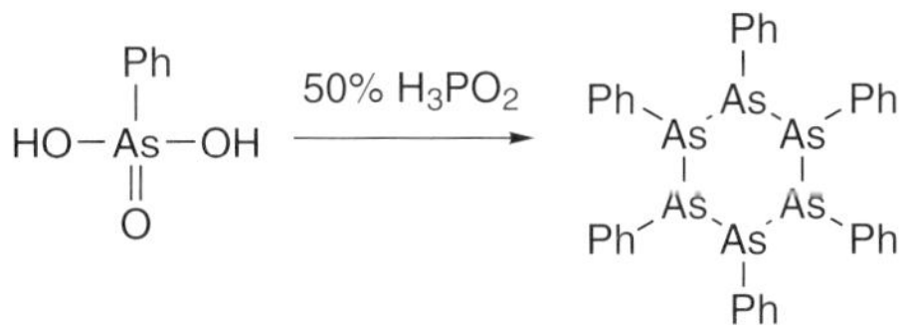

Scheme 9.4 Synthesis of hexaphenylcyclohexaarsine (*cyclo*-$(PhAs)_6$).

Cyclooligophosphines are homocyclic compounds of which the backbone is constructed with only phosphorus atoms. Although interesting behavior, owing to their ability to form complexes with metal carbonyls and transition metal salts, has been shown [56–58], there are few reports concerning reactions of cyclooligophosphines with organic compounds [46, 56]. The methyl-substituted cyclooligophosphine was synthesized by the reduction of methyldichlorophosphine with lithium [59]. It is reported that the five-membered ring structure, pentamethylcyclopentaphosphine, *cyclo*-$(MeP)_5$, is mainly formed by this method and exclusively obtained by distillation under reduced pressure.

Cyclooligostibines are homocyclic compounds with rings built exclusively of antimony atoms, and have been well characterized [60–64]. However, the reactivity of cyclooligostibines was reported in only a few papers [60, 61]. The phenyl-substituted cyclooligostibine, hexaphenylcyclohexastibine (*cyclo*-$(PhSb)_6$), was synthesized by reduction of phenylantimony dichloride using bis(cyclopentadienyl) cobalt(II) [65, 66], Non-bulky substituents (R = Me, Et, Pr, Ph, Tol, etc.) give rise to highly flexible antimony homocycles with equilibria between rings of different size or polymers [67]. Cyclic hexamers, R_6Sb_6 (R = Ph, Tol), exist in crystalline phases.

9.5 Poly(Vinylene-Arsine)s

Poly(vinylene-arsine)s have been produced by a radical reaction between the cyclic organoarsenic compounds, such as *cyclo*-$(MeAs)_5$ or *cyclo*-$(PhAs)_6$, and mono-subsitututed acetylenic compounds using AIBN as a radical initiator (Scheme 9.5) [68]. This is a novel type of radical alternating copolymerization, in which the arsenic ring compound collapses and the arsenic unit is incorporated into the polymer backbone with no arsenic–arsenic bonds. Therefore, we have named this copolymerization system "ring-collapsed radical alternating copolymerization (RCRAC)". In the ^1H NMR spectrum of **2a** derived from *cyclo*-$(MeAs)_5$ with phenylacetylene (**1a**), the integral ratio of two peaks in a vinyl region confirmed that the trans-isomer was obtained predominantly. Analysis of **2a** by ^{13}C NMR spectroscopy showed only one sharp resonance for the methyl carbon, suggesting that the arsenic in **2a** existed in a trivalent state and no arsenic–arsenic bond or oxidized arsenic was present. When using *cyclo*-$(PhAs)_6$ with **1a**, the corresponding poly(vinylene-arsine) **3a** was obtained

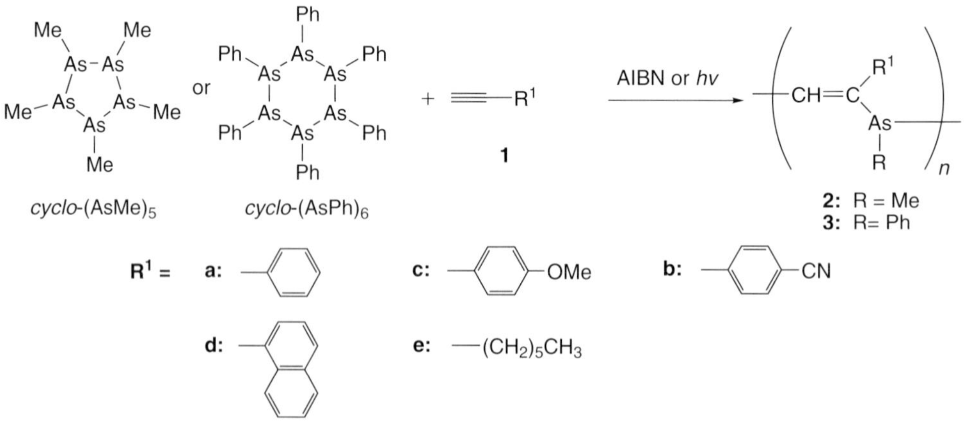

Scheme 9.5 Ring-collapsed radical alternating copolymerization (*RCRAC*) of cyclic organoarsenic compounds with mono-subsitututed acetylenic compounds.

Table 9.1 Results of polymerization of cyclooligoarsines and acetylenic compounds.

Run	Arsenic monomer	1	x[a]	Initiator[d]	Solvent (°C)	M_w[b]	M_n[b]	M_w/M_n[b]	Yield (%)[c]
1	*cyclo*-$(AsMe)_5$	1a	5.0	AIBN	Benzene (reflux)	48 700	11 500	4.3	46
2[e]	*cyclo*-$(AsMe)_5$	1a	5.0	hυ	Benzene (r.t.)	11 100	3400	303	48
3	*cyclo*-$(PhMe)_6$	1a	6.0	AIBN	Benzene (reflux)	5600	3900	1.4	35
4	*cyclo*-$(PhMe)_6$	1a	3.0	AIBN	Benzene (reflux)	5000	3500	1.4	34
5	*cyclo*-$(PhMe)_6$	1a	1.0	AIBN	Benzene (reflux)	5600	3800	1.5	trace
6	*cyclo*-$(PhMe)_6$	1a	12.0	AIBN	Benzene (reflux)	7300	4100	1.8	17
7	*cyclo*-$(PhMe)_6$	1b	6.0	AIBN	Benzene (reflux)	15 100	8400	1.8	29
8	*cyclo*-$(PhMe)_6$	1c	6.0	AIBN	Benzene (reflux)	5100	3700	1.4	22
9	*cyclo*-$(PhMe)_6$	1d	6.0	AIBN	Benzene (reflux)	9900	7600	1.3	33
10	*cyclo*-$(AsMe)_5$	1a	1.7	no	$CHCl_3$ (25 °C)	13 500	5200	2.6	17
11	*cyclo*-$(AsMe)_5$	1a	3.4	No	$CHCl_3$ (25 °C)	14 000	5200	2.7	42
12	*cyclo*-$(AsMe)_5$	1a	5.0	No	$CHCl_3$ (25 °C)	23 300	11 100	2.1	64
13	*cyclo*-$(AsMe)_5$	1a	7.5	no	$CHCl_3$ (25 °C)	30 100	13 100	2.3	52

a) Molar ratio of **1** to the cyclic organoarsenic compounds.
b) GPC ($CHCl_3$). Polystyrene standards.
c) Isolated yields after reprecipitation into hexane based on the molar amount of 1.
d) 3 mol% of AIBN was employed.
e) Xenon lamp.

as a white powder. Both the polymers were readily soluble in common organic solvents such as THF, chloroform and benzene. Several results of the polymerization are summarized in Table 9.1.

Polymer **2a** was also obtained by irradiation of a benzene solution of *cyclo*-$(MeAs)_5$ and **1a** with a xenon lamp at room temperature (Table 9.1, Run 2). We carried out the copolymerization in different feed ratios ($x = 1$–12) of the two monomers, *cyclo*-$(PhAs)_6$ and **1a** (Table 9.1, Run 3–6). In Run 4, the polymerization with 3 equiv of **1a** to *cyclo*-$(PhAs)_6$ gave a polymer which had almost the same molecular weight and the same structure. The polymerization with an equivalent of **1a** to *cyclo*-$(PhAs)_6$ also gave a polymer with almost the same molecular weight and the same structure.

The following reaction mechanism was proposed for the radical alternating copolymerization (Scheme 9.6) [55, 68]. First, AIBN cleaves arsenic–arsenic bonds of the cyclooligoarsines to produce arsenic radicals. Second, the homolysis of the other arsenic–arsenic bonds proceeds spontaneously due to their instability caused by the destruction of the quite stable five- or six-membered ring structures. In competition with this reaction, an arsenic radical added to an acetylenic compound to give a vinyl radical. Next, the vinyl radical reacted immediately with the arsenic–arsenic bond or with the arsenic radical to form a new carbon–arsenic bond. Although the carbon–arsenic bond formation seems to result in loss of the growing radicals and chain growth stops, the labile arsenic–arsenic bonds in the product lead to easy homolysis to produce a new arsine radical, and the chain growth then restarts. In this manner, repeating production of the arsine radicals by arsenic–arsenic bond cleavage and their addition to the acetylenic compound lead to a polymer with a

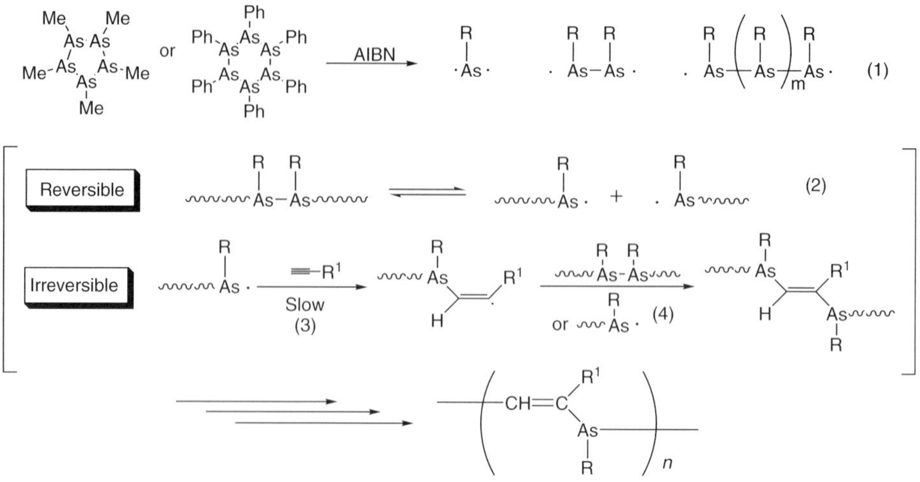

Scheme 9.6 Proposed reaction mechanism of RCRAC of cyclic organoarsenic compounds with acetylenic compounds.

simple main-chain structure, poly(vinylene-arsine). Since the vinyl radicals are more unstable and reactive than the arsenic radicals, formation of the vinyl radicals (Scheme 9.6 Eq. 3) should be relatively slower than creation of the arsine radical (Eq. 2) and of a carbon–arsenic bond (Eq. 4). One of the propagating radicals in this copolymerization system is the vinyl radical, which may not cause recombination due to its low concentration. The vinyl radical may not react with the acetylenic compound to produce a new vinyl radical because of its instability, but it reacts with the arsenic radical or the arsenic–arsenic bond. The other propagating radical is the arsenic radical which cannot cause disproportionation as a termination reaction. A recombination of the arsenic radical generates an arsenic–arsenic bond which is easily cleaved under the benzene refluxing condition to reproduce the propagating arsenic radicals.

The aromatic stabilization of the vinyl radical is essential for the copolymerization. The more effectively aromatic stabilization acts on the vinyl radical, the more smoothly the copolymerization progresses [69]. When using 4-cyanophenylacetylene **1b**, the polymerization proceeded to reach nearly quantitative conversion. The copolymerization of *cyclo*-$(PhAs)_6$ with **1b** or 1-naphthylacetylene **1d** was faster than that with **1a** because the aromatic stabilization of the vinyl radical by 4-cyanophenyl or naphthyl group is stronger than that by the phenyl group. The copolymerization of *cyclo*-$(PhAs)_6$ with 4-methoxyphenylacetylene **1c** was slightly slower than that with **1a**, suggesting that electrostatic effect of the substituent might affect the reaction between the arsenic radical and the carbon–carbon triple bond. The arsenic radical may prefer to react with electron-accepting monomers rather than electron-donating monomers due to the lone pair on the arsenic atom. Both the conjugative and the electrostatic effects of the 4-cyanophenyl group led to the fastest consumption of **1b** during the copolymerization with *cyclo*-$(PhAs)_6$ among the acetylenic monomers employed here.

The methyl-substituted cyclooligoarsine, *cyclo*-$(MeAs)_5$, afforded the corresponding poly(vinylene-arsine) by copolymerization with **1a** using water as a reaction medium. The copolymerization was carried out under a nitrogen atmosphere using water, *cyclo*-$(MeAs)_5$, **1a**, sodium lauryl sulfate as an emulsifier, and potassium peroxodisulfate as a water-soluble radical initiator. The obtained polymer was purified by reprecipitation into *n*-hexane to yield the corresponding poly (vinylene-arsine) as a yellow-white powder (yield 27%). The number-average molecular weight (M_n) of the copolymer was 4700 with a polydispersity index (M_w/M_n) of 4.1. ^{1}H-NMR and ^{13}C-NMR spectra were employed to support the structure of the polymer. The fact that the copolymerization proceeded in the presence of water excludes the possibility that ionic species contribute to the propagation of the copolymerization.

The arsenic ring compound with *cyclo*-$(MeAs)_5$ reacted with **1a** in chloroform at 25 °C in the absence of any radical initiators or catalysts to produce the corresponding poly(vinylene-arsine) in moderate yield [55]. The 1 : 1 alternating structure of MeAs and **1a** is controlled very strictly under a wide variety of copolymerization conditions. Various solvents could be used to afford copolymers with number-average molecular weights of more than ten thousand in moderate yields at 25 °C. While the excess amount of **1a** (Table 9.1, run 13) was found to remain unchanged after completion of the reaction, the excess amount of MeAs (Table 9.1, run 10 and 11) decreased the molecular weights and yields of the obtained polymers in comparison with those in the equivalent case (Table 9.1, run 12). Since almost all the propagating radicals are the arsenic radical, even when excess **1a** was employed, the copolymerization in run 13 proceeded comparably with the equivalent case. When a smaller amount of **1a** than that of MeAs was employed, all the **1a** was consumed before sufficient growth of the copolymer chain. Thus, the copolymerization in runs 10 and 11 yielded alternating copolymer with low molecular weights, and, hence, low yields after removal of the oligomers by reprecipitation.

The ring-collapsed radical alternating copolymerization (*RCRAC*) of *cyclo*-$(MeAs)_5$ with aliphatic acetylenes in the presence of AIBN or irradiation from an incandescent lamp has been achieved to produce poly(vinylene-arsine)s with no aromatic substitution, while *cyclo*-$(PhAs)_6$ did not copolymerize with aliphatic acetylenes [70]. The present results show that acetylenic comonomers with a variety of substitution are available for *RCRAC* with *cyclo*-$(MeAs)_5$. The progress of the copolymerization of *cyclo*-$(MeAs)_5$ with aliphatic acetylenes was much slower than that with PA due to the lack of conjugative substitution to the ethynyl group.

9.6 Poly(Vinylene-Phosphine)s

A conjugated polymer containing phosphorus atom in the main chain, poly(vinylene-phosphine) (**4a**), was synthesized by *RCRAC* of *cyclo*-$(MeP)_5$ and **1a** (Scheme 9.7) [71]. The ^{1}H, ^{13}C, and ^{31}P NMR spectroscopies suggested that trans-isomers of diphosphaalkene units were mainly formed in the main chain.

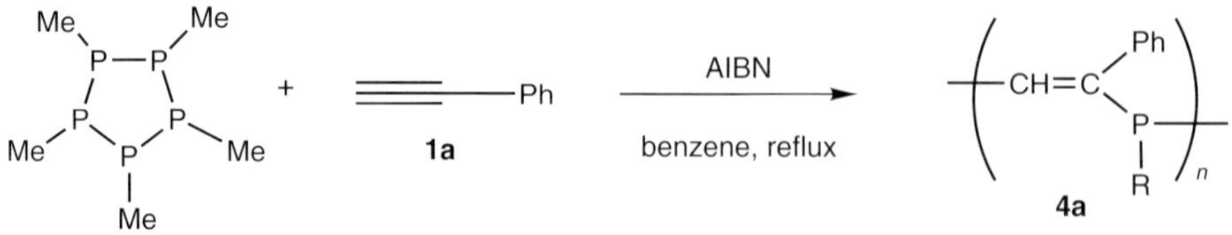

Scheme 9.7 RCRAC of *cyclo*-$(MeP)_5$ with phenylacetylene (**1a**).

No RCRAC of *cyclo*-$(MeP)_5$ with **1a** without AIBN proceeded even when heating to 80 °C or in the bulk. This result suggests that AIBN would be required for RCRAC of *cyclo*-$(MeP)_5$, in contrast with RCRAC of *cyclo*-$(MeAs)_5$. A phosphorus–phosphorus bond is stronger than an arsenic–arsenic bond, and the phosphorus–phosphorus bond in *cyclo*-$(MeP)_5$ cannot cleave spontaneously. The radical reaction of *cyclo*-$(MeP)_5$ with 4-cyanophenylacetylene **1b** and dimethyl acetylenedicarboxylate **1f** gave only oligomeric products. Radicals of lighter elements are expected to have less ability for the capture of vinyl radicals. The capture of the vinyl radicals is hardly achieved in RCRAC of *cyclo*-$(MeP)_5$ with electron-deficient acetylenes which provide less reactive vinyl radicals. Comparing these results with those of *cyclo*-$(MeAs)_5$ the capture ability of the vinyl radical is seen to be lower for *cyclo*-$(MeP)_5$ than for *cyclo*-$(MeAs)_5$ to provide the corresponding poly(vinylene-phosphine) even though the reaction of *cyclo*-$(MeP)_5$ with **1b** or **1c** proceeds.

9.7 Poly(Vinylene-Stibine)s

RCRAC of dimethyl acetylenedicarboxylate **1f** and *cyclo*-$(PhSb)_6$ in the presence of a catalytic amount of AIBN in tetrahydrofuran at 70 °C provided a corresponding poly (vinylene-stibine) **5f** with number-average molecular weight 4300 (Scheme 9.8) [72]. The resulting polymers are the first organic polymers with trivalent antimony in their backbones. Polymer **5f** was readily soluble in common organic solvents such as $CHCl_3$, THF, DMF, and toluene.

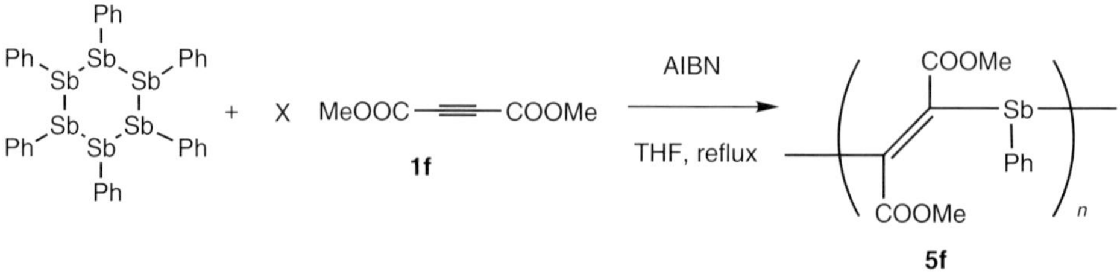

Scheme 9.8 RCRAC of *cyclo*-$(PhSb)_6$ and **1f**.

Although RCRAC of *cyclo*-$(PhSb)_6$ and **1f** gave polymer **5f**, no polymerization proceeded when *cyclo*-$(PhAs)_6$ was used in the RCRAC system with **1f**. When the phenylacetylene derivatives (**1a–1c**) were used, the resulting arsenic polymers had larger number-average molecular weights and higher yields than the corresponding antimony polymers. Monosubstituted acetylenic monomers, phenylacetylene **1a**, 4–cyanophenylacetylene **1b** and 4-methoxyphenylacetylene **1c**, however, gave low molecular weight poly(vinylene-stibine)s. These differences between the case of *cyclo*-$(PhAs)_6$ and *cyclo*-$(PhSb)_6$ are derived from the reactivity of the corresponding radicals. The effects of conjugation or electron-withdrawing of the substituent of the acetylenic monomer promoted RCRAC. This result suggested that the arsenic radical reacts preferentially with electron-accepting monomers due to the lone pair on the arsine atom. However, an excessive conjugative or electrostatic effect of the substitution of an acetylenic monomer such as **1f** decreased the reactivity of the vinyl radicals formed from the acetylenic compounds toward the arsenic radical due to increased stability of the vinyl radicals. The arsenic radicals have more reactivity toward the ethynyl groups than the antimony radicals and the antimony radicals have more reactivity toward the vinyl radicals.

9.8 Periodic Terpolymerization of Cyclooligoarsine, Cyclooligostibine, and an Acetylenic Compound

The different reactivity of pnictogen radicals made it possible to construct the periodic vinylene–arsine–vinylene–stibine backbone. The ring-collapsed radical terpolymerization of *cyclo*-$(MeAs)_5$, *cyclo*-$(PhSb)_6$ and dimethyl acetylenedicarboxylate **1f** provided a periodic terpolymer containing both antimony atom and arsenic atom in the polymer chain (Scheme 9.9) [73]. With a 1/5 : 1/6 : 2 ratio of *cyclo*-$(MeAs)_5$: *cyclo*-$(PhSb)_6$: **1f**, the obtained polymer has a 1 : 1 : 2 structure of the arsine, stibine and vinylene units. Employing excess *cyclo*-$(MeAs)_5$ did not affect the periodic structure of

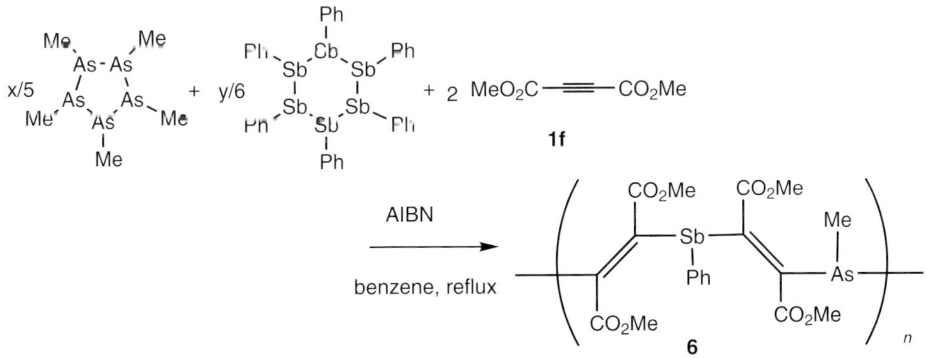

x/5 (MeAs)5 + y/6 (PhSb)6 + 2 MeO_2C—≡—CO_2Me (**1f**) —AIBN, benzene, reflux→ **6**

Scheme 9.9 Periodic terpolymerization of cyclooligoarsine, cyclooligostibine, and acetylenic compound.

the terpolymerization. On the other hand, employing excess *cyclo*-$(PhSb)_6$ resulted in a terpolymer with more stibine units. The arsenic radical was more reactive towards **1f** forming the vinyl radical, while the antimony radical was more reactive towards the vinyl radical forming the terpolymer. After the arsenic radicals were consumed, the antimony radical also reacted with **1f**, thereby affecting the periodic structure of the terpolymerization. There are some similar reports about the reactivity of Group 16 elements. Sulfur radicals attack unsaturated bonds faster than selenic radicals [74] and selenic radicals react with carbon radicals [75]. This is the first example to provide an A–B–A–C type polymer by a simple radical reaction, thus opening a new area employing a novel method to make polymers with periodic backbones.

9.9 Stability

The methyl-substituted poly(vinylene-arsine) **2a** showed a 10% weight losses at 265 °C (under N_2) and 205 °C (under air), and for the phenyl-substituted one (**3a**) the corresponding losses occurred at 284 °C (under N_2) and 250 °C (under air). The poly(vinylene-arsine) with phenyl-substitution (**3a**) was thermally more stable than that with methyl-substitution (**2a**). The glass transition temperatures (T_g) of **2a** and **3a** were 58.2 and 92.9 °C, respectively, determined by DSC analysis. Both the copolymers are stable in the solid state at room temperature. No decrease in molecular weight or change of structure was observed, even after exposing them to air for several months.

After **3a** was stirred vigorously with 30% H_2O_2 at 60 °C for 3 h, the ^{1}H-NMR spectrum and the GPC curve of **3a** were the same as those for the starting polymer. These results suggest that the trivalent state of arsenic in the main chain was insensitive to the oxidation.

Poly(vinylene-phosphine) **4a** is reasonably air- and moisture-stable in the solid state. No change was observed in the GPC trace and the ^{31}P NMR spectrum after **4a** in the solid state was exposed to air for several days. On the other hand, when polymer **4a** was left in $CHCl_3$ for 12 h, a new peak appeared around δ 25 ppm in the ^{31}P NMR spectrum, which suggests the oxidation of the phosphine unit. The peak of the vinyl proton became smaller in the ^{1}H NMR spectrum, suggesting decomposition of the vinylene-phosphine structure. After being exposed to air in the solution state for several days, the polymer decomposed to show no polymeric peaks in a GPC trace. Polymer **4a** was reacted with elemental sulfur in dichloromethane to yield phosphine sulfide, of which the ^{31}P NMR spectrum showed a peak at δ 39.8 ppm and no peak around δ −28 ppm. In this case, the GPC trace exhibited no significant change compared to that before the sulfuration.

Polymer **5f** was air-stable in the solid state at room temperature. No changes in the molecular weight and the structure were observed in the GPC trace and the ^{1}H-NMR spectrum after exposing **5f** to air for several days. The GPC analysis and the ^{1}H-NMR spectrum showed no change after leaving **5f** in an inert atmosphere for several months. Although no oxidation peak was observed in the cyclic voltammetric measurement of **5f**, this polymer was oxidized by 2,3-dichloro-5,6-dicyano-1,

4-benzoquinone (DDQ). When DDQ was added to the $CHCl_3/CH_3CN$ solution of polymer **5f**, the solution color turned to dark red. The UV spectrum of the solution of **5f** and DDQ shows major maxima at 550 and 589 nm, assignable to the anion radical of DDQ. The intensities at 550 and 589 nm were saturated at an amount of [DDQ]/[repeating unit of **5f**] = 0.4–0.5. This result showed that half of the antimony atoms in the polymer can be oxidized in the present condition. This may be due to an electrostatic repulsion of the adjacent oxidized antimony unit.

9.10 Optical Properties

The electronic structures of the poly(vinylene-arsine)s were studied by UV–vis and fluorescence spectroscopies. In the UV–vis analyzes was observed not only strong absorption in the UV region, derived from the π–π^* transition of the benzene ring, but also a small absorption in the visible region, due to the n–π^* transition in the main chain. The polymers **3a–d** showed fluorescent properties. The emission was observed in the visible blue–purple region with peaks at 443 and 466 nm. The emission peak maximum was independent of the concentration of polymer **3b**. In the excitation spectrum of **3b** monitored at 470 nm, the absorption was not observed in the shorter wavelength region but in the longer wavelength region with a peak at 394 nm. This means that the absorption of **3b** in the higher energy region and the absorption in the lower energy region originate from different absorbing species; π–π^* and n–π^* transitions. The emission of **3b** results only from the absorption of the latter transition. Each emission peak maximum of **3a–c** was red-shifted in the order **3c** < **3a** < **3b**, which coincides with the order of strength of the electron-withdrawing properties of the substituents. This indicates that the donor–acceptor (arsenic atom with a lone pair and vinylene unit with an electron-accepting group) repeating units made the band gap narrower and resulted in the lower energy of the emission. The optical properties of the poly(vinylene-arsine)s were tuned by changing the substituents on the acetylenic compounds. In the case of **3d**, the naphthyl group itself has a fluorescent property which is stronger than that derived from the n–π^* transition in the main chain of the polymer, and, therefore, the emission peak due to the main chain was hidden. The absorption peaks in the excitation spectra of **3a–d** showed the same order as the emission peaks for the same reason discussed above.

The poly(vinylene-arsine)s with no aromatic substitution (**2e**) also showed emission in the visible blue–violet region with a peak at 450 nm from irradiation at 350 nm. In the excitation spectrum of **2e** monitored at 450 nm, the absorption peak was observed only in the region where the wavelength was more than 310 nm. Both the emission and excitation peaks of **2e** were blue-shifted in comparison with those of the polymer which was synthesized by the copolymerization of *cyclo*-$(MeAs)_5$ with **1a**. The substituent groups of the acetylenes affected the electronic structure of the main chain of the poly(vinylene-arsine)s.

The UV–vis absorption spectrum of **4a** in chloroform showed a small absorption in the visible region derived from the n–π^* transition in the main chain, as well as strong absorption in the UV derived from the aromatic groups. The fluorescence spectrum of a dilute chloroform solution of **4a** showed an emission peak at 470 nm, which derived from the n–π^* transition along the polymer backbone. The bathochromic shift observed in the excitation spectrum of **4a** compared with that of **2a** might be attributed to the conformation of the phosphorus or arsenic atom in the main chain. The s-character of the lone pair on a phosphorus atom is weaker than that of an arsenic atom. This might lead to increased delocalization between the lone pair electron and the π electron of the vinylene unit.

The lower energy absorption edge was located at around 440 nm. No emission was observed when **5f** was excited at 350 or 400 nm, while the poly(vinylene-arsine)s showed emission. As the atomic number of the pnictogen increases, the s character of the lone pair is enhanced. In polymer **5f**, the lone pairs are very localized on the antimony element because of the stronger s-character. We assume interaction between the lone pair electron and the π electron of the vinylene unit decreases compared to the poly(vinylene-arsine)s.

9.11
Coordination Ability of Poly(Vinylene-Arsine)s Towards Transition Metal Ions

The introduction of transition metal complexes into conjugated polymer backbones provides new dimensions of novel materials with improved electrical, magnetic and catalytic properties. The poly(vinylene-arsine)s are expected to form polymer complexes with various kinds of transition metals due to the lone pair on the arsenic atom [76]. Poly(vinylene-arsine)–cobalt complexes have been prepared because the coordination of the compounds of tertiary arsines with cobalt ion has been well studied providing background information for this research [77–80]. The polymer metal complexes were prepared by adding an acetone solution of **2a** or **3a** to an acetone solution of cobalt(II) chloride hexahydrate at different feed ratios. Polymer **3a** formed no cobalt complex upon addition of the acetone solution of the polymer to the acetone solution of cobalt(II) chloride hexahydrate due to its bulky structure. On the other hand, polymer **2a** formed the **2a**–Co complexes at different cobalt salt-polymer ratios due to its less bulky structure (Scheme 9.10).

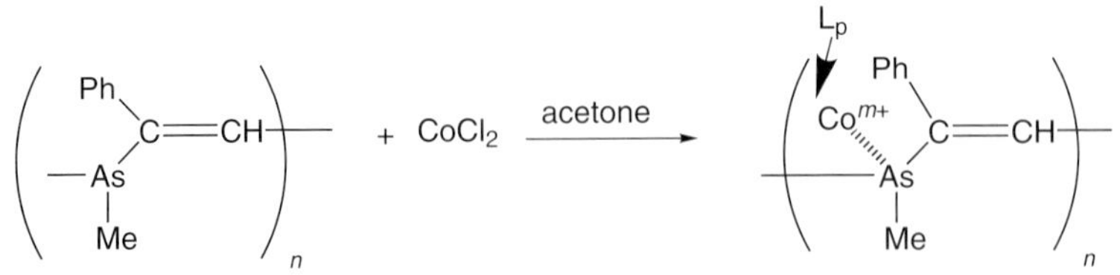

Scheme 9.10 Preparation of poly(vinylene-arsine)-cobalt complex.

As in the usual case, the bivalent complexes of the reaction run corresponding to a feed ratio of 1 : 1 [$CoCl_2$]/[repeating unit of **2a**] oxidized readily in air to give very stable complexes of trivalent cobalt. On the other hand, the bivalent complexes of the reaction runs corresponding to feed ratios of 1 : 10 and 1 : 4 both exhibited unusual stability against air oxidation owing to the higher concentration of the polymer in the reaction mixture.

The UV–vis spectrum of the complex of the run corresponding to the feed ratio of 1 : 1, showed a metal-to-ligand charge transfer (MLCT) absorption peak at 370 nm. This absorption suggests that the bivalent complexes of this reaction run oxidized in air to give very stable complexes of trivalent cobalt. On the other hand, no MLCT band is observed in the UV–vis spectra of the complexes of the runs corresponding to the feed ratios of 1 : 10 and 1 : 20 in chloroform.

Complexations of the poly(vinylene-arsine)s with ruthenium ions were carried out (Scheme 9.11). A half equimolar amount of *cis*-$(bpy)_2RuCl_2{\cdot}2H_2O$ relative to the vinylene-arsine structure and **2a** were stirred in refluxing deaerated $CHCl_3$/ethanol (1/1). The solvent of the resulting mixture was reduced by evaporation and the resulting solution was poured into a methanol solution of NH_4PF_6 to yield a polymer complex. The complexation of **2a** resulted in a red solid (**2a**–Ru complex), soluble in CH_2Cl_2 and acetonitrile, and insoluble in benzene. The ^1H NMR spectrum of the **2a**–Ru complex suggested that 19% of the arsine units were coordinated with ruthenium ions. Polymer **3a** resulted in no or little complexation with ruthenium ions.

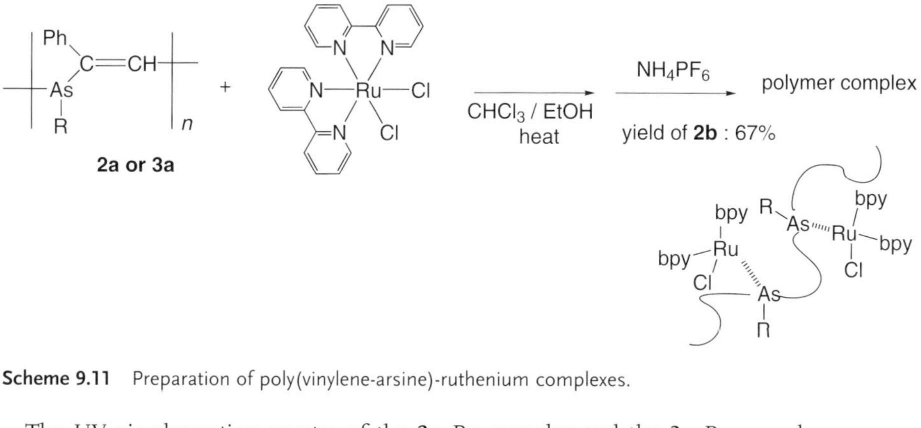

Scheme 9.11 Preparation of poly(vinylene-arsine)-ruthenium complexes.

The UV–vis absorption spectra of the **2a**–Ru complex and the **3a**–Ru complex showed absorption peaks at around 470 nm, derived from metal-to-ligand charge transfer (MLCT) as well as a strong absorption due to the aromatic groups in the UV region. This result also indicates the incorporation of the ruthenium complexes in the polymer. Cyclic voltammetry (CV) of a **2a**–Ru complex in CH_2Cl_2 showed a reversible oxidation peak derived from the oxidation of Ru(II) ions. The potential difference between the oxidation peak and the half oxidation peak was 40 mV, which suggests little interaction between the ruthenium ions when around 20% of the arsine units are coordinated to the ruthenium ions.

The study of metal nanoparticles is one of the most exciting fields due to its potential applications towards electronics, nonlinear optics, and catalysis. One of the widely used methods for the stabilization of metal nanoparticles is steric stabilization by polymers, which is achieved by surrounding the metal center. The polymers covering the metal particles prevent close contact with other particles. The length and nature of the polymer play a key role in proper control of the size of the metal nanoparticles.

The reduction of Ag ions in DMF took place spontaneously in the presence and the absence of **2a** or **3a** at 100 °C. When no polymer was present in the system, the reduction led to silver deposition (Scheme 9.12a). On the other hand, in the presence of **2a** or **3a** in a 1 : 1 ratio with $AgNO_3$, colloidal solutions were obtained (Scheme 9.12b). This means that poly(vinylene-arsine)s stabilized the silver nanoparticles. The UV–vis absorption spectra of the colloidal solutions indicated the formation of silver nanoparticles with a surface plasmon absorption band at 400 nm. Polymer **2a** is a better stabilizer for the silver nanoparticles in DMF owing to its less bulky structure than that of **3a**.

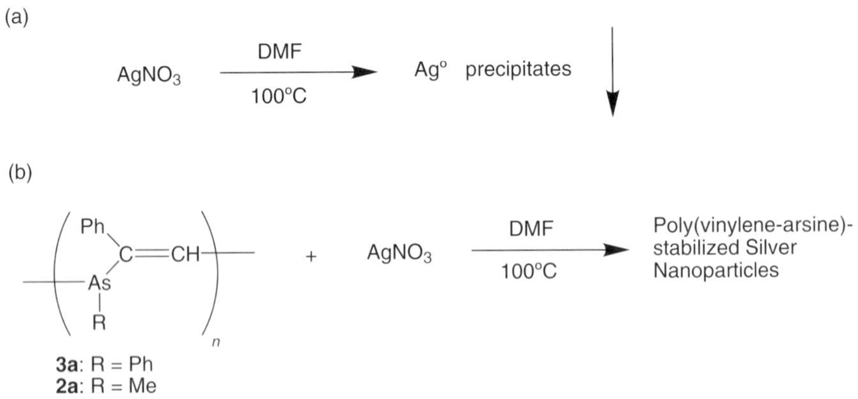

(a)

$AgNO_3$ —DMF, 100°C→ Ag^o precipitates ↓

(b)

[–As(R)–C(Ph)=CH–]$_n$ + $AgNO_3$ —DMF, 100°C→ Poly(vinylene-arsine)-stabilized Silver Nanoparticles

3a: R = Ph
2a: R = Me

Scheme 9.12 Reduction of Ag^+ ions in DMF in the absence (a) and presence (b) of poly(vinylene-arsine)s.

9.12
Cross-Linked Poly(Vinylene-Arsine)s

Conjugated cross-linked polymer networks have been studied because of their unique properties, such as environmental responses [81] and enhanced conductivities [82, 83]. We synthesized cross-linked poly(vinylene-arsine)s by the radical terpolymerization of the cyclooligoarsine, **1a**, and *p*- or *m*-diethynylbenzene (DVB) (Scheme 9.13) [84]. If each polymer chain with the structure of the vinylene-arsine is cross-linked by another, the electrical structure would be affected due to the enhancement of the interchain interaction, and thus more extension of the conjugation length would be expected. In addition, closely packed structures might improve the thermal stability of the poly(vinylene-arsine)s.

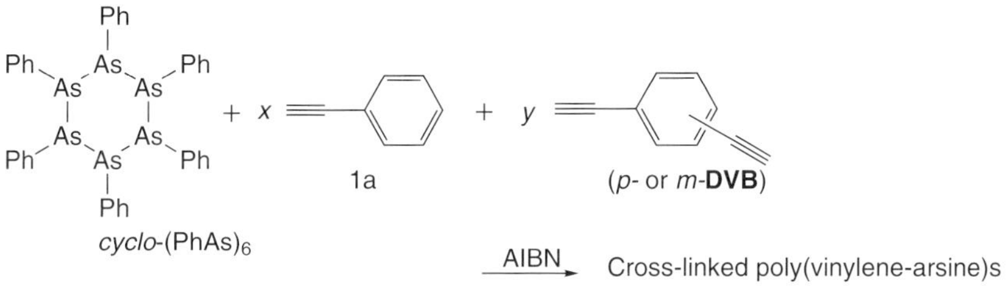

cyclo-(PhAs)$_6$ + x 1a + y (p- or m-**DVB**) $\xrightarrow{\text{AIBN}}$ Cross-linked poly(vinylene-arsine)s

Scheme 9.13 Synthesis of cross-linked poly(vinylene-arsine)s by the radical terpolymerization of cyclooligoarsine, phenylacetylene, and *p*- or *m*-diethynylbenzene.

We carried out terpolymerization at different compositions of *cyclo*-(PhAs)$_6$, **1a**, and *p*- or *m*-DVB with a catalytic amount of AIBN in refluxing benzene to investigate their effects on the polymerization behavior and the polymer properties. The obtained polymers, except the polymer of run 4, were soluble in common organic solvents such as benzene, toluene, chloroform, and DMF. The monomer feed ratios, the molecular weights, yields, and solubility of the obtained polymers are summarized in Table 9.2. In run 4 where we employed **1a** and *p*-DVB with a molar ratio of 1 : 1, the reaction mixture became heterogeneous due to deposition of the insoluble polymeric material within 2 h after the terpolymerization was initiated.

To investigate the thermal properties of the obtained cross-linked polymers, DSC analyses were carried out under nitrogen atmosphere. The values of the glass transition temperatures (T_g) are also listed in Table 9.2. As expected, the T_g value increases with the content of the cross-linker, diethynylbenzene, and when the feed

Table 9.2 Results of radical terpolymerization of *cyclo*-(PhMe)$_6$, **1a**, and *p*- or *m*-diethynylbenzene.

Run	DVB	Feed Ratio 1a/DVB	M_w a)	M_w/M_n a)	T_g b) (°C)	Emission c) λ_{max}/nm	Excitation c) λ_{max}/nm
1		6.0/0	3100	1.8	92.9	437	375
2	*p*-	5.0/0.5	3700	1.9	92.9	470	392
3	*p*-	3.0/1.5	11 800	4.2	101.7	473	394
4	*p*-	2.0/2.0	ND d)	ND	104.9	ND d)	ND
5	*m*-	3.0/1.5	7800	2.9	99.2	465	392
6	*m*-	2.0/2.0	8000	3.1	103.7	464	391

a) Determined by GPC analysis (DMF, polystyrene standards).
b) Determined by DSC measurement.
c) Measured in chloroform at 25 °C.
d) ND = not determined.

ratio of the *p*- and *m*-DVB was the same (Table 9.2, run 3 *vs.* run 5 and run 4 *vs.* run 6) the polymers from *p*-DVB showed higher T_g than those from *m*-DVB. This might be due to an increase in the weight average molecular weights of the cross-linked polymers by increased efficiency of the cross-linking agents. All of the resulting polymers were stable in the solid state at room temperature. No decrease in the molecular weight or change in the structure was observed, even after exposing them to air for several months.

In the UV–vis spectra measured in dilute chloroform solutions, the cross-linked polymers exhibited not only strong absorption in the UV region derived from the π–π^* transition but also small absorptions in the visible region as observed in the spectrum of the linear poly(vinylene-arsine). The lower energy absorption was attributable to a delocalized n–π^* transition in the backbone of the polymer. The absorption edges of the obtained polymers were located around 500 nm.

The fluorescent emission spectrum of the polymer in run 3 showed an emission in the blue region with a peak at 473 nm. The location of the emission and excitation peak maxima are summarized in Table 9.2 runs 2–6. The peaks of the cross-linked polymers were red-shifted compared to those of the linear poly(vinylene-arsine) (Table 9.2, run 1). The favorable interchain interaction of the poly(vinylene-arsine) moiety by the cross-linker and the extension of the effective conjugation length through the conjugated cross-linking bond resulted in the narrower band gap. The longer extension of the conjugation path by the *p*-substituted cross-linker than by the *m*-substituted one (Table 9.2, run 3 *vs.* run 5 and run 4 *vs.* run 6) may be derived from the more efficient resonance effect of the conjugated *p*-substitution.

We made polymer films by letting a few drops of the polymer solutions in chloroform, except for that of run 4, dry on slip glasses at room temperature. All the terpolymers formed films. We compared the degree of difficulty of peeling the films off the slip glasses, and found this to increase in the following order: polymer film of run 1 $\approx$ that of run 2 $<$ that of run 3 and that of run 5 $<$ that of run 6. This is the order of increasing concentration of diethynylbenzene in the copolymerization. Thus, incorporating diethynylbenzene in the copolymerization yields polymers which exhibit improved film-forming properties. The polymers with more cross-linking points possess the larger weight-average molecular weights, and, hence, the more improved processabilities.

9.13 Conclusion

The development of new polymerization methods for heteroatom-containing unsaturated polymers is of considerable interest, not only scientifically but also for the preparation of useful materials with unusual properties. The simplest type of heteroatom-containing unsaturated polymer is heteroarom-including poly(acetylene)s. However, no synthetic procedures have been reported for such polymers except poly(vinylene sulfide), which was prepared by polycondensation of sodium sulfide and *trans*-1,2-dichloroethylene. Synthesis of novel arsenic-containing polymers, poly(vinylene-arsine)s, has been achieved by ring-collapsed radical alternating copolymerization (*RCRAC*) of arsenic homocycles with phenylacetylenic derivatives. Though further investigations of the mechanism are needed, such as EPR analysis, the results of the studies here strongly suggest that the copolymerization proceeds via radical propagating species. Comparing the results of RCRAC of *cyclo*-$(PhAs)_6$ and *cyclo*-$(PhSb)_6$, the arsenic radicals were more reactivite toward ethynyl groups and the antimony radicals had more reactivity toward vinyl radicals.

The poly(vinilene-arsine, phosphine, and -stibine)s showed unique optical properties. In UV–vis analyses of these polymers, there was observed not only strong absorption in the UV region derived from the π–π^* transition of the benzene ring, but also small absorption in the visible region, due to the n–π^* transition in the main chain. The poly(vinylene-arsine)s showed fluorescent properties, which were tunable by changing the substituent of the acetylenic compound. The s-character of the lone pair on a phosphorus atom is weaker than that of an arsenic atom. The bathochromic shift observed in the excitation spectrum of **4a** compared with that of **2a** might be attributed to the conformation of phosphorus or arsenic atom in the main chain. On the other hand, polymer **5f** showed no emission, since the lone pairs are more localized on antimony due to the stronger s-character. We also showed that the poly (vinylene-arsine)s can coordinate with transition metal ions and metal nanoparticles. These polymer–metal complexes would be interesting materials since several studies of organoarsine–transition metal complexes have been reported to show attractive properties.

References

1 Sundararaman, A., Victor, M., Varughese, R., and Jäkle, F. (2005) *J. Am. Chem. Soc.*, **127**, 13748–13749.

2 Jin, Z. and Lucht, B.L. (2005) *J. Am. Chem. Soc.*, **127**, 5586–5595.

3 Smith, R.C. and Protasiewicz, J.D. (2004) *J. Am. Chem. Soc.*, **126**, 2268–2269.

4 Tsang, C.W., Yam, M., and Gates, D.P. (2003) *J. Am. Chem. Soc.*, **125**, 1480–1481.

5 Wright, V.A. and Gates, D.P. (2002) *Angew. Chem., Int. Ed.*, **41**, 2389–2392.

6 Matsumi, N. and Chujo, Y. (2008) *Polym. J.*, **40**, 77–89.

7 Ikeda, Y., Ozaki, M., and Arakawa, T. (1983) *J. Chem. Soc., Chem. Commun.*, **24**, 1518–1519.

8 Ikeda, Y., Ozaki, M., Arakawa, T., Takahashi, A., and Kambara, S. (1984) *Polym. Commun.*, **25**, 79–80.

9 Ikeda, Y., Ozaki, M., and Arakawa, T. (1985) *Mol. Cryst. Liq. Cryst.*, **118**, 431–434.

10 Smith, R.C., Chen, X., and Protasiewicz, J.D. (2003) *Inorg. Chem.*, **42**, 5468–5470.

11 Smith, R.C. and Protasiewicz, J.D. (2004) *Eur. J. Inorg. Chem.*, **126**, 998–1006.

12 Wright, V.A., Patrick, B.O., Schneider, C., and Gates, D.P. (2006) *J. Am. Chem. Soc.*, **128**, 8836–8844.

13 Smith, R.C. and Protasiewicz, J.D. (2004) *J. Am. Chem. Soc.*, **126**, 2268–2269.

14 Lucht, B.L. and Onge, N.O.St. (2000) *Chem. Comm.*, 2097–2098.

15 Lucht, B.L. and Onge, N.O.St. (2002) *J. Organomet. Chem.*, **653**, 167–176.

16 Mao, S.S.H. and Tilley, T.D. (1997) *Macromolecules*, **30**, 5566–5569.

17 Morisaki, Y., Aiki, Y., and Chujo, Y. (2003) *Macromolecules*, **36**, 2594–2597.

18 Vanderark, L.A., Clark, T.J., Rivard, E., Manners, I., Slootweg, C.J., and Lammertsma, K. (2006) *Chem. Comm.*, 3332–3333.

19 Krannich, L.K., and Watkins, C.L. (1994) in *Encyclopedia of Inorganic Chemistry*, vol. **1** (ed. R.B. King), John Wiley & Sons Ltd, Chichester, England, p. 200 and references therein.
20 Kraft, M.Y. (1960) *Dokl. Akad. Nauk SSSR*, **131**, 1342–1344.
21 Kallenbach, L.R., Irgolic, K.J., and Zingaro, R.A. (1970) *Eur. Polym. J.*, **6**, 479–485.
22 Carraher, C.E. Jr. and Moon, W.G. (1976) *Eur. Polym. J.*, **12**, 329–331.
23 Karak, N. and Maiti, S. (1996) *J. Polym. Mater.*, **13**, 179–190.
24 Carraher, C.E. Jr. and Hedlund, L.J. (1980) *J. Macromol. Soc. Chem.*, **A14**, 713–728.
25 Carraher, C.E. Jr., Naas, M.D., Giron, D.J., and Cerutis, D.R. (1983) *J. Macromol. Sci.-Chem.*, **A19**, 1101–1120.
26 Carraher, C.E. Jr. and Blaxall, H.S. (1979) *Angew. Makromol. Chem.*, **83**, 37–45.
27 Karak, N. and Maiti, S. (1997) *J. Polym. Mater.*, **14**, 71–78.
28 Adrova, N.A., Koton, M.M., and Prokhorova, L.K. (1964) *Vysokomol. Soedin.*, 9–10.
29 Chujo, Y., Tomita, I., Hashiguchi, Y., Tanigawa, H., Ihara, E., and Saegusa, T. (1991) *Macromolecules*, **24**, 345–348.
30 Matsumi, N., Naka, K., and Chujo, Y. (1998) *J. Am. Chem. Soc.*, **120**, 5112–5113.
31 Heiba, E.I. and Dessau, R.M. (1967) *J. Org. Chem.*, **32**, 3837–3840.
32 Back, T.G. and Krishna, M.V. (1988) *J. Org. Chem.*, **53**, 2533–2536.
33 Ogawa, A., Yokoyama, H., Masawaki, T., Kambe, N., and Sonoda, N. (1991) *J. Org. Chem.*, **56**, 5721–5723.
34 Ogawa, A., Takami, N., Sekiguchi, M., Yokoyama, H., Kuniyasu, H., Ryu, I., and Sonoda, N. (1991) *Chem. Lett.*, **3**, 2241–2242.
35 Tzschach, A. and Baensch, S. (1971) *J. Prakt. Chem.*, **313**, 254–358.
36 Ogawa, A., Ogawa, I., Obayashi, R., Umezu, K., Doi, M., and Hirao, T. (1999) *J. Org. Chem.*, **64**, 86–92.
37 Kawaguchi, S., Nagata, S., Shirai, T., Tsuchii, K., Nomoto, A., and Ogawa, A. (2006) *Tetrahedron Lett.*, **47**, 3919–3922.
38 Sato, A., Yorimitsu, H., and Oshima, K. (2005) *Angew. Chem. Int. Ed.*, **44**, 1694–1696.
39 Dodds, D.L., Haddow, M.F., Orpen, A.G., Pringle, P.G., and Woodward, G. (2006) *Organometallics*, **25**, 5937–5945.
40 Smith, L.R. and Mills, J.L. (1975) *J. Organomet. Chem.*, **84**, 1–15.
41 Krannich, L.K. and Watkins, C.L. (1994) in *Encyclopedia of Inorganic Chemistry*, vol. **1** (ed. R.B. King), John Wiley and Sons Ltd, Chichester, p. 19.
42 Michaelis, A. and Schulte, C. (1881) *Chem. Ber.*, **14**, 912–914.
43 Ehrlich, P. (1907) *Lancet*, **173**, 351–353.
44 Sennyey, G., Mathey, F., Fischer, J., and Mischler, A. (1983) *Organometallics*, **2**, 298–304.
45 Thiollet, G. and Mathey, F. (1979) *Tetrahedron Lett.*, **20**, 3157–3158.
46 Schmidt, U., Boie, I., Osterroht, C., Schoer, R., and Grutzmacher, H.-F. (1968) *Chem. Ber.*, **101**, 1381–1397.
47 DiMaio, A.-J. and Rheingold, A.L. (1990) *Chem. Rev.*, **90**, 169–190.
48 Power, P.P. (1998) *J. Chem. Soc., Dalton Trans.*, 2939–2951.
49 Patai, S. (ed.) (1994) *The Chemistry of Organic Arsenic, Antimony and Bismuth Compounds*, John Wiley & Sons Inc., New York.
50 Elmes, P.S., Middleton, S., and West, B.O. (1970) *Aust. J. Chem.*, **23**, 1559–1570.
51 Reesor, J.W.B. and Wright, G.F. (1957) *J. Org. Chem.*, **22**, 382–385.
52 Kraft, M.Y. and Katyshkina, V.V. (1949) *Dokl. Akad. Nauk. SSSR*, **66**, 207–209.
53 Rheingold, A.L., Lewis, J.E., and Bellama, J.M. (1973) *Inorg. Chem.*, **12**, 2845–2850.
54 Takeda, K. and Shiraishi, K. (1998) *Phys. Rev. B*, **57**, 6989–6997.
55 Umeyama, T., Naka, K., and Chujo, Y. (2004) *Macromolecules*, **37**, 5952–5958.
56 Baudler, M. and Glinka, K. (1987) in *The Chemistry of Inorganic Homo- and Heterocycles*, vol. **2** (eds I. Haiduc and D.B. Sowerby), Academic Press, London, p. 423.
57 Baudler, M. (1982) *Angew. Chem. Int. Ed. Engl.*, **21**, 492–512.
58 Maier, L. (1972) in *Organic Phosphorus Compounds*, vol. **1** (eds G.M. Kosolapoff and L. Maier), Wiley-Interscience, New York, p. 339.
59 Baudler, M. and Hammerström, K. (1965) *Z. Naturforsch.*, **B20**, 810–811.

60 Ateş, M., Breunig, H.J., Ebert, K.H., Gülec, S., Kaller, R., and Dräger, M. (1992) *Organometallics*, **11**, 145–150.
61 Breunig, H.J., Ghesner, I., and Loak, E. (2001) *Organometallics*, **20**, 1360–1364.
62 Breunig, H.J., Ebert, K.H., Gülec, S., and Probst, J. (1995) *Chem. Ber.*, **128**, 599–603.
63 Breunig, H.J. and Soltani-Neshan, A. (1984) *J. Organomet. Chem.*, **262**, C27–C29.
64 Mourad, Y., Mugnier, Y., Breunig, H.J., and Ateş, M. (1991) *J. Organomet. Chem.*, **406**, 323–329.
65 Nunn, M., Sowerby, D.B., and Wesolec, D.M. (1983) *J. Organomet. Chem.*, **251**, C45–C46.
66 Breunig, H.J., Ebert, K.H., Probst, J., Mourad, Y., and Mugnier, Y. (1996) *J. Organomet. Chem.*, **514**, 149–152.
67 Breunig, H.J. and Rösler, R. (2000) *Chem. Soc. Rev.*, **29**, 403–410.
68 Naka, K., Umeyama, T., and Chujo, Y. (2002) *J. Am. Chem. Soc.*, **124**, 6600–6603.
69 Umeyama, T., Naka, K., Nakahashi, A., and Chujo, Y. (2004) *Macromolecules*, **37**, 1271–1275.
70 Umeyama, T., Naka, K., and Chujo, Y. (2004) *J. Polym. Sci., Polym. Chem.*, **42**, 3604–3611.
71 Naka, K., Umeyama, T., Nakahashi, A., and Chujo, Y. (2007) *Macromolecules*, **40**, 4854–4858.
72 Naka, K., Nakahashi, A., and Chujo, Y. (2006) *Macromolecules*, **39**, 8257–8262.
73 Nakahashi, A., Naka, K., and Chujo, Y. (2007) *Macromolecules*, **40**, 1372–1376.
74 Russell, G.A. and Tashtoush, H. (1983) *J. Am. Chem. Soc.*, **105**, 1398–1399.
75 Ito, O. (1983) *J. Am. Chem. Soc.*, **105**, 850–853.
76 Nakahashi, A., Bravo, M., Naka, K., and Chujo, Y. (2009) *J. Phys.: Conf. Ser.*, **184**, 012021.
77 McAuliffe, C.A. and Levason, W. (1979) *Phosphine, Arsine, and Stibine Complexes of the Transition Elements*, Elsevier, Amsterdam, New York.
78 Feltham, R.D., Metzger, H.G., and Silverthorn, W. (1968) *Inorg. Chem.*, **7**, 2003–2006.
79 Feltham, R.D. and Silverthorn, W. (1968) *Inorg. Chem.*, **7**, 1154–1158.
80 Nyholm, R.S. (1950) *J. Chem. Soc.*, 2071–2078.
81 Chen, L., Kim, B.-S., Nishino, M., Gong, J.P., and Osada, Y. (2000) *Macromolecules*, **33**, 1232–1236.
82 Rebourt, E., Pepin-Donat, B., and Dinh, E. (1995) *Polymer*, **36**, 399–412.
83 Sixou, B., Pepin-Donat, B., and Nechtschein, M. (1997) *Polymer*, **38**, 1581–1587.
84 Umeyama, T., Naka, K., Bravo, M., Nakahashi, A., and Chujo, Y. (2004) *Polym. Bull.*, **52**, 191–199.

10
Synthetic Strategies to Conjugated Main-Chain Metallopolymers

Andreas Wild, Andreas Winter, Martin D. Hager, and Ulrich S. Schubert

10.1
Introduction

The successful free radical polymerization of vinyl ferrocene more than 50 years ago was the starting point for the broad field of metallopolymers [1]. Starting from this specific example a broad variety of different metallopolymers were successfully synthesized and investigated. These great efforts in scientific research made metallopolymers one of the fastest developing fields in macromolecular chemistry. As simple as is the definition of metallopolymers, that is, all organic polymers which contain metals, as diverse and sometimes complex are the structures of the synthesized polymers. The great interest in these materials is mainly attributed to their special properties, which represent a combination of the physical, electronic and optical properties of the organic polymer as well as the physical, electronic and optical properties of the incorporated metal complexes. For this reason π-conjugated oligomers and polymers, one of the youngest polymer classes which were discovered around 30 years ago [2], are very interesting building units for combination with metal complexes. These (semiconducting) polymers feature outstanding electrical and optical properties; due to these properties they are also considered as synthetic metals. They have found applications in many fields, like polymer light emitting diodes (PLEDs) [3] and solar cells [4]. Therefore the introduction of the "real" metal complexes into synthetic metals facilitates the tuning of these special electronic and optical properties and introduces completely new properties. It is important to note that, in particular in the field of conjugated polymers, the introduction of metal complexes can also lead to undesired properties which require a "decoupling" of the metal complex and the conjugated oligomer/polymer. The fruitful combination of metal complexes and conjugated polymers may lead to many applications [5–7] like in organic light emitting diodes (OLEDs) [8, 9], in solar cells [10], and in photonic crystals [11], depending on the structure of the respective polymer. The backbone of a conjugated polymer and the respective metal complexes can, in principle, be combined in three different ways (Figure 10.1) [12]. Type I metallopolymers feature metal complexes which are connected by a nonconjugated linker to a conjugated

Conjugated Polymer Synthesis. Edited by Yoshiki Chujo

ISBN: 978-3-527-32267-1

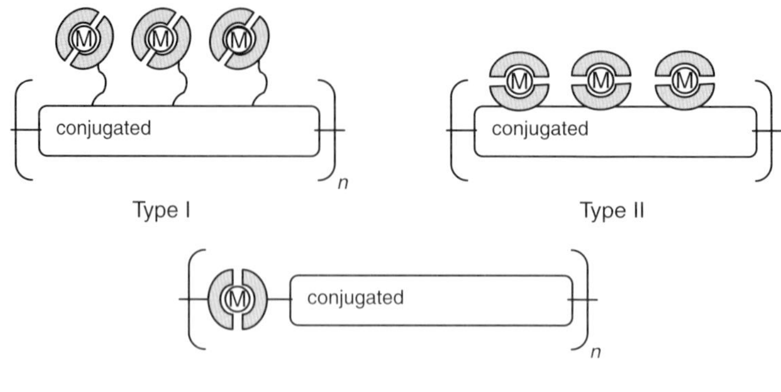

Figure 10.1 Schematic representation of the three different types of conjugated metallopolymers.

backbone. Because of the larger distance between the polymer and the complexes no, or only weak, interactions between both can be observed. Direct coupling of the metal to the conjugated backbone, that is, the polymer can act as a ligand for the metal and a conjugated spacer connects the polymer and ligand, leads to an increased influence of the metal complex on the properties of the conjugated backbone (Type II). Type III conjugated polymers, the title compounds of this chapter; consist of conjugated oligomers/polymers which are connected by metal complexes. Their coordination can be distinguished between two extremes: a covalent and therefore irreversible coordination and a reversible supramolecular coordination.

The metal coordination influences the properties and possible synthetic strategies of the metallopolymers. Generally, three main routes can be used to synthesize conjugated polymers which contain metal complexes in their main chain (Figure 10.2).

Route A involves the step-growth reaction of a conjugated oligomer/polymer which bears two ligands and acts as a bridging ligand connecting the appropriate metal ions. Due to the attractive experimental protocol, that is, the ditopic conjugated

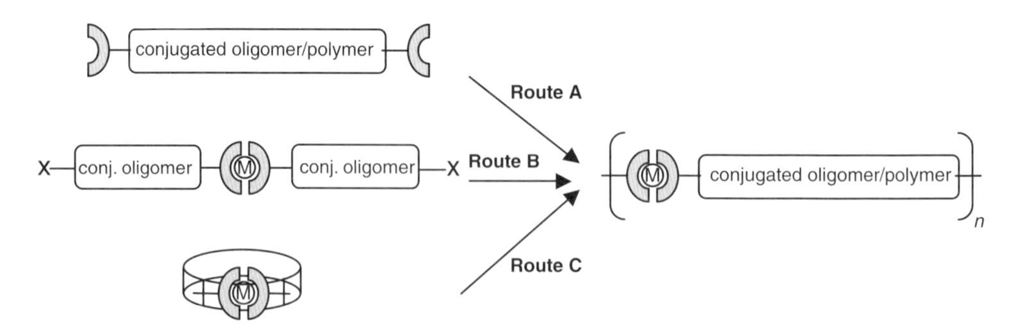

Figure 10.2 Three main synthetic routes to main-chain conjugated metallopolymers. Route A: polymerization by complexation, Route B: polymerization by coupling of the conjugated spacer, Route C: ring-opening polymerization.

ligand and a metal salt are reacted in a suitable solvent, this method is applied frequently. However, the complexation constant must be high enough to form high molar mass polymers. In Route B the conjugated oligomer/polymer which connects the metal complexes is formed during the polymerization. For this purpose, the metal complexes have to be stable under the polymerization conditions. In principle, the well-known polymerization techniques (e.g., electropolymerization, C–C coupling reactions) for conjugated polymers can be applied. However, this technique is not as versatile compared to Route A. Both Routes A and B involve step growth reactions which require exact stoichiometric ratios and high monomer conversions to obtain high molar masses. Route A is mainly influenced by the monomer concentration [M] and the stability constant of the complex *K*. The degree of polymerization is: $DP \sim K [M]^{1/2}$. Therefore high concentrations of the monomers and high stability constants favor the formation of high molar mass metallopolymers (Figure 10.3) [13]. Moreover, these systems are mostly dynamic, that is, reversible assembly and disassembly processes can occur.

The obtainable molar masses of polymers which are synthesized according to Route B are determined by the Carother's equation. Therefore an exact stoichiometry and a high yield of the coupling reaction are necessary to obtain high molar mass polymers.

Route C is based on cyclic monomers. The cycle is opened and the linear polymer is formed. However, the size of the conjugated segment is very limited. Therefore larger conjugated segments are not accessible with this method because of the rigidity of most conjugated oligomers and polymers.

The previously introduced synthetic routes were applied to a large variety of different conjugated metallopolymers. In the following the synthesis of different metallopolymers of Type III with metal complexes in the main chain will be discussed. Due to the broad field of different metallopolymers the main focus will be on conjugated metallopolymers with terpyridines, porphyrins and metallaynes.

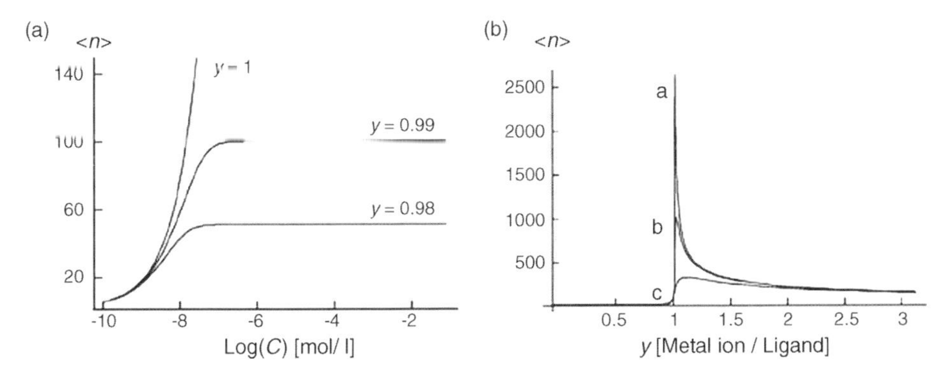

Figure 10.3 Average number of repeat units <*n*> in an assembly (in arbitrary units) as: (a) a function of concentration for different stoichiometries, and (b), a function of stoichiometry, *y*, for different monomer concentrations: a) 10^{-3} mol l^{-1}; b) 10^{-4} mol l^{-1}; c) 10^{-5} mol l^{-1}) [13].

10.2
π-Conjugated Polymers with Terpyridine Units and Other Tridentate Ligands as Part of the Main Chain

A highly versatile ligand which can be used for metallopolymers is the 2,2′:6′,2″-terpyridine (tpy). This tridentate ligand forms stable complexes with many transition metals (e.g., iron, cobalt, nickel, ruthenium, zinc) [14]. The relatively high complex stability and the high binding constants are prerequisites for linear metallopolymers with high molar masses. Due to the supramolecular interaction between the metal ion and the coordinating ligand (terpy) these metallopolymers are called metallo-supramolecular polymers [15]. Most of these polymers feature a reversible, and therefore dynamic, coordination behavior.

The first selected example is deviant from the classical metallopolymers which contain terpyridine complexes, but it demonstrates the great opportunities which metallopolymers offer. The synthesis of conjugated diblock copolymers is not trivial. Two different poly(phenylene-vinylene)s (PPVs) – the acceptor **1** and the donor block **2** – are functionalized at one end with a terpyridine. The acceptor block **1** was reacted with $RuCl_3 \cdot 3\,H_2O$. As known from low molar mass terpyridines a Ru(III) precursor **2** was formed. Subsequently, the monofunctionalized Ru(III)-complex **2** was activated with silver(I) ions and the second block **3** was added; as a result the diblock copolymer **4** was formed [14]. In this case the formed *bis*-terpyridine ruthenium complex plays only a structural role (Scheme 10.1) [16].

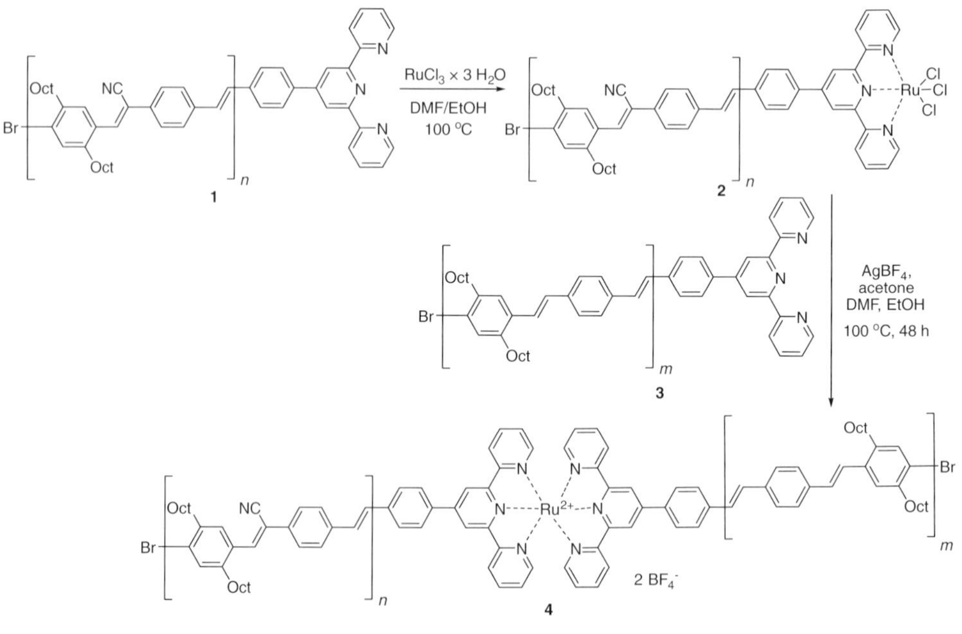

Scheme 10.1 Schematic representation of a PPV diblockcopolymer **4** which is connected by a $Ru(tpy)_2$ complex.

The classical $M(tpy)_2$ metallopolymers contain one metal complex in each repeating unit and the terpyridines are connected by a conjugated oligomer (Scheme 10.2). Other terpyridine-like tridentate ligands, for example, 2,6-*bis*(1′-methylbenzimidazolyl)pyridines, were also utilized for the synthesis of different metallopolymers. The synthetic strategies can be derived from this general structure. When Route A is applied the synthesized ditopic terpyridine ligands are polymerized via complexation. The full spectrum of synthetic methods for the synthesis of conjugated oligomers and polymers has been utilized up to now to synthesize appropriate ligands [17, 18]. Metal *bis*-terpyridine complexes with functional groups were used in Route B for the synthesis of metallopolymers. For this purpose, well-known coupling reactions can be used. The method of choice depends on several factors and is not always obvious.

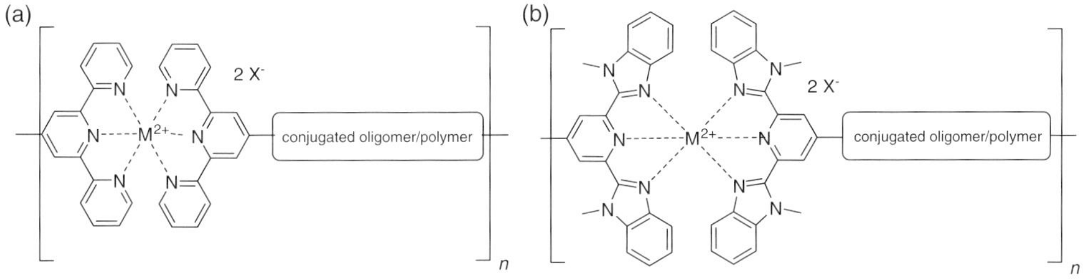

(a) 2 X⁻ conjugated oligomer/polymer n (b) 2 X⁻ conjugated oligomer/polymer n

Scheme 10.2 Schematic representation of metallopolymers containing $M(tpy)_2^{n+}$ binding units (a) and 2,6-*bis*(1′-methylbenzimidazolyl)pyridine (b) ($M = Ru^{2+}$, Fe^{2+}, Ni^{2+}, Zn^{2+}; $X^- = PF_6^-$, halogen, OTf^-).

An early example by Rehahn and coworkers shows the comparison of the two possible synthetic strategies A and B (Scheme 10.3) [19]. For the polymerization by complexation a ditopic ligand is required.

An oligo(phenylene) (OPP) **5**, which was functionalized with two terpyridines, was synthesized by the Suzuki reaction of 4-bromophenyl-terpyridine and a substituted phenylene diboronic acid. This ligand was used for the complexation of ruthenium ions. An experimental protocol which was also applied for the synthesis of low molar mass ruthenium complexes was applied. Ruthenium(III)chloride was activated with $AgBF_4$ in acetone, the ligand was added and the metallopolymer **6** could be obtained by *in situ* reduction to ruthenium (II).

The Strategy B should, in principle, lead also to the same metallopolymer but Rehahn and coworkers could not obtain a polymer when the ruthenium complex of the 4-bromo-phenylterpyridine **7** was reacted with the substituted phenylene *bis*-boronic acid **8**. Only oligomers could be obtained due to many side reactions.

Therefore it is obvious, that most of the described metallopolymers which feature conjugated oligomers like OPPs, OPVs, OPEs were synthesized according to Strategy A, in particular in the case of zinc metallopolymers. The zinc complexes of both ligands are relatively weak, but the polymers feature interesting optical properties. Complexes of stronger binding metals, for example, ruthenium, iron, nickel, feature a characteristic d–π* metal-to-ligand charge-transfer transition which leads to

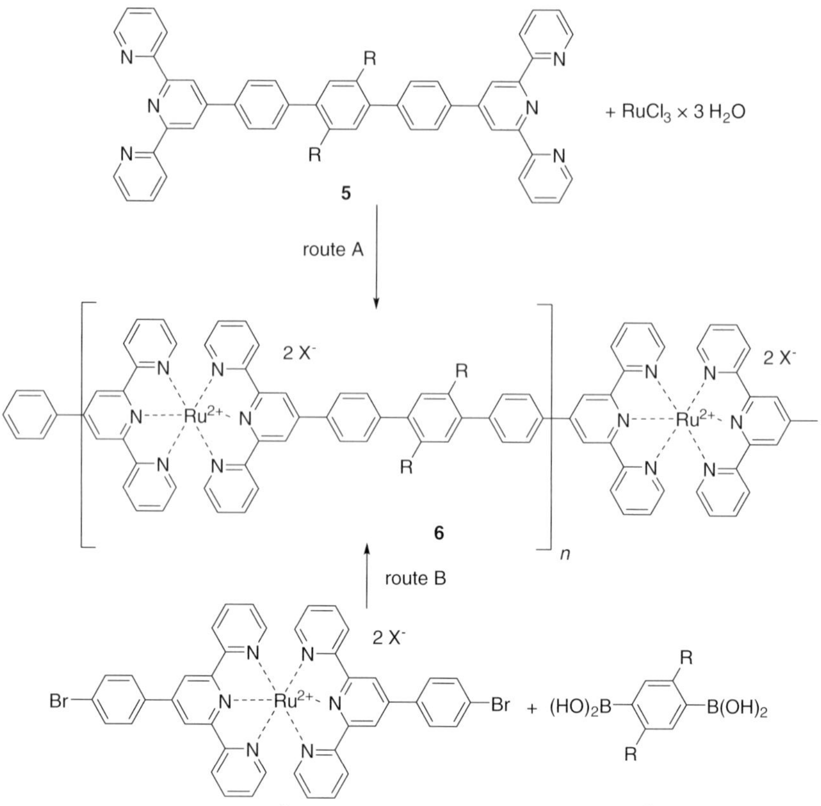

Scheme 10.3 Schematic representation of the two synthetic strategies towards metallopolymer **6** containing an OPP.

the characteristic colors of the complexes [20]. However, the optical properties of the conjugated oligomer/polymer are changed by this transition, for example, the fluorescence quantum yield is decreased.

OPEs with two tpy endgroups based on fluorene moieties were synthesized in a step-wise fashion by the Sonogashira crosscoupling reaction and these ligands were complexed with zinc(II) ions. The resulting metallo-polymers still feature interesting optical properties, like fluorescence, which are hardly influenced by the complexation. This makes them interesting candidates for PLEDs; as a consequence, their electroluminescence was investigated [21, 22]. Due to the small influence of the zinc complexes on the optical properties of the conjugated oligomers the only possibility to tune absorption and emission is the size and the substitution of the conjugated oligomers. Different conjugated oligomers with terpyridines were synthesized and the structure–optical property relationship of the different oligomers was investigated (Scheme 10.4) [9]. Standard synthetic methods, the Horner–Wadsworth–Emmons (HWE) reaction and the Sonogashira crosscoupling reaction, were applied for the synthesis of the OPVs and OPEs.

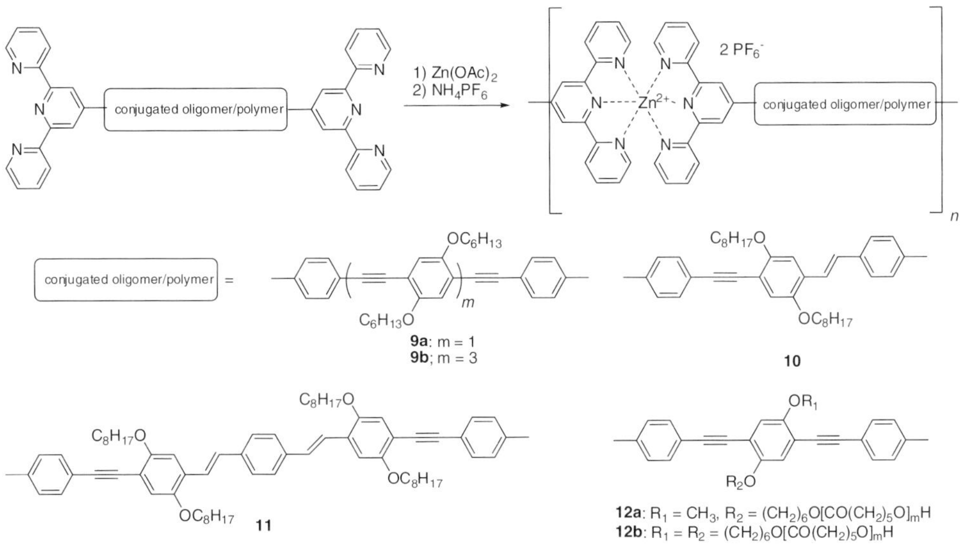

Scheme 10.4 Schematic representation of different $Zn(tpy)_2$ metallopolymers.

A major drawback of many metallopolymers is their rather low solubility. Therefore solubilizing groups were often attached to the conjugated oligomer. Polymeric sidegroups increase strongly the solubility of metallopolymers. Hence a conjugated OPE with two terpyridine endgroups was functionalized with poly(ε-caprolactone) (pCL) which led to an increased solubility (Scheme 10.4) [23]. Two conjugated *bis*terpyridines bearing one and two hydroxyl groups, respectively, were synthesized by the Sonogashira reaction. Subsequently, the free hydroxyl groups were used as initiators for the ring-opening polymerization of ε-caprolactone. It is noteworthy that even these polymeric ligands with molar masses from 5000 to 10 000 g mol^{-1} could be used for the complexation of zinc(II) ions, resulting in metallo-polymers with enhanced solubility and excellent film forming ability.

The discussed examples are all based on monodisperse, well-defined conjugated *bis*terpyridines. Weder and coworkers synthesized conjugated polydisperse macroligands which bear two 2,6-*bis*(1′-methylbenzimidazolyl)pyridines as endgroups [24, 25]. This ligand can be used for the complexation of metal ions in a similar manner as the terpyridine. The standard protocols for PPE polycondensation by the Sonogashira reaction were applied using $Pd(PPh_3)_4$/CuI in toluene/diisopropylamine. To control the size of the oligomers the ratio of the diethynyl **14** and dihalogen compound **15** were adjusted. An excess of the acetylene functionalized ligand **13** was added toward the end of the reaction to obtain a high degree of functionalization (Scheme 10.5). The macroligands **16** could be used for the polymerization by complexation with Zn(II) and Fe(II) ions. This synthetic approach enables the formation of high molar mass conjugated polymers starting from a good soluble conjugated prepolymer. This is advantageous because the good solubility is not always available for classical high molar mass conjugated polymers.

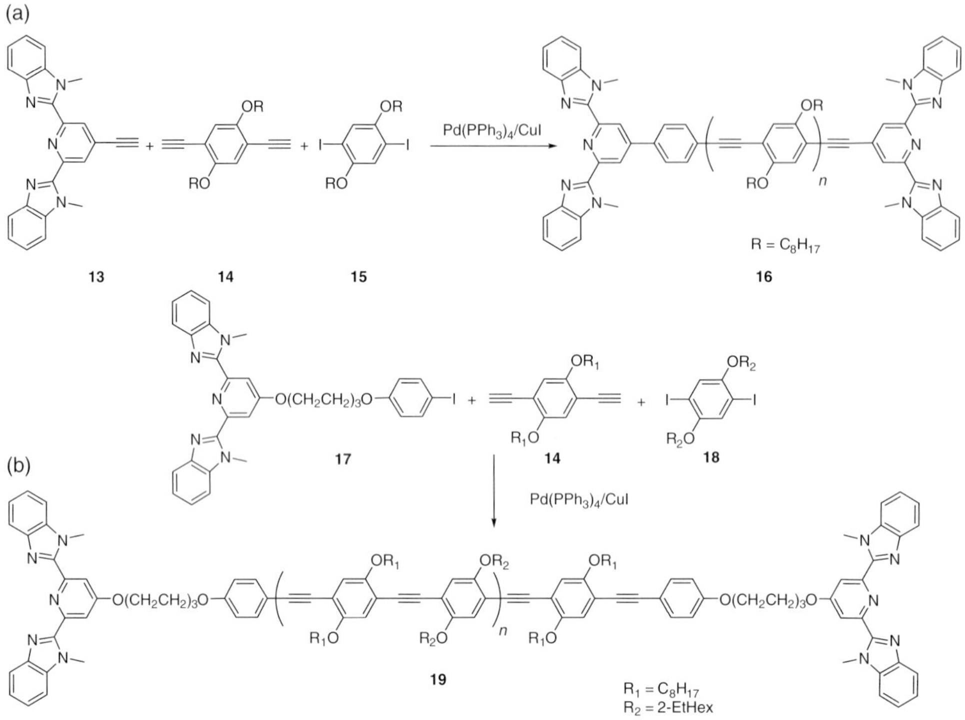

Scheme 10.5 Schematic representation of the synthesis of the conjugated macroligand **16** (a) and the decoupled macroligand **19** (b).

The complexation of iron ions led to drastic changes in the optical properties. An additional absorption – the MLCT band of the iron complex – occurred and the fluorescence was completely quenched in contrast to the zinc analogues. Weder and coworkers also synthesized a ligand in which the conjugated oligomer is connected via an aliphatic hexamethylene spacer (Scheme 10.5) [26].

The synthetic protocol was comparable to the previously discussed example. The ligand **17** was added to the polymerization reaction as an endcapper. Due to the decoupling of the π-conjugated system of the oligomer and the metal complex, the optical properties of the oligomer **19** were changed only slightly upon complexation.

All the above discussed selected examples utilize Route A for the synthesis of different metallo-polymers. However, only a few examples which are based on Route B (coupling of functionalized metal complexes) are described in the literature. In particular, oligothiophene-containing metallopolymers represent here an exception. Ruthenium(II) **20a** and osmium(II) *bis*-terpyridine complexes **20b** were synthesized with thiophene and *bis*thiophene functionalized terpyridines. Electropolymerization, a standard polymerization method for polythiophenes, was used to synthesize the corresponding metallopolymers (**21a**,**b**) (Scheme 10.6) [27–29].

A solution of the ruthenium complex **20a** in dichloromethane was used for the electropolymerization, the polymer **21a** was deposited on the platinum electrode as

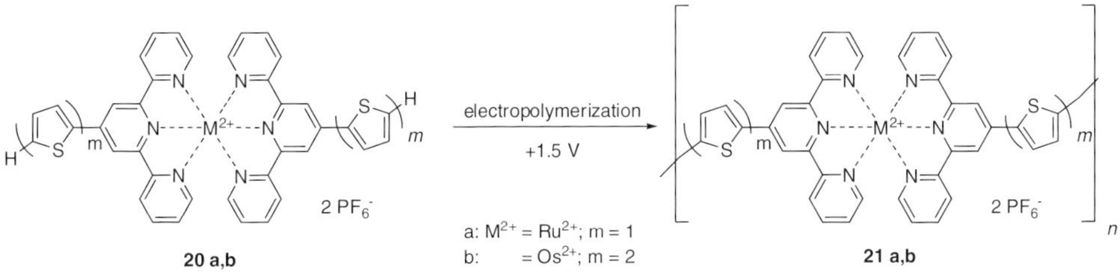

Scheme 10.6 Schematic representation of the electropolymerization of ruthenium and osmium complexes.

a red film. Analogous osmium complexes **20b** could be polymerized in acetonitrile solutions.

Due to the rigid rod structure of conjugated terpyridines, the ring opening polymerization cannot be applied for the synthesis of metallopolymers.

10.3
π-Conjugated Polymers with Porphyrin Units as Part of the Main Chain

In recent years, π-conjugated polymer semiconductors with donor–acceptor (D–A) architectures have attracted considerable attention, since their rich electro-optical properties make them promising candidates for potential applications in the fields of organic light-emitting diodes (OLEDs), organic field-effect transistors (OFETs) and photovoltaic cells [30–67]. In the fabrication of semiconductor photovoltaic devices, conjugated polymers have many advantages compared to inorganic semiconductors, such as low cost, and the suitability for solution processing and large area application [68]. A careful design and selection of the donor (D) and the acceptor (A) moiety enables tuning of the intramolecular charge transfer (ICT) between the D and A subunits towards the desired optical and electronic properties in such conjugated D–A polymer semiconductors [45–63]. For the above mentioned optical applications, for example, the position of the fundamental absorption band (i.e., the optical gap) is an essential parameter governing the emission of OLEDs and the energy threshold in photovoltaics and light detectors [69].

Besides fullerene C_{60}, featuring six delocalized π-electrons, as a three-dimensional acceptor [70–74], porphyrinoid systems have gained much attention. The rich and extensive absorption features thereof guarantee increased absorption cross sections and an efficient use of the solar spectrum [75]. Therefore, porphyrins are the most frequently employed building blocks in organic photonics [76–81] and electronics [82–89]. Porphyrin-containing arrays have been studied intensively and their utilization in artificial photosynthesis [90–94] and molecular devices [95–98] has been elaborated. Additionally, porphyrin derivatives feature a large third-order non-linear optical response, strong two-photon absorption and potential application in photodynamic therapy (PDT) [99–102].

The incorporation of porphyrins into the main backbone of π-conjugated polymers is of the utmost interest, since porphyrins are well-known as the natural photosynthetic reaction center in green plants. Therefore, much effort has been devoted to the design and synthesis of porphyrin-containing macromolecules, including dendrimers [103–109], linear oligomers [110–117] and polymers, such as the conjugated polymers **22–24** (Scheme 10.7) [99, 111, 118–129].

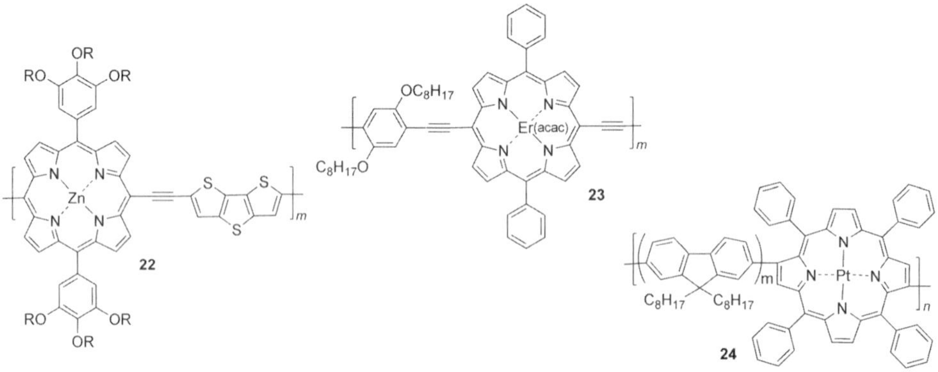

Scheme 10.7 Schematic representation of conjugated metallopolymers with porphyrinoids as part of the main-chain [118, 130, 131].

The synthetic approaches to polymers with porphyrin-units as parts of the conjugated main-chain mirror the chemistry toolbox known for the preparation of π-conjugated polymers, including organic condensation and transition-metal catalyzed coupling reactions [3]. As a common strategy the porphyrin moieties are first synthesized and subsequently utilized as monomers in the polymerization step. The preparation of functionalized porphyrins, suited for synthetic coupling in macromolecular chemistry, has, therefore, been an active area of research in the past [99, 132–140]. By far the most common approach in the preparation of porphyrin monomers is the protocol introduced by Lindsey *et al.* [141]. A selected example for the synthesis of such a functionalized porphyrin is summarized in Scheme 10.8 [118]. According to this, the porphyrin core (**25**) is synthesized in an acid-catalyzed condensation of dipyrromethane with an aldehyde derivative, followed by oxidation with 2,3-dichloro-5,6-dicyano-1,4-benzoquinone (DDQ). After subsequent modification (e.g., bromination with NBS to yield **26**) the functionalized zinc porphyrin (**27**) can be obtained by coordination reaction with zinc acetate. Generally, this reaction sequence can be carried out conveniently on a large scale and yields the desired porphyrin monomers in high yields.

Only a few examples deal with the *homo*polymerization of porphyrin derivatives. In particular, poor solubility, difficult separation and demanding characterization of the products are the crucial issues constraining a more detailed investigation of such materials [83, 114, 117, 134, 142–146]. However, Aratani *et al.* showed that the selective Ag^{I}-catalyzed *meso–meso* coupling of Zn^{II} 5,15-diarylporphyrins **28** represents a versatile tool to obtain monodisperse porphyrin rods (**29**) with a high degree of

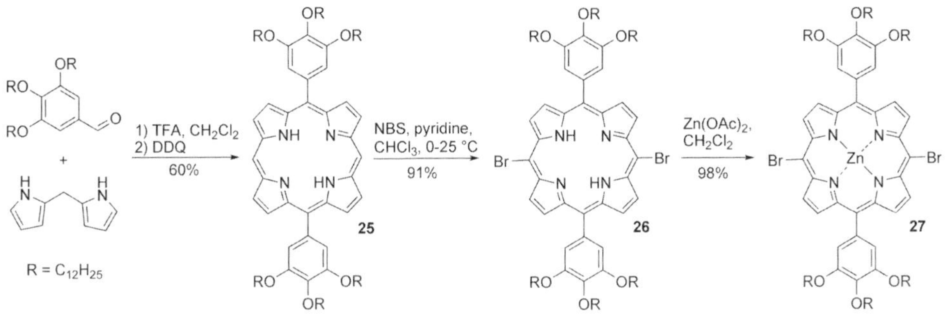

Scheme 10.8 Schematic representation of the synthesis of a functionalized porphyrin monomer [118].

polymerization of up to 130 (Scheme 10.9) [113]. The products could be purified by preparative size-exclusion chromatography (SEC) and the molar masses (up to 133 000 Da) were determined by MALDI-TOF mass spectrometry.

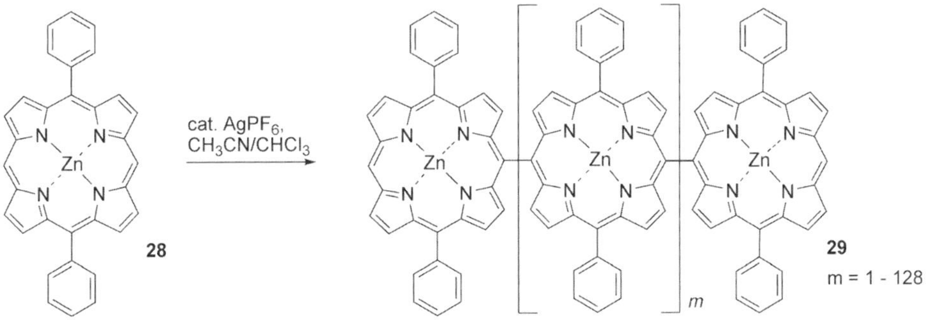

Scheme 10.9 Schematic representation of the synthesis of macromolecular porphyrin-based rods [113].

Due to the ease of film-formation and the potential in catalysis and sensor technology, porphyrins have also been electro-polymerized onto electrode surfaces [147, 148].

In a contribution by Lin *et al.*, a porphyrin-derivative (**30**) with one reactive group (i.e., aniline-ethynyl) was used to generate polymer films on platinum and indium tin oxide (ITO) surfaces (Figure 10.4) [149]. In contrast to the frequently used *tetra*-substituted porphyrins [150], here a higher variability in structure and electro-optical properties is ensured. Electropolymerization of **30** was carried out, monitored and controlled by cyclic voltammetry (CV). As shown by UV–vis absorption spectroscopy, the electropolymerization proceeded in a non-destructive manner. Atomic force microscopy (AFM) proved that the porphyrin films consist of nanoscale dots. The formation of such dots might be of relevance in terms of catalysis and sensor capabilities [151]. The *meso–meso* bridging of porphyrins by acetylene- or butadiyne-spacer moieties offers the opportunity to obtain oligo- (**31**) and polyporphyrin wires

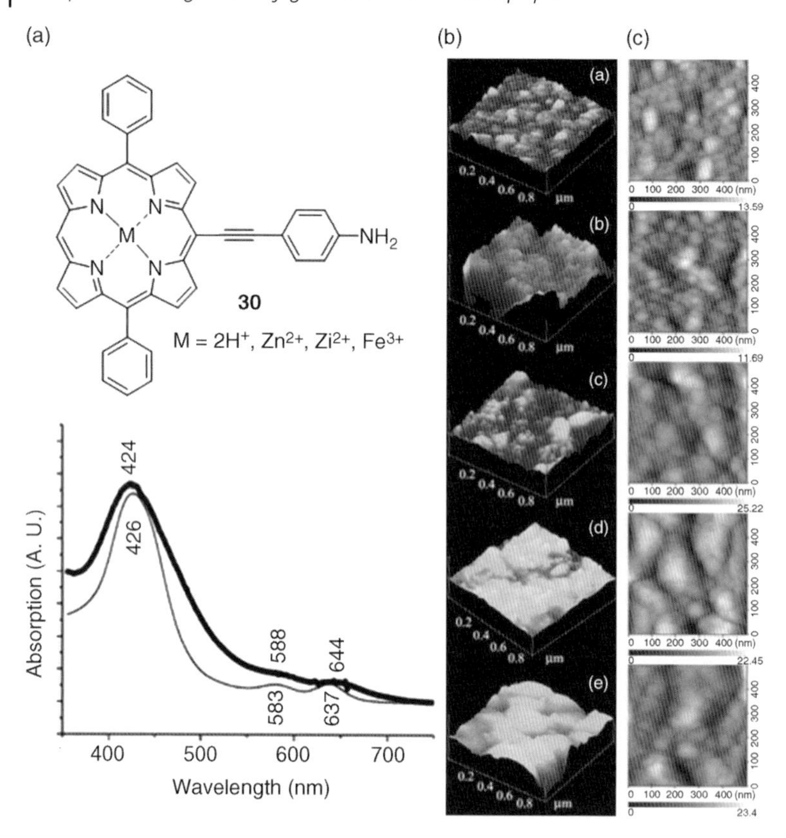

Figure 10.4 A, Schematic representation of the aniline-ethynyl porphyrins (**30**) above UV–vis absorption spectra of Fe-**30** in CH_2Cl_2 (thin line) and after electropolymerization on ITO (bold line). B, 3-D and C, 2-D AFM images of (a) bare ITO, (b) H_2-**30**, (c) Ni-**30**, (d) Zn-**30** and (e) Fe-**30** films (Reproduced with permission from [149]. Copyright 2005 The Royal Society of Chemistry.)

(**33**) featuring efficient charge transfer over long distances [99, 114, 152]. Thus, these types of materials are highly promising candidates for future applications in molecular electronics and solar energy harvesting. However, the time-consuming multi-step preparation of functionalized oligoporphyrins, such as the α-C_{60} ω-ferrocene modified derivative **31** (Figure 10.5) [152], remains as a major drawback.

Therefore, the group of Anderson utilized the straightforward (*homo*)polymerization of a *bis*ethynyl-functionalized porphyrin (**32**) to synthesize a polyporphyrin with butadiyne linkages between the porphyrin centers (Scheme 10.10) [153]. The Glaser-type polymerization conditions were derived from earlier work by the group of Swager [154]. Applying SEC, a molar mass (M_n) of 83 700 Da, corresponding to a number-average degree of polymerization of about 50, and a polydispersity index (PDI) of 1.88 were determined. In solution, the conjugation length of these rod-shaped molecules is limited by the low rotation barrier of one porphyrin with respect to its neighbors [155]. It was shown that the conformation of these polymers can be

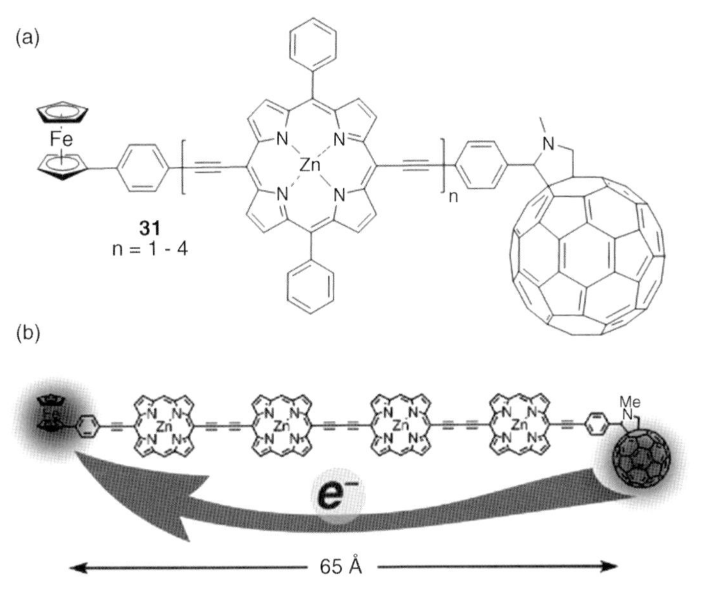

Figure 10.5 (a) Schematic representation of the α-C_{60}-ω-ferrocene-functionalized oligoporphyrins **31**. (b) Schematic representation of the long distance energy transfer in **31** (Reproduced from [152]. Copyright 2007 American Chemical Society.)

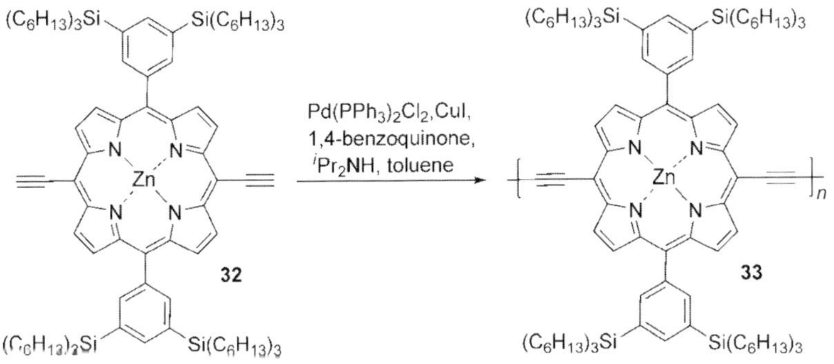

$(C_6H_{13})_3Si$ $Si(C_6H_{13})_3$
Zn
32
$Pd(PPh_3)_2Cl_2$, CuI, 1,4-benzoquinone, $^iPr_2NH$, toluene
33

Scheme 10.10 Schematic representation of the synthesis of the polyporphyrin derivative **33** [153].

altered by the addition of bidentate ligands (e.g., 4,4′-bipyridine, bpy) leading to the formation of double-strand ladder-shaped assemblies **34** (Figure 10.6). In these ladder-type structures the effective conjugation length could be increased remarkably, amplifying the two-photon absorption, nonlinear refraction and charge carrier mobility along the molecular wires by a factor of ten [101, 125, 153, 156, 157].

Since the first observation of electroluminescence from a poly(*p*-phenylenevinylene) (PPV)-containing device almost 20 years ago [158], extensive efforts have been made to further develop the chemical structure and optoelectronic properties of this special class of π-conjugated polymers [30–38]. As already pointed out, in particular the combination of donor- (D) and acceptor-type units (A) within a conjugated

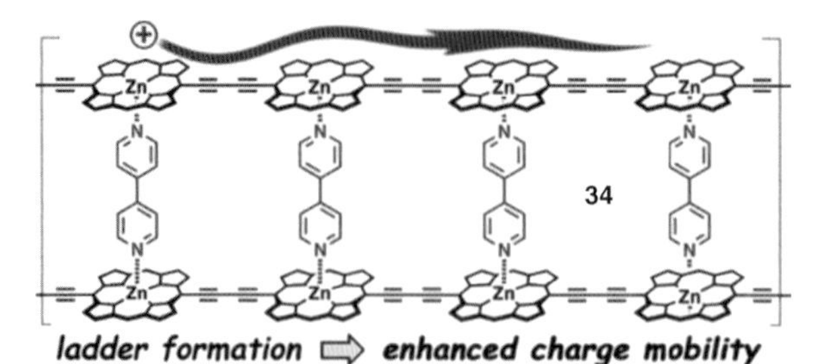

Figure 10.6 Schematic representation of the double-strand ladder-type polyporphyrin **34** (Reproduced with permission from [153]. Copyright 2007 American Chemical Society.)

polymer is of importance for the design of new molecular electronic devices [45–63]. Thus, a variety of systems where porphyrin units were introduced into the main-chain of PPV-type polymers has been reported (Scheme 10.11) [120, 127, 128, 159, 160].

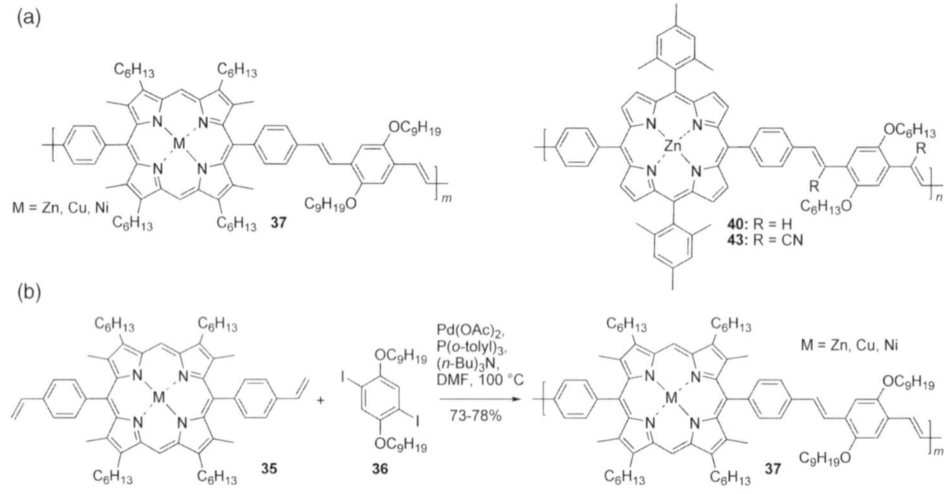

(a)

M = Zn, Cu, Ni; **37**; **40**: R = H; **43**: R = CN

(b)

35 + **36** → Pd(OAc)$_2$, P(o-tolyl)$_3$, (n-Bu)$_3$N, DMF, 100 °C, 73-78% → **37** (M = Zn, Cu, Ni)

Scheme 10.11 (a) Schematic representation of different types of conjugated porphyrin-*p*-phenylenevinylene copolymers [127, 128, 160]. (b) Schematic representation of the synthesis of the metalloporphyrin-containing PPV **37** by Heck reaction [128].

The Pd(0)-catalyzed Heck cross-coupling reaction was used by Bao *et al.* for the copolymerization of a distyrylporphyrin monomer **35** with 1,4-diodo-2,5-dinonoxybenzene **36** and palladium(II) acetate/tri-*o*-tolylphosphite as the catalytic system (Scheme 10.11) [128, 159]. It is noteworthy that under the reaction conditions utilized (i.e., DMF, 100 °C) no exchange of the coordinated metal ions by the Pd^{II} species could be observed. Molar masses (M_n) in the range from 8300 to 46 000 Da, as determined by SEC, were obtained.

Besides this modern Pd(0)-catalysed cross-coupling reaction, well-known condensation reactions – such as the Gilch [127], Wittig [120, 160] and Knoevenagel [160] reactions – have also been used for the synthesis of similar metalloporphyrin-containing PPVs. Two representative examples for the Gilch and Knoevenagel polycondensations (polymers **40** and **43**), respectively, are summarized in Scheme 10.12. In comparison, the aforementioned Heck reaction appears to be somehow superior to the conventional organic condensation reactions, since higher yields and better control over the polydispersity index could be achieved by that route. Furthermore, a porphyrin-containing PPV with additional C_{60} molecules in the side chains (**44**) was prepared by a Wittig co-condensation in moderate yields ($M_n = 5100–7900$ Da; PDI $= 1.78–2.47$) [120]. The investigation of the electro-optical properties revealed that the material is a promising candidate for photoinduced electron-transfer systems.

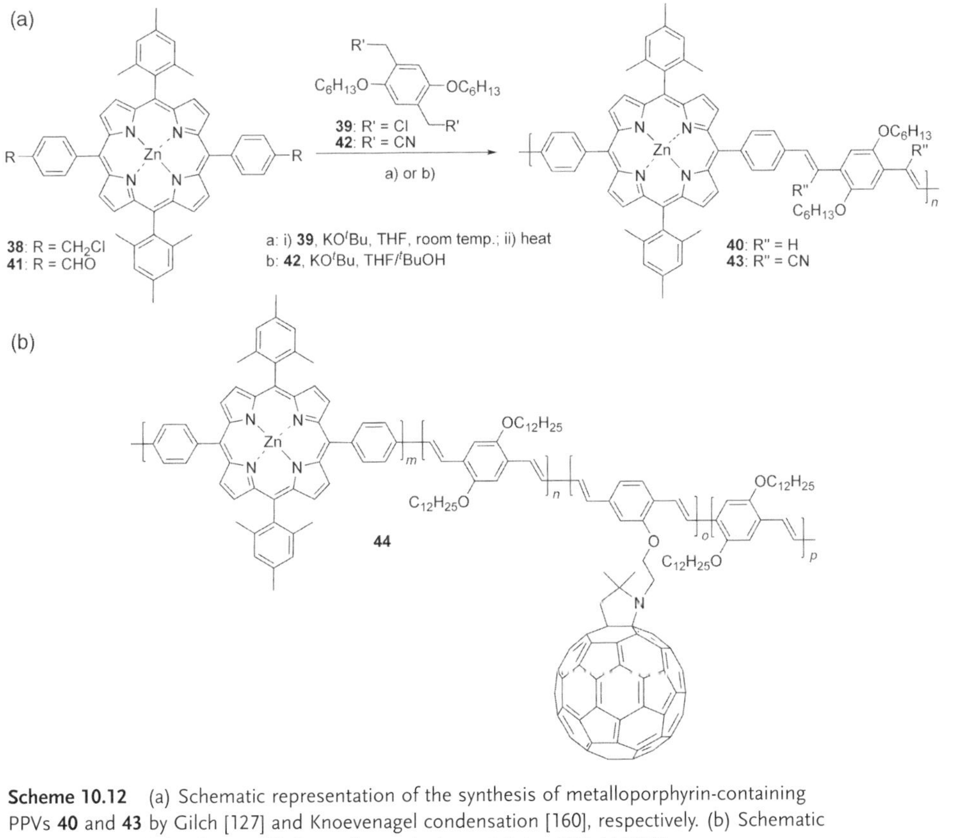

Scheme 10.12 (a) Schematic representation of the synthesis of metalloporphyrin-containing PPVs **40** and **43** by Gilch [127] and Knoevenagel condensation [160], respectively. (b) Schematic representation of the C_{60}-modified metalloporphyrin-containing PPV **44** [120].

Two different types of metalloporphyrin-containing poly(*p*-phenylenethylyne)s (PPEs) have also been prepared by Sonogashira cross-coupling reactions: PPE-type polymers with one central porphyrin unit (**45**) and conjugated PPE-type polymers with alternating porphyrin and dialkoxybenzene units (**46**) (Scheme 10.13).

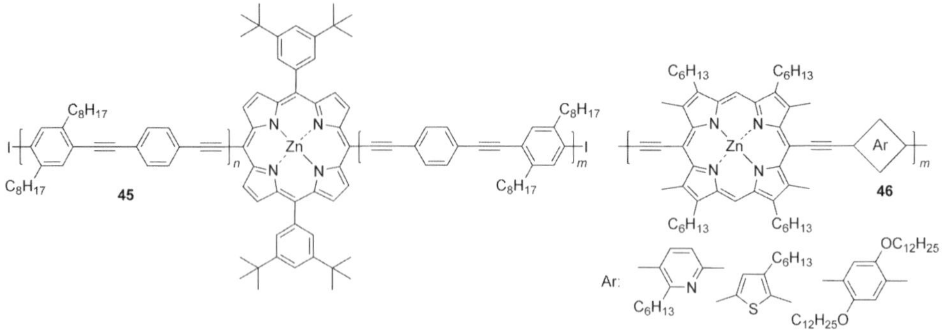

Scheme 10.13 Schematic representation of PPE-type polymers **45** and **46** containing porphyrin units [123, 161].

The synthetic strategy applied by Krebs and coworkers allowed incorporation of only one zinc-porphyrin moiety into the conjugated backbone (**45**). Here, complete suppression of the PPE-based emission in light-emitting devices was observed with exclusive near-infrared emission due to electroluminescence pumping from the conjugated PPE-segments to the central porphyrin core [161].

For the synthesis of the PPE-type polymer with alternating porphyrin and dialkoxybenzene units (**46**) a synthetic approach, similar to the previously discussed Heck coupling, was applied. Utilizing $Pd(PPh_3)_4$ and CuI as the catalytic system, polymers with narrow polydispersity indices (PDI values 1.2–1.4) and molar masses (M_n) from 4800 to 8300 Da were obtained [123].

In addition, the electro-optical properties of this type of conjugated metallopolymers could be further tuned by subsequent exchange of the metal ion coordinated to the porphyrin units [130]. Introducing lanthanide ions into the coordination sphere of the porphyrin rings, new metallopolymers, such as the erbium-porphyrin derivative **23** (see Scheme 10.7), could be obtained. Remarkably red-shifted absorption and near-infrared photoluminescence were reported for these materials.

Due to their photochemical behavior and photoconductive properties, chromophores such as fluorene, carbazole, and perylene units have been widely used as building blocks in the field of π-conjugated polymers [3, 30, 31]. The combination of these structures with electro- and photo-active porphyrin units is, therefore, a growing field of interest. Utilizing the palladium-catalyzed Sonogashira [162] or Suzuki cross-coupling reaction [131, 162, 163], various types of Zn- (**47–49**, see Scheme 10.14) and Pt-containing metallopolymers (**24**, see Scheme 10.7) have been reported recently.

The Sonogashira coupling ($Pd(PPh_3)_4$/CuI as catalyst) towards metallopolymers **47** and **48** was carried out with a *bis*(iodophenyl)-functionalized Zn-porphyrin and diethynyl-substituted fluorene and carbazole, respectively. The polymers were obtained in good yields (75–90%) with molar masses (M_n) in the range 10 500 to

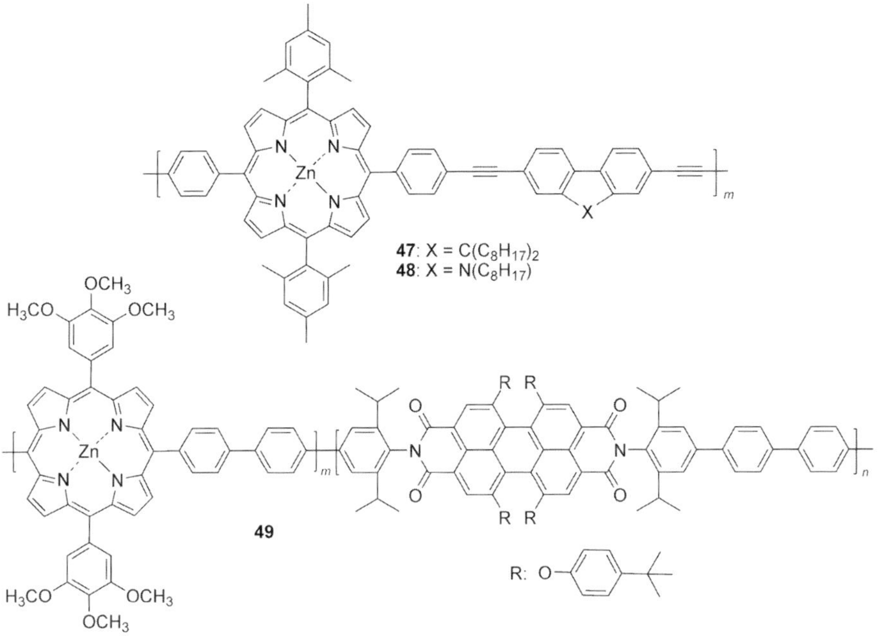

Scheme 10.14 Schematic representation of fluorene- (**47**), carbazole- (**48**) and perylene-containing metalloporphyrin co-polymers (**49**) [162, 163].

133 500 Da (PDI = 1.72–2.74). Thin films of good homogeneity and stability were obtained from **47** and **48** [162].

The metallopolymers containing the Pt-porphyrin unit (**24**, see Scheme 10.7) were synthesized by Suzuki coupling ($PdOAc_2$ and $P(cyclohexyl)_3$ as the catalyst system). At the end of the reaction, all terminal groups were end-capped to avoid quenching effects or the formation of excimers via the bromine or boronic ester functionalities [131]. Polymers with molar masses (M_n) up to 40 000 Da (PDI = 2.1–2.9) were obtained. Furthermore, polymers **24** were also prepared by copolymerization of the metal-free porphyrin, followed by postpolymerization metallation with $PtCl_2$. All polymers showed red photoluminescence, originating from the triplet excited states of the Pt-porphyrin centers. However, the first approach appeared to be preferable, since a much more efficient energy transfer – in solution and in the solid state – was observed for the polymers derived via that route.

The Ni(0)-mediated polycondensation of a dibromoaryl-subsituted porphyrin was utilized by Yamamoto *et al.* to prepare a series of pyridine-, thiophene- and phenyl-bridged Zn^{II}-porphyrins [123]. However, only for the thiophene-derivative could high molar masses (M_n = 37 900 Da) be achieved, and good solubility and processability of the material was reported.

The application of similar thiophene derivatives, promoting ordered π-stacked structures and high hole mobility, in OFETs has been reported in recent years [164–166]. Huang *et al.* introduced dithienothiophene (DDT) and diethynyl-DDT into the polyporphyrin main chain by Stille and Sonogashira coupling reactions, respectively [118]. Reduced steric hindrance, extended conjugation, enhanced absorption and improved charge transport properties were reported for the polymers **22** and **51**. In the case of the Stille product **51** (Scheme 10.15), a significantly higher molar mass ($M_n = 46\,500$ Da) was found, mainly due to the reduced probability of undesired side-reactions (e.g., Glaser coupling) in comparison to the Sonogashira product **22**. Also a higher thermal stability was observed for **51**.

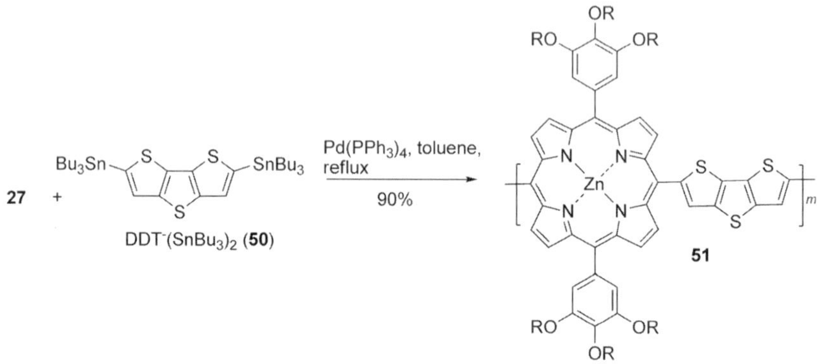

Scheme 10.15 Schematic representation of the synthesis of the DDT-containing metalloporphyrin polymer **51** [118].

10.4 Rigid-Rod Polymetallaynes

Rigid-rod transition metal acetylide polymers represent a class of metallopolymers, which are, due to their unique structures and optoelectronic properties, technologically and scientifically of potential importance [167–170]. In 1975 Hagihara reported the synthesis of oligomeric, [171] and, two years later, polymeric Pt(II) and Pd(II) acetylides [172–175]. Starting from that discovery the chemistry of, in particular Group 10, metal-containing polymetallaynes has become a very active field of research. These materials may combine the ease in processing and special mechanical properties of polymers with the tailor-made electronic and optical properties of functional organic molecules [176, 177]. In recent years, especially Wong *et al.* have invested significant efforts to optimize Pt(II)-acetylide polymers with respect to potential opto-electronic applications [178–183].

The general structure of a d^{10} metal acetylide polymer is depicted in Figure 10.7a, consisting of a metal ion M, the conjugated spacer group R and the auxiliary ligand L (in most cases a phosphine-type ligand, such as PR_3).

By changing one of these structural elements (i.e., spacer, metal or auxiliary ligand), the physical as well as the functional properties of the material can be varied over a wide range, by using for example (R)-1,1′-bi-(2-naphthol) bridges, introducing

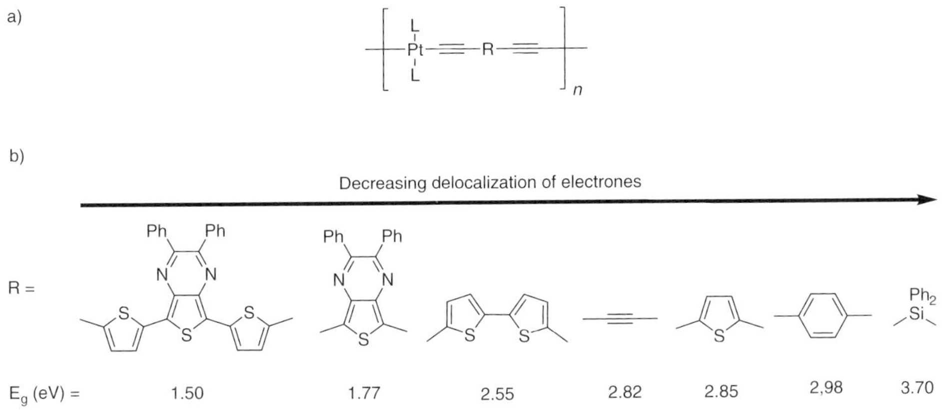

a)

b)

Decreasing delocalization of electrones

E_g (eV) = 1.50 1.77 2.55 2.82 2.85 2,98 3.70

Figure 10.7 (a) Schematic representation of the general structure of a metal acetylide polymer; (b) trend in optical band gaps for platinum(II) polyynes with different spacer groups [170, 180].

chirality [184] or a carboxylic acid-containing spacer system, introducing water solubility [185]. (Figure 10.7b) [186].

As known from conjugated oligomers and polymers, the optical properties are strongly dependent on the number of repeating units. It was shown, for oligomers and polymers of Pt-DEBP (4,4′-*bis*(ethynyl)-biphenyl), that extension of the conjugation length, leads to an optical redshift in absorption [187]. Another very efficient way to tune the bandgap of the resulting polymer is the modulation of the spacer unit. In metal acetylide polymers, π-conjugation is preserved through the metal site, as a result of mixing between the frontier orbitals of the metal and the conjugated ligands. The resultant optical gap in these polymers is determined by a number of factors. The fundamental optical transition is between the mixed 5d/π and 6d/π* orbitals of the metal/ligand conjugated system [188, 189]. The degree of mixing depends on the overlap between ligand and metal orbitals and, thus, will vary from ligand to ligand. Additionally, d^{10} metal ions are electron rich in comparison to the conjugated ligands and should, therefore, act as an electron donor within the polymer [190]. The resulting push–pull system will lower the overall optical gap of the polymer, depending on the acceptor strength of the spacer unit [189, 191]. Therefore, two main approaches were used to achieve low bandgap materials: (i) the use of push–pull polymers with alternating electron-rich and electron-poor units and (ii) the introduction of ethylenic linkages between the aryl rings or ladder type systems, to hinder angular rotation between them [192]. Figure 10.7b shows the optical bandgap of selected metallopolyynes.

There are basically three synthetic routes for the synthesis of d^{10} metal alkynyl polymers. The most common way is still the one introduced by Hagihara in 1975 [171]. Generally, a copper(I)-catalyzed dehydrohalogenation of metal halides in an amine as solvent leads to rigid-rod polymers with molar masses (M_w) larger than 10^6 Da (Scheme 10.16a) [169]. It was shown that the catalyst is needed to ensure efficient intramolecular transfer of the alkynyl ligand to Pt(II) [187]. The amine acts as both, acid acceptor and solvent.

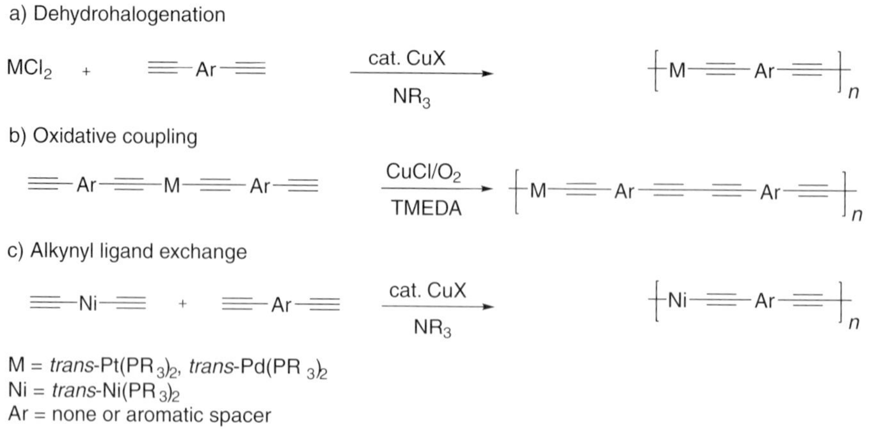

a) Dehydrohalogenation

MCl_2 + ≡–Ar–≡ —(cat. CuX, NR_3)→ [–M–≡–Ar–≡–]$_n$

b) Oxidative coupling

≡–Ar–≡–M–≡–Ar–≡ —($CuCl/O_2$, TMEDA)→ [–M–≡–Ar–≡–≡–Ar–≡–]$_n$

c) Alkynyl ligand exchange

≡–Ni–≡ + ≡–Ar–≡ —(cat. CuX, NR_3)→ [–Ni–≡–Ar–≡–]$_n$

M = *trans*-$Pt(PR_3)_2$, *trans*-$Pd(PR_3)_2$
Ni = *trans*-$Ni(PR_3)_2$
Ar = none or aromatic spacer

Scheme 10.16 (a)–(c) Schematic representation of the general methods for the synthesis of sigma alkynyl polymers.

The second method involves oxidative homocoupling of the *bis*(terminalalkynyl) metal complexes in the presence of a catalytic amount of cuprous halides, with O_2 as the oxidizing agent (Scheme 10.16b) [168, 193, 194]. The basic principle of this reaction is known as Glaser coupling, and, in a modified way, using TMEDA, as Hay's coupling [194]. Since there is only one monomer species required, high degrees of polymerization are obtained by the Hay's coupling. In contrast, for the dehydrohalogenation reactions, the molar masses depend on the exact ratio of the used monomers. Usually, the two above-mentioned methods work well for Pt, are less well for Pd, and not at all for Ni, due to the instability of dihalonickel complexes in amines and dialkynylnickel complexes under oxidizing conditions [193]. Therefore, a third method has to be used to synthesize the analogous nickel-containing compounds (Scheme 10.16c) [195]. An alkynyl-ligand exchange process, using amines as solvents, leads to the desired nickel-containing metallopolymer and can also be used for the synthesis of polymers containing mixed alkynyl ligands and/or different d^{10} metals in the backbone [173, 196]. Detailed overviews of the synthesized Pt(II)-alkynyl polymers and their physical and optical properties are available in the literature [167, 170, 197].

In the early 1990s Lewis and coworkers developed a method to synthesize metallo alkynyl polymers using a metathesis reaction of trimethyltin derivatives with a platinum salt, first described by Lappert [198]. In this strategy *bis*-trimethylstannyl acetylides were reacted with an equimolar amount of metal halide, in the presence of copper iodine as catalyst (Scheme 10.17a). Owing to the absence of amines this strategy can be extended to the Group 8 and 9 metals, namely iron [199], ruthenium [200–202], osmium, rhodium [203, 204], but can also be used for d^{10} metals: platinum [205, 206], palladium [203] and nickel [204].

The so-called "extended one-pot" (EOP) polymerization route was introduced by Lo Sterzo *et al.* and offers the possibility of performing the three-step reaction starting from aromatic diiodo-compounds in one-pot. The aromatic diacetylides and,

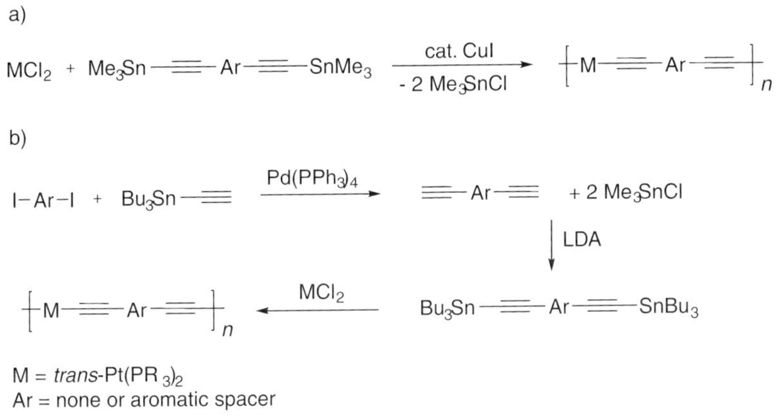

a)

MCl_2 + Me_3Sn—≡—Ar—≡—$SnMe_3$ —(cat. CuI, - 2 Me_3SnCl)→ [M—≡—Ar—≡]$_n$

b)

I–Ar–I + Bu_3Sn—≡ —($Pd(PPh_3)_4$)→ ≡—Ar—≡ + 2 Me_3SnCl

↓ LDA

[M—≡—Ar—≡]$_n$ ←(MCl_2)— Bu_3Sn—≡—Ar—≡—$SnBu_3$

M = *trans*-$Pt(PR_3)_2$
Ar = none or aromatic spacer

Scheme 10.17 Synthesis of polyyne by trimethylstannyl metathesis: (a) Classical route, (b) extended one-pot route.

subsequently, the aromatic ditributyltinacetylides are generated and finally converted into the ethynylaromatic-ethynylmetallo copolymers (Scheme 10.17b) [207–209].

The trimethylstannyl-route has also been used for the synthesis of Co(III) polymers, although 3,9-dimethyl-4,8-diazaundecane-2,10-dione dioxime is needed to stabilize the cobalt-center [210]. Marder *et al.* reported, at the same time, a way to synthesize rhodium-containing polymers through the direct oxidative addition of terminal alkynes to [L_4RhMe], with the loss of methane. [211, 212].

The introduction of an iron center into a metallopolyyne backbone was achieved by palladium-catatalyzed polycondensation between 1,1′-diiodoferrocene and various diethynyl homo- and hetero-aromatic groups [213, 214]. In addition, Sahoo published, in 1999, a procedure to synthesize another metallocene-polyyne, containing Zr atoms in the conjugated backbone (Scheme 10.18) [215]. The conversion of [Zr$(Cp^*)_2Cl_2$] with Li$(C{\equiv}C)_m$Li (**53a**, **b**; m = 1 or 2) yielded two polymers (**54a**, **b**), showing limited solubility in common organic solvents.

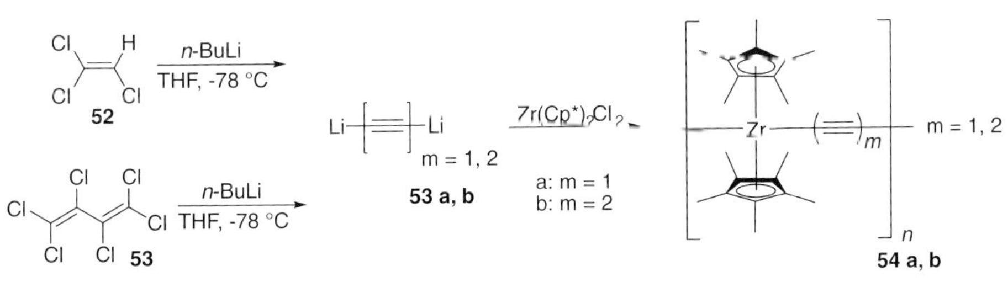

52 —(n-BuLi, THF, -78 °C)→ Li[≡]Li, m = 1, 2 (**53 a, b**)

53 —(n-BuLi, THF, -78 °C)→

—($Zr(Cp^*)_2Cl_2$)→ [Zr—(≡)$_m$]$_n$, m = 1, 2 (**54 a, b**)

a: m = 1
b: m = 2

Scheme 10.18 Schematic representation of the synthesis of a zirconocene-acetylene polymer.

Since the initial reports by the group of Puddephatt [216–221] there is a growing interest in the synthesis of gold(I)-acetylides, oligomers as well as polymers [222]. Gold(I) ions are well suited for forming rigid-rod polymers, tending to possess a coordination number of two with linear stereochemistry, but can also form trigonal-planar, or tetrahedral complexes [222]. A unique feature of gold chemistry is the ability of gold(I) to form gold–gold secondary bonding interactions, called aurophilic interactions, having a bond energy similar to a hydrogen–carbon bond [223]. When reacting the linear di-gold compounds [AuC≡C–Ar–C≡CAu]$_n$ with a range of *para*-substituted diisocyanoarenes, insoluble polymers were obtained (Scheme 10.19a). This low solubility was at least partially explained by crosslinking by Au–Au interactions [219].

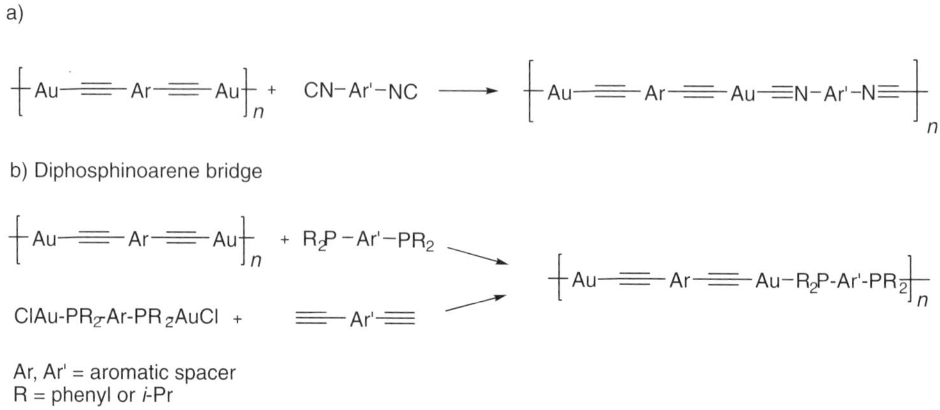

a)

[Au—≡—Ar—≡—Au]$_n$ + CN–Ar'–NC ⟶ [Au—≡—Ar—≡—Au—≡N–Ar'–N≡]$_n$

b) Diphosphinoarene bridge

[Au—≡—Ar—≡—Au]$_n$ + R_2P–Ar'–PR_2 ⟶ [Au—≡—Ar—≡—Au–R_2P-Ar'-PR_2]$_n$

ClAu-PR_2-Ar-PR_2AuCl + ≡—Ar'—≡ ⟶

Ar, Ar' = aromatic spacer
R = phenyl or *i*-Pr

Scheme 10.19 Schematic representation of the synthesis of gold(I)-acetylide polymers by (a) diisocyanoarene and (b) diphosphinarene bridges.

The earlier attempt to synthesize gold(I)-based acetylene polymers using either the reaction of a linear gold(I) diacetylide precursor complex [AuC≡C–Ar–C≡CAu]$_n$ with a diphosphine or the reaction of a dinuclear gold(I) diphosphine complex [ClAu-PR_2–Ar–PR_2AuCl], with a diethynylarene in the presence of a base, resulted in only partly soluble polymers (Scheme 10.19) [217]. Because of the incorporation of P atoms, the polymers cannot be strictly linear. Furthermore, the conjugation will involve p_π–p_π (CC), p_π–d_π (CP), and d_π–d_π (PAu) bonding effects and, therefore, is more complex and probably weaker than with, for example, diisocyanide bridges. The yellow color of the polymers, compared to their colorless monomeric analogs, suggests that there is still some electron delocalization in these diphosphine-bridged polymers [217]. The first anionic polygold(I)acetylide was synthesized by the reaction of PPN[Au(acac)$_2$] (PPN = *bis*(triphenylphosphine)iminium) with stoichiometric amounts of 1,3-diethynylbenzene [224]. Later two anionic heteronuclear Pt(II)-Au(I) σ-acetylide polymers were synthesized [225]. All these polymers suffer from their poor solubility in common organic solvents.

Similar to the gold(I) metallopolyynes, the low solubility is one of the major problems for their isoelectronic mercury(II) analogs. In 2003 Wong *et al.* reported the first examples of soluble, high molar mass and well characterized Hg^{II} polyyne polymers, using a fluorine spacer unit [226]. The synthesis was performed in a one-pot procedure of 9,9-dialkyl-2,7-diethynylfluorenes (**55**) with $HgCl_2$ in a methanolic NaOH solution (Scheme 10.20a).

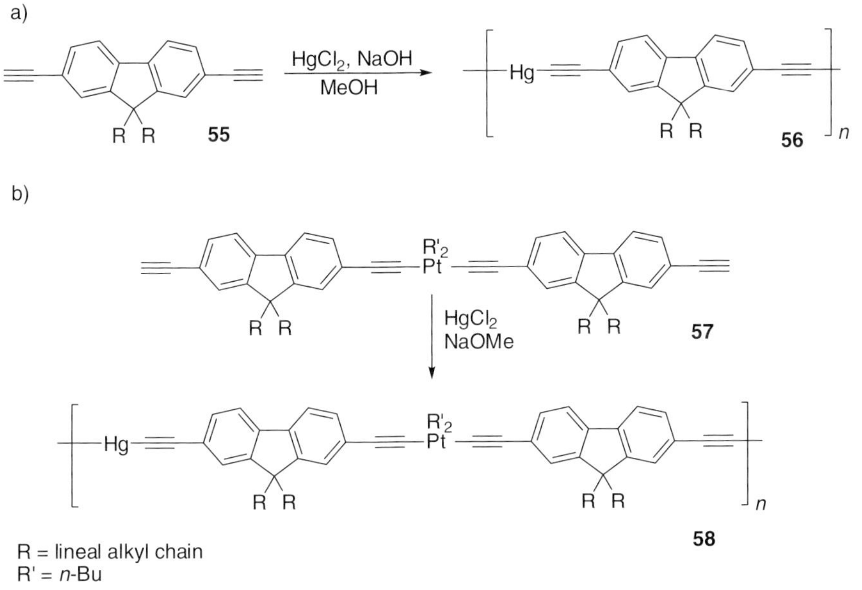

Scheme 10.20 Schematic representation of the synthesis of (a) Hg(II)-acetylide polymers and (b) Hg(II)/Pt(II) heteronuclear copolymers.

Recently, Wong and coworkers published the first rigid-rod heteronuclear Pt(II)/Hg(II) σ-alkynyl polymer (Scheme 10.20b) [182, 183]. The synthesis was carried out, in analogy to the above-mentioned conditions, starting from the terminal diacetylenic Pt complex (**57**) using $HgCl_2$ and NaOMe in a MeOH/CH_2Cl_2 mixture, to obtain the Pt/Hg copolymer (**58**) in good yield as an off-white solid [182, 183].

10.5 Conclusion and Outlook

During the last decades there have been many achievements in the synthesis of metallopolymers, in particular π-conjugated metallopolymers. The expanding synthetic toolbox for metallopolymers is accompanied by a growing interest in the special properties of these polymers. Due to the need for a growing understanding of structure–property relationships of π-conjugated metallopolymers the development of new synthetic pathways will still remain a challenge for the next decades.

This chapter focuses on the already explored synthetic routes for metallopolymers which contain terpyridines (and terpyridine analog ligands), porphyrins and metallaynes within the polymer main-chain. Thus, this is only a small selection from many different polymer classes. Nevertheless, the main synthetic routes for metallopolymers can be illustrated and explained on the basis of these three representative polymer classes.

As pointed out in this chapter three main routes have been developed for the synthesis of these polymers which can be distinguished by whether the metal complex is formed before or after the polymerization step. First, π-conjugated ditopic ligands can be used for polymerization by complexation; the metal complex is formed during polymerization. Secondly, a metal complex with π-conjugated substituents is polymerized by standard polymerization techniques (e.g., Sonogashira cross-coupling, Horner–Wadsworth–Emmons reaction) which are well established for conjugated polymers. Various metallopolymers have been synthesized by these synthetic routes.

However, all synthetic routes have one thing in common, as they are "uncontrolled" polymerizations. The molar mass cannot be controlled and the stoichiometry as well as the concentration is very important for the reaction. Therefore, the reproducibility of polymerizations is often a problem. However, the resulting polymers feature interesting properties and can be used for emerging applications [11].

The challenge for the future is the development of "controlled" polymerization techniques which are suitable for the synthesis of π-conjugated main-chain metallopolymers. Future achievements in the synthesis will surely open the door for a wide range of potential applications of these interesting materials.

References

1 Arimoto, F.S. and Haven, A.C. Jr. (1955) Derivatives of dicyclopentadienyliron. *J. Am. Chem. Soc.*, **77** (23), 6295–6297.

2 Shirakawa, H., Louis, E.J., MacDiarmid, A.G., Chiang, C.K., and Heeger, A.J.J. (1977) Synthesis of electrically conducting organic polymers: halogen derivatives of polyacetylene, $(CH)_x$. *J. Chem. Soc. Chem. Commun.* (16), 578–580.

3 Grimsdale, A.C., Chan, K.L., Martin, R.E., Jokisz, P.G., and Holmes, A.B. (2009) Synthesis of light-emitting conjugated polymers for applications in electroluminescent devices. *Chem. Rev.*, **109** (3), 897–1091.

4 Günes, S., Neugebauer, H., and Sariciftci, N.J. (2007) Conjugated polymer-based organic solar cells. *Chem. Rev.*, **107** (4), 1324–1338.

5 Williams, K.A., Boydston, A.J., and Bielawski, C.W. (2007) Main-chain organometallic polymers: synthetic strategies, applications, and perspectives. *Chem. Soc. Rev.*, **36** (31), 729–744.

6 Whittell, G.R. and Manners, I. (2007) Metallopolymers: new multifunctional materials. *Adv. Mater.*, **19** (21), 3429–3468.

7 Marin, V., Holder, E., Hoogenboom, R., and Schubert, U.S. (2007) Functional ruthenium(II)- and iridium(III)-containing polymers for potential electro-optical applications. *Chem. Soc. Rev.*, **36** (4), 618–635.

8 Wong, W.-Y., Zhou, G.-J., He, Z., Cheung, K.-Y., Ng, A.M.-C., Djurisic, A.B., and Chan, W.-K. (2008) Organometallic polymer light-emitting diodes derived from a platinum(II) polyyne containing the bithiazole ring. *Macromol. Chem. Phys.*, **209** (13), 1319–1332.

9 Winter, A., Friebe, C., Chiper, M., Hager, M.D., and Schubert, U.S. (2009) Self-assembly of π-conjugated bis (terpyridine) ligands with zinc(II) ions: new metallosupramolecular materials for optoelectronic applications. *J. Polym. Sci. Polym. Chem.*, **47** (16), 4083–4098.

10 Wang, X.-Z., Wong, W.-Y., Cheung, K.-Y., Fung, M.-K., Djurisic, A.B., and Chan, W.-K. (2008) Polymer solar cells based on very narrow-bandgap polyplatinynes with photocurrents extended into the near-infrared region. *Dalton Trans.*, (40), 5484–5494.

11 Eloi, J.-C., Chabanne, L., Whitell, G.R., and Manners, I. (2008) Metallopolymers with emerging applications. *Mater. Today*, **11** (4), 28–36.

12 Wolf, M.O. (2001) Transition-metal-polythiophene hybrid materials. *Adv. Mater.*, **13** (8), 545–553.

13 Vriese, V.A. and Kurth, D.G. (2008) Soluble dynamic coordination polymers as a paradigm for materials science. *Coordin. Chem. Rev.*, **252**, 199–211.

14 Schubert, U.S., Hofmeier, H., and Newkome, G.R. (eds) (2006) *Modern Terpyridine Chemistry*, Wiley-VCH Verlag GmbH, Weinheim, Germany.

15 Schubert, U.S., Newkome, G.R., and Manners, I. (eds) (2006) *Metal-Containing and Metallo-Supramolecular Polymers and Materials*, American Chemical Society, Washington, D.C.

16 Duprez, V., Biancardo, M., Spanggaard, H., and Krebs, F.C. (2005) Synthesis of conjugated polymers containing terpyridine-ruthenium complexes: photovoltaic applications. *Macromolecules*, **38** (25), 10436–10448.

17 Han, F.-S., Higuchi, M., and Kurth, D.G. (2008) Synthesis of π-conjugated, pyridine ring functionalized bis-terpyridines with efficient green, blue, and purple emission. *Tetrahedron*, **64** (38), 9108–9116.

18 Winter, A., Egbe, D.A.M., and Schubert, U.S. (2007) Rigid π-conjugated mono, bis- and tris(2,2′:6′,2″-terpyridine)s. *Org. Lett.*, **9** (12), 2345–2348.

19 Kelch, S. and Rehahn, M. (1999) Synthesis and properties in solution of rodlike, 2′,2′:6′,2″-terpyridine-based ruthenium(II) coordination polymers. *Macromolecules*, **32** (18), 5818–5828.

20 Sauvage, J.P., Collin, J.P., Chambron, J.C., Guillerez, S., Coudret, C., Balzani, V., Barigelletti, F., De Cola, L., and Flamigni, L. (1994) Ruthenium(II) and osmium(II) bis(terpyridine) complexes in covalently-linked multicomponent systems: synthesis, electrochemical behavior, absorption spectra, and photochemical and photophysical properties. *Chem. Rev.*, **94** (4), 993–1019.

21 Chen, Y.-Y., Tao, Y.-T., and Lin, H.-C. (2006) Novel self-assembled metallo-homopolymers and metallo-*alt*-copolymer containing terpyridyl zinc(II) moieties. *Macromolecules*, **39** (25), 8559–8566.

22 Chen, Y.-Y. and Lin, H.-C. (2007) Synthesis and characterization of light-emitting main-chain metallo-polymers containing bis-terpyridyl ligands with various lateral substituents. *J. Polym. Sci. Polym. Chem.*, **45** (15), 3243–3255.

23 Winter, A., Friebe, C., Hager, M.D., and Schubert, U.S. (2008) Advancing the solid state properties of metallo-supramolecular materials: poly (ε-caprolactone) modified π-conjugated bis(terpyridine)s and their Zn(II) based metallo-polymers. *Macromol. Rapid Commun.*, **29** (21), 1679–1686.

24 Knapton, D., Rowan, S.J., and Weder, C. (2006) Synthesis and properties of metallo-supramolecular poly(*p*-phenylene ethynylene)s. *Macromolecules*, **39** (2), 651–657.

25 Burnworth, M., Knapton, D., Rowan, S.J., and Weder, C. (2007) Metallo-supramolecular polymerization: a route to easy-to-process organic/inorganic hybrid materials. *J. Inorg. Organomet. Polym. Mater.*, **17** (1), 91–103.

26 Burnworth, M., Mendez, J.D., Schroeter, M., Rowan, S.J., and Weder, C. (2008) Decoupling optical properties in metallo-supramolecular poly(*p*-phenylene ethynylene)s. *Macromolecules*, **41** (6), 2157–2163.

27 Hjelm, J., Constable, E.C., Figgemeier, E., Hagfeldt, A., Handel, R., Housecroft, C.E., Mukhtar, E., and Schofield, E. (2002) A rod-like polymer containing

{Ru(terpy)$_2$} units prepared by electrochemical coupling of pendant thienyl moieties. *Chem. Commun.* (3), 284–285.

28 Hjelm, J., Handel, R.W., Hagfeldt, A., Constable, E.C., Housecroft, E.C., and Forster, R.J. (2003) Conducting polymers containing in-chain metal centers: homogeneous charge transport through a quaterthienyl-bridged {Os(tpy)$_2$} polymer. *J. Phys. Chem. B*, **107** (38), 10431–10439.

29 Hjelm, J., Handel, R.W., Hagfeldt, A., Constable, E.C., Housecroft, E.C., and Forster, R.J. (2004) Electropolymerisation dynamics of a highly conducting metallopolymer: poly-[Os(4′-(5-(2,2′-bithienyl))-2,2′:6′,2″-terpyridine)$_2$]$^{2+}$. *Electrochem. Commun.*, **6** (2), 193–200.

30 Müllen, M. and Scherf, U. (eds) (2006) *Organic Light Emitting Devices*, Wiley-VCH Verlag GmbH, Weinheim.

31 Yersin, H. (eds) (2008) *Highly Efficient OLEDs with Phosphorescent Materials*, Wiley-VCH Verlag GmbH, Weinheim.

32 Jenekhe, S.A. (2007) The special issue on organic electronics and optoelectronics. *Chem. Rev.*, **107** (4), 923–1386.

33 Jenekhe, S.A. (2004) The special issue on organic electronics. *Chem. Mater.*, **16** (23), 4381–4846.

34 Jenekhe, S.A., Lu, L., and Alam, M.M. (2001) New conjugated polymers with donor-acceptor architectures: synthesis and photophysics of carbazole-quinoline and phenothiazine-quinoline copolymers and oligomers exhibiting large intramolecular charge transfer. *Macromolecules*, **34** (21), 7315–7324.

35 Van Mullekom, H.A.M., Vekemans, J.A.J.M., Havinga, E.E., and Meijer, E.W. (2001) Developments in the chemistry and band gap engineering of donor-acceptor substituted conjugated polymers. *Mater. Sci. Eng.*, **R32** (1), 1–40.

36 Bunz, U.H.F. (2000) Poly (aryleneethynylene)s: Syntheses, properties, structures, and applications. *Chem. Rev.*, **100** (4), 1605–1644.

37 Lee, B.-L. and Yamamoto, T. (1999) Syntheses of new alternating CT-type copolymers of thiophene and pyrido[3,4-b]pyrazine units: their optical and electrochemical properties in comparison with similar CT copolymers of thiophene with pyridine and quinoxaline. *Macromolecules*, **32** (5), 1375–1382.

38 Yamamoto, T., Zhou, Z.H., Kanbara, T., Shimura, M., Kizu, K., Maruyama, T., Nakamura, Y., Fukuda, T., Lee, B.-L., Ooba, N., Tomaru, S., Kurihara, T., Kaino, T., Kubota, K., and Sasaki, S. (1996) π-Conjugated donor-acceptor copolymers constituted of π-excessive and π-deficient arylene units. Optical and electrochemical properties in relation to CT structure of the polymer. *J. Am. Chem. Soc.*, **118** (43), 10389–10399.

39 Wu, P.-T., Bull, T., Kim, F.S., Luscombe, C.K., and Jenekhe, S.A. (2009) Organometallic donor-acceptor conjugated polymer semiconductors: tunable optical, electrochemical, charge transport, and photovoltaic properties. *Macromolecules*, **42** (3), 671–681.

40 Wild, A., Egbe, D.A.M., Birckner, E., Cimrova, V., Baumann, R., Grummt, U.-W., and Schubert, U.S. (2009) Anthracene- and thiophene-containing MEH-PPE-PPVs: synthesis and study of the effect of the aromatic ring position on the photophysical and electrochemical properties. *J. Polym. Sci. Polym. Chem.*, **47** (9), 2243–2261.

41 Zhu, Y., Yen, C.-T., Jenekhe, S.A., and Chen, W.-C. (2004) Poly (pyrazinoquinoxaline)s: new n-type conjugated polymers that exhibit highly reversible reduction and high electron affinity. *Macromol. Rapid Comm.*, **25** (21), 1829–1834.

42 Hoppe, H. and Sariciftci, N.S. (2008) Polymer solar cells. *Adv. Polym. Sci.*, **214** (Photoresponsive Polymers II), 1–86.

43 Thompson, B.C. and Frèchet, J.M.J. (2008) Polymer-fullerene composite solar cells. *Angew. Chem. Int. Ed.*, **47** (1), 58–77.

44 Scheblykin, I.G., Yartsev, A., Pullerits, T., Gulbinas, V., and Sundström, V.J. (2007) Excited State and Charge Photogeneration Dynamics in Conjugated Polymers. *J. Phys. Chem. B*, **111** (23), 6303–6321.

45 Blouin, N., Michaud, A., Gendron, D., Wakim, S., Blair, E., Neagu-Plesu, R.,

Belletête, M., Dorucher, G., Tao, Y., and Leclerc, M. (2008) Toward a rational design of poly(2,7-carbazole) derivatives for solar cells. *J. Am. Chem. Soc.*, **130** (2), 732–742.

46 Admassie, S., Inganäs, O., Mammo, W., Perzon, E., and Andersson, M.R. (2006) Electrochemical and optical studies of the band gaps of alternating polyfluorene copolymers. *Synthetic Met.*, **156** (7-8), 614–623.

47 Sonmez, G., Shen, C.K.F., Rubin, Y., and Wudl, F. (2005) The unusual effect of bandgap lowering by C60 on a conjugated polymer. *Adv. Mater.*, **17** (7), 897–900.

48 Campos, L.M., Tontcheva, A., Günes, S., Sonmez, G., Neugebauer, H., Sariciftci, N.S., and Wudl, F. (2005) Extended photocurrent spectrum of a low band gap polymer in a bulk heterojunction solar cell. *Chem. Mater.*, **17** (16), 4031–4033.

49 Alam, M.M. and Jenekhe, S.A. (2004) n-Type conjugated dendrimers: convergent synthesis, photophysics, electroluminescence, and use as electron-transport materials for light-emitting diodes. *Chem. Mater.*, **16** (23), 4657–4666.

50 Svensson, M., Zhang, F., Veenstra, S.C., Verhees, W.J.H., Hummelen, J.C., Kroon, J.M., Inganäs, O., and Andersson, M.R. (2003) High-performance polymer solar cells of an alternating polyfluorene copolymer and a fullerene derivative. *Adv. Mater.*, **15** (12), 988–991.

51 Alam, M.M. and Jenekhe, S.A. (2001) Nanolayered heterojunctions of donor and acceptor conjugated polymers of interest in light emitting and photovoltaic devices: photoinduced electron transfer at polythiophene/polyquinoline interfaces. *J. Phys. Chem. B*, **105** (13), 2479–2482.

52 Jenekhe, S.A. and Yi, S. (2000) Efficient photovoltaic cells from semiconducting polymer heterojunctions. *Appl. Phys. Lett.*, **77** (17), 2635–2637.

53 Yu, G., Gao, J., Hummelen, J.C., Wudl, F., and Heeger, A.J. (1995) Polymer photovoltaic cells: enhanced efficiencies via a network of internal donor-acceptor heterojunctions. *Science*, **270** (5243), 1789–1791.

54 Hancock, J.M., Gifford, A.P., Zhu, Y., Lou, Y., and Jenekhe, S.A. (2006) n-Type conjugated oligoquinoline and oligoquinoxaline with triphenylamine endgroups: efficient ambipolar light emitters for device applications. *Chem. Mater.*, **18** (20), 4924–4932.

55 Wu, W.-C., Liu, C.-L., and Chen, W.-C. (2006) Synthesis and characterization of new fluorene-acceptor alternating and random copolymers for light-emitting applications. *Polymer*, **47** (2), 527–538.

56 Thompson, B.C., Madrigal, L.G., Pinto, M.R., Kang, T.-S., Schanze, K.S., and Reynolds, J.R. (2005) Donor-acceptor copolymers for red- and near-infrared-emitting polymer light-emitting diodes. *J. Polym. Sci. Polym. Chem.*, **43** (7), 1417–1431.

57 Kulkarni, A.P., Zhu, Y., and Jenekhe, S.A. (2005) Quinoxaline-containing polyfluorenes: synthesis, photophysics, and stable blue electroluminescence. *Macromolecules*, **38** (5), 1553–1563.

58 Ego, C., Marsitzky, D., Becker, S., Zhang, J., Grimsdale, A.C., Müllen, K., MacKenzie, J.D., Silva, C., and Friend, R.H. (2003) Attaching perylene dyes to polyfluorene: three simple, efficient methods for facile color tuning of light-emitting polymers. *J. Am. Chem. Soc.*, **125** (2), 437–443.

59 Pang, H., Skabara, P., Crouch, J.D., Duffy, W., Heeney, M., McCulloch, I., Coles, S.J., Horton, P.N., and Hursthouse, M.B. (2007) Structural and electronic effects of 1,3,4-thiadiazole units incorporated into polythiophene chains. *Macromolecules*, **40** (18), 6585–6593.

60 Zhu, Y., Champion, R.D., and Jenekhe, S.A. (2006) Conjugated donor-acceptor copolymer semiconductors with large intramolecular charge transfer: synthesis, optical properties, electrochemistry, and field effect carrier mobility of thienopyrazine-based copolymers. *Macromolecules*, **39** (25), 8712–8719.

61 Chen, M., Crispin, X., Perzon, E., Andersson, M.R., Pullerits, T., Andersson, M., Inganäs, O., and Berggren, M. (2005) High carrier

mobility in low band gap polymer-based field-effect transistors. *Appl. Phys. Lett.*, **87** (25), 252105/1–252105/3.

62 Yamamoto, T., Yasuda, T., Sakai, Y., and Aramaki, S. (2005) Ambipolar field-effect transistor (FET) and redox characteristics of a π-conjugated thiophene/1,3,4-thiadiazole CT-type copolymer. *Macromol. Rapid Commun.*, **26** (15), 1214–1217.

63 Yamamoto, T., Kokubo, H., Kobashi, M., and Sakai, Y. (2004) Alignment and field-effect transistor behavior of an alternative π-conjugated copolymer of thiophene and 4-alkylthiazole. *Chem. Mater.*, **16** (23), 4616–4618.

64 Wong, W.-Y. (2008) Metallopolyyne polymers as new functional materials for photovoltaic and solar cell applications. *Macromol. Chem. Phys.*, **209** (1), 14–24.

65 Wong, W.-Y., Wang, X.-Z., He, Z., Djurišić, A.B., Yip, C.T., Cheung, K.-Y., Wang, H., Mak, C.S.K., and Chan, W.-K. (2007) Response to "on the efficiency of polymer solar cells". *Nature Mater.*, **6** (10), 704–705.

66 Wong, W.-Y., Wang, X.-Z., He, Z., Djurišić, A.B., Cheung, K.Y., Yip, C.T., Ng, A.M.-C., Xi, Y.Y., Mak, C.S.K., and Chan, W.-K. (2007) Tuning the absorption, charge transport properties, and solar cell efficiency with the number of thienyl rings in platinum-containing poly(aryleneethynylene)s. *J. Am. Chem. Soc.*, **129** (46), 14372–14380.

67 Wong, W.-Y. (2005) Recent advances in luminescent transition metal polyyne polymers. *J. Inorg. Organomet. Polym. Mater.*, **15** (2), 197–219.

68 Tekin, E., Egbe, D.A.M., Kranenburg, J.M., Ulbricht, C., Rathgeber, S., Birckner, E., Rehmann, N., Meerholz, K., and Schubert, U.S. (2008) Effect of side chain length variation on the optical properties of PPE-PPV hybrid polymers. *Chem. Mater.*, **20** (8), 2727–2735.

69 Pedersen, T.G., Lynge, T.B., Kristensen, P.K., and Johansen, P.M. (2005) Theoretical study of conjugated porphyrin polymers. *Thin Solid Films*, **477** (1-2), 182–186.

70 Mizuseki, H., Niimura, K., Majmuder, C., Belosludov, R.V., Farajian, A.A., and Kawazoe, Y. (2003) Theoretical study of donor-spacer-acceptor structure molecule for stable molecular rectifier. *Mol. Cryst. Liq. Cryst.*, **406**, 205–211.

71 Gust, D., Moore, T.A., and Moore, A.L. (2001) Mimicking photosynthetic solar energy transduction. *Acc. Chem. Res.*, **34** (1), 40–48.

72 Reed, C.A. and Bolskar, R.D. (2000) Discrete fulleride anions and fullerenium cations. *Chem. Rev.*, **100** (3), 1075–1119.

73 Guldi, D.M. and Prato, M. (2000) Excited-state properties of C60 fullerene derivatives. *Acc. Chem. Res.*, **33** (10), 695–703.

74 Martin, N., Sanchez, L., Illescas, B., and Perez, I. (1998) C60-Based electroactive organofullerenes. *Chem. Rev.*, **98** (7), 2527–2547.

75 Kadish, K.M., Smith, K.M., and Guilard, R. (eds) (1999) *The Porphyrin Handbook*, Academic Press, New York.

76 Li, L., Tedeschi, C., Kurth, D.G., and Moehwald, H. (2004) Synthesis of a pyrene-labeled polyanion and its adsorption onto polyelectrolyte hollow capsules functionalized for electron transfer. *Chem. Mater.*, **16** (4), 570–573.

77 Hasobe, T., Imahori, H., Kamat, P.V., Ahn, T.K., Kim, S.K., Kim, D., Fujimoto, A., Hirakawa, T., and Fukuzumi, S. (2005) Photovoltaic cells using Composite nanoclusters of porphyrins and fullerenes with gold nanoparticles. *J. Am. Chem. Soc.*, **127** (4), 1216–1228.

78 Itoh, T., Yano, K., Kajino, T., Itoh, S., Shibata, Y., Mino, H., Miyamoto, R., Inada, Y., Iwai, S., and Fukushima, Y. (2004) Nanoscale organization of chlorophyll *a* in mesoporous silica: efficient energy transfer and stabilized charge separation as in natural photosynthesis. *J. Phys. Chem. B*, **108** (36), 13683–13687.

79 Kumble, R., Palese, S., Lin, V.S.Y., Therien, M.J., and Hochstrasser, R.M. (1998) Ultrafast dynamics of highly conjugated porphyrin arrays. *J. Am. Chem. Soc.*, **120** (44), 11489–11498.

80 Choi, M.-S., Aida, T., Yamazaki, T., and Yamazaki, I. (2001) A large dendritic multiporphyrin array as a mimic of the

bacterial light-harvesting antenna complex: molecular design of an efficient energy funnel for visible photons. *Angew. Chem. Int. Ed.*, **40** (17), 3194–3198.

81 Fujitsuka, M., Okada, A., Tojo, S., Takei, F., Onitsuka, K., Takahashi, S., and Majima, T. (2004) Rapid exciton migration and fluorescent energy transfer in helical polyisocyanides with regularly arranged porphyrin pendants. *J. Phys. Chem. B*, **108** (32), 11935–11941.

82 Nagata, T., Osuka, A., and Maruyama, K. (1990) Synthesis and optical properties of conformationally constrained trimeric and pentameric porphyrin arrays. *J. Am. Chem. Soc.*, **112** (8), 3054–3059.

83 Wagner, R.W. and Lindsey, J.S. (1994) A photonic molecular wire. *J. Am. Chem. Soc.*, **116** (21), 9759–9760.

84 Cho, H.S., Rhee, H., Song, J.K., Min, C.K., Takase, M., Aratani, N., Cho, S., Osuka, A., Joo, T., and Kim, D. (2003) Excitation energy transport processes of porphyrin monomer, dimer, cyclic trimer, and hexamer probed by ultrafast fluorescence anisotropy decay. *J. Am. Chem. Soc.*, **125** (19), 5849–5860.

85 Kim, D. and Osuka, A. (2003) Photophysical properties of directly linked linear porphyrin Arrays. *J. Phys. Chem. A*, **107** (42), 8791–8816.

86 Takei, F., Hayashi, H., Onitsuka, K., Kobayashi, N., and Takahashi, S. (2001) Helical chiral polyisocyanides possessing porphyrin pendants: determination of helicity by exciton-coupled circular dichroism. *Angew. Chem. Int. Ed.*, **40** (21), 4092–4094.

87 De Witte, P.A.J., Castriciano, M., Cornelissen, J.J.L.M., Scolaro, L.M., Nolte, R.M.J., and Rowan, A.E. (2003) Helical polymer-anchored porphyrin nanorods. *Chem.-Eur. J.*, **9** (8), 1175–1781.

88 Jiang, B., Yang, S., and Jones, W.E. Jr. (1997) Conjugated porphyrin polymers: control of chromophore separation by oligophenylenevinylene bridges. *Chem. Mater.*, **9** (10), 2031–2034.

89 Wolffs, M., Hoeben, F.J.M., Beckers, E.H.A., Schenning, A.P.H.J., and Meijer, E.W. (2005) Sequential energy and electron transfer in aggregates of tetrakis [oligo(*p*-phenylenevinylene)] porphyrins and C60 in water. *J. Am. Chem. Soc.*, **127** (39), 13484–13485.

90 Balaban, T.S. (2005) Tailoring porphyrins and chlorins for self-assembly in biomimetic artificial antenna systems. *Acc. Chem. Res.*, **38** (8), 612–623.

91 Imahori, H. (2004) Giant multiporphyrin arrays as artificial light-harvesting antennas. *J. Phys. Chem. B*, **108** (20), 6130–6143.

92 Sakai, H., Masada, Y., Onuma, H., Takeoka, S., and Tsuchida, E. (2004) Reduction of methemoglobin *via* electron transfer from photoreduced flavin: restoration of O_2-binding of concentrated hemoglobin solution coencapsulated in phospholipid vesicles. *Bioconjug. Chem.*, **15** (5), 1037–1045.

93 Uno, H., Masumoto, A., and Ono, N. (2003) Hexagonal columnar porphyrin assembly by unique trimeric complexation of a porphyrin dimer with π-π stacking: remarkable thermal behavior in a solid. *J. Am. Chem. Soc.*, **125** (40), 12082–12083.

94 Hasobe, T., Imahori, H., Kamat, P.V., and Fukuzumi, S. (2003) Quaternary self-organization of porphyrin and fullerene units by clusterization with gold nanoparticles on SnO_2 electrodes for organic solar cells. *J. Am. Chem. Soc.*, **125** (49), 14962–14963.

95 Park, M., Cho, S., Yoon, Z.S., Aratani, N., Osuka, A., and Kim, D. (2005) Single molecule spectroscopic investigation on conformational heterogeneity of directly linked zinc(II) porphyrin arrays. *J. Am. Chem. Soc.*, **127** (43), 15201–15206.

96 Fujitsuka, M., Hara, M., Tojo, S., Okada, A., Troiani, V., Solladie, N., and Majima, T. (2005) Fast exciton migration in porphyrin-functionalized polypeptides. *J. Phys. Chem. B*, **109** (1), 33–35.

97 Langford, S.J. and Yann, T. (2003) Molecular logic: a half-subtractor based on tetraphenylporphyrin. *J. Am. Chem. Soc.*, **125** (37), 11198–11199.

98 Yoon, D.H., Lee, S.B., Yoo, K.H., Kim, J., Lim, J.K., Aratani, N., Tsuda, A., Osuka, A., and Kim, D. (2003) Electrical conduction through linear porphyrin

arrays. *J. Am. Chem. Soc.*, **125** (36), 11062–11064.

99 Anderson, H.L., Martin, S.J., and Bradley, D.D.C. (1994) Synthesis and 3rd-order nonlinear-optical properties of a conjugated porphyrin polymer. *Angew. Chem. Int. Ed.*, **33** (6), 655–657.

100 Rath, H., Sankar, J., PrabhuRaja, V., Chandrashekar, T.K., Nag, A., and Goswami, D. (2005) Core-modified expanded porphyrins with large third-order nonlinear optical response. *J. Am. Chem. Soc.*, **127** (33), 11608–11609.

101 Drobizhev, M., Stepanenko, Y., Rebane, A., Wilson, C.J., Screen, T.E.O., and Anderson, H.L. (2006) Strong cooperative enhancement of two-photon absorption in double-strand conjugated porphyrin ladder arrays. *J. Am. Chem. Soc.*, **128** (38), 12432–12433.

102 MacDonald, I.J. and Dougherty, T.J. (2001) Basic principles of photodynamic therapy. *J. Porphyrins Phthalocyanines*, **5** (2), 105–129.

103 Duncan, T.V., Ghoroghchian, P.P., Rubtsov, I.V., Hammer, D.A., and Therien, M.J. (2008) Ultrafast excited-state dynamics of nanoscale near-infrared emissive polymersomes. *J. Am. Chem. Soc.*, **130** (30), 9773–9784.

104 Fei, Z.P., Han, Y., and Bo, Z.S. (2008) Synthesis of porphyrin-embedded dendronized polymers by Suzuki polycondensation. *J. Polym. Sci. Polym. Chem.*, **46** (12), 4030–4037.

105 Imaoka, T., Tanaka, R., Arimoto, S., Sakai, M., Fujii, M., and Yamamoto, K. (2005) Probing stepwise complexation in phenylazomethine dendrimers by a metallo-porphyrin core. *J. Am. Chem. Soc.*, **127** (40), 13896–13905.

106 Collis, G.E., Campbell, W.M., Officer, D.L., and Burrell, A.K. (2005) The design and synthesis of porphyrin/oligiothiophene hybrid monomers. *Org. Biomol. Chem.*, **3** (11), 2075–2084.

107 Choi, M.S., Aida, T., Yamazaki, T., and Yamazaki, I. (2002) Dendritic multiporphyrin arrays as light-harvesting antennae: effects of generation number and morphology on intramolecular energy transfer. *Chem.-Eur. J.*, **8** (12), 2668–2678.

108 Nishioka, T., Tashiro, K., Aida, T., Zheng, J.Y., Kinbara, K., Saigo, K., Sakamoto, S., and Yamaguchi, K. (2000) Molecular design of a novel dendrimer porphyrin for supramolecular fullerene/dendrimer hybridization. *Macromolecules*, **33** (25), 9182–9184.

109 Officer, D.L., Burrell, A.K., and Reid, D.C.W. (1996) Building large porphyrin arrays: pentamers and nonamers. *Chem. Commun.* (14), 1657–1658.

110 Jiu, T.G., Li, Y.J., Gan, H.Y., Li, Y.L., Liu, H.B., Wang, S., Zhou, W.D., Wang, C.R., Li, X.F., Liu, X.F., and Zhu, D.B. (2007) Synthesis of oligo(*p*-phenylene vinylene)-porphyrin-oligo(*p*-phenylene vinylene) triads as antenna molecules for energy transfer. *Tetrahedron*, **63** (1), 232–240.

111 Nielsen, K.T., Spanggaard, H., and Krebs, F.C. (2005) Synthesis, light harvesting, and energy transfer properties of a zinc porphyrin linked poly (phenyleneethynylene). *Macromolecules*, **38** (4), 1180–1189.

112 Tsuda, A. and Osuka, A. (2001) Fully conjugated porphyrin tapes with electronic absorption bands that reach into infrared. *Science*, **293** (527), 79–82.

113 Aratani, N., Osuka, A., Kim, Y.H., Jeong, D.H., and Kim, D. (2000) Extremely long, discrete meso-meso-coupled porphyrin arrays. *Angew. Chem. Int. Edit.*, **39** (8), 1458–1462.

114 Anderson, H.L. (1999) Building molecular wires from the colours of life: conjugated porphyrin oligomers. *Chem. Commun.* (23), 2323–2330.

115 Seth, J., Palaniappan, V., Wagner, R.W., Johnson, T.E., Lindsey, J.S., and Bocian, D.F. (1996) Soluble synthetic multiporphyrin arrays. Part 3. Static spectroscopic and electrochemical probes of electronic communication. *J. Am. Chem. Soc.*, **118** (45), 11194–11207.

116 Wagner, R.W., Johnson, T.E., and Lindsey, J.S. (1996) Soluble synthetic multiporphyrin arrays. Part 1. Modular design and synthesis. *J. Am. Chem. Soc.*, **118** (45), 11166–11180.

117 Anderson, H.L. (1994) Conjugated porphyrin ladders. *Inorg. Chem.*, **33** (5), 972–981.

118 Huang, X.B., Zhu, C.L., Zhang, S.M., Li, W.W., Guo, Y.L., Zhan, X.W., Liu, Y.Q., and Bo, Z.Z. (2008) Porphyrin-dithienothiophene π-conjugated copolymers: synthesis and their applications in field-effect transistors and solar cells. *Macromolecules*, **41** (19), 6895–6902.

119 Takeuchi, M., Fujikoshi, C., Kubo, Y., Kaneko, K., and Shinkai, S. (2006) Conjugated polymers complexed with helical porphyrin oligomers create micron-sized ordered structures. *Angew. Chem. Int. Ed.*, **45** (33), 5494–5499.

120 Huang, C.S., Wang, N., Li, Y.L., Li, C.H., Li, J.B., Liu, H.B., and Zhu, D.B. (2006) A new class of conjugated polymers having porphyrin, poly(*p*-phenylenevinylene), and fullerene units for efficient electron transfer. *Macromolecules*, **39** (16), 5319–5325.

121 Hou, Q., Zhang, Y., Li, F.Y., Peng, J.B., and Cao, Y. (2005) Red electrophosphorescence of conjugated organoplatinum(II) polymers prepared via direct metalation of poly(fluorene-*co*-tetraphenylporphyrin) copolymers. *Organometallics*, **24** (19), 4509–4518.

122 Krebs, F.C., Hagemann, O., and Spanggaard, H. (2003) Directional synthesis of a dye-linked conducting homopolymer. *J. Org. Chem.*, **68** (6), 2463–2466.

123 Yamamoto, T., Fukushima, N., Nakajima, H., Maruyama, T., and Yamaguchi, I. (2000) Synthesis and chemical properties of π-conjugated zinc porphyrin polymers with arylene and aryleneethynylene groups between zinc porphyrin units. *Macromolecules*, **33** (16), 5988–5994.

124 Hecht, S., Ihre, H., and Frechet, J.M.J. (1999) Porphyrin core star polymers: synthesis, modification, and implication for site isolation. *J. Am. Chem. Soc.*, **121** (39), 9239–9240.

125 Taylor, P.N. and Anderson, H.L. (1999) Cooperative self-assembly of double-strand conjugated porphyrin ladders. *J. Am. Chem. Soc.*, **121** (49), 11538–11545.

126 Jiang, B., Yang, S.-W., Barbini, D.C., and Jones, W.E. Jr. (1998) Synthesis of soluble conjugated metalloporphyrin polymers with tunable electronic properties. *Chem. Commun.* (2), 213–214.

127 Jiang, B. and Jones, W.E. Jr. (1997) Synthesis and characterization of a conjugated copolymer of poly (phenylenevinylene) containing a metalloporphyrin incorporated into the polymer backbone. *Macromolecules*, **30** (19), 5575–5581.

128 Bao, Z.N., Chen, Y.M., and Yu, L.P. (1994) New metalloporphyrin containing polymers from the Heck coupling reaction. *Macromolecules*, **27** (16), 4629–4631.

129 Anderson, H.L. (1994) Supramolecular orientation of conjugated porphyrin oligomers in stretched polymers. *Adv. Mater.*, **6** (11), 834–836.

130 Pizzoferrato, R., Ziller, T., Paolesse, R., Mandoj, F., Micozzi, A., Ricci, A., and Lo Sterzo, C. (2006) Optical properties of novel Er-containing co-polymers with emission at 1530 nm. *Chem. Phys. Lett.*, **426** (1–3), 124–128.

131 Zhuang, W.L., Zhang, Y., Hou, Q., Wang, L., and Cao, Y. (2006) High-efficiency, electrophosphorescent polymers with porphyrin-platinum complexes in the conjugated backbone: synthesis and device performance. *J. Polym. Sci. Polym. Chem.*, **44** (13), 4174–4186.

132 Wagner, R.W., Lindsey, J.S., Seth, J., Palaniappan, V., and Bocian, D.F. (1996) Molecular optoelectronic gates. *J. Am. Chem. Soc.*, **118** (16), 3996–3997.

133 Bedioui, F., Devynck, J., and Biedcharreton, C. (1995) Immobilization of metalloporphyrins in electropolymerized films – design and applications. *Acc. Chem. Res.*, **28** (1), 30–36.

134 Lin, V.S.Y., Dimagno, S.G., and Therien, M.J. (1994) Highly conjugated, acetylenyl bridged porphyrins - New models for light-harvesting antenna systems. *Science*, **264** (5162), 1105–1111.

135 Prathapan, S., Johnson, T.E., and Lindsey, J.S. (1993) Building-block synthesis of porphyrin light-harvesting arrays. *J. Am. Chem. Soc.*, **115** (16), 7519–7520.

136 Anderson, H.L. and Sanders, J.K.M. (1990) Amine-template-directed

synthesis of cyclic porphyrin oligomers. *Angew. Chem. Int. Ed.*, **29** (12), 1400–1403.

137 Daunert, S., Wallace, S., Florido, A., and Bachas, L.G. (1991) Anion-selective electrodes based on electropolymerized porphyrin films. *Anal. Chem.*, **63** (17), 1676–1679.

138 Nishide, H., Kawakami, H., Suzuki, T., Azechi, Y., Soejima, Y., and Tsuchida, E. (1991) Effect of polymer matrix on the oxygen diffusion via a cobalt porphyrin fixed in a membrane. *Macromolecules*, **24** (23), 6306–6309.

139 Wamser, C.C., Bard, R.R., Senthilathipan, V., Anderson, V.C., Yates, J.A., Lonsdale, H.K., Rayfield, G.W., Friesen, D.T., Lorenz, D.A., Stangle, G.C., Vaneikeren, P., Baer, D.R., Ransdell, R.A., Golbeck, J.H., Babcock, W.C., Sandberg, J.J., and Clarke, S.E. (1989) Synthesis and photoactivity of chemically asymmetric polymeric porphyrin films made by interfacial polymerization. *J. Am. Chem. Soc.*, **111** (22), 8485–8491.

140 Kamachi, M., Cheng, X.S., Kida, T., Kajiwara, A., Shibasaka, M., and Nagata, S. (1987) Synthesis of new polymers containing porphyrins in their side-chains – radical polymerizations of 5-[4-(acryloyloxy)phenyl]-10,15,20-triphenylporphyrin and 5-[4-(methacryloyloxy)phenyl]-10,15,20-triphenylporphyrin. *Macromolecules*, **20** (11), 2665–2669.

141 Lindsey, J.S., Prathapan, S., Johnson, T.E., and Wagner, R.W. (1994) Porphyrin building blocks for modular construction of bioorganic model systems. *Tetrahedron*, **50** (30), 8941–8968.

142 Sugiura, K., Tanaka, H., Matsumoto, T., Kawai, T., and Sakata, Y. (1999) A mandala-patterned bandanna-shaped porphyrin oligomer, $C_{1244}H_{1350}N_{84}$-$Ni_{20}O_{88}$, having a unique size and geometry. *Chem. Lett.* (11), 1193–1194.

143 Graca, M., Vicente, H., Jaquinod, L., and Smith, K.M. (1999) Oligomeric porphyrin arrays. *Chem. Commun.* (18), 1771–1782.

144 Mongin, O., Papamicae, C., Hoyler, N., and Gossauer, A. (1998) Modular synthesis of benzene-centered porphyrin trimers and a dendritic porphyrin hexamer. *J. Org. Chem.*, **63** (16), 5568–5580.

145 Mak, C.C., Bampos, N., and Sanders, J.K.M. (1998) Metalloporphyrin dendrimers with folding arms. *Angew. Chem. Int. Ed.*, **37** (21), 3020–3023.

146 Wasielewski, M.R. (1992) Photoinduced electron-transfer in supramolecular systems for artificial photosynthesis. *Chem. Rev.*, **92** (3), 435–461.

147 Griveau, S., Albin, V., Pauporte, T., Zagal, J.H., and Bedioui, F. (2002) Comparative study of electropolymerized cobalt porphyrin and phthalocyanine based films for the electrochemical activation of thiols. *J. Mater. Chem.*, **12** (2), 225–232.

148 Allietta, N., Pansu, R., Bied Charreton, C., Albin, V., Bedioui, F., and Devynck, J. (1996) New conducting polymers: preparation and spectroscopic properties of zinc-porphyrin and anthraquinone-coated electrodes. *Synth. Met.*, **81** (2-3), 205–210.

149 Lin, C.Y., Hung, Y.C., Liu, C.M., Lo, C.F., Lin, Y.C., and Lin, C.L. (2005) Synthesis, electrochemistry, absorption and electro-polymerization of aniline-ethynyl metalloporphyrins. *Dalton Trans.* (2), 396–401.

150 Kadish, K.M., Smith, K.M., and Guilard, G. (eds) (2000) *The Porphyrin Handbook*, vol. **8**, Academic Press, New York, pp. 1–97.

151 See e.g.: Subramanian, V., Wolf, E.E., and Kamat, P.V. (2004) Catalysis with TiO_2/gold nanocomposites. Effect of metal particle size on the Fermi level equilibration. *J. Am. Chem. Soc.*, **126** (15), 4943–4950.

152 Winters, M.U., Dahlstedt, E., Blades, H.E., Wilson, C.J., Frampton, M.J., Anderson, H.L., and Albinsson, B. (2007) Probing the efficiency of electron transfer through porphyrin-based molecular wires. *J. Am. Chem. Soc.*, **129** (14), 4291–4297.

153 Grozema, F.C., Houarner-Rassin, C., Prins, P., Siebbeles, L.D.A., and Anderson, H.L. (2007) Supramolecular control of charge transport in molecular wires. *J. Am. Chem. Soc.*, **129** (44), 13370–13371.

154 Williams, V.E. and Swager, T.M. (2000) An improved synthesis of poly(*p*-phenylenebutadiynylene)s. *J. Polym. Sci. Pol. Chem.*, **38** (S1), 4669–4676.

155 Winters, M.U., Karnbratt, J., Eng, M., Wilson, C.J., Anderson, H.L., and Albinsson, B. (2007) Photophysics of a butadiyne-linked porphyrin dimer: influence of conformational flexibility in the ground and first singlet excited state. *J. Phys. Chem. C*, **111** (19), 7192–7199.

156 Screen, T.E.O., Thorne, J.R.G., Denning, R.G., Bucknall, D.G., and Anderson, H.L. (2003) Two methods for amplifying the optical nonlinearity of a conjugated porphyrin polymer: transmetallation and self-assembly. *J. Mater. Chem.*, **13** (11), 2796–2808.

157 Screen, T.E.O., Thorne, J.R.G., Denning, R.G., Bucknell, D.G., and Anderson, H.L. (2002) Amplified optical nonlinearity in a self-assembled double-strand conjugated porphyrin polymer ladder. *J. Am. Chem. Soc.*, **124** (33), 9712–9713.

158 Burroughes, J.H., Bradley, D.D.C., Brown, A.R., Marks, R.N., Mackay, K., Friend, R.H., Burns, P.L., and Holmes, A.B. (1990) Light-emitting-diodes based on conjugated polymers. *Nature*, **347** (6293), 539–541.

159 Lee, Y.G., Liang, Y.Y., and Yu, L.P. (2006) The Heck polycondensation for functional polymers. *Synlett* (18), 2879–2893.

160 Jiang, B., Yang, S.-W., Niver, R., and Jones, W.E. Jr. (1998) Metalloporphyrin polymers bridged with conjugated cyano-substituted stilbene units. *Synth, Met.*, **94** (2), 205–210.

161 Nielsen, K.T., Spanggaard, H., and Krebs, F.C. (2004) Dye linked conjugated homopolymers: using conjugated polymer electroluminescence to optically pump porphyrin-dye emission. *Displays*, **25** (5), 231–235.

162 Zhao, J.L., Li, B.S., and Bo, Z.S. (2006) Synthesis and photocurrent response of porphyrin-containing conjugated polymers. *Chin. Sci. Bull.*, **51** (11), 1287–1295.

163 Wang, N., Lu, F.S., Huang, C.S., Li, Y.L., Yuan, M.J., Liu, X.F., Liu, H.B., Gan, L.B., Jiang, L., and Zhu, D.B. (2006) Construction of diads and triads copolymer systems containing perylene, porphyrin, and/or fullerene blocks. *J. Polym. Sci. Polym. Chem.*, **44** (20), 5863–5874.

164 Hunziker, C., Zhan, X., Losio, P.A., Figi, H., Kwon, O.P., Barlow, S., Guenter, P., and Marder, S.R. (2007) Highly ordered thin films of a bis(dithienothiophene) derivative. *J. Mater. Chem.*, **17** (47), 4972–4979.

165 Sun, Y.M., Ma, Y.W., Liu, Y.Q., Lin, Y.Y., Wang, Z.Y., Wang, Y., Di, C.G., Xiao, K., Chen, X.M., Qiu, W.F., Zhang, B., Yu, G., Hu, W.P., and Zhu, D.B. (2006) High-performance and stable organic thin-film transistors based on fused thiophenes. *Adv. Func. Mater.*, **16** (3), 426–432.

166 Xiao, K., Liu, Y.Q., Qi, T., Zhang, W., Wang, F., Gao, J.H., Qiu, W.F., Ma, Y.Q., Cui, G.L., Chen, S.Y., Zhan, X.W., Yu, G., Qin, J.G., Hu, W.P., and Zhu, D.B. (2005) A highly π-stacked organic semiconductor for field-effect transistors based on linearly condensed pentathienoacene. *J. Am. Chem. Soc.*, **127** (38), 13281–13286.

167 Manners, I. (ed). (2004) *Synthetic Metal-Containing Polymers*, Wiley-VCH Verlag GmbH, Weinheim, Germany.

168 Nguyen, P., Gomez-Elipe, P., and Manners, I. (1999) Organometallic polymers with transition metals in the main chain. *Chem. Rev.*, **99** (6), 1515–1548.

169 Long, N.J. and Williams, C.K. (2003) Metal alkynyl sigma complexes: synthesis and materials. *Angew. Chem. Int. Ed.*, **42** (23), 2586–2617.

170 Wong, W.Y. and Ho, C.L. (2006) Di-, oligo and polymetallaynes: syntheses, photophysics, structures and applications. *Coord. Chem. Rev.*, **250** (19-20), 2627–2690.

171 Fujikura, Y., Sonogashira, K., and Hagihara, N. (1975) Preparation and UV spectra of some oligomer-complexes composed of platinum group metals and conjugated poly-yne systems. *Chem. Lett.* (10), 1067–1070.

172 Sonogashira, K., Takahashi, S., and Hagihara, N. (1977) New extended chain polymer – poly[trans-bis(tri-*N*-

butylphosphine)platinum 1,4-butadiynediyl]. *Macromolecules*, **10** (4), 879–880.

173 Sonogashira, K., Kataoka, S., Takahashi, S., and Hagihara, N. (1978) Studies of poly-yne polymers containing transition-metals in main chain: Part III. Synthesis and characterization of a poly-yne polymer containing mixed metals in main chain. *J. Organomet. Chem.*, **160** (1), 319–327.

174 Sonogashira, K., Fujikura, Y., Yatake, T., Toyoshima, N., Takahashi, S., and Hagihara, N. (1978) Syntheses and properties of cis-dialkynyl and trans-dialkynyl complexes of platinum(II). *J. Organomet. Chem.*, **145** (1), 101–108.

175 Takahashi, S., Kariya, M., Yatake, T., Sonogashira, K., and Hagihara, N. (1978) Studies of poly-yne polymers containing transition-metals in main chain: Part II. Synthesis of poly[trans-bis(tri-normal-butylphosphine)platinum 1,4-butadiynediyl] and evidence of a rodlike structure. *Macromolecules*, **11** (6), 1063–1066.

176 Wong, W.Y. (2005) Metallated molecular materials of fluorene derivatives and their analogues. *Coordin. Chem. Rev.*, **249** (9-10), 971–997.

177 Liu, L., Wong, W.Y., Shi, J.X., and Cheah, K.W. (2006) Exploring 9-arylcarbazole moiety as the building block for the synthesis of photoluminescent group 10-12 heavy metal diynes and polyynes with high-energy triplet states. *J. Polym. Sci. Polym. Chem.*, **44** (19), 5588–5607.

178 Wong, W.Y., Zhou, G.J., He, Z., Cheung, K.Y., Ng, A.M.C., Djurisic, A.B., and Chan, W.K. (2008) Organometallic polymer light-emitting diodes derived from a platinum(II) polyyne containing the bithiazole ring. *Macromol. Chem. Phys.*, **209** (13), 1320–1332.

179 Wong, W.Y., Wang, X.Z., Zhang, H.L., Cheung, K.Y., Fung, M.K., Djurisic, A.B., and Chan, W.K. (2008) Synthesis, characterization and photovoltaic properties of a low-bandgap platinum(II) polyyne functionalized with a 3,4-ethylenedioxythiophene-benzothiadiazole hybrid spacer. *J. Organomet. Chem.*, **693** (24), 3603–3612.

180 Wang, X.Z., Wong, W.Y., Cheung, K.Y., Fung, M.K., Djurisic, A.B., and Chan, W.K. (2008) Polymer solar cells based on very narrow-bandgap polyplatinynes with photocurrents extended into the near-infrared region. *Dalton Trans.* (40), 5484–5494.

181 Wong, W.Y. (2008) Metallopolyyne polymers as new functional materials for photovoltaic and solar cell applications. *Macromol. Chem. Phys.*, **209** (1), 14–24.

182 Zhou, G.J., Wong, W.Y., Ye, C., and Lin, Z.Y. (2007) Optical power limiters based on colorless di-, oligo-, and polymetallaynes: highly transparent materials for eye protection devices. *Adv. Func. Mater.*, **17** (6), 963–975.

183 Zhou, G.J., Wong, W.Y., Lin, Z.Y., and Ye, C. (2006) White metallopolyynes for optical limiting/transparency trade-off optimization. *Angew. Chem. Int. Ed.*, **45** (37), 6189–6193.

184 Onitsuka, K., Harada, Y., Takei, F., and Takahashi, S. (1998) Synthesis of transition metal-poly(yne) polymer possessing chiral acetylene bridges. *Chem. Commun.* (6), 643–644.

185 Haskins-Glusac, K., Pinto, M.R., Tan, C.Y., and Schanze, K.S. (2004) Luminescence quenching of a phosphorescent conjugated polyelectrolyte. *J. Am. Chem. Soc.*, **126** (45), 14964–14971.

186 Jang, W.D. (2005) Synthesis of dendrimer based polymeric and macrocyclic complexes with a platinum-acetylide π-conjugated organometallic core. *Macromol. Res.*, **13** (4), 334–338.

187 Fratoddi, I., Battocchio, C., Groia, A.L., and Russo, M.V. (2007) Nanostructured polymetallaynes of controlled length: synthesis and characterization of oligomers and polymers from 1,1′-bis-(ethynyl)-4,4′-biphenyl bridging Pt(II) or Pd(II) centers. *J. Polym. Sci. Polym. Chem.*, **45** (15), 3311–3329.

188 Beljonne, D., Wittmann, H.F., Kohler, A., Graham, S., Younus, M., Lewis, J., Raithby, P.R., Khan, M.S., Friend, R.H., and Bredas, J.L. (1996) Spatial extent of the singlet and triplet excitons in transition metal-containing poly-ynes. *J. Chem. Phys.*, **105** (9), 3868–3877.

189 Younus, M., Kohler, A., Cron, S., Chawdhury, N., Al-Mandhary, M.R.A., Khan, M.S., Lewis, J., Long, N.J., Friend, R.H., and Raithby, P.R. (1998) Synthesis, electrochemistry, and spectroscopy of blue platinum(II) polyynes and diynes. *Angew. Chem. Int. Ed.*, **37** (21), 3036–3039.

190 Wilson, J.S., Kohler, A., Friend, R.H., Al-Suti, M.K., Al-Mandhary, M.R.A., Khan, M.S., and Raithby, P.R. (2000) Triplet states in a series of Pt-containing ethynylenes. *J. Chem. Phys.*, **113** (17), 7627–7634.

191 Havinga, E.E., Tenhoeve, W., and Wynberg, H. (1993) Alternate donor-acceptor small-band-gap semiconducting polymers – polysquaraines and polycroconaines. *Synth. Met.*, **55** (1), 299–306.

192 Roncali, J. (1997) Synthetic principles for bandgap control in linear π-conjugated systems. *Chem. Rev.*, **97** (1), 173–205.

193 Takahashi, S., Murata, E., Sonogashira, K., and Hagihara, N. (1980) Studies on poly-yne polymers containing transition-metals in the main chain: Part IV. Polymer synthesis by oxidative coupling of transition metal-bis(acetylide) complexes. *J. Polym. Sci. Polym. Chem.*, **18** (2), 661–669.

194 Hay, A.S. (1962) Oxidative coupling of acetylenes. Part II. *J. Org. Chem.*, **27** (9), 3320–3321.

195 Sonogashira, K., Ohga, K., Takahashi, S., and Hagihara, N. (1980) Studies of poly-yne polymers containing transition-metals in the main chain: Part VI. Synthesis of nickel-poly-yne polymers by alkynyl ligand exchange using a copper(I) catalyst. *J. Organomet. Chem.*, **188** (2), 237–243.

196 Takahashi, S., Ohyama, Y., Murata, E., Sonogashira, K., and Hagihara, N. (1980) Studies of poly-yne polymers containing transition-metals in the main chain: Part V. Design of metal arrangement in polymer backbone. *J. Polym. Sci. Polym. Chem.*, **18** (1), 349–353.

197 Manners, I. and Abd-El-Aziz, A.S. (eds) (2007) *Frontiers in Transition Metal-Containing Polymers*, Wiley-VCH Verlag GmbH, Weinheim, Germany.

198 Cardin, C.J., Cardin, D.J., and Lappert, M.F. (1977) Unsaturated sigma-hydrocarbyl transition-metal complexes. Part 2. Synthesis and reactions of vinyl-platinum complexes and a comparison with analogous fluorovinyl and alkynyl complexes. *J. Chem. Soc. Dalton.Trans.* (8), 767–779.

199 Johnson, B.F.G., Kakkar, A.K., Khan, M.S., and Lewis, J. (1991) Synthesis of novel rigid rod iron metal containing polyyne polymers. *J. Organomet. Chem.*, **409** (3), C12–C14.

200 Davies, S.J., Johnson, B.F.G., Lewis, J., and Raithby, P.R. (1991) Synthesis of mononuclear and oligomeric ruthenium (II) acetylides. *J. Organomet. Chem.*, **414** (2), C51–C53.

201 Khan, M.S., Kakkar, A.K., Ingham, S.L., Raithby, P.R., Lewis, J., Spencer, B., Wittmann, F., and Friend, R.H. (1994) Synthesis and electronic-structure of rigid-rod octahedral Ru-sigma acetylide complexes. *J. Organomet. Chem.*, **472** (1-2), 247–255.

202 Faulkner, C.W., Ingham, S.L., Khan, M.S., Lewis, J., Long, N.J., and Raithby, P.R. (1994) Ruthenium(II) sigma-acetylide complexes – monomers, dimers and polymers. *J. Organomet. Chem.*, **482** (1-2), 139–145.

203 Davies, S.J., Johnson, B.F.G., Khan, M.S., and Lewis, J. (1991) Synthesis of monomeric and oligomeric bis(acetylide) complexes of platinum and rhodium. *J. Chem. Soc., Chem. Commun.* (3), 187–188.

204 Khan, M.S., Davies, S.J., Kakkar, A.K., Schwartz, D., Lin, B., Johnson, B.F.G., and Lewis, J. (1992) Synthesis of monomeric, oligomeric and polymeric sigma-acetylide complexes of platinum, palladium, nickel and rhodium. *J. Organomet. Chem.*, **424** (1), 87–97.

205 Johnson, B.F.G., Kakkar, A.K., Khan, M.S., Lewis, J., Dray, A.E., Friend, R.H., and Wittmann, F. (1991) Synthesis and optical spectroscopy of platinum-metal-containing di-acetylenic and tri-acetylenic polymers. *J. Mater. Chem.*, **1** (3), 485–486.

206 Lewis, J., Khan, M.S., Kakkar, A.K., Johnson, B.F.G., Marder, T.B., Fyfe, H.B., Wittmann, F., Friend, R.H., and Dray,

A.E. (1992) Di-acetylenic, pseudo-di-acetylenic and pseudo-tetra-acetylenic polymers of platinum – synthesis, characterization and optical-spectra. *J. Organomet. Chem.*, **425** (1-2), 165–176.

207 Lo Sterzo, C. (1999) The wonder of palladium catalysis: from carbon-carbon to metal-carbon bond formation. An opportunity of getting astonishment from reality. *Synlett* (11), 1704–1722.

208 Antonelli, E., Rosi, P., Lo Sterzo, C., and Viola, E. (1999) A convenient short cut from aromatic iodides to alkynylstannanes and their use for the straightforward preparation of polyacetylene and polymetallaacetylene polymers. *J. Organomet. Chem.*, **578** (1-2), 210–222.

209 Pizzoferrato, R., Berliocchi, M., Di Carlo, A., Lugli, P., Venanzi, M., Micozzi, A., Ricci, A., and Lo Sterzo, C. (2003) Improvement of the extended one-pot (EOP) procedure to form poly (aryleneethynylene)s and investigation of their electrical and optical properties. *Macromolecules*, **36** (7), 2215–2223.

210 Khan, M.S., Pasha, N.A., Kakkar, A.K., Raithby, P.R., Lewis, J., Fuhrmann, K., and Friend, R.H. (1992) Synthesis and optical spectroscopy of monomeric and polymeric cobalt sigma-acetylide complexes. *J. Mater. Chem.*, **2** (7), 759–760.

211 Fyfe, H.B., Mlekuz, M., Zargarian, D., Taylor, N.J., and Marder, T.B. (1991) Synthesis of mononuclear, dinuclear and oligomeric rigid-rod acetylide complexes of rhodium, and the molecular-structure of [Rh$(PMe_3)_4$(CC-*p*-C_6H_4-CC)Rh $(PMe_3)_4$]. *J. Chem. Soc., Chem. Commun.* (3), 188–190.

212 Marder, T.B., Lesley, G., Yuan, Z., Fyfe, H.B., Chow, P., Stringer, G., Jobe, I.R., Taylor, N.J., Williams, I.D., and Kurtz, S.K. (1991) Transition-metal acetylides for nonlinear optical-properties. *ACS Symp. Ser.*, **455**, 605–615.

213 Yamamoto, T., Morikita, T., Maruyama, T., Kubota, K., and Katada, M. (1997) Poly (aryleneethynylene) type polymers containing a ferrocene unit in the π-conjugated main chain. Preparation, optical properties, redox behavior, and Mossbauer spectroscopic analysis. *Macromolecules*, **30** (18), 5390–5396.

214 Vorotyntsev, M.A. and Vasilyeva, S.V. (2008) Metallocene-containing conjugated polymers. *Adv. Colloid Interface Sci.*, **139** (1-2), 97–149.

215 Sahoo, P.K. and Swain, S.K. (1999) Synthesis of zirconocene-acetylene and zirconocene-diacetylene polymer. *J. Polym. Sci. Polym. Chem.*, **37** (21), 3899–3902.

216 Jia, G.C., Puddephatt, R.J., Vittal, J.J., and Payne, N.C. (1993) Rigid-rod compounds – monomeric and oligomeric complexes with gold(I) centers bridged by (isocyanoaryl)acetylides. *Organometallics*, **12** (2), 263–265.

217 Jia, G.C., Puddephatt, R.J., Scott, J.D., and Vittal, J.J. (1993) Organometallic polymers with gold(I) centers bridged by diphosphines and diacetylides. *Organometallics*, **12** (9), 3565–3574.

218 Jia, G.C., Payne, N.C., Vittal, J.J., and Puddephatt, R.J. (1993) (Isocyanoaryl) acetylides as bridging ligands in rigid-rod polymers – mononuclear and oligonuclear gold(I) complexes. *Organometallics*, **12** (12), 4771–4778.

219 Irwin, M.J., Jia, G.C., Payne, N.C., and Puddephatt, R.J. (1996) Rigid-rod polymers and model compounds with gold(I) centers bridged by diisocyanides and diacetylides. *Organometallics*, **15** (1), 51–57.

220 Irwin, M.J., Vittal, J.J., Yap, G.P.A., and Puddephatt, R.J. (1996) Linear gold(I) coordination polymers: a polymer with a unique sine wave conformation. *J. Am. Chem. Soc.*, **118** (51), 13101–13102.

221 Irwin, M.J., Vittal, J.J., and Puddephatt, R.J. (1997) Luminescent gold(I) acetylides: from model compounds to polymers. *Organometallics*, **16** (15), 3541–3547.

222 Puddephatt, R.J. (2008) Macrocycles, catenanes, oligomers and polymers in gold chemistry. *Chem. Soc. Rev.*, **37** (9), 2012–2027.

223 Pyykko, P. (2004) Theoretical chemistry of gold. *Angew. Chem. Int. Edit.*, **43** (34), 4412–4456.

224 Vicente, J., Chicote, M.T., Abrisqueta, M.D., and Alvarez-Falcon, M.M. (2002)

The first anionic arenediethynylgold(I) complexes. *J. Organomet. Chem.*, **663** (1-2), 40–45.

225 Vicente, J., Chicote, M.T., Alvarez-Falcon, M.M., and Jones, P.G. (2005) Platinum (II) and mixed platinum(II)/gold(I) sigma-alkynyl complexes. The first anionic sigma-alkynyl metal polymers. *Organometallics*, **24** (11), 2764–2772.

226 Wong, W.Y., Liu, L., and Shi, J.X. (2003) Triplet emission in soluble mercury(II) polyyne polymers. *Angew. Chem. Int. Ed.*, **42** (34), 4064–4068.

11
Helical Polyacetylene Prepared in a Liquid Crystal Field

Kazuo Akagi

11.1
Introduction

Polyacetylene is a simple linear conjugated macromolecule and is representative of conducting polymers [1]. Pristine polyacetylene is a typical semiconductor, but its electrical conductivity can be varied by over 14 orders of magnitude through doping [2]. The maximum conductivity reported to date is more than 10^5 S cm^{-1} [3], which is comparable to copper and gold. Strong interchain interaction gives rise to a fibrillar crystal consisting of rigidly π-stacked polymer chains [4]. This structure makes polyacetylene infusible and insoluble in any kind of solvent. Therefore, the solid-state structure and morphology of polyacetylene are determined during polymerization, which is not the case for substituted polyacetylenes [5]. The fibril morphology of polyacetylene film is randomly oriented, as is usually encountered in ordinary polymers, which depresses the inherent one-dimensionality of this polymer. Several polymerization methods for the macroscopic alignment of the polymer have been developed to achieve higher electrical conductivity with its anisotropic nature [3, 6, 7]. The introduction of a liquid crystal (LC) group into the side chain is one approach to align the polymer under an external force such as shear stress, rubbing, or a magnetic field [8–10]. On the other hand, the use of nematic LC as a solvent gave us directly aligned polyacetylene with the aid of gravity flow or a magnetic force field [11, 12].

It has generally been accepted that polyacetylene has a planar structure, for both the cis- and trans-forms, due to the π-conjugation between the sp^2 hybridized carbon atoms in the polymer chain [1, 4]. If it were possible to modify such a planar structure of polyacetylene into a helical one [13], one might expect novel magnetic and optical properties [14]. Here, we report on the polymerization of acetylene in an asymmetric reaction field constructed with chiral nematic LCs, and show that it is possible to synthesize polyacetylene films formed by helical chains and fibrils [15]. The polymerization mechanism giving the helical structure from primary to higher-order and hierarchical spiral morphology is discussed.

Conjugated Polymer Synthesis. Edited by Yoshiki Chujo

ISBN: 978-3-527-32267-1

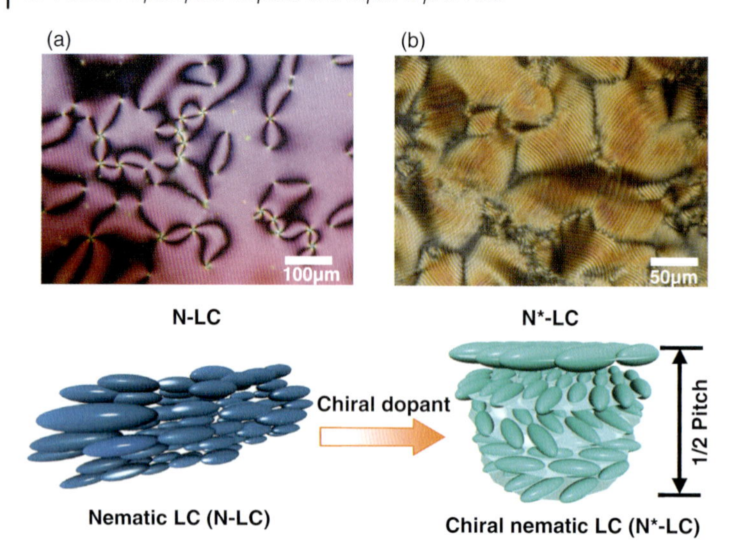

Figure 11.1 Chiral nematic LC induced by addition of a chiral dopant into nematic LC. Schlieren texture (a) and fingerprint texture (b) are observed for nematic and chiral nematic LCs, respectively, with a polarized optical microscope.

11.2
Chiral Dopants and Chiral Nematic LCs

The chiral nematic LC to be used as an asymmetric solvent is prepared by adding a small amount of chiral compound, as a chiral dopant, to nematic LC (Figure 11.1). The formation of chiral nematic LC is recognized when the Schlieren texture characteristic of nematic LC changes into a striated Schlieren or a fingerprint texture when viewed with a polarized optical microscope (POM). The distance between the striae corresponds to a half helical pitch of the chiral nematic LC. Note that, as the degree of twist in the chiral nematic LC increases, the helical pitch observed with POM decreases.

The helical pitch of the chiral nematic LC can be adjusted by two methods: changing either the concentration or the twisting power of the chiral dopant [16]. However, the mesophase temperature region of the chiral nematic LC is affected by changing the concentration of the chiral dopant. That is, it becomes narrow as the concentration increases and, finally, the mesophase is destroyed when the concentration approaches a critical value [17]. In this study, owing to the limitation of the concentration method, an alternative approach of utilizing a chiral compound with a large twisting power was adopted. Axially chiral binaphthyl derivatives were used as chiral dopants since they have been reported to possess greater twisting power than asymmetric carbon-containing chiral compounds [15c].

(*R*)- and (*S*)-1,1′-bi-naphthyl-2,2′-di-[para-(trans-4-*n*-pentylcyclohexyl)phenoxy-1-hexyl]ether were synthesized through the Williamson etherification reactions of chiroptical (*R*)-(+)- and (*S*)-(−)-1,1′-bi-2-naphthols, respectively, with phenylcyclo-

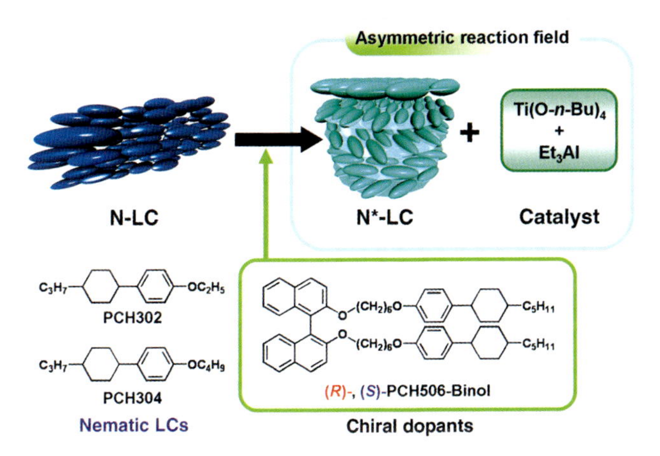

Figure 11.2 Construction of asymmetric reaction field for acetylene polymerization by dissolving Ziegler–Natta catalyst, Ti(O-*n*-Bu)$_4$–AlEt$_3$, in the chiral nematic LC. The chiral nematic LC includes an axially chiral binaphthyl derivative, (*R*)- or (*S*)-PCH506-Binol.

hexyl derivatives. Hereafter, the products will be abbreviated as (*R*)- and (*S*)-PCH506-Binol (Figure 11.2). The substituent is composed of a phenylcyclohexyl (PCH) moiety, an *n*-pentyl group (containing 5 carbon atoms), and a hexamethylene chain linked with an ether-type oxygen atom, [-$(CH_2)_6O$-, 06], and thus abbreviated as PCH506.

To prepare an induced chiral nematic LC, 5 to 14 wt% of (*R*)- or (*S*)-PCH506-Binol was added as a chiral dopant to the equimolar mixture of the nematic LCs 4-(trans-4-*n*-propylcyclohexyl)ethoxybenzene (PCH302) and 4-(trans-4-*n*-propylcyclohexyl)butoxybenzene (PCH304). The PCH506 substituent group in the (*R*)- and (*S*)-PCH506-Binol enhances the miscibility between the nematic LC mixture and the binaphthyl derivative used as the chiral dopant. Use of a similar substituent with a shorter methylene spacer, such as PCH503 or a normal alkyl substituent, gave insufficient miscibility, yielding no chiral nematic phase.

In polarizing optical micrographs of the mixture of PCH302, PCH304, and (*R*)-PCH506-Binol (abbreviated as *R*-1) and that of PCH302, PCH304, and (*S*)-PCH506-Binol (abbreviated as *S*-1), a striated Schlieren or fingerprint texture characteristic of chiral nematic LC phases was observed (Figure 11.1). Cholesteryl oleyl carbonate is known to be a left-handed chiral nematic LC, and therefore it is available as a standard LC for a miscibility test with the present LCs. The miscibility test method is based on the observation of the mixing area between the chiral nematic LC and the standard LC with POM, where the screw direction of the standard LC is known. If the screw direction of the chiral nematic LC is the same as that of the standard LC, the mixing area will be continuous, otherwise, it will be discontinuous (shown as a Schlieren texture of the nematic LC).

As shown in Figure 11.3, the mixture of (*R*-1)-chiral nematic LC and cholesteryl oleyl carbonate lost the striae characteristic of a chiral nematic phase with the POM, yielding instead features corresponding to an ordinary nematic phase. In contrast, the mixture of (*S*-1)-chiral nematic LC and cholesteryl oleyl carbonate showed no

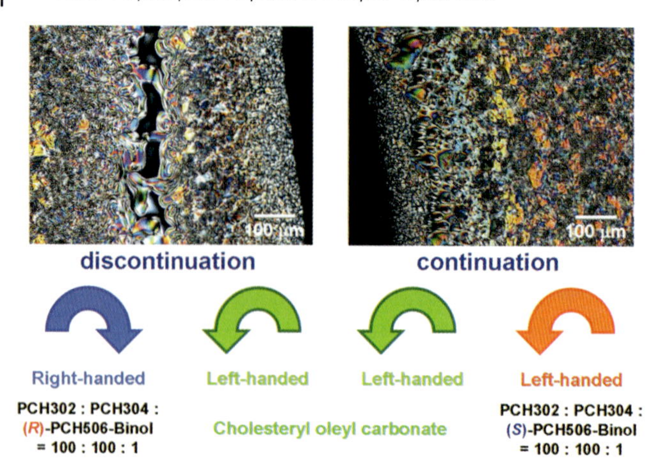

Figure 11.3 Miscibility test between the chiral nematic LC induced by (*R*)- or (*S*)- PCH506-Binol and the standard LC, cholesteryl oleyl carbonate, of the left-handed screw direction.

change in optical texture, keeping a chiral nematic phase. The results demonstrate that the screw directions of the (*R*-1)- and (*S*-1)-chiral nematic LCs are opposite to and the same as that of cholesteryl oleyl carbonate, respectively. Namely, they are right-handed and left-handed chiral nematic LCs, respectively.

11.3
Acetylene Polymerization in Chiral Nematic LC

First, it should be noted that although each component (PCH302 or PCH304) shows a LC phase, the LC temperature region is very narrow, that is, less than 1 to 2 °C. This is not suitable for acetylene polymerization in a nematic or chiral nematic LC reaction field because the exothermal heat evoked during acetylene polymerization would raise the temperature inside a Schlenk flask, and easily convert the LC phase into an isotropic one. Hence, the LC mixture is prepared by mixing two equimolar LC components. In the LC mixture, the nematic–isotropic temperature, T_{N-I}, and the crystalline–nematic temperature, T_{C-N}, might be raised and lowered, respectively. In fact, the mixture exhibited the LC phase in the region from 20 to 35 °C. Subsequently, the change of T_{N-I} upon addition of Ti(O-*n*-Bu)$_4$–AlEt$_3$ catalyst was examined by DSC measurement. Taking into account the effect of supercooling on LCs, the catalyst solution consisting of the LC mixture and the chiral dopant was found available for room temperature polymerization ranging from 5 to 25 °C. This sufficiently wide temperature region enabled us to perform the acetylene polymerization in the N*-LC phase.

The Ziegler–Natta catalyst consisting of Ti(O-*n*-Bu)$_4$ and AlEt$_3$ was prepared using the (*R*-1)- or (*S*-1)-chiral nematic LC as a solvent (Figure 11.2). The concentration of Ti(O-*n*-Bu)$_4$ was 15 mmol l^{-1}, and the mole ratio of the cocatalyst to catalyst, [AlEt$_3$]/[Ti(O-*n*-Bu)$_4$], was 4.0. The catalyst solution was aged for 0.5 h at room temperature.

During the aging, the chiral nematic LC containing the catalyst showed no noticeable change in optical texture and only a slight lowering of the transition temperature by 2 to 5 °C: The transition temperature between the solid and chiral nematic phases was 16 to 17 °C, and that between the chiral nematic and isotropic ones was 30 to 31 °C. No solidification as a result of supercooling was observed down to −7 °C. Thus, the (*R*-1)- and (*S*-1)-chiral nematic LCs are confirmed to be chemically stable with the catalyst. These LCs can therefore be used as asymmetric solvents for acetylene polymerization.

Acetylene gas of six-nine grade (99.9999% purity) was used without further purification. The apparatus and procedure employed were similar to those in earlier studies [12c,d] except for the polymerization temperature. Here, the polymerization temperature was kept between 17 and 18 °C to maintain the chiral nematic phase, by circulating cooled ethanol through an outer flask enveloping the Schlenk flask. The initial acetylene pressure was 11.6 to 22.6 Torr and the polymerization time was 10 to 43 min. After polymerization, the polyacetylene films were carefully stripped off the container and washed with toluene several times under argon gas at room temperature. The films were dried through vacuum pumping on a Teflon sheet and stored in a freezer at −20 °C.

11.4 Characterization of Helical Polyacetylene Film

SEM images of polyacetylene films show that multidomains of spiral morphology are formed (Figure 11.4a), and each domain is composed of a helical structure of a bundle of fibrils in the direction of a one-handed screw (Figure 11.4b). The multi-domain type fibril morphology of polyacetylene seems to replicate that of the chiral nematic LC during the interfacial acetylene polymerization.

Closer observation of SEM images indicates that helical polyacetylenes synthesized in the (*R*-1)- and (*S*-1)-chiral nematic LCs form twisted bundles of fibrils and even twisted fibrils in the left- and right-handed directions, respectively (Figure 11.5). This result implies that the screw direction of helical polyacetylene is controllable by choosing the helicity, that is, optical configuration of the chiral dopant, if the chiral nematic LC induced by the chiral dopant is employed as an asymmetric polymerization solvent. Moreover, it is very interesting that the screw directions of the bundle and fibrils are opposite to those of the (*R*-1)- and (*S*-1)-chiral nematic LCs used as solvents (see Figure 11.3). This is an unexpected and even surprising result, requiring a sensible interpretation, which will be discussed later.

The helical pitch of the chiral nematic LC depends on the helical twisting power of the chiral dopant as well as on its concentration and optical purity. This means that the helical pitch of the polyacetylene chain can also be varied by changing the helical twisting power of the chiral dopant. Another axially chiral dopant, (*R*)- or (*S*)-6,6′-PCH506-2,2′-Et-Binol, abbreviated as *R* or *S*-2, (Scheme 11.1) [18] gave a shorter helical pitch of chiral nematic LC by 0.3 μm than that of the corresponding (*R*)- or (*S*)-PCH506-Binol. Acetylene polymerizations using these sorts of highly twisted chiral

(a)

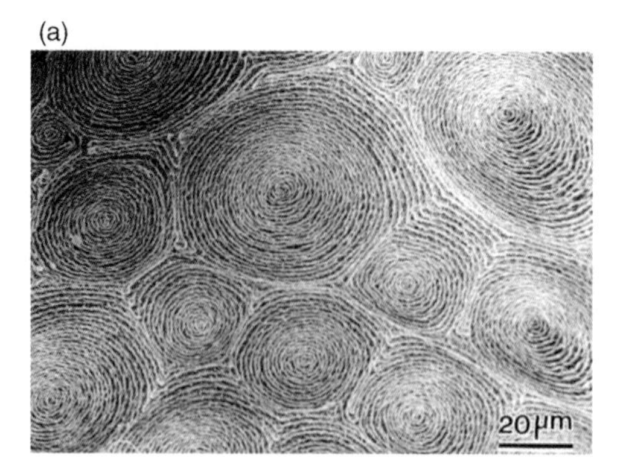

(b)

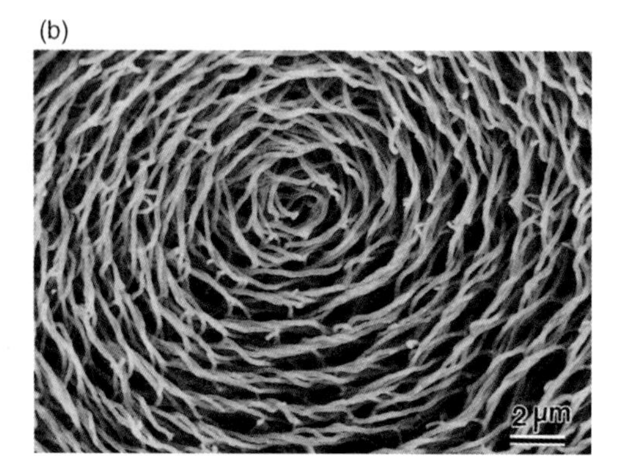

Figure 11.4 SEM photographs showing hierarchical spiral morphologies of helical polyacetylene film synthesized in chiral nematic LCs including (*R*)-6,6'-PCH506-2,2'-Et-Binol. Hierarchical spiral morphologies of helical polyacetylene film. (a) scanning electron microscope (SEM) photograph of multi-domain type spiral morphology and (b) left-handed twisted bundles of fibrils in a domain.

nematic LCs, designated (*R*-2)- and (*S*-2)-chiral nematic LCs, afforded clearer spiral morphologies consisting of helical bundles of fibrils (Figure 11.6). That is, these bundles are aligned parallel to each other in the microscopic regime and form spiral morphologies in the macroscopic regime. It is noteworthy that the hierarchical higher order structures observed in Figure 11.6 resemble the helical self-assembled microstructure of biological molecules such as lipids [19], but they, as well as those of Figure 11.4, are rarely formed in synthetic polymers. This indicates the validity of chiral nematic LC as a template polymerization medium for controlling a higher order structure of synthetic polymers.

The bundles of fibrils for helical polyacetylenes synthesized in the (*R*-2)- and (*S*-2)-chiral nematic LCs have left- and right-handed screw directions, respectively. The screw directions of helical polyacetylene are opposite to those of the corresponding (*R*-2)- and (*S*-2)-chiral nematic LCs whose directions are confirmed to be right-handed and left-handed, respectively, through the miscibility test with cholesteryl oleyl carbonate. This is the same situation as in the case of the (*R*-1)- and (*S*-1)-chiral nematic LCs including (*R*)- and (*S*)-PCH506-Binol.

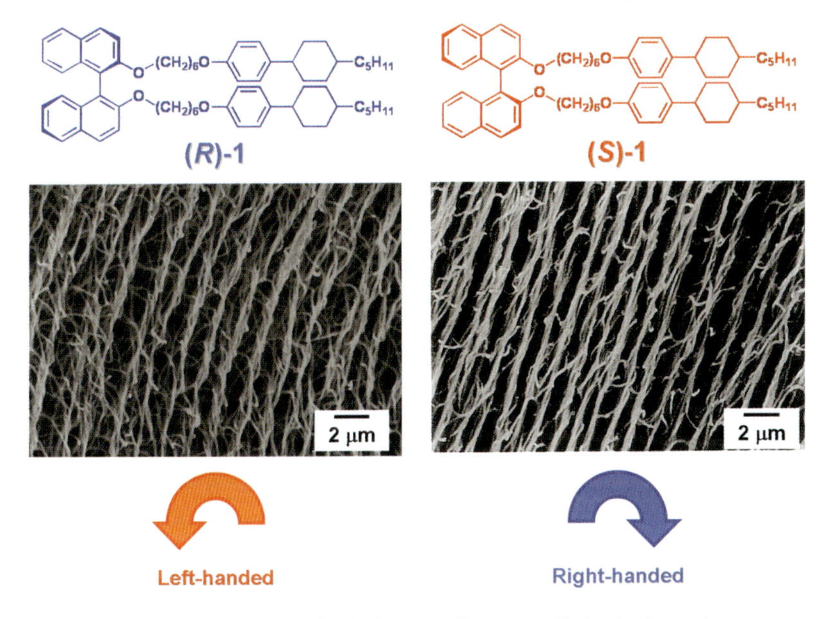

Figure 11.5 SEM photographs of helical polyacetylene films synthesized in chiral nematic LCs including (*R*)- and (*S*)-PCH506-Binol. The left- and right-handed screw directions of helical polyacetylenes are determined by the chirality of the chiral dopants with *R*- and *S*-configurations, respectively.

It has been shown so far that the polyacetylene chains propagate along the director (an averaged direction for the LC molecules within a domain) of the chiral nematic LC. Since the helical axis of polyacetylene is parallel to the polyacetylene chain, and the director of the chiral nematic LC is perpendicular to the helical axis of the chiral

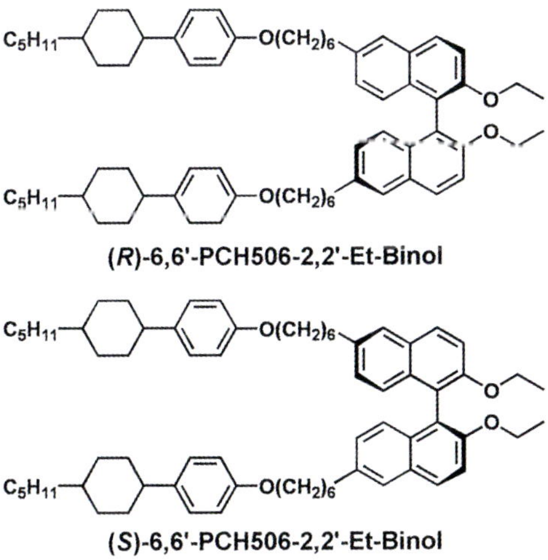

C_5H_{11} — $O(CH_2)_6$

(*R*)-6,6'-PCH506-2,2'-Et-Binol

C_5H_{11} — $O(CH_2)_6$

(*S*)-6,6'-PCH506-2,2'-Et-Binol

Scheme 11.1 Axially chiral binaphthyl derivatives used as chiral dopants, (*R*)- and (*S*)-6,6′-PCH506-2,2′-Et-Binol.

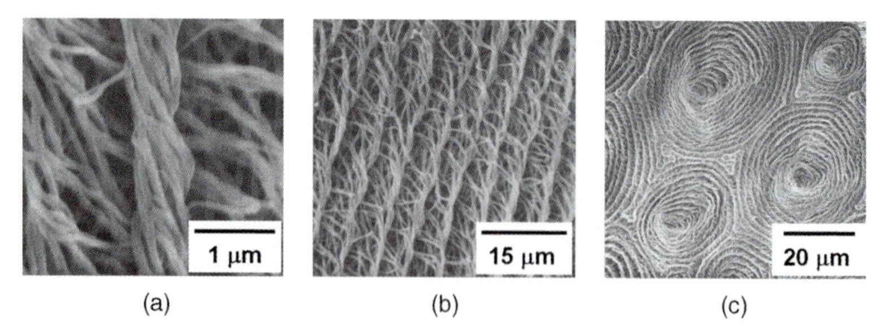

Figure 11.6 SEM photographs showing hierarchical spiral morphologies of helical polyacetylene film synthesized in chiral nematic LCs including (*R*)- 6,6′-PCH506-2,2′-Et-Binol. The fibrils of helical polyacetylene are gathered to form helical bundles of fibrils (a). The helical bundles are aligned parallel to each other in the microscopic regime (b), and they finally form spiral morphologies in the macroscopic regime (c).

nematic LC, the helical axis of polyacetylene is perpendicular to that of the chiral nematic LC. Taking these aspects into account, one can describe a plausible mechanism for interfacial acetylene polymerization in the chiral nematic LC, as shown in Figure 11.7. In the case of a right-handed chiral nematic LC, for example, the polyacetylene chain would propagate with a left-handed manner, starting from the catalytic species, but not with a right-handed one. This is because the polyacetylene chains with the screw direction opposite to that of the chiral nematic LC could propagate along the LC molecules, but those with the same direction as that of the chiral nematic LC would encounter LC molecules, making propagation stereospecifically impossible. The detailed mechanism of acetylene polymerization in the chiral nematic LC must be clarified.

In the circular dichroism (CD) spectra of the polyacetylene thin films synthesized with the (*R*-2)- and (*S*-2)-chiral nematic LCs, positive and negative Cotton effects are

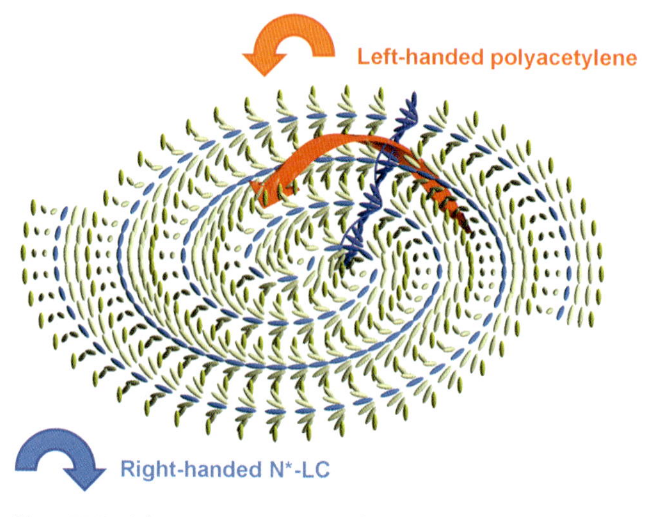

Figure 11.7 Schematic representation of a plausible mechanism for acetylene polymerization in a chiral nematic LC. The helical polyacetylene with the left-handed screw direction (red arrow) grows starting from the catalytic species in the right-handed chiral nematic LC.

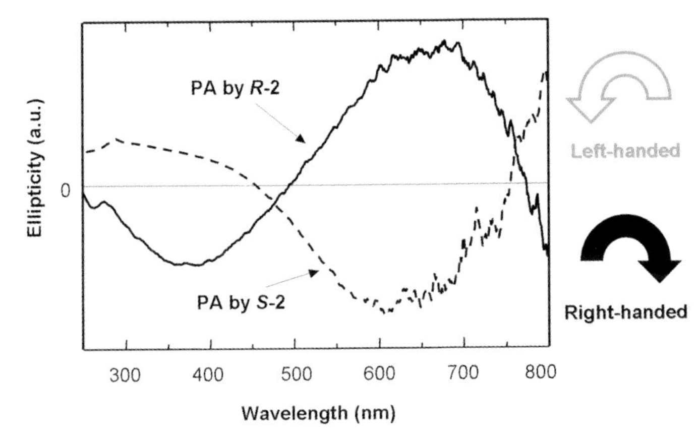

Figure 11.8 Circular dichroism (CD) spectra of helical polyacetylene films. The polyacetylene films synthesized in the (*R*-2)- and (*S*-2)-chiral nematic LCs including (*R*)- and (*S*)-6,6′-PCH506-2,2′-Et-Binol are designated as "PA by *R*-2" and "PA by *S*-2," respectively.

observed, respectively, in the region from 450 to 800 nm corresponding to the $\pi \rightarrow \pi^*$ transition of the polyacetylene chain (Figure 11.8), despite the absence of chiroptical substituents in the side chains. This indicates that the polyacetylene chain itself is helically twisted. It is evident that the above Cotton effect is not due to the chiral dopant [(*R*)- or (*S*)- 6,6′-PCH506-2,2′-Et-Binol] because the Cotton effect of the chiral dopant is only observed at shorter wavelengths such as 240 to 340 nm.

From the results mentioned above, it can be stated that left- and right-handed helical polyacetylene chains are formed in (*R*)- and (*S*)-chiral nematic LCs, respectively, and that these helical chains are bundled through van der Waals interactions to form helical fibrils with screw directions opposite those of the chiral nematic LCs. The bundles of fibrils further form the spiral morphology with various sizes of domains (Figure 11.9).

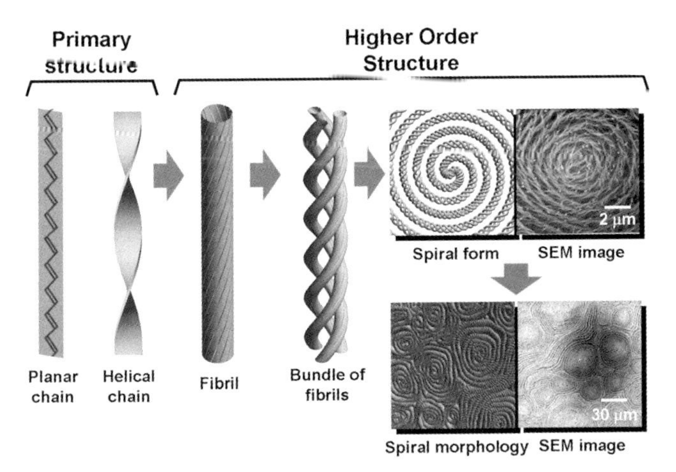

Figure 11.9 Hierarchical helical structures from primary to higher order in helical polyacetylene.

The dihedral angle between neighboring unit cells, (–CH = CH–), of the helical polyacetylene was estimated to be from 0.02° to 0.23°. Although such a very small dihedral angle might allow us to regard the present polyacetylene as an approximately planar structure, the polymer is definitely twisted in a one-handed direction with the non-zero dihedral angle. The present helical polyacetylene films have high trans-contents of 90%, and become highly conductive upon iodine doping. In fact, the electrical conductivities of the doped films are $1.5 \sim 1.8 \times 10^3\,S\,cm^{-1}$ at room temperature, which are comparable in order to those of metals. The iodine-doped polyacetylene showed the same Cotton effect as that of non-doped polyacetylene although the CD peak was slightly shifted to shorter wavelengths. This indicates that the helical structure is preserved, even after iodine doping. Furthermore, CD and X-ray diffraction measurements showed that the helical structure was also preserved after heating to 150 °C (which corresponds to the isomerization temperature from the cis- to the trans-form). The most stable structure of polyacetylene is the planar one. However, since polyacetylene is actually insoluble and infusible, the helical structure formed during polymerization can be preserved, even if it is washed with toluene or heated below the isomerization temperature. In other words, the insolubility and infusibility of polyacetylene are indispensable for preserving its metastable helical structure.

Here, it is worthwhile to emphasize the following experimental results. (i) Acetylene polymerization was carried out under a nematic LC environment using the equal weighted mixture of PCH302 and PCH304, but without a chiral dopant. The polyacetylene film showed neither helical morphology in SEM photographs nor the Cotton effect in CD spectra: The morphology observed was composed of fibrils that were locally aligned owing to spontaneous orientation of the LC solvent. This has also been confirmed in previous studies [12b–e]. (ii) Next, acetylene polymerization was performed in toluene and a small amount (less than 10%) of a chiral dopant. The synthesized polyacetylene showed the usually encountered randomly oriented fibrillar morphology, but not a helical one. At the same time, the polyacetylene showed no Cotton effect in the CD spectrum, although the characteristic very broad absorption band due to the $\pi \rightarrow \pi^*$ transitions in the conjugated polyene chain was observed in the region of 450 to 800 nm. (iii) The acetylene polymerization at 35 to 40 °C, where the LC mixture including the chiral dopant was isotropic, produced polyacetylene with no helical morphology. These results demonstrate that a chiral nematic LC environment is essential for producing helical polyacetylene.

11.5
Summary

Helical polyacetylene with a super-hierarchical structure was synthesized in an asymmetric reaction field consisting of chiral nematic LC. The chiral nematic LC was prepared by adding a chiroptical binaphthyl derivative as a chiral dopant to a mixture of two nematic LCs. Acetylene polymerizations were carried out using the catalyst $Ti(O\text{-}n\text{-}Bu)_4$–$AlEt_3$ dissolved in the chiral nematic LC solvent. The polyacetylene film consisted of a right- or left-handed helical structure of fibrils, as seen in scanning electron micrographs. The Cotton effect was observed in the region of the $\pi \rightarrow \pi^*$ transition of the polyacetylene chain in circular dichroism spectra. The high electrical conductivities of $1.5 \sim 1.8 \times 10^3\ \mathrm{S\,cm^{-1}}$ after iodine doping and the chiral helicity of the present films should make them suitable for applications with novel electromagnetic properties, such as an induced solenoid magnetism. Macroscopic alignment of helical polyacetylene has also been successfully carried out in order to prepare samples for the examination of its electromagnetic and optical properties, which will be presented in the near future.

Finally, it is worth noting that, by using the chiral nematic LC as an asymmetric polymerization solvent, helix formation is possible not only for polyacetylene but also for π-conjugated polymers without chiroptical substituents in the side chains. In fact, very recently, other kinds of spiral morphology containing conjugated polymers such as polybithiophene, polyethylenedioxythiophene derivatives, and phenylene-thiophene copolymers [20] have been synthesized through chemical or electrochemical polymerization in chiral nematic LC [21].

Acknowledgments

The author is grateful to Dr. Mutsumasa Kyotani for his generous and constructive help in measuring the SEM photographs. The author also thanks Dr. Taizo Mori and Dr. Munju Goh for their helpful cooperation in syntheses, polymerizations, spectroscopic measurements, and structural analyses. This work was supported by a Grant-in-Aid for Science Research in a Priority Area, "Super-Hierarchical Structures" (No. 446), and that for Science Research (S) (No. 20225007), from the Ministry of Education, Culture, Sports, Science and Technology, Japan.

References

1 (a) Skotheim, T.A. (ed.) (1986) *Handbook of Conducting Polymers*, Marcel Dekker, New York, (b) Akagi, K. and Shirakawa, H. (1996) Chapter 8, in *The Polymeric Materials Encyclopedia. Synthesis, Properties and Applications* (editor-in-Chief J.C. Salamone), CRC Press, p. 5315; (c) Nalwa, H.S. (ed.) (1997) *Handbook of Organic Conductive Molecules and Polymers*, John Wiley and Sons Inc., New York, (d) Akagi, K. and Shirakawa, H. (1998) Chapter 28, in *Electrical and Optical Polymer Systems, Fundamentals, Methods, and Applications* (eds D.L. Wise, G.E. Wnek, D.J. Trantolo, T.M. Cooper and J.D. Gresser), Marcel Dekker, New York, p. 983; (e) Akagi, K. (2007) Chapter 3, in *Handbook of Conducting Polymers, Third Edition, Conjugated Polymers*, 3rd edn (eds T.A. Skotheim and J.R. Reynolds), (eds Skotheim, T.A. and Reynolds, J.R.), CRC Press, New York, p. 3.

2 (a) Shirakawa, H., Louis, E., MacDiarmid, A.G., Chiang, C.K., and Heeger, A.J. (1977) *J. Chem. Soc. Chem. Commun.*, 578; (b) Chiang, C.K., Fincher, C.R., Park, Y.W., Heeger, A.J., Shirakawa, H., Louis, E.J., Gau, S.C., and MacDiarmid, A.G. (1977) *Phys. Rev. Lett.*, **39**, 1098.

3 (a) Naarrmann, H. and Theophilou, N. (1987) *Synth. Met.*, **22**, 1; (b) Tsukamoto, J., Takahashi, A., and Kawasaki, K. (1990) *Jpn. J. Appl. Phys.*, **29**, 125.

4 (a) Shirakawa, H. and Ikeda, S. (1971) *Polym. J.*, **2**, 231; (b) Ito, T., Shirakawa, H., and Ikeda, J. (1974) *Polym. Sci., Polym.*

Chem. Ed., **12**, 11; (c) Tanabe, Y., Kyotani, H., Akagi, K., and Shirakawa, H. (1995) *Macromolecules*, **28**, 4173.

5 For instance: Shirakawa, H., Masuda, T., and Takeda, T. (1994) Chapter 17, in *The Chemistry of Triple-Bonded Functional Groups*, vol. 2 (Suppl. C2) (ed. S. Patai), John Wiley and Sons Ltd, Chichester, UK.

6 (a) Akagi, K., Suezaki, M., Shirakawa, H., Kyotani, H., Shimomura, M., and Tanabe, Y. (1989) *Synth. Met.*, **28**, D1; (b) Cao, Y., Smith, P., and Heeger, A.J. (1991) *Polymer*, **32**, 1210.

7 (a) Shibahara, S., Yamane, M., Ishikawa, K., and Takezoe, H. (1998) *Macromolecules*, **31**, 3756; (b) Scherman, O.A. and Grubbs, R.H. (2001) *Synth. Met.*, **124**, 431; (c) Scherman, O.A., Rutenberg, I.M., and Grubbs, R.H. (2003) *J. Am. Chem. Soc.*, **125**, 8515; (d) Schuehler, D.E., Williams, J.E., and Sponsler, M.B. (2004) *Macromolecules*, **37**, 6255; (e) Gu, H., Zheng, R., Zhang, X., and Xu, B. (2004) *Adv. Mater.*, **16**, 1356.

8 (a) Oh, S.Y., Akagi, K., Shirakawa, H., and Araya, K. (1993) *Macromolecules*, **26**, 6203; (b) Oh, S.Y., Ezaki, R., Akagi, K., and Shirakawa, H. (1993) *J. Polym. Sci., Part A: Polym. Chem.*, **31**, 2977; (c) Akagi, K. and Shirakawa, H. (1996) *Macromol. Symp.*, **104**, 137; (d) Kuroda, H., Goto, H., Akagi, K., and Kawaguchi, A. (2002) *Macromolecules*, **35**, 1307; (e) Goto, H., Dai, X., Ueoka, T., and Akagi, K. (2004) *Macromolecules*, **37**, 4783.

9 (a) Jin, S.H., Choi, S.J., Ahn, W., Cho, H.N., and Choi, S.K. (1993) *Macromolecules*, **26**, 1487; (b) Choi, S.K., Gal, Y.S., Jin, S.H., and Kim, H.K. (2000) *Chem. Rev.*, **100**, 1645; (c) Tang, B.Z., Kong, X., Wan, X., Peng, H., Lam, W.Y., Feng, X.D., and Kwok, H.S. (1998) *Macromolecules*, **31**, 2419; (d) Kong, X., Lam, J.W.Y., and Tang, B.Z. (1999) *Macromolecules*, **32**, 1722; (e) Geng, J., Zhao, X., Zhou, E., Li, G., Lam, J.W.Y., and Tang, B.Z. (2003) *Polymer*, **44**, 8095.

10 (a) Koltzenburg, S., Wolff, D., Stelzer, F., Springer, J., and Nuyken, O. (1998) *Macromolecules*, **31**, 9166; (b) Ting, C.H., Chen, J.T., and Hsu, C.S. (2002) *Macromolecules*, **35**, 1180; (c) Schenning, A.P.H.J., Fransen, M., and Meijer, E.W. (2002) *Macromol. Rapid Commun.*, **23**, 265; (d) Stagnaro, P., Conzatti, L., Costa, G., Gallot, B., and Valenti, B. (2003) *Polymer*, **44**, 4443.

11 (a) Araya, K., Mukoh, A., Narahara, T., and Shirakawa, H. (1984) *Chem. Lett.*, 1141; (b) Akagi, K., Shirakawa, H., Araya, K., Mukoh, A., and Narahara, T. (1987) *Polym. J.*, **19**, 185; (c) Akagi, K., Katayama, S., Shirakawa, H., Araya, K., Mukoh, A., and Narahara, T. (1987) *Synth. Met.*, **17**, 241; (d) Akagi, K., Katayama, S., Ito, M., Shirakawa, H., and Araya, K. (1989) *Synth. Met.*, **28**, D51; (e) Sinclair, M., Moses, D., Akagi, K., and Heeger, A.J. (1988) *Phys. Rev. B: Condens. Matter*, **38**, 10724.

12 (a) Montaner, A., Rolland, M., Sauvajol, J.L., Galtier, M., Almairac, R., and Ribet, J.L. (1988) *Polymer*, **29**, 1101; (b) Coustel, N., Foxonet, N., Ribet, J.L., Bernier, P., and Fischer, J.E. (1991) *Macromolecules*, **24**, 5867.

13 Bozovic, I. (1987) *Mod. Phys. Lett. B*, **1**, 81.

14 (a) Suh, D.S., Kim, T.J., Aleshin, A.N., Park, Y.W., Piao, G., Akagi, K., Shirakawa, H., Han, J.S., Qualls, S.Y., and Brooks, J.S. (2001) *J. Chem. Phys.*, **114**, 7222; (b) Aleshin, A.N., Lee, H.J., Park, Y.W., and Akagi, K. (2004) *Phys. Rev. Lett.*, **93**, 196601; (c) Lee, H.J., Jin, Z.X., Aleshin, A.N., Lee, J.Y., Goh, M.J., Akagi, K., Kim, Y.S., Kim, D.W., and Park, Y.W. (2004) *J. Am. Chem. Soc.*, **126**, 16722.

15 (a) Akagi, K., Piao, G., Kaneko, S., Sakamaki, K., Shirakawa, H., and Kyotani, M. (1998) *Science*, **282**, 1683; (b) Akagi, K., Piao, G., Kaneko, S., Higuchi, I., Shirakawa, H., and Kyotani, M. (1999) *Synth. Met.*, **102**, 1406; (c) Akagi, K., Higuchi, I., Piao, G., Shirakawa, H., and Kyotani, M. (1999) *Mol. Cryst. Liq. Cryst.*, **332**, 463; (d) Akagi, K., Guo, S., Mori, T., Goh, M., Piao, G., and Kyotani, M. (2005) *J. Am. Chem. Soc.*, **127**, 14647; (e) Goh, M.J., Kyotani, M., and Akagi, K. (2006) *Curr. Appl. Phys.*, **6**, 948; (f) Akagi, K. (2007) *Polym. Int.*, **56**, 1192; (g) Goh, M., Matsushita, T., Kyotani, M., and Akagi, K. (2007) *Macromolecules*, **40**, 4762; (h) Mori, T., Kyotani, M., and Akagi, K. (2008) *Macromolecules*, **41**, 607; (i) Goh, M., Kyotani, M., and Akagi, K. (2007) *J. Am. Chem. Soc.*, **129**, 8519; (j) Mori, T.,

Sato, T., Kyotani, M., and Akagi, K. (2009) *Macromolecules*, **42**, 1817; (k) Kyotani, M., Matsushita, S., Nagai, T., Matsui, Y., Shimomura, M., Kaito, A., and Akagi, K. (2008) *J. Am. Chem. Soc.*, **130**, 10880.

16 Gottarelli, G., Mariani, P., Spada, G.P., Samori, B., Forni, A., Solladie, G., and Hibert, M. (1983) *Tetrahedron*, **39**, 1337.

17 (a) Semenkova, G.P., Kutulya, L.A., Shkol'nikova, N.I., and Khandrimailova, T.V. (2001) *Kristallografiya*, **46**, 128; (b) Guan, L. and Zhao, Y. (2001) *J. Mater. Chem.*, **11**, 1339; (c) Hatoh, H. (1994) *Mol. Cryst. Liq. Cryst. Sci. Technol., Sect. A*, **250**, 1; (d) Lee, H. and Labes, M.M. (1982) *Mol. Cryst. Liq. Cryst.*, **84**, 137.

18 (a) Kanazawa, K., Higuchi, I., and Akagi, K. (2001) *Mol. Cryst. Liq. Cryst.*, **364**, 825; (b) Goh, M.J., Kyotani, M., and Akagi, K. (2006) *Curr. Appl. Phys.*, **6**, 948; (c) Goh, M. and Akagi, K. (2008) *Liq. Cryst.*, **35**, 953.

19 For, instance, Schnur, J.M. (1993) *Science*, **262**, 1669 and references therein.

20 (a) Osaka, I., Nakamura, A., Inoue, Y., and Akagi, K. (2001) *Trans. Mater. Res. Soc. Jpn.*, **27**, 567; (b) Yorozuya, S., Osaka, I., Nakamura, A., Inoue, Y., and Akagi, K. (2003) *Synth. Met.*, **135–136**, 93; (c) Oh-e, M., Yokoyama, H., Yorozuya, S., Akagi, K., Belkin, M.A., and Shen, Y.R. (2004) *Phys. Rev. Lett.*, **93**, 267402; (d) Goto, H. and Akagi, K. (2005) *Angew. Chem. Int. Ed.*, **44**, 4322.

21 (a) Goto, H. and Akagi, K. (2004) *Macromol. Rapid Commun.*, **25**, 1482; (b) Goto, H. and Akagi, K. (2005) *Macromolecules*, **38**, 1091; (c) Goto, H., Nomura, N., and Akagi, K. (2005) *J. Polym. Sci. Part A: Polym. Chem.*, **43**, 4298.

Index

Conjugated Polymer Synthesis. Edited by Yoshiki Chujo

ISBN: 978-3-527-32267-1

x

y

z